Review of Gross Anatomy

Text and Illustrations by

BEN PANSKY, Ph.D., M.D.
Professor of Anatomy,
Medical College of Ohio at Toledo,
Toledo, Ohio

Review of
Gross Anatomy

FIFTH EDITION

Macmillan Publishing Company
New York

Collier Macmillan Canada, Inc.
Toronto

Collier Macmillan Publishers
London

Macmillan Publishing Company
866 Third Avenue, New York, New York 10022

Collier Macmillan Canada, Inc.

Collier Macmillan Publishers • London

Library of Congress Cataloging in Publication Data

Pansky, Ben.
   Review of gross anatomy.

   Bibliography: p. 565
   Includes index.
   1. Anatomy, Human—Outlines, syllabi, etc.
2. Anatomy, Surgical and topographical—Outlines,
syllabi, etc. I. Title. [DNLM: 1. Anatomy—Atlases.
QS 17 P196r]
QM31.P28 1984   611   83-25539
ISBN 0-02-390650-2

          Printing: 1 2 3 4 5 6 7 8   Year 4 5 6 7 8 9 0 1 2

This fifth edition is dedicated to my dearly beloved wife, **Julie,** who has spent 20 of our 30 years of marriage giving me the patient inspiration, encouragement, and love needed to illustrate and write the various editions of the book; to my son, **Jon,** who grew up with "the little red book"; and to the book itself, which has become such an integral part of our lives over these many years.

Preface to the Fifth Edition

This book has been written for all who require a visual comprehension and review of the fundamentals of anatomy. The presentation is concise, simplified, and in outline form. The text is complete, functionally oriented, and clinically informative to stress the need for understanding the essentials of anatomy. Where repetition occurs, it does so to reemphasize particular points and to help the continuity of study. With only a few exceptions, illustrations appear on right-hand pages and corresponding text on opposite left-hand pages. The terminology used throughout is an anglicized version of the standard nomenclature (*Nomina Anatomica*, 5th ed.), approved at the Eleventh International Congress of Anatomists, held at Mexico City in 1980.

Most students and teachers will agree that anatomy is a study of a three-dimensional body, best learned at the dissection table. Visualization and palpation play major roles in the understanding of the fundamentals of anatomy. Thus, this fifth edition has maintained its tradition and been written around its pictures rather than the latter planned around the text. It contains well over 1000 original linecut illustrations that have been conceived to give a simplified, multifaceted, three-dimensional aspect and visualization of the beauty and function of the body that are rarely found in brief, or review-type, texts.

As the student progresses from region to region—from the head and neck to the back, to the upper extremity, to the thorax, to the abdomen and pelvis, and, last, to the lower extremity—he/she can fully appreciate the continuity between regions.

The illustrations demonstrate an overlapping of the structures, so as to enable the reader to move from one region to the next more easily. In addition, black-and-white or gray tabs are printed in the margins of left-hand pages to facilitate rapid location of each of the major units within the text.

The body has been discussed from its superficial layers to its deep structures, except for the osteology. Because the bones form the framework and lend themselves to the attachment of soft parts, one may find that if they are studied first, the relationships of soft body parts are often more clearly and easily understood.

By extracting information from within the living organism, the student and practitioner of medicine are better able to describe and define normal and disease states. Aside from detailed dissection of the body and the use of the stethoscope and ophthalmoscope, x-ray and imaging technology (discovered at the close of the nineteenth century) has grown rapidly. Nuclear medicine and ultrasonography were introduced in the 1950s. Computed tomography (CT), position emission tomography, digital radiography, and nuclear magnetic resonance (NMR) became available in the 1970s. A modern textbook of anatomy would be incomplete without some discussion of radiography, computed tomography, and nuclear magnetic resonance. Thus, many new radiographs have been added to this fifth edition, along with a series of CT scans through the head, thorax, abdomen, and pelvis. Also included is an NMR illustration of cranial anatomy.

All units of the book have been thoroughly reviewed, and many have been revised or rewritten. New chapters, with accompanying new figures and plates, have been added in order to clarify, simplify, and make the regions more understandable.

Through years of teaching gross anatomy to a great variety of students, the author has become well aware of the fact that the subject, quickly learned and memorized, is just as easily forgotten and, therefore, requires a constant and almost endless review. Yet, "time" is the student's adversary, and his/her many duties are often so overwhelming that a repeated reading of a long and detailed textbook is impossible. Accordingly, the author is hopeful that this fifth edition of *Review of Gross Anatomy* will continue to fill an obvious gap in present-day medical literature.

The author would like to express his appreciation to the many teachers, colleagues, and, above all, students whose comments and suggestions have helped to maintain the accuracy and quality of this book. I am particularly grateful to John Doman, who photographed the models used in depicting areas of surface anatomy; Ms. Faye Keen, Library Media Technical Assistant in the Department of Radiology of the Medical College of Ohio, for many of the radiologic prints used in this edition, including the CT scans; to Dr. A. J. Christoforidis, former Chairman and Professor of Radiology at the Medical College of Ohio (he presently holds a similar position at Ohio State University Medical Center), for supplying me with many of the radiographs and CT scans; and to Dr. Prem C. Chadnani, Associate Professor and Acting Chairman of Radiology at the Medical College of Ohio, who supplied many of the additional radiographs and gave his time to help me sort out the details of the radiographs and CT scans. And a special thanks to my editor, Ms. Joan C. Zulch, whose encouragement, patience, and understanding have seen me through all five editions of this book.

B.P.

Contents

Unit Two. BACK

Unit Three. UPPER EXTREMITY

Unit Four. THORAX

Unit Five. ABDOMEN AND PELVIS

Unit Six   LOWER EXTREMITY

Appendixes

UNIT ONE

Head and Neck

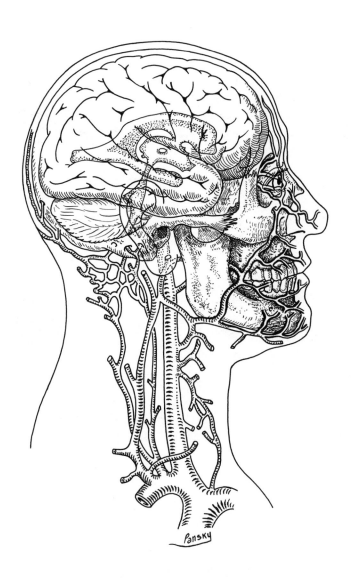

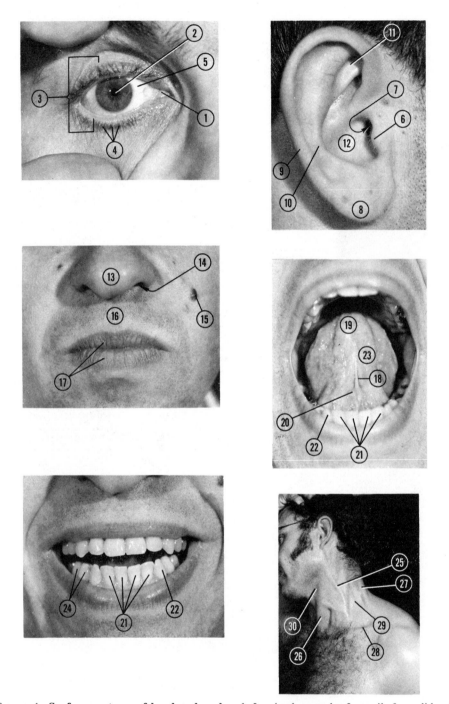

FIGURE 1. **Surface anatomy of head and neck.** *1*, Lacrimal caruncle; *2*, pupil; *3*, eyelids; *4*, eyelashes; *5*, conjunctiva; *6*, tragus; *7*, external auditory meatus; *8*, lobule; *9*, helix; *10*, antihelix; *11*, triangular fossa; *12*, concha; *13*, tip of nose; *14*, nares; *15*, nevus; *16*, philtrum; *17*, lips; *18*, frenulum; *19*, tip of tongue; *20*, sublingual caruncle; *21*, incisors; *22*, canine; *23*, fimbriated fold; *24*, premolars; *25*, external jugular vein; *26*, sternocleidomastoid m.; *27*, trapezius m.; *28*, clavicle; *29*, posterior triangle; *30*, carotid triangle.

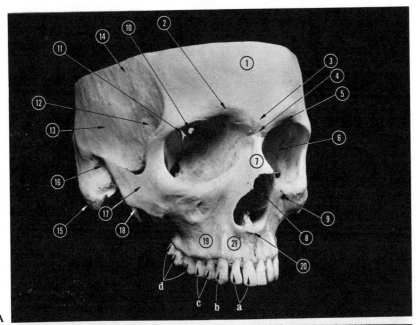

A

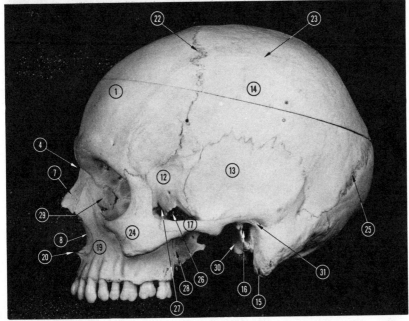

B

FIGURE 2. **A. Anterolateral aspect of the skull; B. Lateral aspect of the skull.** *1*, Frontal bone; *2*, superciliary arch; *3*, glabella; *4*, nasion; *5*, superior orbital notch; *6*, orbit; *7*, nasal bone; *8*, piriform (nasal) aperture; *9*, infraorbital foramen; *10*, optic foramen; *11*, superior orbital fissure; *12*, sphenoid bone (greater wing); *13*, temporal bone; *14*, parietal bone; *15*, mastoid process; *16*, external auditory meatus; *17*, zygomatic arch; *18*, styloid process; *19*, maxillary bone; *20*, anterior nasal spine; *21*, canine fossa. *a*, Incisors; *b*, canine; *c*, premolars; *d*, molars; *22*, coronal suture; *23*, temporal lines; *24*, zygomatic bone; *25*, lambdoidal suture; *26*, infratemporal fossa; *27*, spine of sphenoid; *28*, lateral pterygoid plate; *29*, lacrimal bone; *30*, tympanic plate; *31*, suprameatal triangle.

1. THE SKULL: GENERAL CONSIDERATIONS

I. **Introduction:** the skull surrounds and protects the brain and brain stem; houses the organs of special sensation for seeing (the eyes), hearing (middle and inner ear), taste (tongue, etc.), and smell (nasal structures); and contains openings that lead into the digestive tract (oral cavity), respiratory tract (nasal and oral cavities), and vertebral foramen (foramen magnum)
   A. IN THE ANATOMIC POSITION, the orbits are directed forward, and the infraorbital margins and the superior margin of the external acoustic meatus are on the same horizontal plane
   B. THE VERTEX is the highest part of the skull
   C. THE FOREHEAD describes the skull's frontal region (brow, metopion, or frons); the occiput or occipital region is at the back of the head; and the temple or side of the head denotes the temporal region above the zygomatic arch

II. **The skull** consists of 2 parts: *the calvaria,* which encloses the brain and brain stem, and *the face,* which is attached to its anteroinferior portion
   A. SKULL refers to the entire bony framework of the head including the mandible
   B. CRANIUM refers to the skull, excluding the mandible and facial bones. It consists of 8 bones: occipital, 2 parietals, frontal, 2 temporals, sphenoid, and ethmoid
   C. CALVARIA refers to the cranial vault or brain case whose roof is called *skullcap* and is formed anteriorly by the frontal bone and posteriorly by parietal and occipital bones, all articulating with each other via *sutures*
      1. The floor of the calvaria (internal surface of the base of the skull) contains the anterior, middle, and posterior cranial fossae
         a. The floor also contains many foramina and thin areas, is fragile and prone to fractures

III. **Most of the bones of the calvaria** consist of inner and outer cortices or tables of compact bone (the inner table being thinner than the outer) which are separated by spongy, cancellous bone or diploë (with red bone marrow)
   A. DIPLOIC VEINS are located between the bony tables of the skull
   B. EMISSARY VEINS connect the intracranial venous sinuses with the diploic veins and with veins on the outside of the cranium. They are valveless, and blood may flow in both directions, but usually flows from the brain to the outside

IV. **Clinical considerations**
   A. TRAUMATIC INJURY to the side of the head where the calvaria is thin may produce fractures with ultimate intracranial bleeding due to tearing of blood vessels lying in the bony grooves of the inner table of bone
   B. THE SKULL SUTURES begin to obliterate (synostosis) at about middle age (30 to 40 years), except for the metopic suture between the frontal bones (begins to fuse up to 2 years of age). The suture is seen in a small percentage of adults
      1. Premature closure of the sutures may lead to deformities:
         a. Scaphocephaly: premature closure of the sagittal suture, resulting in a narrow, long, wedge-shaped skull
         b. Plagiocephaly: premature closure of the lambdoid or coronal sutures on one side only, resulting in a twisted, asymmetric skull
         c. Oxycephaly (acrocephaly or turricephaly): premature closure of the coronal suture, resulting in a high, towerlike skull
   C. FRACTURES
      1. Linear: resembling a line, most frequent type and occurring at point of contact
      2. Contrecoup: fracture of skull on opposite side of impact area
      3. Closed or simple: fracture with intact overlying scalp and/or mucous membrane
      4. Open or compound: with laceration of overlying scalp and/or mucous membrane
      5. Comminuted: fracture with bone fragmentation
      6. Depressed: fracture with inward displacement of a part of the calvaria
      7. Expressed: fracture with outward displacement of a part of the cranium
      8. Diastatic: separation of cranial bones at a suture

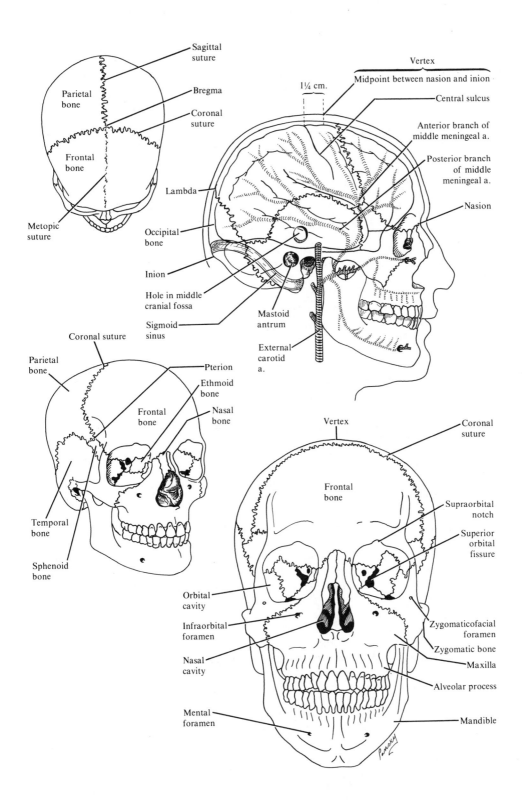

Sagittal suture

Parietal bone

Bregma

Coronal suture

Frontal bone

Metopic suture

Lambda

Occipital bone

Inion

Hole in middle cranial fossa

Sigmoid sinus

1¼ cm.

Vertex

Midpoint between nasion and inion

Central sulcus

Anterior branch of middle meningeal a.

Posterior branch of middle meningeal a.

Nasion

Mastoid antrum

External carotid a.

Coronal suture

Parietal bone

Pterion

Ethmoid bone

Frontal bone

Nasal bone

Temporal bone

Sphenoid bone

Vertex

Coronal suture

Frontal bone

Supraorbital notch

Superior orbital fissure

Orbital cavity

Infraorbital foramen

Nasal cavity

Zygomaticofacial foramen

Zygomatic bone

Maxilla

Alveolar process

Mental foramen

Mandible

2. ANTERIOR VIEW OF THE SKULL (NORMA FRONTALIS), MANDIBLE, AND HYOID BONE

I. **Bones of the skull:** anterior part of the cranial vault and face
A. FRONTAL: frontal tuber or eminence, superciliary arch with supraorbital notch or foramen, supraorbital margin, glabella, zygomatic process, and nasal part
B. NASAL
C. NASAL SEPTUM
D. MAXILLA: facial surface, incisive and canine fossae, infraorbital foramen, nasal spine, nasal notch, and alveolar, zygomatic, and frontal processes
E. ZYGOMATIC (malar): frontal and temporal processes and infraorbital margin

II. **Sutures of the skull**
A. FRONTONASAL: frontal and two nasal bones; nasion is the midpoint of the suture
B. ZYGOMATICOMAXILLARY: zygomatic and maxillary bones
C. FRONTOZYGOMATIC: zygomatic and frontal bones
D. INTERMAXILLARY: between maxillary bones
E. METOPIC: between frontal bones (may or may not be present in the adult skull)

III. **Hyoid:** not a skull bone; consists of 5 parts: body, 2 greater, and 2 lesser cornua
A. BODY: quadrangular, convex ventrally (anteriorly)
B. GREATER CORNUA: long, projecting backward from body; joined to body in a synostosis
    1. The tip of the greater horn is found midway between the laryngeal prominence and mastoid process and is a landmark for the lingual artery
C. LESSER CORNUA: short, conical processes projecting upward and backward from body; base attached by fibrous tissue (syndesmosis) at line of union of body and greater cornu (horn)
D. HYOID BONE is at the level of the 3rd cervical vertebra

IV. **Mandible:** made up of a body and 2 rami. Largest, strongest bone of the face
A. BODY: convex anteriorly; fused at the symphysis menti by 2 years of age
    1. Externally: mental protuberance with mental tubercles on either side; mental foramen; oblique line; groove for facial artery
    2. Internally: mental spines, mylohyoid line, alveolar border, submandibular and sublingual fossae
B. RAMI form angles with the body
    1. Condyloid process, with rounded head for articulation
    2. Constricted neck
    3. Coronoid process: thin, separated from condyloid process by a wide mandibular notch and is pointed at its superior end
    4. On inner (medial) surface: mandibular foramen leading into the mandibular canal; the lingula; and the mylohyoid groove

V. **Clinical considerations**
A. A SKULL FRACTURE may be due to a blow on the head by a solid object, crushing of the head between 2 solid objects, or impaction of the moving head against a stationary solid object
    1. If bone distortion is sufficient, fracture results.
B. FRACTURES OF THE MANDIBLE
    1. Open type: bone fragments break the mucous membrane of the mouth
    2. Closed type: vertical breaks between ramus and body of mandible. The ramus fragment is pulled upward by temporalis and masseter muscles, and medially by the medial pterygoid muscle; the body fragment is pulled downward by gravity and the geniohyoid muscle and backward by digastric and mylohyoid muscles
    3. Displacement of bone fragments at fracture lines may result in malocclusion
C. WITH AGING AND LOSS OF TEETH, bone of the mandibular alveoli is absorbed, the shape of the face changes, foramina come to lie closer to the upper border of bone, and their nerves may be exposed, leading to injury and pain

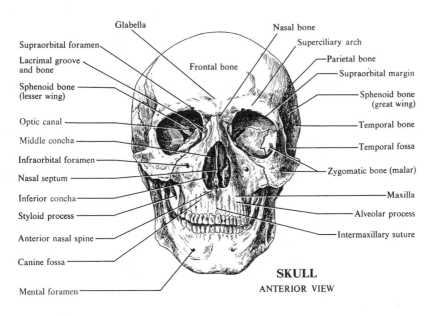

Glabella

Nasal bone

Supraorbital foramen

Superciliary arch

Lacrimal groove and bone

Parietal bone

Frontal bone

Supraorbital margin

Sphenoid bone (lesser wing)

Sphenoid bone (great wing)

Optic canal

Temporal bone

Middle concha

Temporal fossa

Infraorbital foramen

Nasal septum

Zygomatic bone (malar)

Inferior concha

Maxilla

Styloid process

Alveolar process

Anterior nasal spine

Intermaxillary suture

Canine fossa

SKULL
ANTERIOR VIEW

Mental foramen

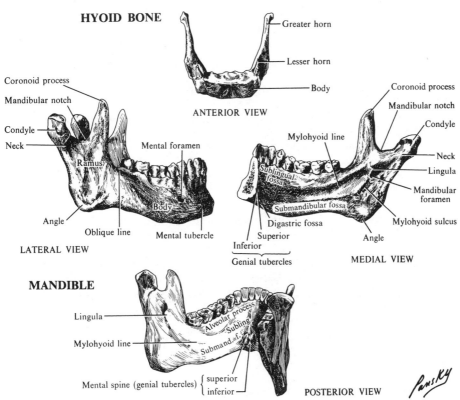

HYOID BONE

Greater horn

Lesser horn

Body

ANTERIOR VIEW

Coronoid process

Mandibular notch

Condyle

Neck

Mental foramen

Ramus

Coronoid process

Mandibular notch

Condyle

Mylohyoid line

Neck

Lingula

Sublingual fossa

Mandibular foramen

Submandibular fossa

Mylohyoid sulcus

Angle

Body

Angle

Oblique line

Mental tubercle

Digastric fossa

Superior

Angle

Inferior

LATERAL VIEW

Genial tubercles

MEDIAL VIEW

MANDIBLE

Lingula

Alveolar process

Subling

Mylohyoid line

Submand. f.

Mental spine (genial tubercles) { superior { inferior

POSTERIOR VIEW

Pansky

–7–

3. LATERAL VIEW OF THE SKULL (NORMA LATERALIS)

I. Bones
   A. PARIETAL: part of roof and sides of cranium. Note superior and inferior temporal lines, and parietal foramen (inconsistent)
   B. FRONTAL (squamous part): anterior and anterolateral part of cranium. Note superciliary arch, the glabella (uniting the arches over the nose), the zygomatic process, and the temporal line
   C. LACRIMAL: anteromedial wall of the orbit; lacrimal sulcus and crest
   D. GREAT WING OF SPHENOID AND LATERAL PTERYGOID PLATE
   E. TEMPORAL
      1. Squamous: flat, forming part of side of cranium; note zygomatic process, articular tubercle, mandibular fossa, and suprameatal crest. Lies outside temporal lobe of the brain
      2. Mastoid: large portion behind ear; mastoid process and foramen are evident
      3. Tympanic: surrounding external auditory meatus; part of wall of tympanic cavity
      4. Styloid process with its vaginal sheath: gives attachment to ligaments and muscles
      5. Zygomatic process of temporal: helps make up zygomatic arch
   F. ZYGOMATIC (cheek bone): note zygomaticofacial and zygomaticotemporal foramina
   G. MAXILLA: note anterior nasal spine and alveolar border; frontal process of bone
   H. OCCIPITAL (squamous portion): external protuberance and superior nuchal line
   I. NASAL BONE
   J. MANDIBLE: head and neck; ramus; angle; body; coronoid process; alveolar part; mental tubercle, foramen, and protuberance

II. Sutures of the skull
   A. CORONAL: frontal and parietal bones
   B. SPHENOFRONTAL: great wing of sphenoid and frontal bones
   C. SPHENOPARIETAL: great wing of sphenoid and parietal bones (*pterion* is the posterior end of the suture)
   D. SQUAMOSAL: squama of temporal and parietal bones
   E. PARIETOMASTOID: parietal and mastoid of temporal bone
   F. LAMBDOIDAL: parietal and occipital bones
   G. OCCIPITOMASTOID: occipital and mastoid of temporal bone; *asterion* is the point at which the lambdoidal and occipitomastoid sutures meet
   H. SPHENOSQUAMOSAL: great wing of sphenoid and squamous temporal bones
   I. TEMPOROZYGOMATIC: processes of zygomatic and temporal bones
   J. FRONTOZYGOMATIC: zygomatic and frontal bones

III. Special features
   A. TEMPORAL FOSSA: deep to zygomatic process of temporal bone; bounded above and behind by temporal lines, in front by frontal and zygomatic bones, laterally by the zygomatic arch, and below by the infratemporal crest
   B. INFRATEMPORAL FOSSA: deep to ramus of mandible; bounded in front by the maxilla, behind by the articular tubercle of the temporal and angular spine of sphenoid, above by great wing of sphenoid, below by the infratemporal crest and alveolar border of the maxilla, and medially by the lateral pterygoid plate
   C. PTERYGOPALATINE FOSSA: bounded above by body of sphenoid, in front by maxilla, behind by pterygoid process and great wing of sphenoid, and medially by palatine bone. Five foramina open into fossa: rotundum, pterygoid canal, pharyngeal canal, sphenopalatine foramen, and pterygopalatine (palatine) canal. In addition, anteriorly, the fossa communicates with the orbit via the inferior orbital fissure, and via the pterygomaxillary fissure connects laterally with the infratemporal fossa
   D. PTERYGOID LAMINA: medial and lateral, suspended from base of sphenoid bone
   E. PTERYGOMAXILLARY FISSURE: between pterygoid process posteriorly and maxillary bone anteriorly. Leads into the pterygopalatine fossa

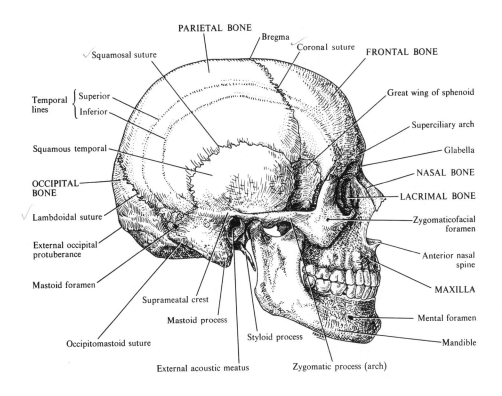

PARIETAL BONE

Bregma

Coronal suture

FRONTAL BONE

Squamosal suture

Great wing of sphenoid

Temporal lines
{ Superior
  Inferior }

Superciliary arch

Squamous temporal

Glabella

NASAL BONE

OCCIPITAL BONE

LACRIMAL BONE

Lambdoidal suture

Zygomaticofacial foramen

External occipital protuberance

Anterior nasal spine

Mastoid foramen

MAXILLA

Suprameatal crest

Mastoid process

Mental foramen

Styloid process

Mandible

Occipitomastoid suture

External acoustic meatus

Zygomatic process (arch)

SKULL
LATERAL VIEW

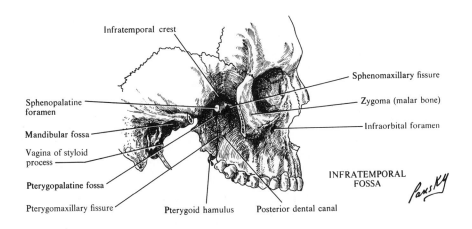

Infratemporal crest

Sphenomaxillary fissure

Zygoma (malar bone)

Sphenopalatine foramen

Infraorbital foramen

Mandibular fossa

Vagina of styloid process

Pterygopalatine fossa

INFRATEMPORAL FOSSA

Pterygomaxillary fissure

Pterygoid hamulus

Posterior dental canal

4. SKULL FROM ABOVE AND BEHIND

I. Bones
A. FRONTAL: usually smooth and convex in adult. In infancy, 2 frontal bones are joined at frontal (metopic) suture. Lower end of this suture, above nose, sometimes seen in adults
B. PARIETAL forms most of superior and lateral cranium; shows parietal foramen (for emissary vein) and tuberosity
C. OCCIPITAL, forming dorsal part of cranium, shows external occipital protuberance with external occipital crest running caudally and the supreme and superior nuchal lines arching laterally from it; the planum occipitale lies above highest line; the inferior nuchal lines run laterally from midpoint of the median crest; and planum nuchale is that part of bone below superior nuchal lines

II. Sutures of skull
A. SAGITTAL: between the parietal bones on either side
B. CORONAL: between frontal and the 2 parietal bones; bregma is that point of union of coronal and sagittal sutures
C. LAMBDOIDAL: between occipital and 2 parietal bones
D. FRONTAL (metopic) may not be visible in adults

III. Fontanelles: portions of the skull not ossified at birth, usually 6 in number
A. ANTERIOR (bregmatic) is largest and diamond-shaped, located at junctions of coronal, sagittal, and frontal sutures. Closes middle of second year
B. POSTERIOR: triangular in shape, is located at union of sagittal and lambdoidal sutures. Closes 2 months after birth
C. SPHENOIDAL: 1 on each side, is a small gap between parietal, great wing of sphenoid, and squamous temporal bones. Closes 1 to 2 months after birth
D. MASTOID: 1 on each side, is a small gap between parietal, temporal, and occipital bones. Closes 1 to 2 months after birth

IV. Clinical considerations
A. FRACTURE LINES follow the paths of least resistance: through the vault between the reinforcing pillars; through the base within the fossae, not at the junctions of the fossae; and along the suture lines. All lines of fracture directed toward the base will converge on the body of the sphenoid bone
B. IN BASAL SKULL FRACTURES, bleeding from the ear is common, and Battle's sign is discoloration of the skin along the course of the posterior auricular artery
C. A RING FRACTURE of the occipital bone about the foramen magnum may occur in injuries in which the base of the skull is jammed against the atlas
D. A DEPRESSED FRACTURE occurs when a piece of bone is displaced inward and presses on or lacerates the brain
E. A LINEAR FRACTURE consists of a single line of fracture without bone displacement
F. A COMPOSITE FRACTURE is one in which the fracture lines branch without bone displacement
G. SKULL FRACTURES may be either comminuted or compound
H. THE MOST COMMON CAUSE OF DEATH from a skull fracture is laceration of the brain; other causes include subdural hematoma, cerebral concussion, extradural hemorrhage, and meningitis
I. WHILE THE FONTANELLES ARE OPEN, the brain is very vulnerable. Care is needed to protect these regions, especially anteriorly, because this fontanelle does not close until the middle of the second year

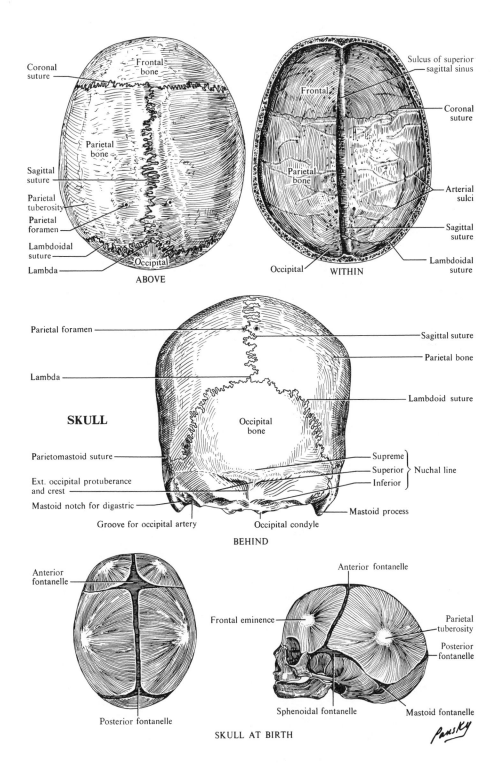

Coronal suture

Frontal bone

Parietal bone

Sagittal suture

Parietal tuberosity

Parietal foramen

Lambdoidal suture

Lambda

Occipital

ABOVE

Sulcus of superior sagittal sinus

Frontal

Coronal suture

Parietal bone

Arterial sulci

Sagittal suture

Lambdoidal suture

Occipital **WITHIN**

SKULL

Parietal foramen

Lambda

Parietomastoid suture

Ext. occipital protuberance and crest

Mastoid notch for digastric

Groove for occipital artery

Occipital bone

Sagittal suture

Parietal bone

Lambdoid suture

Supreme
Superior } Nuchal line
Inferior

Mastoid process

Occipital condyle

BEHIND

Anterior fontanelle

Posterior fontanelle

Anterior fontanelle

Frontal eminence

Parietal tuberosity

Posterior fontanelle

Sphenoidal fontanelle

Mastoid fontanelle

SKULL AT BIRTH

Pansky

5. BASAL VIEW OF THE SKULL

I. Bones
A. PALATINE PROCESSES OF MAXILLAE
   1. Incisive fossa and foramen, behind incisor teeth
B. HORIZONTAL PLATE OF PALATINE: this, with A above, forms hard palate. Located here are the greater and lesser palatine foramina and posterior nasal spine
C. VOMER
D. MEDIAL PTERYGOID PLATE: scaphoid fossa located on lateral side of base; hamulus, at lower extremity
E. LATERAL PTERYGOID PLATE (process or lamina)
F. GREATER WING OF SPHENOID contains foramina ovale, rotundum, spinosum; angular spine; and sulcus or groove for auditory tube
G. TEMPORAL BONE: mandibular and jugular fossae, styloid process, stylomastoid foramen, mastoid process with mastoid notch and occipital groove, tympanomastoid fissure, carotid canal, foramen lacerum, inferior tympanic canaliculus, and mastoid canaliculus
H. BASILAR PART OF OCCIPITAL shows pharyngeal tubercle
I. LATERAL PART OF OCCIPITAL: condyles, condylar fossa and canal, hypoglossal canal, jugular process
J. SQUAMOUS PART OF OCCIPITAL: external occipital protuberance with the superior nuchal line, external occipital crest with the inferior nuchal line running laterally from its middle, and planum nuchale

II. Foramina (see p. 14): only those not given elsewhere will be listed below

Name	Bone	Contents
Incisive canals	Palatine process of maxilla	Nasopal. n. Ant. brs. descending palat. a. & v.
Greater palatine	Palatine	Great. palat. n. Descend. palat. vessels
Lesser palatine	Palatine	Lesser palatine nn. & aa.
Stylomastoid	Temporal	Facial n. Stylomastoid a.
Carotid	Temporal	Int. carotid a. Carotid plexus
Condyloid	Occipital	Vein from transverse sinus
Tympanic canaliculus	Temporal	Tympanic br. of IX
Mastoid canaliculus	Temporal	Auricular br. of X
Tympanomastoid fissure	Temporal	Auricular br. of X

III. Special features
A. CHOANAE: posterior openings of nasal cavity into nasopharynx
B. PHARYNGEAL CANALS: just lateral to base of vomer
C. EMISSARY SPHENOIDAL FORAMEN: at base of lateral lamina just medial to the foramen ovale (contains emissary vein)
D. FORAMEN LACERUM: between basilar part of occipital bone, petrous part of temporal and sphenoid. Filled with cartilage and transmits small vessels, nerves, and lymphatics

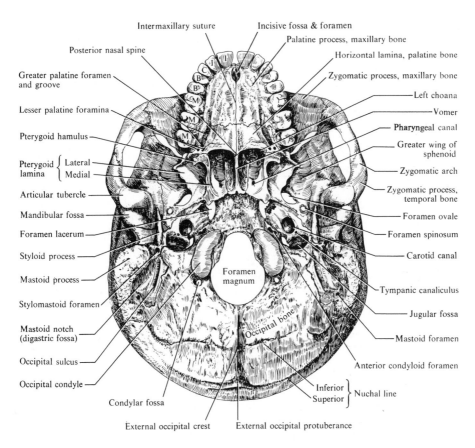

Intermaxillary suture

Incisive fossa & foramen

Posterior nasal spine

Palatine process, maxillary bone

Greater palatine foramen and groove

Horizontal lamina, palatine bone

Zygomatic process, maxillary bone

Lesser palatine foramina

Left choana

Vomer

Pterygoid hamulus

Pharyngeal canal

Pterygoid { Lateral
lamina { Medial

Greater wing of sphenoid

Zygomatic arch

Articular tubercle

Zygomatic process, temporal bone

Mandibular fossa

Foramen ovale

Foramen lacerum

Foramen spinosum

Styloid process

Carotid canal

Mastoid process

Foramen magnum

Tympanic canaliculus

Stylomastoid foramen

Occipital bone

Jugular fossa

Mastoid notch (digastric fossa)

Mastoid foramen

Occipital sulcus

Anterior condyloid foramen

Occipital condyle

Inferior } Nuchal line
Superior }

Condylar fossa

External occipital crest

External occipital protuberance

BASE OF SKULL

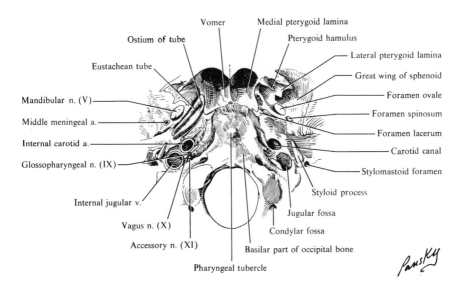

Vomer

Medial pterygoid lamina

Ostium of tube

Pterygoid hamulus

Eustachean tube

Lateral pterygoid lamina

Great wing of sphenoid

Mandibular n. (V)

Foramen ovale

Middle meningeal a.

Foramen spinosum

Internal carotid a.

Foramen lacerum

Glossopharyngeal n. (IX)

Carotid canal

Stylomastoid foramen

Internal jugular v.

Styloid process

Vagus n. (X)

Jugular fossa

Accessory n. (XI)

Condylar fossa

Basilar part of occipital bone

Pharyngeal tubercle

Netter / Pansky

6. SKULL INTERIOR: THE CRANIAL FOSSAE AND FORAMINA

I. Anterior cranial fossa
A. BOUNDARIES: anteriorly and laterally by squamous frontal bone, the ethmoid bone in the midline, and posteriorly by the posterior margin of lesser wing of sphenoid and anterior margin of chiasmatic sulcus
B. FLOOR: orbital plate of frontal bone, cribriform plate of ethmoid bone, the crista galli, and body and lesser wing of sphenoid bone
C. SPECIAL FEATURES: frontal crest, foramen cecum, crista galli, olfactory groove, ant. and post. ethmoidal foramina, and grooves for anterior meningeal vessels
   1. Crista galli: serves as anterior attachment for falx cerebri (dura)

II. Middle cranial fossa
A. BOUNDARIES: anterior: lesser wings of sphenoid, ant. clinoid processes, and ant. margin of chiasmatic groove; posterior: sup. angles of petrous bone and dorsum sellae; lateral: squamous temporal, petrous bone, great wing of sphenoid, and parietal bones
B. FLOOR: squamous temp., petrous bone, great wing of sphenoid, and sella turcica
C. SPECIAL FEATURES: sella turcica, posterior clinoid processes, carotid sulcus (groove), carotid canal, superior orbital fissure, optic foramina, foramen rotundum, ovale, spinosum, and lacerum, arcuate eminence, tegmen tympani of petrous temporal, hiatus for greater petrosal n., grooves for greater and lesser petrosal nerves, and grooves for branches of middle meningeal arteries

III. Posterior fossa
A. BOUNDARIES: anterior: dorsum sellae, basal occipital, and crest of petrous bone; lateral: parietal bone; posterior: squamous occipital bones
B. FLOOR: occipital and temporal bones
C. SPECIAL FEATURES: grooves for superior petrosal sinus, foramen magnum, hypoglossal canal, condylar canal, petro-occipital fissure, jugular foramen, internal acoustic (auditory) meatus, vestibular aqueduct, internal occipital crest, grooves for the transverse and sigmoid sinuses, and mastoid foramen

IV. Foramina or other openings and their principal contents

Name	Fossa	Contents
Cribriform	Ant.	Olfactory nerve fibers
Ant. ethmoid	Ant.	Ant. ethmoid vessels and nerve
Post. ethmoid	Ant.	Post. ethmoid vessels and nerve
Foramen cecum	Ant.	Origin of sup. sagittal venous sinus
Optic	Mid.	Optic n. and ophthalmic a.
Sup. orbital fissure	Mid.	III, IV, VI, ophthalmic V nn.; sympathetic nn.; ophthalmic vv.
Rotundum	Mid.	Maxillary n.
Ovale	Mid.	Mandibular n., access. meningeal a.
Spinosum	Mid.	Middle meningeal a.
Hiatus for greater petrosal n.	Mid.	Greater petrosal n., petrosal br. middle meningeal a.
Carotid canal	Mid.	Internal carotid a. and sympathetic nn.
Magnum	Post.	Spinal cord, accessory n., vertebral aa., ant. and post. spinal aa.
Jugular	Post.	Inf. petrosal and trans. sinuses, meningeal brs. of occip. and ascend. pharyngeal aa.; IX, X, XI nn.
Hypoglossal	Post.	XII n., meningeal br. of ascend. pharyngeal a.
Int. aud. meatus	Post.	VII, VIII nn.; labyrinthine (int. auditory) a.

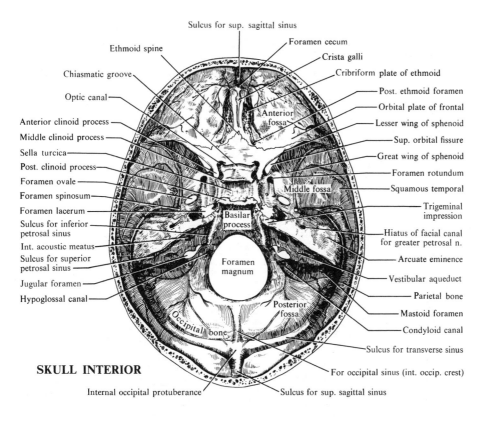

Sulcus for sup. sagittal sinus

Ethmoid spine

Foramen cecum

Crista galli

Chiasmatic groove

Cribriform plate of ethmoid

Optic canal

Post. ethmoid foramen

Orbital plate of frontal

Anterior
fossa

Lesser wing of sphenoid

Anterior clinoid process

Sup. orbital fissure

Middle clinoid process

Great wing of sphenoid

Sella turcica

Foramen rotundum

Post. clinoid process

Middle fossa

Squamous temporal

Foramen ovale

Foramen spinosum

Trigeminal
impression

Foramen lacerum

Basilar
process

Sulcus for inferior
petrosal sinus

Hiatus of facial canal
for greater petrosal n.

Int. acoustic meatus

Arcuate eminence

Foramen
magnum

Sulcus for superior
petrosal sinus

Vestibular aqueduct

Jugular foramen

Parietal bone

Hypoglossal canal

Mastoid foramen

Posterior
fossa

Condyloid canal

Occipital bone

Sulcus for transverse sinus

SKULL INTERIOR

For occipital sinus (int. occip. crest)

Internal occipital protuberance

Sulcus for sup. sagittal sinus

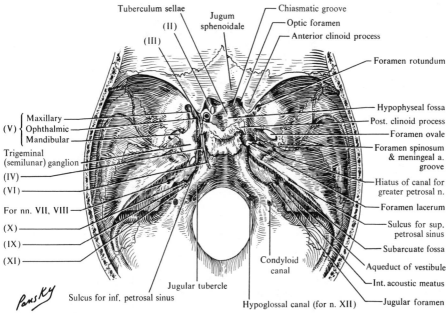

Tuberculum sellae

Jugum
sphenoidale

Chiasmatic groove

(II)

Optic foramen

(III)

Anterior clinoid process

Foramen rotundum

Maxillary

Hypophyseal fossa

(V) { Ophthalmic

Post. clinoid process

Mandibular

Foramen ovale

Trigeminal
(semilunar) ganglion

Foramen spinosum
& meningeal a.
groove

(IV)

(VI)

Hiatus of canal for
greater petrosal n.

For nn. VII, VIII

Foramen lacerum

(X)

Sulcus for sup.
petrosal sinus

(IX)

Subarcuate fossa

(XI)

Condyloid
canal

Aqueduct of vestibule

Int. acoustic meatus

Jugular tubercle

Jugular foramen

Sulcus for inf. petrosal sinus

Hypoglossal canal (for n. XII)

Pansky

-15-

7. CUTANEOUS NERVES AND DERMATOMES OF THE HEAD

I. **Introduction:** the innervation of the skin of the face is largely via the 3 branches of the trigeminal nerve (cranial nerve V). The skin over the angle of the mandible, anterior to and behind the ear, the anterior neck, and the back of the head and neck are innervated from the cervical plexus (C2–4). The trigeminal nerve is the largest of the 12 cranial nerves and is the principal general sensory nerve to the head, particularly the face (it is the motor nerve of muscles of mastication). Its sensory cell bodies lie in the trigeminal (semilunar or gasserian) ganglion and mesencephalic nucleus (midbrain). The peripheral processes of the unipolar cells of the ganglion form the ophthalmic (CN V1), the maxillary (CN V2), and sensory part of the mandibular nerve (CN V3).

II. **Cutaneous nerves of scalp**
   A. AREA IN FRONT OF EAR
      1. Rostral and medial: supratrochlear and supraorbital branches of ophthalmic (V)
      2. Lateral: zygomaticotemporal branch of maxillary (V), auriculotemporal branch of mandibular (V)
   B. AREA BEHIND EAR
      1. Lateral: great auricular and small occipital nerves from cervical plexus C2 and C3
      2. Posterior and medial: great occipital nerve—the posterior division of C2; least (3rd) occipital nerve—posterior division of C3

III. **Cutaneous nerves of face**
   A. NOSE: medially, supratrochlear, infratrochlear and nasal branches of ophthalmic (V); laterally, infraorbital branch of maxillary (V)
   B. EYE
      1. Upper lid: supratrochlear, infratrochlear, supraorbital, and lacrimal branches of ophthalmic (V)
      2. Lower lid: infratrochlear of ophthalmic (V), infraorbital and zygomaticofacial branches of maxillary (V)
   C. SKIN OVER UPPER JAW AND ZYGOMATIC AREA: infraorbital and zygomaticofacial branches of maxillary (V)
   D. SKIN OVER BUCCINATOR: buccal branch of mandibular (V)
   E. OVER LOWER JAW (front to back): mental, buccal and auriculotemporal branches of mandibular (V), great auricular of cervical plexus, C2 and C3

IV. **Cutaneous nerves of auricle (external ear)**
   A. LATERAL ASPECT: upper part, auriculotemporal branch of mandibular (V); lower, great auricular branch of cervical plexus C2 and C3
   B. MEDIAL ASPECT: upper part, lesser occipital from cervical plexus, C2 and C3; lower part, great auricular branch of cervical plexus, C2 and C3

V. **Cutaneous nerves of neck**
   A. POSTERIOR (from cranial to caudal): third occipital and posterior divisions of C2–C6
   B. LATERAL: great auricular and lesser occipital branches of cervical plexus, C2 and C3; intermediate supraclavicular branch of cervical plexus, C3 and C4
   C. ANTERIOR: transverse cervical branch of cervical plexus, C2 and C3; medial supraclavicular branch of cervical plexus, C3 and C4

VI. **Dermatomes,** a term applied to those areas supplied by spinal nerves. Taking the back of the head and the neck as a unit, keeping in mind that the first cervical nerve has no cutaneous branch, one begins at the highest point with C2 and descends to the lowest point at the back of the root of the neck, with C6. On the back of the neck, the posterior divisions of these nerves have direct cutaneous branches. The lateral and anterior aspects, receiving fibers from C2–C4, are carried through branches of the cervical plexus

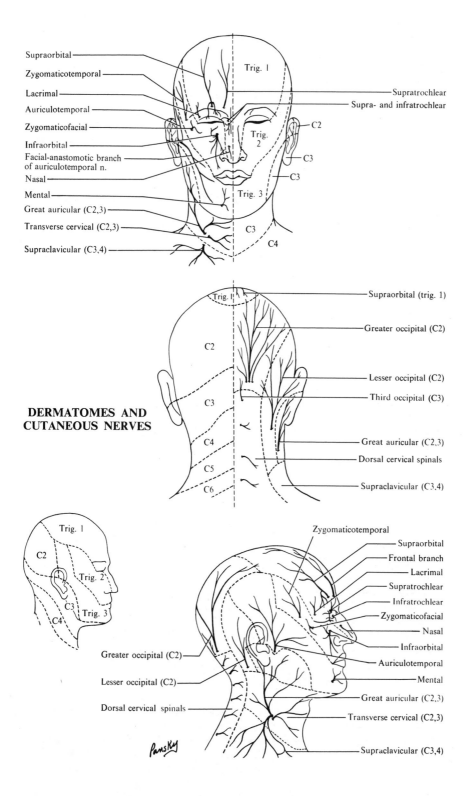

Supraorbital
Zygomaticotemporal
Lacrimal
Auriculotemporal
Zygomaticofacial
Infraorbital
Facial-anastomotic branch
of auriculotemporal n.
Nasal
Mental
Great auricular (C2,3)
Transverse cervical (C2,3)
Supraclavicular (C3,4)

Trig. 1
Trig. 2
Trig. 3

Supratrochlear
Supra- and infratrochlear
C2
C3
C3
C3
C4

**DERMATOMES AND
CUTANEOUS NERVES**

Trig. 1
C2
C3
C4
C5
C6

Supraorbital (trig. 1)
Greater occipital (C2)
Lesser occipital (C2)
Third occipital (C3)
Great auricular (C2,3)
Dorsal cervical spinals
Supraclavicular (C3,4)

Trig. 1
C2
Trig. 2
C3
Trig. 3
C4

Zygomaticotemporal
Supraorbital
Frontal branch
Lacrimal
Supratrochlear
Infratrochlear
Zygomaticofacial
Nasal
Infraorbital
Auriculotemporal
Mental
Great auricular (C2,3)
Transverse cervical (C2,3)
Supraclavicular (C3,4)

Greater occipital (C2)
Lesser occipital (C2)
Dorsal cervical spinals

Pansky

−17−

8. THE SCALP

I. Structure: composed of 5 layers

A. SKIN(S): thick, with many close-set hair follicles and their associated sebaceous and sweat glands. Firmly joined to next deeper layer

B. SUBCUTANEOUS TISSUE (C), SUPERFICIAL FASCIA: thick; strong with fiber bundles woven together, with fat interspaced
   1. Contains superficial vessels and nerves in abundance
   2. Hair follicles of skin project into this layer

C. MUSCULOAPONEUROTIC (A): represents the deep fascia (aponeurosis epicranialis)
   1. In forehead and occipital regions: the frontalis and occipitalis muscles are located here. In temporal region, auricular muscles are also in this layer
   2. Galea aponeurotica: a dense, thin, fibrous sheet that unites the frontal and occipital muscles of cranial vault

D. SUBAPONEUROTIC LAYER (L): very loose and scanty. Contains a few small vessels. The nature of this layer permits easy movement of layers A–C, which act as a unit, over the next layer

E. PERICRANIUM (P): the periosteum of the bones. Except at sutures, is poorly fixed to bone by connective tissue fibers (Sharpey's fibers)

If the above letters, *S C A L P*, are put together, they form the very word that each layer helps create.

II. Arteries of the scalp

A. OCCIPITAL: from external carotid artery to back of head

B. POSTERIOR AURICULAR: from external carotid artery to ear and scalp in posterior temporal region

C. SUPERFICIAL TEMPORAL: one of terminal branches of external carotid artery to lateral part of scalp in front of ear

D. SUPRAORBITAL: from ophthalmic artery to skin of forehead and rostral area of scalp

E. SUPRATROCHLEAR: from ophthalmic artery to most rostral and medial parts of scalp

III. Cutaneous nerves

A. ROSTRALLY: the supratrochlear and supraorbital branches of ophthalmic (V) supply regions corresponding to arteries of same name

B. LATERALLY: auriculotemporal branch of mandibular (V) follows superficial temporal artery

C. POSTERIORLY: the greater occipital (C2) accompanies branches of occipital artery

D. POSTEROLATERALLY: the lesser occipital from cervical plexus (C2, 3); in general, covers same areas as posterior auricular artery

IV. Clinical considerations

A. DUE TO LOOSENESS of subaponeurotic layer, large amounts of blood can form enormous hematomas after blows on head

B. FOR SAME REASON, INFECTIONS entering this layer may also spread widely by way of:
   1. Emissary veins (see p. 24), which pass through bones directly to dura, can carry infections to meninges
   2. Diploic veins (see p. 24), located between bony tables of the skull

C. DUE TO THE THICKNESS AND EXTENT of the fibrous tissue of the second layer and extensive vascular anastomoses, wounds of the scalp tend to bleed profusely

D. THE PERICRANIUM possesses little osteogenic capacity as compared to most periosteum, and except over the sutures it is rather loosely attached to the bones of the calvaria

E. IT IS THE PULSATING DISTENTION of the arteries of the scalp that accounts for most of the pain in migraine headache

F. SEBACEOUS CYSTS (wens): obstruction of ducts of sebaceous glands associated with hair follicles and retention of secretions

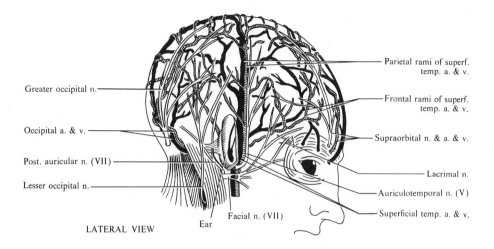

Greater occipital n.

Occipital a. & v.

Post. auricular n. (VII)

Lesser occipital n.

Parietal rami of superf. temp. a. & v.

Frontal rami of superf. temp. a. & v.

Supraorbital n. & a. & v.

Lacrimal n.

Auriculotemporal n. (V)

Superficial temp. a. & v.

Facial n. (VII)

Ear

LATERAL VIEW

SCALP VESSELS AND NERVES

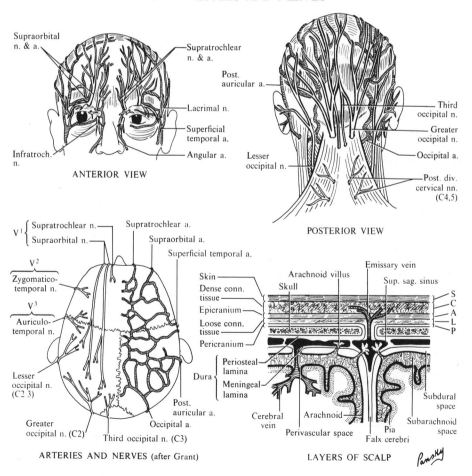

Supraorbital n. & a.

Supratrochlear n. & a.

Post. auricular a.

Lacrimal n.

Superficial temporal a.

Infratroch. n.

Angular a.

ANTERIOR VIEW

Third occipital n.

Greater occipital n.

Occipital a.

Post. div. cervical nn. (C4,5)

Lesser occipital n.

POSTERIOR VIEW

V^1 { Supratrochlear n.
Supraorbital n.

V^2

Zygomatico-temporal n.

V^3

Auriculo-temporal n.

Lesser occipital n. (C2,3)

Greater occipital n. (C2)

Third occipital n. (C3)

Supratrochlear a.

Supraorbital a.

Superficial temporal a.

Post. auricular a.

Occipital a.

ARTERIES AND NERVES (after Grant)

Skin

Dense conn. tissue

Epicranium

Loose conn. tissue

Pericranium

Dura { Periosteal lamina
Meningeal lamina

Cerebral vein

Perivascular space

Arachnoid villus

Skull

Emissary vein

Sup. sag. sinus

S
C
A
L
P

Subdural space

Subarachnoid space

Arachnoid

Pia

Falx cerebri

LAYERS OF SCALP

Pansky

9. SUPERFICIAL AND DEEP VEINS

    I. Supratrochlear vein: from middle scalp and forehead, near midline. Descends to medial angle of eye, where it joins supraorbital vein to form angular vein and communicates with branches of superficial temporal vein

   II. Supraorbital vein: from scalp and forehead. Sends branch through supraorbital notch and continues to join frontal vein, forming angular vein

  III. Angular vein: runs caudad at side of nose to level of lower orbit, where it becomes anterior facial vein. Receives tributaries from sides of nose and communicates, through nasofrontal vein at medial angle of orbit, with superior ophthalmic vein

  IV. Anterior facial vein: continuation of angular v. caudally. Runs beneath facial muscles, over masseter muscle, crosses mandible, runs caudally and posteriorly under platysma muscle to join ant. branch of retromandibular v., forming common facial v. Has communication through deep facial v. with pterygoid plexus. Receives tributaries from eyelids, lips, cheeks, and masseter muscles. In neck receives submental, submandibular, and palatine branches

   V. Superficial temporal vein: from side of head and scalp. Branches communicate with I and II, above. Is joined by middle temporal vein from temporalis muscle. Crosses zygomatic arch to enter parotid gland where it joins maxillary vein to form retromandibular vein. Tributaries from parotid gland, external ear, side of face

  VI. Pterygoid plexus: network between temporalis and pterygoid muscles. Receives middle meningeal, sphenopalatine, deep temporal, alveolar, palatine, and muscular veins. Communicates with the anterior facial vein through inferior orbital fissure via the inferior orbital vein and with cavernous sinus through veins in foramina ovale and lacerum

 VII. Maxillary vein: short vein draining pterygoid plexus and joining superficial temporal to form posterior facial vein

VIII. Retromandibular vein: formed by union of superficial temporal and maxillary veins. Descends in parotid, lateral to ext. carotid a. but deep to facial n. Divides into 2 branches
   A. ANTERIOR, running rostrally to join ant. facial vein to form common facial vein
   B. POSTERIOR, which is joined by post. auricular vein to form external jugular vein

  IX. Posterior auricular vein: from side of head, behind ear, descends behind auricle and joins posterior branch of posterior facial to form external jugular vein

   X. Occipital vein: from back of head to deep cervical and vertebral veins

  XI. External jugular vein: formed by VIIIB and IX above, in parotid gland at level of angle of mandible. Descends vertically to enter subclavian triangle, where it pierces fascia and terminates in subclavian vein. Communicates with internal jugular vein; receives trans. cervical, trans. scapular, and anterior jugular veins

 XII. Posterior external jugular vein: from occipital region and upper posterior side of neck to external jugular vein

XIII. Anterior jugular vein: from submandibular region, descends near midline to enter external jugular vein. Receives some drainage from larynx and thyroid. Above sternum, veins of 2 sides are joined through venous jugular arch

XIV. Clinical considerations: infections on the face (above mouth) can be carried into the cavernous dural sinus via the facial, angular, and sup. ophthalmic vv.

SUPERFICIAL AND DEEP VEINS

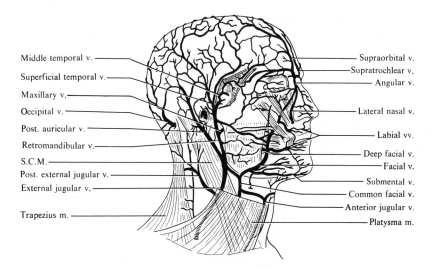

Middle temporal v.
Superficial temporal v.
Maxillary v.
Occipital v.
Post. auricular v.
Retromandibular v.
S.C.M.
Post. external jugular v.
External jugular v.
Trapezius m.

Supraorbital v.
Supratrochlear v.
Angular v.
Lateral nasal v.
Labial vv.
Deep facial v.
Facial v.
Submental v.
Common facial v.
Anterior jugular v.
Platysma m.

SUPERFICIAL VEINS

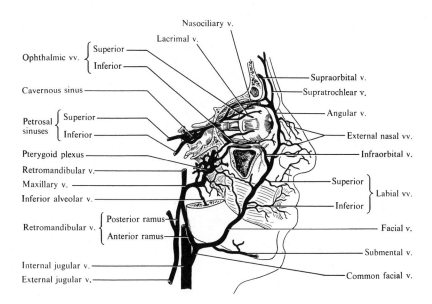

Nasociliary v.
Lacrimal v.

Ophthalmic vv. { Superior
Inferior

Cavernous sinus

Petrosal sinuses { Superior
Inferior

Pterygoid plexus
Retromandibular v.
Maxillary v.
Inferior alveolar v.

Retromandibular v. { Posterior ramus
Anterior ramus

Internal jugular v.
External jugular v.

Supraorbital v.
Supratrochlear v.
Angular v.
External nasal vv.
Infraorbital v.
Superior } Labial vv.
Inferior
Facial v.
Submental v.
Common facial v.

FACIAL AND OPHTHALMIC VEINS

– 21 –

10. SUPERFICIAL LYMPHATICS OF HEAD AND NECK

I. Lymphatic vessels
A. LYMPHATICS OF SCALP: three drainage areas: frontal, ending in anterior auricular and parotid nodes; temporal and parietal, ending in parotid and retroauricular nodes; occipital, ending in occipital and deep cervical nodes
B. LYMPHATICS OF EXTERNAL EAR drain into pre- and retroauricular nodes and superficial and deep cervical nodes
C. LYMPHATICS OF FACE
    1. Eyelids and conjunctiva: to submandibular and parotid nodes
    2. Cheek: to parotid and submandibular nodes
    3. Side of nose, upper lip, and lateral lower lip: to submandibular nodes
    4. Medial lower lip: into submental nodes
    5. Temporal and infratemporal fossae: into deep facial and deep cervical nodes

II. Lymph nodes of head
A. OCCIPITAL: back of head close to edge of trapezius. Afferents from scalp at back of head; efferents to superior deep cervical
B. RETROAURICULAR: at insertion of sternocleidomastoid on mastoid. Afferents from posterior temporal and parietal regions; efferents to superior deep cervicals
C. PREAURICULAR: in front of tragus of ear. Afferents from pinna and temporal region; efferents to superior deep cervical
D. PAROTID: two sets, either embedded in gland or just below it. Afferents from root of nose, eyelids, anterior temporal region and external auditory meatus; efferents to superior deep cervical
E. FACIAL: three sets: infraorbital, buccal, and mandibular. Afferents from eyelids, conjunctiva, skin of nose, nasal mucosa, and cheek; efferents to submandibular nodes

III. Lymph nodes of neck
A. SUBMANDIBULAR: under body of mandible. Afferents from cheek, nose, upper lip, lower lip, facial and submental nodes; efferents to superior deep cervical
B. SUBMENTAL: between anterior bellies of digastric. Afferents from central lower lip, floor of mouth; efferents to submandibular and deep cervical
C. SUPERFICIAL CERVICAL: along external jugular vein. Afferents from ear and parotid region; efferents to superior deep cervical
D. DEEP CERVICAL: along carotid sheath
    1. Superior, under sternocleidomastoid muscle, along accessory nerve and internal jugular vein. Afferents from back of head and neck, tongue, larynx, thyroid, palate, nose, esophagus, and all nodes except inferior deep cervical; efferents to inferior deep cervical nodes and jugular trunk
    2. Inferior, extending below border of sternocleidomastoid, close to subclavian vein. Afferents from dorsum of scalp and neck, superficial pectoral region, part of arm, and superior deep cervical; efferents join efferents of superior deep nodes to form jugular trunk

IV. Jugular trunk terminates on right, at junction of internal jugular and subclavian veins; on left, in thoracic duct

V. Clinical considerations
A. Since lymph nodes act as filters for lymph, and both infectious material and metastatic cells are carried in lymph, one must know the position as well as the drainage areas of node groups. In case of cancer, the swelling of regional nodes may be the first sign of a malignancy in some deep-lying structure

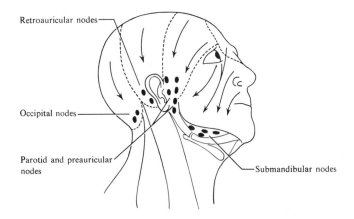

Retroauricular nodes

Occipital nodes

Parotid and preauricular
nodes

Submandibular nodes

LYMPH DRAINAGE OF HEAD AND NECK

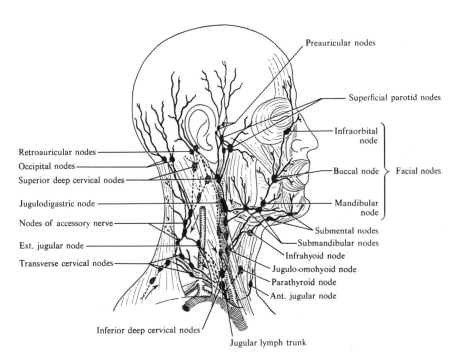

Preauricular nodes

Superficial parotid nodes

Infraorbital
node

Retroauricular nodes

Occipital nodes

Superior deep cervical nodes

Buccal node

Facial nodes

Jugulodigastric node

Nodes of accessory nerve

Mandibular
node

Ext. jugular node

Submental nodes

Transverse cervical nodes

Submandibular nodes

Infrahyoid node

Jugulo-omohyoid node

Parathyroid node

Ant. jugular node

Inferior deep cervical nodes

Jugular lymph trunk

PRINCIPAL NODE GROUPS AND DRAINAGE

Pansky

11. THE DIPLOIC AND EMISSARY VEINS

I. The diploic veins are endothelium-lined channels found between the inner and outer tables of the calvaria. The meningeal veins, diploic veins, veins of the scalp, and dural venous sinuses do not have valves, and there is free communication between these venous systems. There are 4 major diploic veins on each side:

A. THE FRONTAL DIPLOIC VEIN drains the anterior portion of the frontal bone into the supraorbital vein and superior sagittal sinus

B. THE ANTERIOR TEMPORAL DIPLOIC VEIN drains the posterior portion of the frontal bone and the anterior portion of the parietal and temporal bones into the sphenoparietal sinus and into the temporal veins via emissary veins in the greater wing of the sphenoid

C. THE POSTERIOR TEMPORAL DIPLOIC VEINS drain the posterior portions of the parietal and temporal bones into the transverse and sigmoid sinuses and the mastoid emissary vein

D. THE OCCIPITAL DIPLOIC VEINS drain the occiput into the confluence of sinuses, transverse sinuses, occipital veins, and the mastoid emissary vein

II. The emissary veins connect the intracranial venous sinuses with veins outside the cranium. Although they are valveless and blood can flow in both directions, the flow is usually away from the brain. Their size and number are variable. There are usually 6 to 10 recognized connections.

A. PARIETAL EMISSARY VEINS, one on each side, pass through parietal foramina and connect the superior sagittal sinus with external veins of the scalp. They may communicate with diploic veins

B. MASTOID EMISSARY VEINS connect each sigmoid sinus via the mastoid foramen with occipital or posterior auricular veins

C. POSTERIOR CONDYLAR EMISSARY VEINS pass through the condylar canals connecting the sigmoid sinus with the suboccipital plexus of veins

D. FRONTAL EMISSARY VEIN (seen in children and in some adults) passes through the foramen cecum connecting the superior sagittal sinus with veins of the frontal sinus and/or those of the nasal cavities

E. OCCIPITAL EMISSARY VEINS

F. SPHENOIDAL EMISSARY VEINS

G. VEINS IN THE HYPOGLOSSAL CANAL

H. VEINS IN THE FORAMEN OVALE

I. VEINS FOLLOWING THE INTERNAL CAROTID ARTERY

J. OPHTHALMIC VEINS: although these are not technically emissary veins, they serve the same dangerous function of connecting outside surfaces with dural sinuses and drain areas likely to become infected

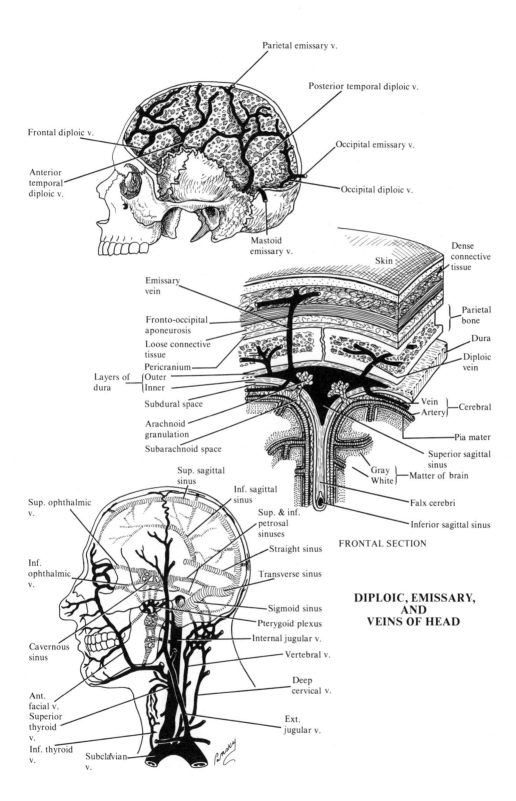

Parietal emissary v.

Posterior temporal diploic v.

Frontal diploic v.

Occipital emissary v.

Anterior
temporal
diploic v.

Occipital diploic v.

Mastoid
emissary v.

Dense
connective
tissue

Skin

Emissary
vein

Parietal
bone

Fronto-occipital
aponeurosis

Dura

Loose connective
tissue

Diploic
vein

Pericranium

Layers of
dura

Outer

Inner

Vein
Artery

Cerebral

Subdural space

Arachnoid
granulation

Pia mater

Subarachnoid space

Superior sagittal
sinus

Sup. sagittal
sinus

Gray
White

Matter of brain

Inf. sagittal
sinus

Falx cerebri

Sup. ophthalmic
v.

Sup. & inf.
petrosal
sinuses

Inferior sagittal sinus

FRONTAL SECTION

Straight sinus

Inf.
ophthalmic
v.

Transverse sinus

Sigmoid sinus

**DIPLOIC, EMISSARY,
AND
VEINS OF HEAD**

Pterygoid plexus

Cavernous
sinus

Internal jugular v.

Vertebral v.

Deep
cervical v.

Ant.
facial v.
Superior
thyroid
v.

Ext.
jugular v.

Inf. thyroid
v.

Subclavian
v.

12. MUSCLES OF FACIAL EXPRESSION—PART I

I. **General considerations:** all are skin (cutaneous) muscles. They lie in superficial fascia and may arise from either fascia or bone. They insert into skin. Many individual muscles are indistinctly separated from those closely adjoining. All are innervated by the facial nerve (VII). They are grouped according to location or area of principal action. Due to a deficiency of deep fascia and the fact that the superficial fascia between the cutaneous attachments of the muscles is loose, lacerations of the face tend to open wide. Thus, suturing of the skin of the face must be done with great care to avoid scarring

II. **Groupings of facial muscles,** with principal actions only
   A. SCALP
      1. Frontalis: raises brows and wrinkles forehead as in surprise or frowning
         a. Arises from the galea aponeurotica and inserts into the skin and dense subcutaneous connective tissue at the level of the eyebrows
      2. Occipitalis: acts with the frontalis in raising the eyebrows
         a. Arises from the lateral 2/3 of the highest nuchal line of the occipital bone, the mastoid part of the temporal bone, and ends in the galea aponeurotica
   B. EAR
      1. Anterior auricular draws ear up and forward
      2. Superior auricular draws ear up
      3. Posterior auricular draws ear back
   C. EYE
      1. Orbicularis oculi
         a. Consists of 3 parts: a thick *orbital part* for closing the eyes tightly to protect against the glare of light (squinting) and dust; a thin *palpebral part* for closing the eyelids lightly to keep the cornea from drying (responsible for blinking); and a *lacrimal part* (deep to the palpebral part and often a part of it) for drawing the eyelids and the lacrimal puncta medially. The latter part encloses the lacrimal canaliculi and passes behind the lacrimal sac (important in control of tears)
         b. When all parts contract, the eyes are firmly closed and adjacent skin is wrinkled. These wrinkles become permanent by age 30–35 and are called ''crow's feet''
         c. Paralysis results in drooping of the lower lid (ectropion) and the spilling of tears (epiphora)
   D. NOSE
      1. Procerus: a small muscle that is continuous with the occipitofrontalis, passing from the forehead over the bridge of the nose where it inserts. It draws the medial angle of the eyebrow downward and produces wrinkles over the bridge of the nose
      2. Depressor septi: runs from the incisor fossa of the maxilla to the mobile part of the nasal septum. It assists the dilator nares in widening the aperture during deep inspiration and also depresses the nasal septum
      3. Nasalis: consists of transverse (compressor naris) and alar (dilator naris) portions
         a. Compressor nares: from the maxilla above the incisor teeth to the dorsum of the nose. Compress the anterior nasal openings
         b. Dilator nares: from the maxilla into the alar cartilages of the nose. They widen the anterior nasal openings. In children, a marked or increased action of these muscles suggests respiratory distress (pneumonia)

III. **Clinical considerations**
   A. DAMAGE TO THE FACIAL NERVE or its branches produces variable amounts of weakness or paralysis of facial muscles, sometimes called ''Bell's palsy.'' Weakness is particularly noticeable about the mobile mouth, where it may be evidenced even in repose by a sagging corner and becomes very obvious as a result of the asymmetry attending an attempt to show the teeth or smile. Upper facial weakness can similarly be brought out by having the patient attempt to frown, raise his eyebrows, or close his eyes tightly.

DEEP AND SUPERFICIAL MUSCLES OF FACIAL EXPRESSION

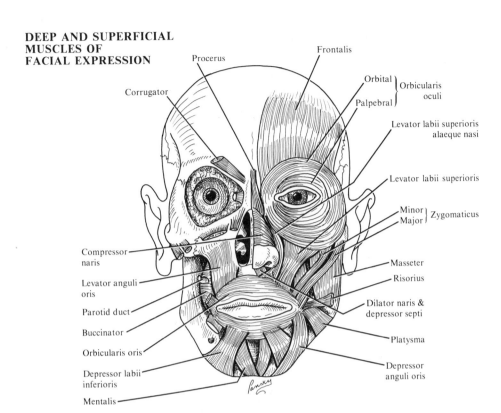

Procerus

Corrugator

Frontalis

Orbital
Palpebral } Orbicularis oculi

Levator labii superioris alaeque nasi

Levator labii superioris

Minor
Major } Zygomaticus

Compressor naris

Levator anguli oris

Parotid duct

Buccinator

Orbicularis oris

Depressor labii inferioris

Mentalis

Masseter

Risorius

Dilator naris & depressor septi

Platysma

Depressor anguli oris

MUSCLE ATTACHMENTS

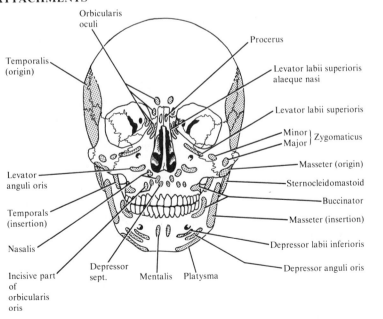

Orbicularis oculi

Temporalis (origin)

Procerus

Levator labii superioris alaeque nasi

Levator labii superioris

Minor
Major } Zygomaticus

Masseter (origin)

Sternocleidomastoid

Buccinator

Masseter (insertion)

Depressor labii inferioris

Depressor anguli oris

Levator anguli oris

Temporals (insertion)

Nasalis

Incisive part of orbicularis oris

Depressor sept.

Mentalis

Platysma

13. MUSCLES OF FACIAL EXPRESSION—PART II

II. Groupings of facial muscles (cont'd)

E. MOUTH

1. Levator labii superioris: descends from maxilla just above the infraorbital foramen (below infraorbital margin) to insert into upper lip. It raises upper lip
2. Levator labii superioris alaeque nasi: from frontal process of maxilla to insert on ala of nose and on lip. It raises the lip and dilates the nares
3. Zygomaticus major: from the zygomatic bone to insert into the orbicularis oris near angle of mouth. It draws the angle of mouth up and back, as in laughing and smiling
4. Zygomaticus minor: an occasional muscle slip almost continuous with the orbicularis oculi and inserts on the angle of the mouth. With 1 and 2 above, it helps form the nasolabial groove. It helps lift the angles of the mouth in smiling and deepen the nasolabial groove in sorrow
5. Levator anguli oris (caninus): from the maxilla below the infraorbital foramen to insert on the corner of the mouth (on a deeper plane than the zygomaticus and levator labii superioris muscles). With the levator labii superioris, alaeque nasi, and zygomaticus minor, it helps express contempt or disdain by raising the angle of the mouth
6. Risorius: extends posterior from corners of the mouth as a small, very superficial muscle which pulls the corners of the mouth laterally, retracting the angle of the mouth. These are the so-called "laughter" muscles
7. Depressor labii inferioris: deep to the depressor anguli oris, arising from the mandible just above the depressor anguli oris, and inserts into the lower lip. It depresses the lower lip, drawing the lip down and back
8. Depressor anguli oris: from the mandible near the attachment of the platysma to insert on the corners of the mouth. It pulls down or depresses the corner of the mouth
9. Mentalis: is named for its location, not its action. It is a small muscle just anterior to the mental foramen, arising from the mandible and inserting just inferior to the incisor teeth. It raises and wrinkles the skin of the chin and pushes up the lower lip
10. Platysma: a thin sheet of muscle in the superficial fascia of the neck, from the pectoral region to the mandible. It retracts and depresses the angle of the mouth
11. Orbicularis oris: completely surrounds the mouth, with no bony attachments. It is a complex muscle with layers, some parts intrinsic to the lips, and others derived from the buccinator, levator anguli oris, depressor anguli oris, and zygomaticus major and minor. It closes the lips, protrudes and purses the lips, and presses the lips to the teeth aiding in articulation and mastication (chewing)
12. Buccinator: from the pterygomandibular ligament and lateral surfaces of the mandible and maxilla to insert into the muscles around the mouth, with fibers crossing to the upper and lower lip. It compresses the cheek, holds food under the teeth in mastication, and is important in blowing when the cheeks are distended with air

B. PARALYSIS OF AN ENTIRE SIDE OF THE FACE is an indication that the facial nerve as a whole has been damaged along its course

C. WEAKNESS RATHER THAN COMPLETE PARALYSIS of a group of muscles typically results from injury to facial nerve branches because of the overlap of their distribution

D. FACIAL INFLAMMATION results in much swelling because of the looseness of the superficial fascia that permits fluid and blood to accumulate in the loose connective tissue

E. IN PARKINSON'S DISEASE OR SYNDROME, emotional changes of facial expression are usually lost, giving the face a masklike expressionless appearance

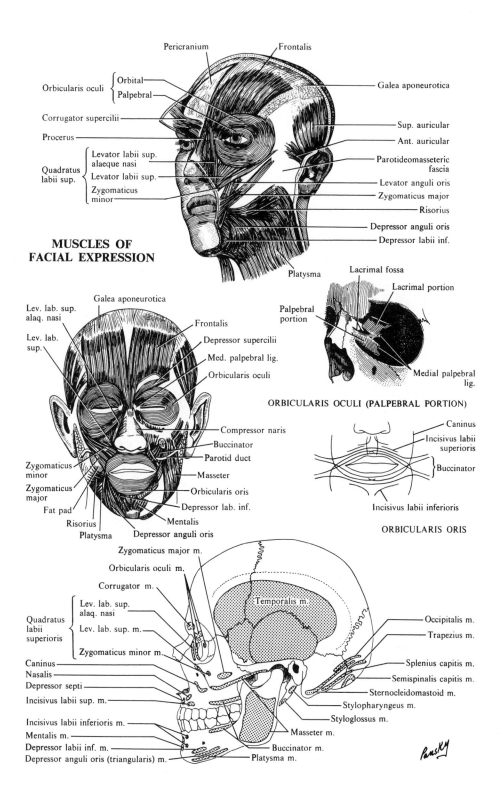

Pericranium · Frontalis

Orbicularis oculi { Orbital · Palpebral }

Galea aponeurotica

Corrugator supercilii

Procerus

Quadratus labii sup. {
Levator labii sup. alaeque nasi
Levator labii sup.
Zygomaticus minor
}

Sup. auricular

Ant. auricular

Parotideomasseteric fascia

Levator anguli oris

Zygomaticus major

Risorius

Depressor anguli oris

Depressor labii inf.

Platysma

MUSCLES OF FACIAL EXPRESSION

Lacrimal fossa

Lacrimal portion

Palpebral portion

Medial palpebral lig.

ORBICULARIS OCULI (PALPEBRAL PORTION)

Galea aponeurotica

Lev. lab. sup. alaq. nasi

Lev. lab. sup.

Frontalis

Depressor supercilii

Med. palpebral lig.

Orbicularis oculi

Compressor naris

Buccinator

Parotid duct

Masseter

Orbicularis oris

Depressor lab. inf.

Mentalis

Zygomaticus minor

Zygomaticus major

Fat pad

Risorius

Platysma

Depressor anguli oris

Caninus

Incisivus labii superioris

Buccinator

Incisivus labii inferioris

ORBICULARIS ORIS

Zygomaticus major m.

Orbicularis oculi m.

Corrugator m.

Quadratus labii superioris {
Lev. lab. sup. alaq. nasi
Lev. lab. sup. m.
Zygomaticus minor m.
}

Caninus

Nasalis

Depressor septi

Incisivus labii sup. m.

Incisivus labii inferioris m.

Mentalis m.

Depressor labii inf. m.

Depressor anguli oris (triangularis) m.

Temporalis m.

Occipitalis m.

Trapezius m.

Splenius capitis m.

Semispinalis capitis m.

Sternocleidomastoid m.

Stylopharyngeus m.

Styloglossus m.

Masseter m.

Buccinator m.

Platysma m.

Pansky

14. MUSCLES OF MASTICATION

I. **General considerations:** all these muscles are inserted upon the mandible and are concerned in the process of biting and chewing. All are innervated by the mandibular division of the trigeminal (V) nerve

II. **Arrangement**
   A. TEMPORALIS: (see p. 40)
      1. Origin: from temporal fascia and entire temporal fossa from temporal lines to infratemporal crest
      2. Insertion: coronoid process and anterior border of ramus of mandible
      3. Action: closes jaw, posterior part retracts jaw
   B. MASSETER: (see p. 40)
      1. Origin
         a. Superficial part: lower border of zygomatic arch
         b. Deep part: posterior and medial side of zygomatic arch
      2. Insertion
         a. Superficial part: angle and lower lateral side of ramus of mandible
         b. Deep part: upper lateral ramus and coronoid process of mandible
      3. Action: closes jaw
   C. MEDIAL PTERYGOID (INTERNAL PTERYGOID)
      1. Origin: medial side of lateral pterygoid plate of sphenoid, pyramidal process of palatine and tuberosity of maxilla
      2. Insertion: lower and posterior part of medial surface of ramus of mandible
      3. Action: closes jaw
   D. LATERAL PTERYGOID (EXTERNAL PTERYGOID)
      1. Origin
         a. Upper part: from lower lateral great wing of sphenoid and infratemporal crest
         b. Lower part: from lateral surface of lateral pterygoid plate
      2. Insertion: neck of mandibular condyle and articular disk and capsule of temporomandibular joint
      3. Action: depresses, protrudes, and moves mandible from side to side

III. **Clinical considerations**
   A. LESIONS OF THE MANDIBULAR DIVISION OF THE TRIGEMINAL NERVE will cause unilateral paralysis of muscles of mastication followed by atrophy. This results in a sunken-in appearance along the ramus of the mandible and above the zygomatic arch
   B. THE MASSETER, TEMPORALIS, AND MEDIAL PTERYGOID MUSCLES are powerful closers of the jaw, accounting for the strength of the bite
   C. THE MASSETER AND TEMPORALIS MUSCLES both abduct (deviate) the jaw to the same side, but the medial pterygoid abducts to the opposite side. Thus, with proper synchronization, these muscles can produce the grinding movement of chewing (the tongue and the buccinator muscle also aid in mastication, but in different ways: the tongue positions the food on the teeth, while the buccinator helps to maintain it there during chewing)
   D. THE POSTERIOR FIBERS OF THE TEMPORALIS MUSCLE are the chief retractor of the mandible and are also responsible for maintaining the resting position of closure of the mouth
   E. ALTERNATING ACTION OF THE LATERAL PTERYGOIDS can move the jaw from side to side, but the important action of these muscles is to act bilaterally and helps to open the mouth

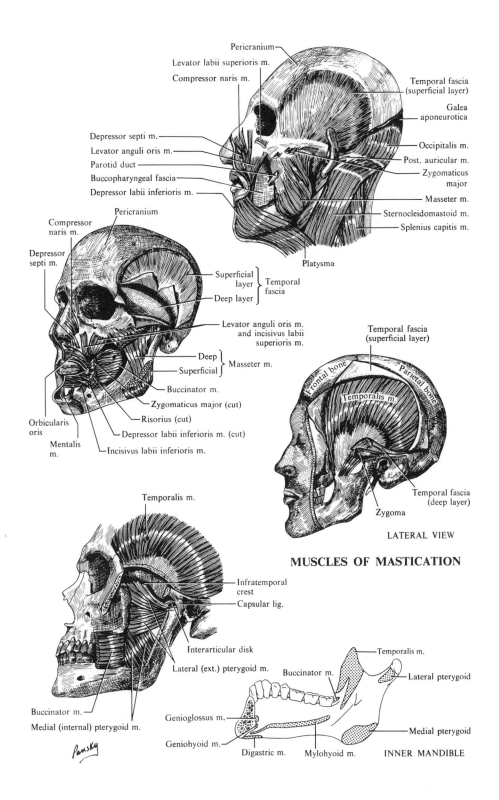

Pericranium
Levator labii superioris m.
Compressor naris m.

Temporal fascia (superficial layer)
Galea aponeurotica

Depressor septi m.
Levator anguli oris m.
Parotid duct
Buccopharyngeal fascia
Depressor labii inferioris m.

Occipitalis m.
Post. auricular m.
Zygomaticus major
Masseter m.
Sternocleidomastoid m.
Splenius capitis m.

Platysma

Compressor naris m.
Depressor septi m.
Pericranium

Superficial layer
Deep layer
} Temporal fascia

Levator anguli oris m. and incisivus labii superioris m.

Deep
Superficial
} Masseter m.

Buccinator m.
Zygomaticus major (cut)
Risorius (cut)
Depressor labii inferioris m. (cut)
Incisivus labii inferioris m.

Orbicularis oris
Mentalis m.

Temporal fascia (superficial layer)

Frontal bone
Parietal bone
Temporalis m.

Temporal fascia (deep layer)
Zygoma

LATERAL VIEW

MUSCLES OF MASTICATION

Temporalis m.

Infratemporal crest
Capsular lig.

Interarticular disk
Lateral (ext.) pterygoid m.

Buccinator m.
Medial (internal) pterygoid m.

Temporalis m.
Lateral pterygoid

Buccinator m.

Genioglossus m.
Geniohyoid m.
Digastric m.
Mylohyoid m.

Medial pterygoid

INNER MANDIBLE

Pansky

15. THE FACIAL (VII) NERVE

I. **Roots:** 2 in number: a large voluntary motor and a smaller mixed sensory and parasympathetic (nervus intermedius)

II. Course

A. LEAVES CRANIAL FOSSA with acoustic nerve (VIII) via internal acoustic meatus

B. LEAVES NERVE VIII to enter facial canal, which passes laterally and then posteriorly along the medial wall of the tympanic cavity above oval window, then turns caudally to exit from skull through the stylomastoid foramen

C. GANGLION: geniculate, located at acute bend of the facial canal

D. BRANCHES
1. Greater petrosal nerve: mixed, with parasympathetic and sensory fibers. Leaves region of geniculate ganglion through *facial hiatus* to run in sulcus on petrous bone to foramen lacerum and then through pterygoid canal to pterygopalatine fossa. Unites with deep petrosal nerve (sympathetic) to form nerve of pterygoid canal (vidian nerve). The parasympathetic fibers synapse in pterygopalatine ganglion whose postganglionic fibers reach the lacrimal gland and glands of nose and palate. Sensory fibers from these same areas merely pass through the pterygopalatine ganglion to reach their cell bodies in the geniculate ganglion
2. Nerve to stapedius muscle: very small, leaves facial nerve in its descending part to supply muscle
3. Chorda tympani: leaves facial nerve in descending part of facial canal, arches upward and rostrally, enters tympanic cavity by crossing its lateral wall, exits petrous bone through petrotympanic fissure and descends in the infratemporal fossa where it joins the posterior border of the lingual nerve (V). Its sensory fibers are those of taste from the anterior 2/3 of tongue, and these follow the general sensory branches of the lingual nerve. The parasympathetic fibers leave the lingual nerve by parasympathetic roots which terminate by synapse in the submandibular ganglion. From the latter, postganglionic glandular rami go to the submandibular and sublingual glands
4. Muscular branches: supply posterior belly of digastric, stylohyoid, and all muscles of facial expression. The latter branches are designated as follows: temporal, zygomatic, buccal, marginal mandibular, and cervical, depending on the region of the face supplied
5. Posterior auricular branch: gives motor fibers to superior and posterior auricular muscles and occipital muscle of scalp. Carries some sensory fibers to external ear

III. Clinical considerations

A. AFTER LEAVING THE STYLOMASTOID FORAMEN, the nerve runs forward in substance of parotid gland. In surgical operations on parotid, this must be noted

B. PARALYSIS: if lesions occur in either the facial nucleus or its peripheral fibers, there is total unifacial paralysis; if lesions are in either the facial area of cerebral cortex or the descending corticobulbar fibers, only muscles below the eye will be paralyzed since those above the eye receive cortical fibers from both sides of the brain. Bell's facial paralysis is sometimes caused by nerve irritations, therefore may be only temporary. Paralysis of muscles of facial expression leads to a lowered eyebrow, inability to blink (conjunctiva becomes ulcerous), inability to dilate nostrils, a one-sided smile, inability to contain food or saliva in mouth, and a distinct droop of face. Symptoms are ipsilateral

C. LOSS OF TASTE on anterior 2/3 of tongue indicates chorda tympani involvement

D. THE EXACT DISTRIBUTION AND FUNCTION OF MANY OF THE SENSORY FIBERS of the facial nerve are not known, but some are thought to be concerned with deep pain from the face; some are apparently distributed to a small part of the soft palate; a few may reach the middle ear cavity; and a few cutaneous fibers are distributed to the skin on the posterior surface of the ear, along with similar fibers from the ninth (IX) and tenth (X) cranial nerves. The best-known sensory fibers in this nerve are those for taste on the anterior 2/3 of the tongue

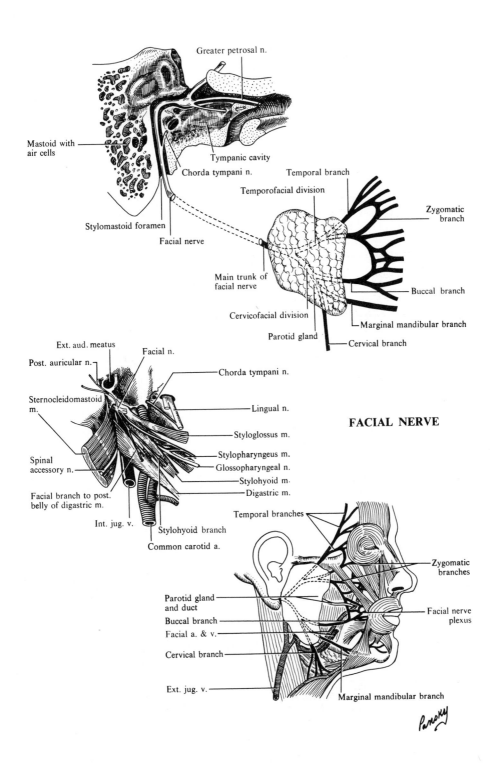

Greater petrosal n.

Mastoid with
air cells

Tympanic cavity

Chorda tympani n.

Temporal branch

Temporofacial division

Zygomatic
branch

Stylomastoid foramen

Facial nerve

Main trunk of
facial nerve

Buccal branch

Cervicofacial division

Marginal mandibular branch

Parotid gland

Cervical branch

Ext. aud. meatus

Post. auricular n.

Facial n.

Chorda tympani n.

Sternocleidomastoid
m.

Lingual n.

FACIAL NERVE

Styloglossus m.

Spinal
accessory n.

Stylopharyngeus m.

Glossopharyngeal n.

Stylohyoid m.

Facial branch to post.
belly of digastric m.

Digastric m.

Int. jug. v.

Stylohyoid branch

Common carotid a.

Temporal branches

Zygomatic
branches

Parotid gland
and duct

Buccal branch

Facial nerve
plexus

Facial a. & v.

Cervical branch

Ext. jug. v.

Marginal mandibular branch

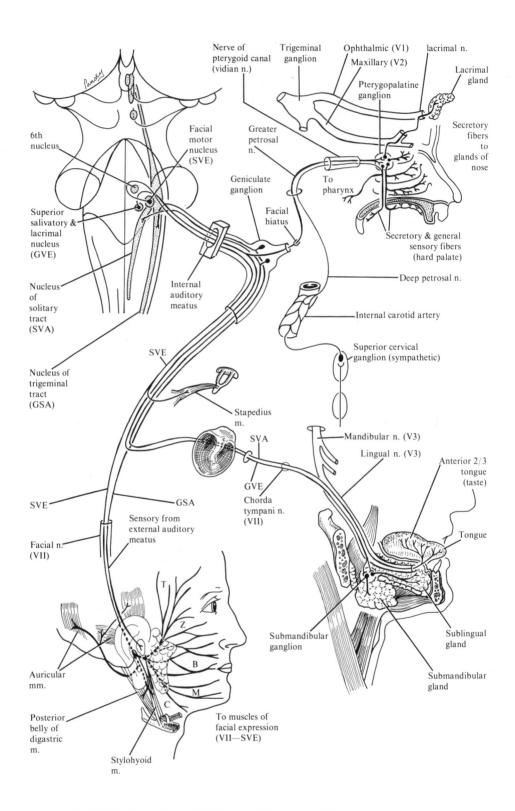

Nerve of pterygoid canal (vidian n.)

Trigeminal ganglion

Ophthalmic (V1)

lacrimal n.

Maxillary (V2)

Lacrimal gland

Pterygopalatine ganglion

Secretory fibers to glands of nose

6th nucleus

Facial motor nucleus (SVE)

Greater petrosal n.

Geniculate ganglion

To pharynx

Superior salivatory & lacrimal nucleus (GVE)

Facial hiatus

Secretory & general sensory fibers (hard palate)

Nucleus of solitary tract (SVA)

Internal auditory meatus

Deep petrosal n.

Internal carotid artery

Nucleus of trigeminal tract (GSA)

SVE

Superior cervical ganglion (sympathetic)

Stapedius m.

SVA

Mandibular n. (V3)

Lingual n. (V3)

Anterior 2/3 tongue (taste)

SVE

GSA

GVE

Chorda tympani n. (VII)

Tongue

Sensory from external auditory meatus

Facial n. (VII)

T

Z

B

M

C

Submandibular ganglion

Sublingual gland

Auricular mm.

Submandibular gland

Posterior belly of digastric m.

To muscles of facial expression (VII—SVE)

Stylohyoid m.

FIGURE 3. **Distribution of facial (VII) nerve (diagrammatic).**

PARALYSIS OF LEFT
FACIAL MUSCLES
(BELL'S PALSY)

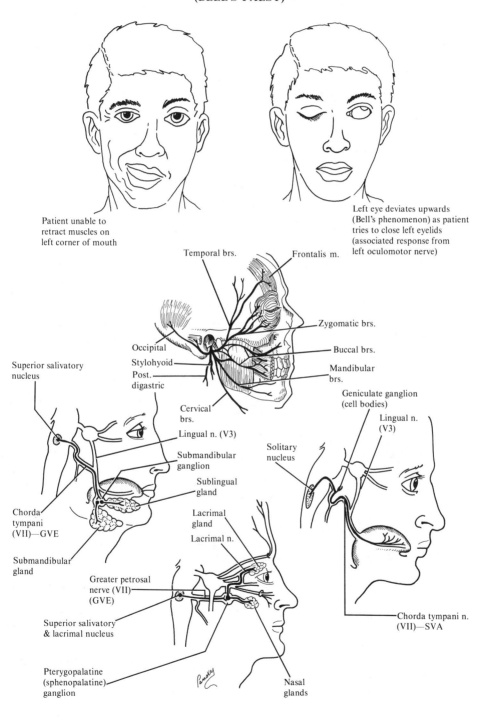

Patient unable to
retract muscles on
left corner of mouth

Left eye deviates upwards
(Bell's phenomenon) as patient
tries to close left eyelids
(associated response from
left oculomotor nerve)

Temporal brs.

Frontalis m.

Zygomatic brs.

Occipital

Buccal brs.

Superior salivatory
nucleus

Stylohyoid

Mandibular
brs.

Post.
digastric

Geniculate ganglion
(cell bodies)

Cervical
brs.

Lingual n.
(V3)

Lingual n. (V3)

Solitary
nucleus

Submandibular
ganglion

Sublingual
gland

Chorda
tympani
(VII)—GVE

Lacrimal
gland

Lacrimal n.

Submandibular
gland

Greater petrosal
nerve (VII)
(GVE)

Chorda tympani n.
(VII)—SVA

Superior salivatory
& lacrimal nucleus

Pterygopalatine
(sphenopalatine)
ganglion

Nasal
glands

FIGURE 4. **Clinical aspects of facial (VII) nerve distribution.**

16. THE PAROTID GLAND

I. **Size:** largest of major salivary glands; weight, 14–28 g

II. **Type:** compound, branched tubuloalveolar gland; purely serous

III. **Location and relations**
  A. ANTERIOR SURFACE: lies against posterior border of ramus of mandible, with rostral projections on both medial and lateral surface of the bone
    1. Medial projection: extension between pterygoid mm. medial to mandible
    2. Lateral projection: lies on external surface of masseter muscle, with small piece, often detached, just below zygomatic arch (accessory part)
  B. POSTERIOR SURFACE: on external auditory meatus and sternocleidomastoid muscle
  C. SUPERFICIAL SURFACE: lobulated, covered by skin, fascia, lymph nodes, and facial branches of great auricular nerve
  D. DEEP SURFACE: has 2 extensions; most posterior part lies on styloid process and its muscles as well as under mastoid and sternocleidomastoid muscle; more cephalic part enters mandibular fossa
    1. Relations: internal and external carotid arteries, internal jugular vein, vagus and glossopharyngeal nerves, and pharyngeal wall

IV. **Capsule:** continuous with deep cervical fascia, the superficial layer being dense and closely united with gland (parotideomasseteric fascia). Between the styloid process and angle of mandible this fascia forms the stylomandibular ligament, which separates parotid from submandibular gland

V. **Duct (Stensen's):** from rostral border of gland, crosses masseter muscle, turns inward to pierce the fat pad of cheek and then the buccinator muscle, to open into mouth opposite second upper molar tooth. Approximately 5 cm long

VI. **Innervaton**
  A. PARASYMPATHETIC: preganglionics arising in the inferior salivatory nucleus pass by way of the tympanic branch of the glossopharyngeal nerve, through the tympanic plexus and into the lesser petrosal nerve. This nerve ends by synapsing in the otic ganglion. Postganglionic fibers enter the mandibular (V) nerve, pass through its auriculotemporal branch to reach the gland (secretomotor)
  B. SYMPATHETICS: preganglionics from the intermediolateral cell column of the upper thoracic cord, through ventral roots and white rami to synapse in the superior cervical ganglion. Postganglionic fibers run in a plexus on the external carotid artery to reach gland (vasomotor)

VII. **Structures embedded in gland**
  A. EXTERNAL CAROTID ENTERS GLAND, gives posterior auricular branch and its terminal branches, superficial temporal and maxillary; the transverse facial artery also arises in gland
  B. VEINS ARE MORE SUPERFICIAL BUT ARE ALSO IN GLAND: maxillary and superficial temporal join to form retromandibular vein in gland; in lower part, latter divides into branches to the common facial and external jugular veins
  C. MOST SUPERFICIAL ARE BRANCHES of facial and auriculotemporal nerves

VIII. **Clinical considerations**
  A. VIRAL INFLAMMATION OF THE PAROTID GLAND (mumps) causes it to swell, resulting in pain on movement of jaw
  B. ABSCESSES OR CYSTS OF THE GLAND may result in pressure on facial nerve

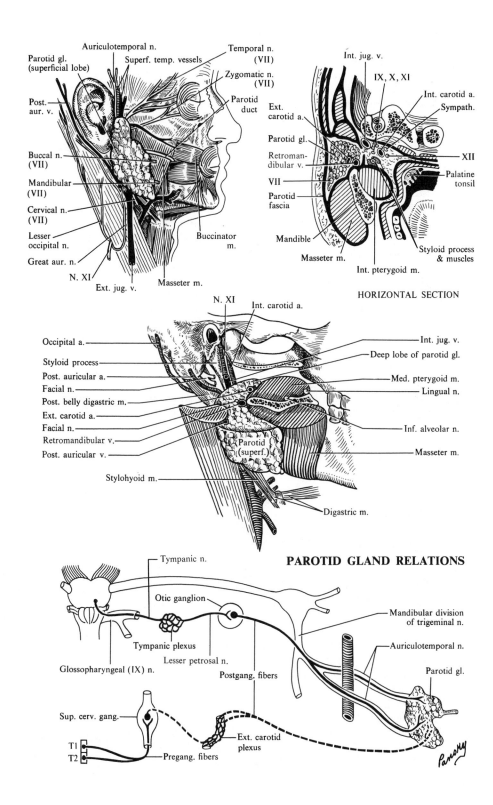

Auriculotemporal n.
Parotid gl.
(superficial lobe)
Superf. temp. vessels
Temporal n.
(VII)
Zygomatic n.
(VII)
Post.
aur. v.
Parotid
duct
Buccal n.
(VII)
Mandibular
(VII)
Cervical n.
(VII)
Lesser
occipital n.
Great aur. n.
N. XI
Ext. jug. v.
Masseter m.
Buccinator
m.

Int. jug. v.
IX, X, XI
Int. carotid a.
Sympath.
Ext.
carotid a.
Parotid gl.
Retroman-
dibular v.
XII
VII
Palatine
tonsil
Parotid
fascia
Mandible
Masseter m.
Int. pterygoid m.
Styloid process
& muscles

HORIZONTAL SECTION

N. XI
Int. carotid a.
Occipital a.
Int. jug. v.
Styloid process
Deep lobe of parotid gl.
Post. auricular a.
Med. pterygoid m.
Facial n.
Lingual n.
Post. belly digastric m.
Ext. carotid a.
Facial n.
Inf. alveolar n.
Retromandibular v.
Post. auricular v.
Parotid
(superf.)
Masseter m.
Stylohyoid m.
Digastric m.

PAROTID GLAND RELATIONS

Tympanic n.
Otic ganglion
Mandibular division
of trigeminal n.
Tympanic plexus
Auriculotemporal n.
Glossopharyngeal (IX) n.
Lesser petrosal n.
Postgang. fibers
Parotid gl.
Sup. cerv. gang.
T1
T2
Pregang. fibers
Ext. carotid
plexus

– 37 –

17. THE TRIGEMINAL (V) NERVE

I. Trigeminal nerve, largest of cranial nerves

A. ROOTS: sensory (portio major), larger; motor (portio minor)

B. GANGLION: semilunar (gasserian), lies in trigeminal impression lateral to the cavernous sinus and internal carotid artery (see p. 112)

C. DIVISIONS

    1. Ophthalmic: all sensory; runs through the lateral border of cavernous sinus and enters orbit through the superior orbital fissure. Distribution: (see p. 146)

    2. Maxillary: all sensory; runs through lower lateral wall of cavernous sinus, under dura to foramen rotundum, crosses pterygopalatine fossa, enters orbit via inferior orbital fissure, and runs in groove on floor of orbit as the *infraorbital nerve*. Distribution: (for branches to skin, see p. 16)

        a. Middle meningeal nerve to dura

        b. Pterygopalatine nerves, greater and lesser palatine nerves (see p. 80)

        c. Posterior superior nasal branch to mucous membrane of nose

        d. Pharyngeal nerve to nasopharynx and auditory tube

        e. Alveolar branches: posterior superior to gums and 3 molar teeth; middle superior to gums and 2 premolar teeth; anterior superior to nasal cavity, canine and incisor teeth. The middle and anterior superior alveolar nerves are branches of the infraorbital nerve

    3. Mandibular: largest; is a mixed nerve (motor and sensory); leaves skull through foramen ovale. This trunk divides into anterior and posterior divisions

        a. Main trunk branches: meningeal to dura; medial pterygoid to medial pterygoid, tensor veli palatini, and tensor tympani muscles

        b. Anterior division (for branches to skin, see p. 16): masseter nerve to masseter muscle; deep temporal nerves (2) to temporalis muscle; lateral pterygoid nerve to lateral pterygoid muscle; and buccal nerve to mucous membrane of mouth

        c. Posterior division: auriculotemporal nerve (see p. 16); lingual nerve (see p. 74); inferior alveolar nerve to mylohyoid and anterior belly of digastric muscles and to all lower teeth

        d. Autonomic connections (not actually fibers of V)

            i. To ophthalmic: sympathetic fibers from the cavernous plexus, to vessels and lacrimal gland of the orbit; parasympathetics (VII) from the pterygopalatine ganglion through the zygomatic branch of the maxillary nerve pass to the lacrimal branch of the ophthalmic nerve to the lacrimal gland

           ii. To maxillary: postganglionic parasympathetics from the pterygopalatine ganglion via the pterygopalatine nerves for vessels and glands of the nasal cavity, palate (palatine and nasal branches of the pterygopalatine nerves)

           iii. To mandibular: postganglionic parasympathetics (IX) from otic ganglion through the auriculotemporal nerve to the parotid gland (see p. 36); to lingual nerve; preganglionic parasympathetic fibers via the chorda tympani nerve leave lingual nerve to synapse in the submandibular ganglion; postganglionics go to the submandibular and sublingual glands

II. Clinical considerations

A. TIC DOULOUREUX: condition of unknown cause associated with excruciating pain, especially along the maxillary and mandibular roots of the trigeminal nerve. It may be relieved by alcohol injections either into the trigeminal ganglion via the foramen ovale or along the root involved as it leaves the skull. It is sometimes necessary to transect the entire sensory part of the nerve between the ganglion and pons

B. WHILE ALL 4 OF THE CRANIAL PARASYMPATHETIC GANGLIA are located close to or on some branch of the 5th nerve, the preganglionic fibers to these ganglia actually come from other nerves: the 3rd, 7th, and 9th. The postganglionic fibers join and are distributed with peripheral branches of the trigeminal nerve

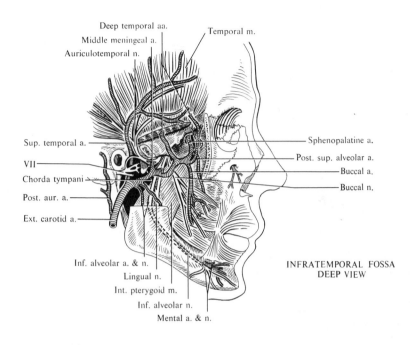

Deep temporal aa.
Middle meningeal a.
Auriculotemporal n.
Temporal m.

Sup. temporal a.
VII
Chorda tympani
Post. aur. a.
Ext. carotid a.

Sphenopalatine a.
Post. sup. alveolar a.
Buccal a.
Buccal n.

Inf. alveolar a. & n.
Lingual n.
Int. pterygoid m.
Inf. alveolar n.
Mental a. & n.

**INFRATEMPORAL FOSSA
DEEP VIEW**

TRIGEMINAL NERVE

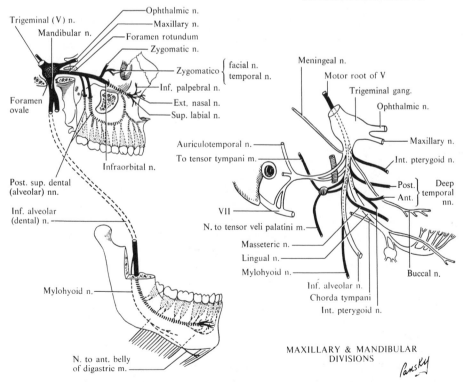

Trigeminal (V) n.
Mandibular n.
Ophthalmic n.
Maxillary n.
Foramen rotundum
Zygomatic n.

Foramen
ovale

Zygomatico { facial n.
temporal n.
Inf. palpebral n.
Ext. nasal n.
Sup. labial n.

Post. sup. dental
(alveolar) nn.

Infraorbital n.

Inf. alveolar
(dental) n.

Mylohyoid n.

N. to ant. belly
of digastric m.

Meningeal n.
Motor root of V
Trigeminal gang.
Ophthalmic n.

Auriculotemporal n.
To tensor tympani m.

Maxillary n.
Int. pterygoid n.

Post. } Deep
Ant. } temporal
nn.

VII
N. to tensor veli palatini m.
Masseteric n.
Lingual n.
Mylohyoid n.

Inf. alveolar n.
Chorda tympani
Int. pterygoid n.

Buccal n.

**MAXILLARY & MANDIBULAR
DIVISIONS**

Pansky

18. THE TEMPOROMANDIBULAR JOINT AND RELATED STRUCTURES

I. Type: combined hinge or ginglymus and diarthrodial (synovial)

II. Movements: opening and closing jaws, protrusion, retraction, lateral displacement

III. Bones: condyle of mandible, the mandibular fossa (posteriorly) and articular tubercle of temporal bone (anteriorly)

IV. Ligaments
A. CAPSULE: from rim of mandibular fossa and articular tubercle to neck of condyloid process of mandible
B. LATERAL (TEMPOROMANDIBULAR): from zygomatic arch to neck of mandible
C. SPHENOMANDIBULAR: from angular spine of sphenoid to lingula of mandible
D. ARTICULAR DISK: divides joint cavity into upper and lower parts; undersurface, concave to conform to shape of mandibular condyle; upper surface, convex to fit into mandibular fossa. Is attached to capsule and to tendon of lateral pterygoid muscle
E. STYLOMANDIBULAR: apex of styloid to angle and posterior border of mandible

V. Muscles acting on joint

Open	Close	Protrude	Retract	Lateral Displacement
Lateral pterygoid	Masseter	Lateral pterygoids (together)	Posterior fibers of temporalis	Lateral pterygoids (individually)
Digastric	Medial pterygoid			
Mylohyoid Geniohyoid	Temporalis			

VI. Important relations
A. MAXILLARY VESSELS pass between sphenomandibular ligament and neck of condyle. At lower level, part of parotid gland and inferior alveolar vessels and nerve lie between this ligament and ramus of mandible
B. STYLOMANDIBULAR LIGAMENT separates submandibular gland from the parotid
C. EXTERNAL CAROTID ARTERY DIVIDES INTO TERMINAL BRANCHES: superficial temporal and maxillary arteries behind neck of mandible
D. INFERIOR ALVEOLAR NERVE RUNS DOWNWARD with its artery and enters mandibular foramen. Gives off the mylohyoid branch just before it enters the bone
   1. Mylohyoid nerve descends in groove on ramus of mandible to reach mylohyoid muscle and runs on its outer surface. Supplies this and anterior belly of digastric muscle
E. LINGUAL NERVE descends parallel to inferior alveolar nerve but is medial and rostral to it, between the medial pterygoid muscle and ramus of mandible

VII. Clinical considerations
A. DISLOCATIONS at this joint can occur. Care should be taken in returning jaw to normal position since the muscles of mastication exert great force
B. BLOWS AGAINST THE JAW may drive the condyle upward and backward, resulting in injury to the external auditory meatus, and may cause bleeding
C. THE PRESSURE produced by the closers of the jaw is normally borne almost entirely by the molar teeth and the synovial membrane of the joint, which extends in part over the articular surfaces. Thus, malocclusion, or any factor that leads to spastic contraction of the muscles (trismus), may cause pain

TEMPEROMANDIBULAR JOINT

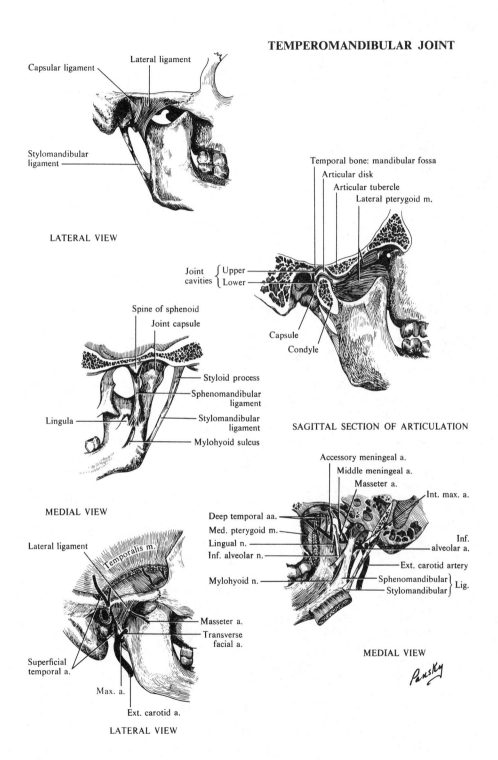

Capsular ligament

Lateral ligament

Stylomandibular ligament

LATERAL VIEW

Temporal bone: mandibular fossa
Articular disk
Articular tubercle
Lateral pterygoid m.

Joint cavities { Upper / Lower

Capsule
Condyle

Spine of sphenoid
Joint capsule

Styloid process
Sphenomandibular ligament
Stylomandibular ligament
Mylohyoid sulcus

Lingula

SAGITTAL SECTION OF ARTICULATION

MEDIAL VIEW

Accessory meningeal a.
Middle meningeal a.
Masseter a.
Int. max. a.

Lateral ligament
Temporalis m.

Deep temporal aa.
Med. pterygoid m.
Lingual n.
Inf. alveolar n.

Inf. alveolar a.

Ext. carotid artery
Sphenomandibular } Lig.
Stylomandibular

Mylohyoid n.

Masseter a.
Transverse facial a.

Superficial temporal a.

Max. a.

Ext. carotid a.

MEDIAL VIEW

Pansky

LATERAL VIEW

19. SUPERFICIAL AND DEEP ARTERIES OF THE FACE

I. Facial
A. ORIGIN: from the external carotid in the carotid triangle of neck
B. COURSE: beneath digastric and stylohyoid muscles, crosses posterior surface of submandibular gland, curves over body of mandible. Proceeds upward and forward to angle of mouth, ascending along side of nose to terminate at medial angle of eye as *angular artery*. In its course, it successively crosses the mandible, the buccinator m., the maxilla, and the levator anguli oris m., but lies deep to the zygomaticus major and levator labii superioris muscles
C. BRANCHES
   1. Cervical: *ascending palatine, tonsillar, glandular* (to submandibular gland), *submental, and muscular*
   2. Facial: *inferior* and *superior labial, lateral nasal, muscular,* and *angular*

II. Superficial temporal artery: smaller of terminal branches of external carotid artery
A. COURSE: ascends in substance of parotid gland, crosses zygomatic process, 5 cm (2 in.) above which it terminates as frontal and parietal branches
B. BRANCHES: *transverse facial, medial temporal, anterior auricular, frontal,* and *parietal*

III. Maxillary artery: the larger of the terminal branches of the external carotid artery, arising posterior to the neck of mandible
A. COURSE: runs forward in parotid gland, then between mandible and sphenomandibular ligament, crosses external pterygoid muscle (either medially or laterally), and enters pterygopalatine fossa
B. PARTS AND THEIR BRANCHES
   1. First part (branches pass through foramina or canals): lies between mandible and sphenomandibular ligament. Branches: *anterior tympanic* to eardrum, *deep auricular* to external acoustic meatus, *middle meningeal* to cranial cavity via foramen ovale, *accessory meningeal* to cranial cavity via foramen ovale, and *inferior alveolar* to mandible, gingivae (gums), and teeth
   2. Second part (supplies muscles): runs medial to ramus of mandible and tendon of temporalis muscle, lying on the external pterygoid muscle, and passes between heads of the latter. Branches: 2 *deep temporals, pterygoid, masseteric,* and *buccal*
   3. Third part (arteries, accompanied by maxillary nerve branches, pass through bony canals or foramina): lies in pterygopalatine fossa lateral to the ganglion. Branches: *posterior superior alveolar, infraorbital, descending palatine, artery of pterygoid canal, pharyngeal,* and *sphenopalatine*
C. SPECIAL FEATURE: it should be noted that there are 5 branches from each part of the maxillary artery. The first 5 reach their destinations by entering foramina, the middle 5 supply soft tissues and use no foramina, and the last 5 also use foramina (the infraorbital or 6th branch is really the termination of the maxillary)

IV. Clinical considerations
A. WOUNDS OF EITHER THE FACE OR SCALP bleed profusely because of multiple anastomoses, both from side to side and front to back
B. THERE ARE ANASTOMOSES IN THE HEAD between branches of the internal and external carotid arteries, especially through the dura mater
C. BRANCHES OF THE POSTERIOR SUPERIOR ALVEOLAR supply the gums, molar and premolar teeth, and maxillary sinus
D. BRANCHES OF THE INFRAORBITAL ARTERY supply orbital structures, maxillary sinus, and via its *anterior superior alveolar branch* the canine and incisor teeth
E. DESCENDING PALATINE ARTERY divides into *greater* and *lesser palatine arteries*
F. THE PHARYNGEAL BRANCH via the pharyngeal canal supplies the pharynx posterior to the opening of the auditory tube
G. THE SPHENOPALATINE ARTERY passes through the sphenopalatine foramen and supplies the lateral wall and septum of nasal cavity

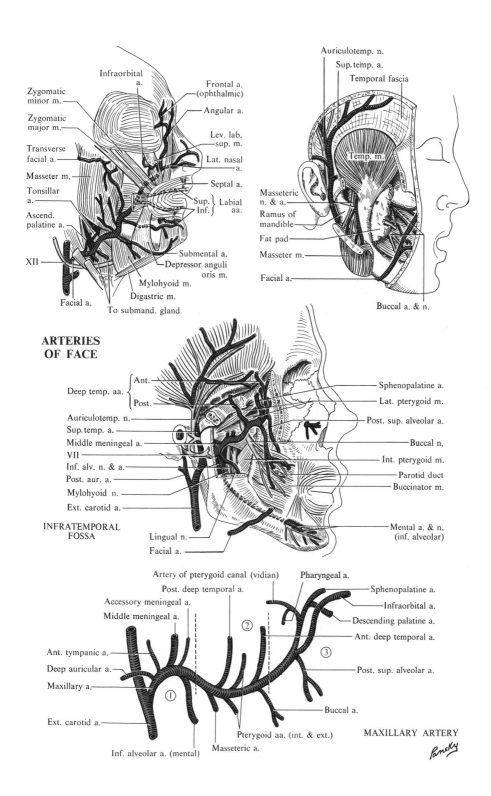

Infraorbital a.
Zygomatic minor m.
Zygomatic major m.
Transverse facial a.
Masseter m.
Tonsillar a.
Ascend. palatine a.
XII
Facial a.
To submand. gland
Digastric m.
Mylohyoid m.
Depressor anguli oris m.
Submental a.
Frontal a. (ophthalmic)
Angular a.
Lev. lab. sup. m.
Lat. nasal a.
Septal a.
Sup. Inf. } Labial aa.

Auriculotemp. n.
Sup. temp. a.
Temporal fascia
Temp. m.
Masseteric n. & a.
Ramus of mandible
Fat pad
Masseter m.
Facial a.
Buccal a. & n.

ARTERIES OF FACE

Deep temp. aa. { Ant. Post.
Auriculotemp. n.
Sup. temp. a.
Middle meningeal a.
VII
Inf. alv. n. & a.
Post. aur. a.
Mylohyoid n.
Ext. carotid a.
Lingual n.
Facial a.

Sphenopalatine a.
Lat. pterygoid m.
Post. sup. alveolar a.
Buccal n.
Int. pterygoid m.
Parotid duct
Buccinator m.
Mental a. & n. (inf. alveolar)

INFRATEMPORAL FOSSA

Artery of pterygoid canal (vidian)
Post. deep temporal a.
Accessory meningeal a.
Middle meningeal a.
Pharyngeal a.
Sphenopalatine a.
Infraorbital a.
Descending palatine a.
Ant. deep temporal a.
Ant. tympanic a.
Deep auricular a.
Maxillary a.
②
③
Post. sup. alveolar a.
Ext. carotid a.
①
Buccal a.
Inf. alveolar a. (mental)
Masseteric a.
Pterygoid aa. (int. & ext.)

MAXILLARY ARTERY

Pansky

20. THE INFRATEMPORAL FOSSA AND ITS CONTENTS

I. Introduction: the infratemporal fossa is an irregular-shaped space lying behind the maxilla, deep to the ramus of the mandible and inferior to the temporal bone

II. Bony boundaries of the fossa

A. LATERAL WALL: ramus of the mandible

B. MEDIAL WALL: lateral pterygoid plate and free border of this plate followed to the foramen ovale

C. ANTERIOR WALL: the infratemporal surface of the maxilla (limited superiorly by the inferior orbital fissure and medially by the pterygomaxillary fissure)

D. POSTERIOR WALL: the anterior surface of the condylar process (head and neck) of the mandible and the styloid process of the temporal bone

E. ROOF: inferior surface of the greater wing of the sphenoid (separated from the temporal fossa by the infratemporal crest)

    1. The foramen ovale in the roof transmits the mandibular division of the trigeminal nerve (V3)

F. INFERIOR BOUNDARY: point where the medial pterygoid muscle inserts into the medial aspect of the mandible near its angle

III. Contents of the fossa

A. LOWER PORTION OF THE TEMPORALIS MUSCLE

B. MEDIAL AND LATERAL PTERYGOID MUSCLES

C. MAXILLARY ARTERY (larger of 2 terminal branches of the external carotid artery) and its branches

D. PTERYGOID PLEXUS OF VEINS

E. MANDIBULAR DIVISION OF TRIGEMINAL NERVE (V3)

    1. Inferior alveolar nerve

    2. Nerve to the mylohyoid muscle

    3. Lingual nerve

    4. Buccal nerve

    5. Auriculotemporal nerve

    6. Nerves to the muscles of mastication

F. CHORDA TYMPANI (cranial VII)

G. OTIC GANGLION (associated with autonomic fibers of cranial IX)

IV. Clinical considerations

A. MANDIBULAR NERVE BLOCK: local anesthesia may be applied to the mandibular nerve as it emerges from the foramen ovale and enters the infratemporal fossa

    1. Extraoral approach: via the mandibular notch, anesthetizing the auriculotemporal, inferior alveolar, lingual, and buccal nerves. All skin areas innervated by the mandibular division of the trigeminal are thus anesthetized

    2. Intraoral approach: through the buccal mucosa and buccinator muscle just medial to the ramus of the mandible, near the mandibular foramen. This anesthetizes the inferior alveolar and lingual nerves and their subdivisions. The areas involved are the body and inferior ramus of the mandible, the mandibular teeth and gingivae, and the mucous membrane of the anterior 2/3 of the tongue

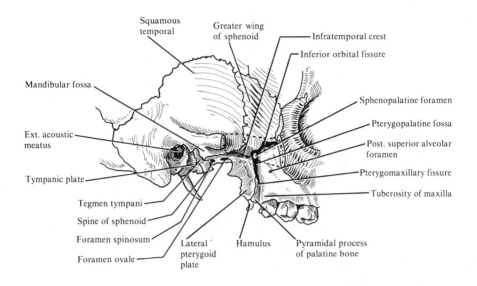

Squamous temporal

Greater wing of sphenoid

Infratemporal crest

Inferior orbital fissure

Mandibular fossa

Sphenopalatine foramen

Pterygopalatine fossa

Ext. acoustic meatus

Post. superior alveolar foramen

Pterygomaxillary fissure

Tympanic plate

Tuberosity of maxilla

Tegmen tympani

Spine of sphenoid

Foramen spinosum

Foramen ovale

Lateral pterygoid plate

Hamulus

Pyramidal process of palatine bone

INFRATEMPORAL REGION

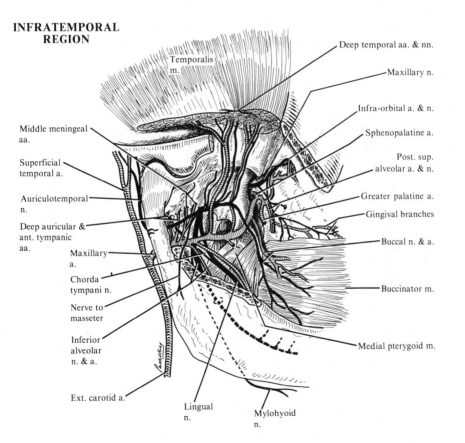

Temporalis m.

Deep temporal aa. & nn.

Maxillary n.

Infra-orbital a. & n.

Middle meningeal aa.

Sphenopalatine a.

Post. sup. alveolar a. & n.

Superficial temporal a.

Auriculotemporal n.

Greater palatine a.

Gingival branches

Deep auricular & ant. tympanic aa.

Buccal n. & a.

Maxillary a.

Chorda tympani n.

Nerve to masseter

Buccinator m.

Inferior alveolar n. & a.

Medial pterygoid m.

Ext. carotid a.

Lingual n.

Mylohyoid n.

21. THE CERVICAL TRIANGLES AND FASCIA

I. Cervical triangles
A. ANTERIOR: midline ventrally; sternocleidomastoid laterally; body of mandible cephalically. This is further subdivided into:
  1. Digastric (submandibular): cephalically, the body of the mandible; anteriorly and below, anterior belly of digastric muscle; posteriorly and below, posterior belly of digastric and stylohyoid muscles
  2. Carotid: cephalically, posterior belly of digastric and stylohyoid muscles; caudally, superior belly of omohyoid muscle; posteriorly, sternocleidomastoid muscle
  3. Submental (suprahyoid): laterally, anterior belly of digastric muscle; caudally, body of the hyoid bone; medially, midline of the neck
  4. Muscular: posteriorly and below, sternocleidomastoid muscle; posteriorly and above, superior belly of omohyoid muscle; medially, midline, from hyoid to sternum
B. POSTERIOR: anteriorly, sternocleidomastoid muscle; posteriorly, trapezius muscle; caudally, clavicle. Further subdivided:
  1. Occipital: posteriorly and anteriorly, as above; caudally, inferior belly of omohyoid
  2. Omoclavicular (subclavian): anteriorly, sternocleidomastoid muscle; cephalically, inferior belly of omohyoid muscle; caudally, clavicle

II. Cervical fascia
A. SUPERFICIAL: covers the ant. and post. triangles but splits to enclose sternocleidomastoid and trapezius. Attachments: posteriorly, ext. occipital protuberance, ligamentum nuchae, spine of C7; cephalically, sup. nuchal line, mastoid process, and mandible; inferiorly, clavicle, manubrium of sternum, acromion, and spine of scapula
B. FASCIA OF THE INFRAHYOID MUSCLES consists of 2 layers: superficial, which encloses the sternohyoid and omohyoid muscles, and a deep, which invests the sternothyroid and thyrohyoid muscles
C. VISCERAL encloses pharynx, larynx, trachea, esophagus, and thyroid. Has 2 parts:
  1. Pretracheal: covers larynx and trachea; splits to enclose thyroid. Cephalically is attached to hyoid bone and thyroid cartilage; laterally is continuous with the next layer; caudally enters the thorax to join fascia of aorta and the pericardium
  2. Buccopharyngeal: covers buccinator muscle and posterior side of esophagus. Cephalically is attached to pharyngeal tubercle and medial pterygoid plate, covers the superior pharyngeal constrictor, and joins the pterygomandibular raphé
D. PREVERTEBRAL encloses vertebral column and its muscles. Covers prevertebral muscles and forms floor of the post. triangle. Attachments: to the transverse processes of the cervical vertebrae laterally; the occipital bone near jugular foramen, the sup. nuchal line, and mastoid process cephalically; continues into mediastinum caudally
  1. Suprapleural membrane (Sibson's fascia): fascia deep to the scalene muscles. Covers cervical pleura

III. Carotid sheath:
an investment of the internal and common carotid arteries, the internal jugular vein, and the vagus nerve. It is adherent to the thyroid sheath and the fascia under the sternocleidomastoid. Cephalically, it is attached to the bone around the jugular foramen and carotid canal; caudally, it continues into the thorax

IV. Fascial spaces
A. RETROPHARYNGEAL: between buccopharyngeal and prevertebral fascia. Above, ends at skull; laterally, is closed by carotid sheath; below, continues into mediastinum
  1. Alar fascia: thin layer of fascia which subdivides this space. It is attached in the midline to the buccopharyngeal fascia; laterally, it joins the carotid sheath
B. SUPRASTERNAL. This is an interval between the layers of the superficial fascia which has split to attach to the posterior and anterior sides of the manubrium

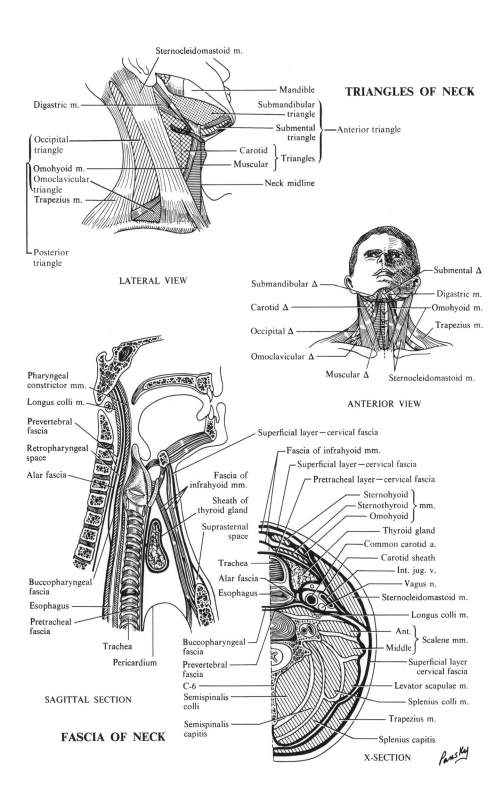

TRIANGLES OF NECK

Sternocleidomastoid m.

Digastric m.

Occipital triangle

Omohyoid m.

Omoclavicular triangle

Trapezius m.

Posterior triangle

Mandible

Submandibular triangle

Submental triangle

Carotid

Muscular

Neck midline

Anterior triangle

Triangles

LATERAL VIEW

Submandibular Δ

Carotid Δ

Occipital Δ

Omoclavicular Δ

Muscular Δ

Submental Δ

Digastric m.

Omohyoid m.

Trapezius m.

Sternocleidomastoid m.

ANTERIOR VIEW

Pharyngeal constrictor mm.

Longus colli m.

Prevertebral fascia

Retropharyngeal space

Alar fascia

Buccopharyngeal fascia

Esophagus

Pretracheal fascia

Trachea

Pericardium

SAGITTAL SECTION

FASCIA OF NECK

Fascia of infrahyoid mm.

Sheath of thyroid gland

Suprasternal space

Trachea

Alar fascia

Esophagus

Buccopharyngeal fascia

Prevertebral fascia

C-6

Semispinalis colli

Semispinalis capitis

Superficial layer—cervical fascia

Fascia of infrahyoid mm.

Superficial layer—cervical fascia

Pretracheal layer—cervical fascia

Sternohyoid
Sternothyroid } mm.
Omohyoid

Thyroid gland

Common carotid a.

Carotid sheath

Int. jug. v.

Vagus n.

Sternocleidomastoid m.

Longus colli m.

Ant.
Middle } Scalene mm.

Superficial layer cervical fascia

Levator scapulae m.

Splenius colli m.

Trapezius m.

Splenius capitis

X-SECTION

22. VEINS OF THE NECK

I. Internal jugular

A. ORIGIN: in jugular fossa, as a continuation of sigmoid sinus

B. COURSE: in carotid sheath, first with internal and then common carotid artery

C. TERMINATION: joins subclavian vein just lateral to sternoclavicular joint to form brachiocephalic vein

    1. A dilation at its origin is called the *superior bulb*

    2. One inch above termination is a pair of valves below which is a dilation, the *inferior bulb*

D. TRIBUTARIES (are variable)

    1. At origin: inferior petrosal sinus and a meningeal vein

    2. Veins from pharyngeal plexus, near angle of jaw

    3. Common facial (formed from anterior branch of retromandibular and facial, see p. 20) enters at level of hyoid bones

    4. Lingual, from tongue, may enter with or just below common facial, drains tongue and sublingual area

    5. Superior thyroid, from upper thyroid gland, enters with or just below common facial

    6. Middle thyroid vein, from the lateral lobe of gland

    7. The right lymphatic duct or, on the left side, the thoracic duct, opens usually into the internal jugular or into the junction between the internal jugular and subclavian vein

II. External jugular, formed by posterior branch of retromandibular vein and posterior auricular vein, terminates in subclavian vein behind middle of clavicle (see p. 20)

A. TRIBUTARIES

    1. Posterior auricular, occipital, retromandibular, and anterior jugular (see p. 20)

    2. Posterior external jugular from cephalic part of back of neck

    3. Transverse scapular from scapular region

    4. Transverse cervical (inconstant) from root of the neck. Most often this ends in subclavian vein but may join the external jugular vein

III. Subclavian

A. ORIGIN: continuation of axillary vein, beginning at lateral border of first rib

B. COURSE: medial, anterior, and slightly inferior to subclavian artery. Its second part is separated from the artery by the anterior scalene muscle

C. TERMINATION: joins internal jugular to form brachiocephalic vein just lateral to sternoclavicular joint

    1. Contains pair of valves just before ending

D. TRIBUTARIES

    1. External jugular (see II, above)

    2. Transverse cervical from root of neck (see II, A4 above)

    3. Thoracoacromial from shoulder and upper pectoral region

IV. Clinical considerations

A. THE LOSS OF A MAJOR VEIN such as the internal jugular is well tolerated because of interconnections between the veins of the two sides, both within and outside the skull

B. WHEN VENOUS PRESSURE INCREASES (heart failure), the external jugular vein becomes prominent throughout its course

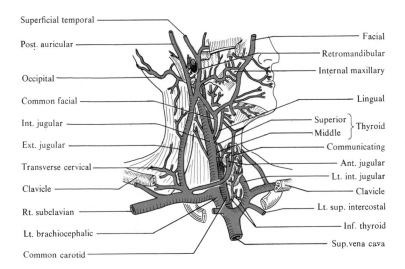

Superficial temporal	Facial
Post. auricular	Retromandibular
	Internal maxillary
Occipital	
Common facial	Lingual
Int. jugular	Superior } Thyroid
Ext. jugular	Middle }
	Communicating
Transverse cervical	Ant. jugular
	Lt. int. jugular
Clavicle	Clavicle
Rt. subclavian	Lt. sup. intercostal
Lt. brachiocephalic	Inf. thyroid
	Sup. vena cava
Common carotid	

DEEP VEINS OF NECK AND HEAD

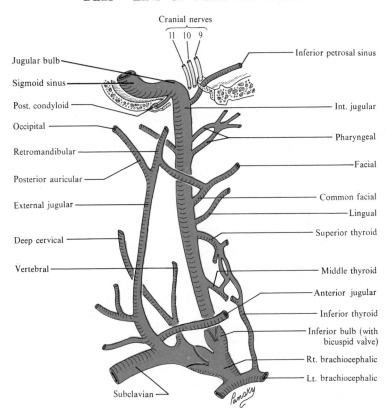

Cranial nerves
11 10 9

Jugular bulb	Inferior petrosal sinus
Sigmoid sinus	
Post. condyloid	Int. jugular
Occipital	Pharyngeal
Retromandibular	Facial
Posterior auricular	
External jugular	Common facial
	Lingual
Deep cervical	Superior thyroid
Vertebral	Middle thyroid
	Anterior jugular
	Inferior thyroid
	Inferior bulb (with bicuspid valve)
	Rt. brachiocephalic
	Lt. brachiocephalic
Subclavian	

23. ARTERIES OF THE HEAD AND NECK

I. **Introduction:** the head and neck receive their arterial supply from the common carotid and its internal and external carotid branches, from the vertebral, and from other branches of the subclavian. Generally, the internal carotid supplies the cranial cavity contents and orbit; the vertebral, the brain and spinal cord; the remaining subclavian branches supply deep structures in the neck and the visceral tube below the middle of the thyroid; and the external carotid supplies everything else

II. **Arteries**

A. COMMON CAROTIDS: left usually arises from aortic arch independently, right branches from the brachiocephalic. Both course superiorly in the carotid sheath with the internal jugular vein and vagus (X) nerve. There are no branches. The common carotids divide into internal and external carotid arteries

   1. Internal carotid artery: continues in carotid sheath and through the base of the skull in an S-shaped curve in the carotid canal. Before dividing into the *middle* and *anterior cerebral arteries,* it gives off:

      a. Branches to the middle ear: caroticotympanic

      b. Branches to the nerves in the cavernous sinus, the trigeminal ganglion, the hypophysis, and the choroid plexus

      c. Ophthalmic artery to the orbit (see p. 144)

   2. Middle cerebral artery: courses laterally in lateral fissure between frontal and temporal lobes of brain and ramifies on lateral surface of cerebrum

   3. Anterior cerebral: runs anteriorly, communicates with mate of opposite side, curls around corpus callosum in the great cerebral fissure giving off branches to medial side of hemispheres and superficial surface. Both cerebrals are terminal, have no collateral connections, and occlude easily

B. VERTEBRAL ARTERY is a branch of the subclavian. Courses superiorly in front of vertebral column and enters transverse process of C6. Runs to atlas in processes, turns medially and posteriorly on posterior arch of atlas, pierces dura, traverses the foramen magnum, courses along the medulla, and joins with the same vessel of the opposite side as the *basilar artery.* Vertebral supplies, along its course, contiguous muscles, vertebral column, spinal cord, meninges, medulla, pons, cerebellum, and cerebrum

   1. Spinal branches enter intervertebral foramina to supply cord

   2. Anterior and posterior spinal arteries descend along cord, forming a single arterial channel anteriorly and double channels posteriorly

   3. Last branch of vertebral is the *posterior inferior cerebellar*

C. BASILAR ARTERY: (see p. 114)

D. POSTERIOR CEREBRAL ARTERIES: (see p. 114)

E. OTHER SUBCLAVIAN ARTERY BRANCHES: *deep cervical* of costocervical trunk; *thyrocervical trunk,* which contributes the *inferior thyroid* which nourishes the thyroid and parathyroid glands and prevertebral muscles, trachea, esophagus, and larynx, by way of ascending cervical, tracheal, esophageal, and inferior laryngeal branches

F. EXTERNAL CAROTID ARTERY: its major branches are

   1. Superior thyroid artery: first branch to surrounding muscles, infrahyoid region, larynx, and thyroid gland

   2. Ascending pharyngeal artery: to pharynx, prevertebral muscles, meninges via jugular and hypoglossal openings, a tympanic branch, and palatine branches to soft palate

   3. Lingual artery: to tongue, suprahyoid region, sublingual gland, and palatine tonsils

   4. Facial artery: to submandibular gland, lips (labial), nose (lateral nasal), facial muscles, and an angular branch to angle of eye

   5. Occipital artery: to sternocleidomastoid muscle, meningeal branches, and to scalp

   6. Posterior auricular artery: to muscles, tympanic membrane, external ear, and posterior scalp

   7. Superficial temporal artery: to parotid, ear, face, zygomatic, and temporal regions

   8. Maxillary artery: (see p. 42)

ARTERIES OF HEAD AND NECK

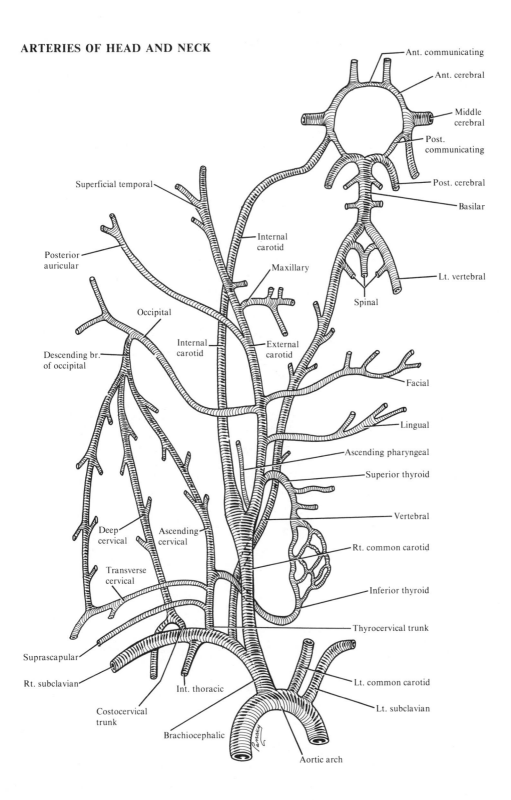

Ant. communicating

Ant. cerebral

Middle cerebral

Post. communicating

Post. cerebral

Basilar

Superficial temporal

Internal carotid

Maxillary

Lt. vertebral

Posterior auricular

Spinal

Occipital

Descending br. of occipital

Internal carotid

External carotid

Facial

Lingual

Ascending pharyngeal

Superior thyroid

Vertebral

Deep cervical

Ascending cervical

Rt. common carotid

Transverse cervical

Inferior thyroid

Thyrocervical trunk

Suprascapular

Rt. subclavian

Int. thoracic

Lt. common carotid

Lt. subclavian

Costocervical trunk

Brachiocephalic

Aortic arch

24. THE ANTEROLATERAL MUSCLES OF THE NECK

Name	Origin	Insertion	Action	Nerve
Platysma	Deltoid and pectoral fascia	Mandible & skin	Depresses angle of mouth, opens jaw	Facial (VII)
Sternocleido-mastoid	Sternum, clavicle	Mastoid process	Bends head to same side, rotates head, raises chin to opposite side, together bend head forward & elevate chin	Accessory (XI), C2, & C3
Digastric				
Ant. belly	Lower border mandible	Body & great cornu of hyoid	Opens jaw, draws hyoid forward	Trigeminal (V)
Post. belly	Mastoid notch of temporal bone	Body & great cornu of hyoid	Draws hyoid back, together raise hyoid	Facial (VII)
Stylohyoid	Styloid process	Body of hyoid	Draws hyoid up & back	Facial (VII)
Mylohyoid (see p. 71)	Mylohyoid line of mandible	Body of hyoid, median raphé	Raises hyoid	Trigeminal (V)
Geniohyoid (see p. 71)	Inf. mental spine behind symphysis	Body of hyoid	Draws hyoid and tongue forward	C1, via ansa cervicalis
Sternohyoid (see p. 71)	Medial end of clavicle, post. manubrium	Body of hyoid	Depresses hyoid	C1–3, via ansa cervicalis
Sternothyroid (see p. 71)	Post. manubrium	Oblique line on lamina of thyroid cartilage	Depresses thyroid cartilage	C1–3, via ansa cervicalis
Thyrohyoid (see p. 62)				
Omohyoid (see p. 71)				
Inf. belly	Scapula, sup. transverse lig.	By tendon to clavicle	Depresses hyoid	C1–3, via ansa cervicalis
Sup. belly	Tendon suspended from clavicle	Body of hyoid bone	Depresses hyoid	C1–3 via ansa cervicalis

Trapezius, levator scapulae, and scalene mm. (see p. 71)

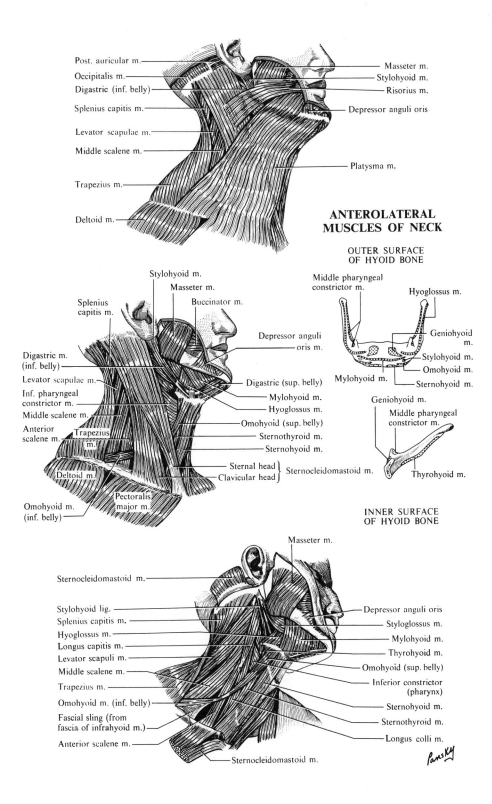

Post. auricular m.

Occipitalis m.

Digastric (inf. belly)

Splenius capitis m.

Levator scapulae m.

Middle scalene m.

Trapezius m.

Deltoid m.

Masseter m.

Stylohyoid m.

Risorius m.

Depressor anguli oris

Platysma m.

ANTEROLATERAL MUSCLES OF NECK

OUTER SURFACE OF HYOID BONE

Stylohyoid m.

Masseter m.

Buccinator m.

Middle pharyngeal constrictor m.

Hyoglossus m.

Splenius capitis m.

Digastric m. (inf. belly)

Levator scapulae m.

Inf. pharyngeal constrictor m.

Middle scalene m.

Anterior scalene m.

Trapezius m.

Deltoid m.

Omohyoid m. (inf. belly)

Pectoralis major m.

Depressor anguli oris m.

Digastric (sup. belly)

Mylohyoid m.

Hyoglossus m.

Omohyoid (sup. belly)

Sternothyroid m.

Sternohyoid m.

Sternal head }
Clavicular head } Sternocleidomastoid m.

Geniohyoid m.

Stylohyoid m.

Omohyoid m.

Sternohyoid m.

Mylohyoid m.

Geniohyoid m.

Middle pharyngeal constrictor m.

Thyrohyoid m.

INNER SURFACE OF HYOID BONE

Masseter m.

Sternocleidomastoid m.

Stylohyoid lig.

Splenius capitis m.

Hyoglossus m.

Longus capitis m.

Levator scapuli m.

Middle scalene m.

Trapezius m.

Omohyoid m. (inf. belly)

Fascial sling (from fascia of infrahyoid m.)

Anterior scalene m.

Sternocleidomastoid m.

Depressor anguli oris

Styloglossus m.

Mylohyoid m.

Thyrohyoid m.

Omohyoid (sup. belly)

Inferior constrictor (pharynx)

Sternohyoid m.

Sternothyroid m.

Longus colli m.

Pansky

25. THE ANTERIOR TRIANGLE OF THE NECK

I. The boundaries of the anterior triangle are formed by midline of the neck (anteromedian line), anterior border of the sternocleidomastoid muscle (posteriorly), and inferior border of the mandible (superiorly).

A. THE DIGASTRIC, STYLOHYOID, AND SUPERIOR BELLY of the omohyoid muscles cross the triangle. They further subdivide it into 4 smaller triangles:

   1. Carotid (superior carotid): bounded by the posterior belly of digastric, superior belly of omohyoid, and anterior border of the sternocleidomastoid
      a. Roof: skin, superficial fascia, platysma muscle, and deep fascia
      b. Floor: thyrohyoid, hyoglossus, middle and inferior pharyngeal constrictor muscles
      c. Contents: bifurcation of common carotid a.; superior thyroid, lingual, facial, occipital, and ascending pharyngeal aa.; internal jugular v.; ansa cervicalis; cervical sympathetic trunk; hypoglossal and superior laryngeal nerves

   2. Muscular (inferior carotid): bounded by superior belly of omohyoid, anterior border of sternocleidomastoid, and anterior median line of neck
      a. Roof: skin, superficial fascia, platysma muscle, and deep fascia
      b. Floor: sternohyoid and sternothyroid muscles
      c. Contents: sternohyoid and sternothyroid mm., thyroid isthmus, larynx, and trachea

   3. Digastric (submandibular): bounded by mandible and two bellies of the digastric muscle
      a. Roof: skin, superficial fascia, platysma muscle, and deep fascia
      b. Floor: mylohyoid and hyoglossus muscles
      c. Divisions: stylomandibular ligament divides it into rostral and posterior parts
      d. Contents:
         i. Rostral part: submandibular gl., facial a. and v., submental a., mylohyoid a. and n.
         ii. Posterior part: parotid gland containing external carotid artery; crossed laterally by facial nerve, internal jugular vein, and vagus nerve

   4. Submental (suprahyoid): bounded by body of hyoid bone and anterior belly of digastric muscle of each side. Extends across anterior median line of neck
      a. Roof: skin, superficial fascia, and deep fascia
      b. Floor: mylohyoid muscle
      c. Contents: 1 or 2 lymph nodes and small tributaries of anterior jugular vein

II. Structures beneath the sternocleidomastoid muscle

A. ANSA CERVICALIS is on the surface of the carotid sheath
B. CAROTID SHEATH containing vagus n., internal jugular v., and common carotid a.
C. CERVICAL SYMPATHETIC TRUNK is embedded in dorsomedial wall of carotid sheath

III. Phrenic nerve: from 4th, but may receive fibers from 3rd and 5th cervical nerves

A. COURSE (in neck): on ventral surface of anterior scalene muscle. Crossed by inferior belly of omohyoid muscle and transverse (superficial) cervical and suprascapular arteries.

IV. Carotid sinus: a slight dilation of proximal part of internal carotid artery which can involve common carotid. Is an *arterial blood-pressure–regulating area*. Innervated mainly by CN IX through carotid sinus nerve. Also supplied by CN X and sympathetic autonomic fibers

V. Carotid body: small, flattened, ovoid tissue mass at bifurcation of common carotid a. Is a *chemoreceptor* that responds to either increased CO_2 tension or decreased O_2 tension in blood. Innervated by carotid sinus nerve of CN IX, CN X, and sympathetics

VI. Branches of external carotid artery

Anterior	Posterior	Ascending	Terminal
Sup. thyroid	Occipital	Ascend. pharyngeal	Superfic. temp.
Lingual	Posterior auricular		Maxillary
Facial			

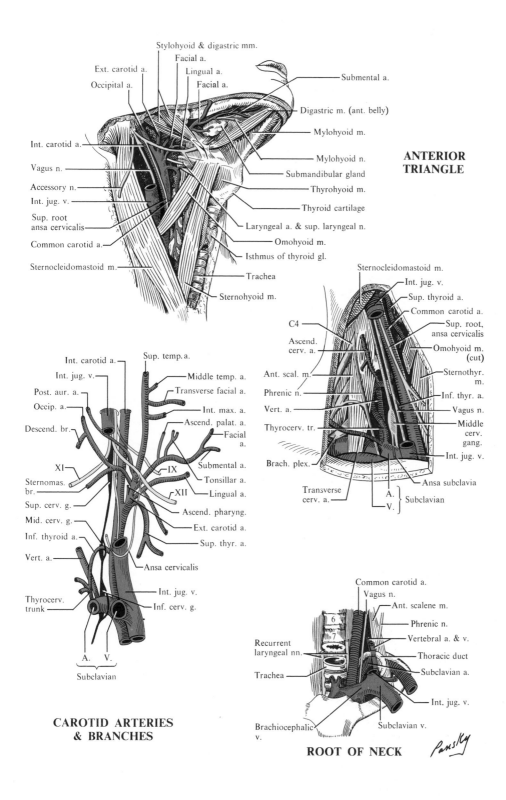

Stylohyoid & digastric mm.
Facial a.
Ext. carotid a.
Lingual a.
Occipital a.
Facial a.
Submental a.

Digastric m. (ant. belly)

Mylohyoid m.

Int. carotid a.
Mylohyoid n.

Vagus n.
Submandibular gland

Accessory n.
Thyrohyoid m.

Int. jug. v.
Thyroid cartilage

Sup. root
ansa cervicalis
Laryngeal a. & sup. laryngeal n.

Common carotid a.
Omohyoid m.

Isthmus of thyroid gl.

Sternocleidomastoid m.
Trachea

Sternohyoid m.

**ANTERIOR
TRIANGLE**

Sternocleidomastoid m.
Int. jug. v.

Sup. thyroid a.

Common carotid a.

C4
Sup. root,
ansa cervicalis

Ascend.
cerv. a.
Omohyoid m.
(cut)

Ant. scal. m.
Sternothyr.
m.

Phrenic n.
Inf. thyr. a.

Vert. a.
Vagus n.

Thyrocerv. tr.
Middle
cerv.
gang.

Brach. plex.
Int. jug. v.

Transverse
cerv. a.
Ansa subclavia

A. } Subclavian
V. }

Int. carotid a.
Sup. temp. a.

Int. jug. v.
Middle temp. a.

Post. aur. a.
Transverse facial a.

Occip. a.
Int. max. a.

Descend. br.
Ascend. palat. a.

Facial
a.

XI
IX
Submental a.

Sternomas.
br.
Tonsillar a.

XII
Lingual a.

Sup. cerv. g.
Ascend. pharyng.

Mid. cerv. g.
Ext. carotid a.

Inf. thyroid a.
Sup. thyr. a.

Vert. a.
Ansa cervicalis

Int. jug. v.

Thyrocerv.
trunk
Inf. cerv. g.

A.   V.

Subclavian

**CAROTID ARTERIES
& BRANCHES**

Common carotid a.
Vagus n.

Ant. scalene m.

Phrenic n.

Recurrent
laryngeal nn.
Vertebral a. & v.

6
Thoracic duct

7
Subclavian a.

Trachea
Int. jug. v.

Brachiocephalic
v.
Subclavian v.

ROOT OF NECK

Pansky

– 55 –

26. ROOT OF NECK

I. **The root of the neck** connects the thoracic cavity contents with those in neck. Structures passing to and from the thoracic cavity and neck course vertically, whereas those passing to and from the thoracic cavity to the upper limb course more horizontally and laterally as they cross the 1st rib. Essential points are:
   A. THE THORACIC INLET is bounded by the manubrium of sternum (anterior), the ribs (lateral), and the vertebral column (posterior)
   B. THE ANTERIOR SCALENE MUSCLE is an important landmark and inserts on the 1st rib
   C. THE DOME OF THE LUNGS extends above the clavicle, well into the neck region
   D. THE ESOPHAGUS lies just anterior to the vertebral column, in the midline
   E. THE TRACHEA is found just anterior to the esophagus

II. **Structures identified at the root of the neck**
   A. ANTERIOR SCALENE MUSCLE: many structures are related to it
   B. STERNOHYOID AND STERNOTHYROID MUSCLES are seen first, just deep to inferior end of the sternocleidomastoid muscle
   C. THE INTERNAL JUGULAR VEIN is located laterally (in the carotid sheath)
   D. THE COMMON CAROTID ARTERY is found medially (in the carotid sheath)
   E. THE VAGUS NERVE is posterior, between the internal jugular vein and carotid artery (in the carotid sheath), just lateral to the trachea and esophagus
   F. TRACHEA AND ESOPHAGUS: midline structures
   G. SUBCLAVIAN VEIN courses superficial to the anterior scalene muscle as it passes over 1st rib on way to root of neck and joins internal jugular to form brachiocephalic vein
   H. SUBCLAVIAN ARTERY courses behind (deep) to anterior scalene muscle on way to upper limb. Its branches are vertebral, thyrocervical trunk, costocervical trunk, and internal thoracic arteries (for details, see p. 58)
   I. PHRENIC NERVE seen on anterior surface of anterior scalene muscle and enters thoracic cavity between the subclavian artery and vein
   J. LYMPHATIC CHANNELS
      1. On the left: the *thoracic duct* arches above left clavicle and lies ventral to subclavian artery, vertebral artery and vein, thyrocervical trunk, and prevertebral fascia, which separates duct from phrenic nerve and anterior scalene muscle. It lies behind left common carotid artery, internal jugular vein, and vagus nerve and terminates at junction of left internal jugular and subclavian veins. Near its termination, it is joined by the *left jugular trunk* (drains left 1/2 of head and neck), *left subclavian trunk* (drains left upper limb), and *left mediastinal trunk* (drains left 1/2 of thorax). These channels may enter the subclavian or brachiocephalic veins independently rather than join the terminal thoracic duct
      2. On the right: the 3 lymph trunks (see above) drain corresponding areas. The right jugular and subclavian trunks may join to form the *right lymphatic duct*
         a. The right lymphatic duct usually ends in the internal jugular vein
         b. The subclavian lymphatic trunk usually ends in the subclavian vein
         c. The mediastinal lymphatic trunk usually ends in the brachiocephalic vein
         d. All the lymphatic drainage of the right side drains the right upper 1/2 of the thorax, neck, and the right extremity
   K. SYMPATHETIC CHAIN AND THE MIDDLE AND INFERIOR GANGLIA: one of the most posteriorly located structures in the root of neck. Both ganglia are related to the vertebral artery (the inferior often combines with the 1st thoracic ganglion to form the *stellate ganglion*). The ganglia both lie behind the apex (dome) of the pleura and lung
   L. RECURRENT LARYNGEAL NERVES: found in the groove between the esophagus and trachea
   M. DOME OF THE PLEURA AND LUNG extends superiorly into neck, being higher than the upper edge of the sternum and 1st rib anteriorly, but posteriorly, it does not reach above the 1st rib level

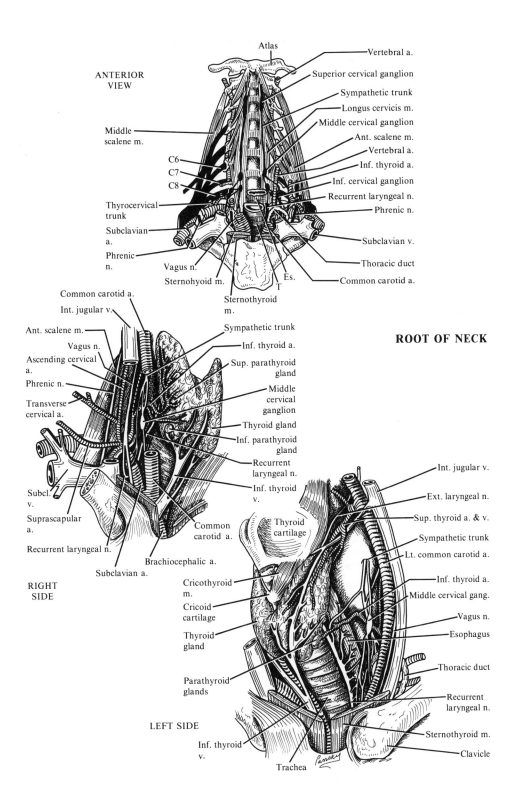

ANTERIOR VIEW

Atlas

Vertebral a.

Superior cervical ganglion

Sympathetic trunk

Longus cervicis m.

Middle cervical ganglion

Ant. scalene m.

Vertebral a.

Inf. thyroid a.

Inf. cervical ganglion

Recurrent laryngeal n.

Phrenic n.

Subclavian v.

Thoracic duct

Common carotid a.

Middle scalene m.

C6
C7
C8

Thyrocervical trunk

Subclavian a.

Phrenic n.

Vagus n.

Sternohyoid m.

Es.

T

Sternothyroid m.

ROOT OF NECK

Common carotid a.

Int. jugular v.

Ant. scalene m.

Vagus n.

Ascending cervical a.

Phrenic n.

Transverse cervical a.

Subcl. v.

Suprascapular a.

Recurrent laryngeal n.

Subclavian a.

Brachiocephalic a.

Common carotid a.

Sympathetic trunk

Inf. thyroid a.

Sup. parathyroid gland

Middle cervical ganglion

Thyroid gland

Inf. parathyroid gland

Recurrent laryngeal n.

Inf. thyroid v.

RIGHT SIDE

Thyroid cartilage

Cricothyroid m.

Cricoid cartilage

Thyroid gland

Parathyroid glands

LEFT SIDE

Inf. thyroid v.

Trachea

Int. jugular v.

Ext. laryngeal n.

Sup. thyroid a. & v.

Sympathetic trunk

Lt. common carotid a.

Inf. thyroid a.

Middle cervical gang.

Vagus n.

Esophagus

Thoracic duct

Recurrent laryngeal n.

Sternothyroid m.

Clavicle

27. THE POSTERIOR TRIANGLE AND SUBCLAVIAN ARTERY

I. Posterior triangle: (for boundaries, see p. 46)
  A. OCCIPITAL DIVISION
    1. Floor: splenius capitis, levator scapulae, posterior and middle scalene muscles
      a. Structures crossing floor: accessory nerve and brachial plexus
    2. Structures piercing roof (see p. 16): lesser occipital, great auricular, cervical cutaneous, and supraclavicular nerves. All emerge just behind the posterior border of the sternocleidomastoid muscle
  B. OMOCLAVICULAR (SUBCLAVIAN) DIVISION
    1. Floor: first rib and first part of serratus anterior muscle
      a. Structures crossing floor: subclavian artery and vein, part of brachial plexus
    2. Structures crossing through: transverse scapular and superficial cervical vessels
    3. Structures piercing roof: external jugular vein

II. Subclavian artery
  A. ORIGIN: on right, from brachiocephalic trunk; on left, from arch of aorta
  B. PARTS: first, from origin to medial border of anterior scalene muscle; second, lies behind anterior scalene muscle; third, extends from lateral border of that muscle to lateral border of first rib
  C. RELATIONS (OTHER THAN MUSCLE OR FASCIA) AND BRANCHES OF PARTS:
    1. Part I: right side
      a. Relations: crossed by internal jugular and vertebral veins, vagus nerve, cardiac branches of the vagus nerve, and the subclavian loop of the sympathetic chain. Behind and below: pleura, sympathetic trunk, and recurrent nerve
      b. Branches: vertebral, thyrocervical trunk, and internal thoracic arteries
    2. Part I: left side
      a. Relations: crossed by vagus, cardiac, and phrenic nerves; common carotid artery; int. jugular and vertebral veins; and origin of brachiocephalic vein. Behind: esophagus, thoracic duct, recurrent nerve, and inferior cervical ganglion. Medial: esophagus, trachea, thoracic duct, and recurrent nerve. Lateral: pleura and lung
      b. Branches: same as right side plus the costocervical trunk
    3. Part II
      a. Relations of the two sides are the same. Behind and below, related to pleura; above is the brachial plexus
      b. Branches: sometimes, descending (transverse) scapular artery; costocervical trunk on the right side
    4. Part III
      a. Relations of the two sides are the same. In front: external jugular vein, venous plexus, transverse scapular vessels, and subclavian vein. Behind: lowest trunk of brachial plexus. Above and lateral: upper and middle trunks of plexus. Below: upper surface of first rib
      b. Branches: descending (transverse) scapular artery

III. Arteries crossing triangle of neck
  A. SUPRASCAPULAR from thyrocervical trunk, which goes over transverse scapular ligament to supraspinous fossa
  B. TRANSVERSE CERVICAL from thyrocervical trunk. It divides into a *superficial* and a *deep* branch (which descends along vertebral border of scapula). It may arise from 2 arteries: a *superficial cervical* from the thyrocervical trunk and a *descending (dorsal) scapular,* from the third part of the subclavian

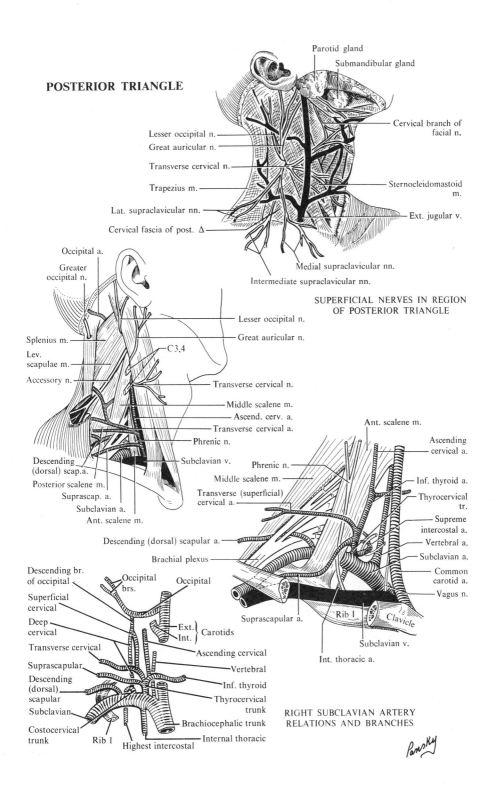

POSTERIOR TRIANGLE

Parotid gland

Submandibular gland

Cervical branch of facial n.

Lesser occipital n.

Great auricular n.

Transverse cervical n.

Trapezius m.

Sternocleidomastoid m.

Lat. supraclavicular nn.

Ext. jugular v.

Cervical fascia of post. Δ

Medial supraclavicular nn.

Intermediate supraclavicular nn.

SUPERFICIAL NERVES IN REGION OF POSTERIOR TRIANGLE

Occipital a.

Greater occipital n.

Lesser occipital n.

Great auricular n.

Splenius m.

Lev. scapulae m.

C 3,4

Accessory n.

Transverse cervical n.

Middle scalene m.

Ascend. cerv. a.

Transverse cervical a.

Phrenic n.

Descending (dorsal) scap.a.

Subclavian v.

Posterior scalene m.

Suprascap. a.

Subclavian a.

Ant. scalene m.

Descending (dorsal) scapular a.

Brachial plexus

Ant. scalene m.

Ascending cervical a.

Phrenic n.

Middle scalene m.

Transverse (superficial) cervical a.

Inf. thyroid a.

Thyrocervical tr.

Supreme intercostal a.

Vertebral a.

Subclavian a.

Common carotid a.

Vagus n.

Suprascapular a.

Rib 1

Clavicle

Subclavian v.

Int. thoracic a.

Descending br. of occipital

Occipital brs.

Occipital

Superficial cervical

Deep cervical

Ext. Int. } Carotids

Transverse cervical

Ascending cervical

Suprascapular

Vertebral

Descending (dorsal) scapular

Inf. thyroid

Thyrocervical trunk

Subclavian

Costocervical trunk

Rib 1

Highest intercostal

Brachiocephalic trunk

Internal thoracic

RIGHT SUBCLAVIAN ARTERY RELATIONS AND BRANCHES

Pansky

−59−

28. THYROID AND PARATHYROID GLANDS

I. Thyroid gland: U-shaped, in neck, on either side of larynx and trachea
A. PARTS: 2 lateral lobes interconnected by an *isthmus*. Weighs about 25 g
B. POSITION, RELATIONS, AND SIZE OF LOBES: varies in shape and size
   1. Apex of lateral lobes reaches junction of middle and lower thirds of thyroid cartilage. Base at level of fifth (or sixth) tracheal ring
   2. Lateral surface: convex; covered from superficial to deep by skin, superficial and deep fascia, sternocleidomastoid muscle, superior belly of omohyoid muscle, sternothyroid and sternohyoid muscles, and visceral (pretracheal) fascia, which forms a capsule
   3. Medial surface: trachea, inferior pharyngeal constrictor muscle, posterior part of cricothyroid muscle, esophagus, superior and inferior thyroid arteries, and the recurrent laryngeal nerve
   4. Posterior border: common carotid artery and parathyroid glands
C. POSITION AND RELATIONS OF ISTHMUS
   1. Connects lower thirds of the lateral lobes at the level of tracheal rings 2 and 3
   2. Covered by skin, fascia, and sternohyoid muscles
   3. A communicating artery between the two superior thyroid arteries runs on its cephalic border, and the inferior thyroid veins are located at its lower border
   4. Pyramidal lobe, if present, is attached to the left side of the superior border
D. ARTERIAL BLOOD SUPPLY
   1. Superior thyroid artery from external carotid artery
   2. Inferior thyroid artery from thyrocervical trunk
   3. Thyroidea ima (sometimes present) from brachiocephalic trunk
E. VENOUS DRAINAGE starts from a plexus on surface of gland and trachea
   1. Superior and middle veins arise in plexus and end in internal jugular vein
   2. Inferior vein terminates in brachiocephalic veins of either side
F. LYMPHATIC DRAINAGE: from a subcapsular plexus to superior deep cervical, pretracheal, paratracheal, retropharyngeal, and inferior deep cervical nodes

II. Parathyroids
A. PARTS: 4 separate glands, arranged into superior and inferior pairs
B. POSITION AND RELATIONS: they lie between the posterior border of thyroid and its capsule; the superior at the level of lower cricoid cartilage; the inferior near lower end of lateral lobes
C. ARTERIAL BLOOD supplied from the superior thyroid artery to the superior parathyroids and from the inferior thyroid artery to inferior pair
D. VEINS AND LYMPHATICS: same as thyroid

III. Special features
A. THE THYROID GLAND secretes thyroxin, which is important in metabolism
B. THE SUPERIOR THYROID ARTERY accompanies the superior thyroid vein, but the inferior thyroid artery is accompanied by the middle thyroid vein. The thyroidea ima, when present, is accompanied by the inferior thyroid vein
C. THE PARATHYROIDS secrete parathormone, important in calcium metabolism

IV. Clinical considerations
A. A PATHOLOGIC ENLARGEMENT of the thyroid gland is termed *goiter*
B. A THYROGLOSSAL DUCT can persist between foramen cecum and pyramidal lobe
C. IN THYROIDECTOMY, great care must be taken to avoid the recurrent laryngeal nerves and to preserve some parathyroid tissue
D. PARATHYROID ATROPHY OR REMOVAL leads to a severe convulsive disorder called *tetany*

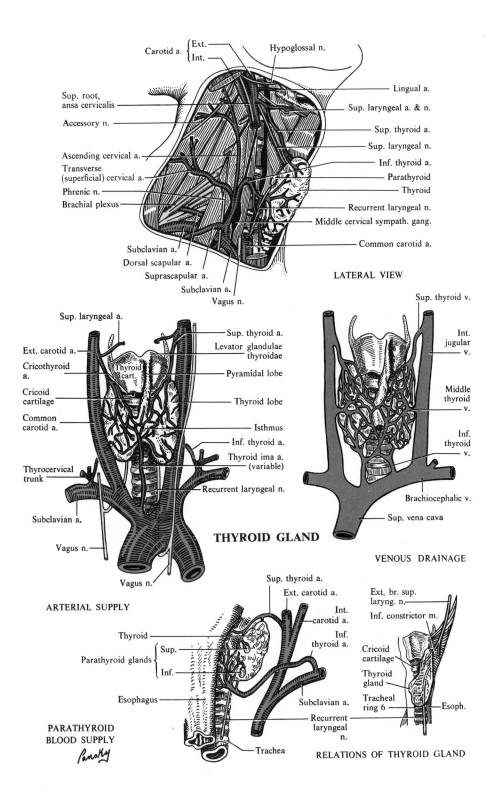

Carotid a. {Ext. Int.

Hypoglossal n.

Sup. root, ansa cervicalis

Accessory n.

Ascending cervical a.

Transverse (superficial) cervical a.

Phrenic n.

Brachial plexus

Subclavian a.

Dorsal scapular a.

Suprascapular a.

Subclavian a.

Vagus n.

Lingual a.

Sup. laryngeal a. & n.

Sup. thyroid a.

Sup. laryngeal n.

Inf. thyroid a.

Parathyroid

Thyroid

Recurrent laryngeal n.

Middle cervical sympath. gang.

Common carotid a.

LATERAL VIEW

Sup. laryngeal a.

Ext. carotid a.

Cricothyroid a.

Cricoid cartilage

Common carotid a.

Thyrocervical trunk

Subclavian a.

Vagus n.

Vagus n.

Thyroid cart.

Sup. thyroid a.

Levator glandulae thyroidae

Pyramidal lobe

Thyroid lobe

Isthmus

Inf. thyroid a.

Thyroid ima a. (variable)

Recurrent laryngeal n.

ARTERIAL SUPPLY

THYROID GLAND

Sup. thyroid v.

Int. jugular v.

Middle thyroid v.

Inf. thyroid v.

Brachiocephalic v.

Sup. vena cava

VENOUS DRAINAGE

Thyroid

Parathyroid glands {Sup. Inf.

Esophagus

PARATHYROID BLOOD SUPPLY

Pansky

Sup. thyroid a.

Ext. carotid a.

Int. carotid a.

Inf. thyroid a.

Subclavian a.

Recurrent laryngeal n.

Trachea

Ext. br. sup. laryng. n.

Inf. constrictor m.

Cricoid cartilage

Thyroid gland

Tracheal ring 6

Esoph.

RELATIONS OF THYROID GLAND

29. MUSCLES OF NECK, PHARYNX, AND TONGUE

I.

Name	Origin	Insertion	Action	Nerve
Buccinator (see p. 28)	Maxilla, mandible, pterygomandibular raphé	Muscles in lips	Compresses cheek, holds food between teeth	Facial (VII)
Geniohyoid	Inf. mental spine	Hyoid bone	Raises hyoid, depresses mand.	Ansa cervicalis (C1)
Genioglossus	Sup. mental spine	Hyoid bone, underside of tongue	Post. protrudes tongue, ant. retracts and depresses tongue	Hypoglossal (XII)
Hyoglossus	Body and great cornu of hyoid	Under lateral surface of tongue	Depresses tongue, draws sides down	Hypoglossal (XII)
Styloglossus	Styloid proc. stylomandib. lig.	Dorsal lat. tongue	Raises and retracts tongue	Hypoglossal (XII)
Palatoglossus	Soft palate	Lat. side of tongue	Elevates post. tongue	Pharyngeal plexus (X, XI?)
Thyrohyoid	Oblique line on thyroid cart.	Lat. side of great cornu of hyoid	Depresses hyoid, raises thyroid cart.	Sup. root, ansa cervicalis (C1–2)
Cricothyroid	Cricoid cart.	Inf. cornu & lower border thyroid cart.	Tilts lamina of cricoid cart. dorsally, tenses vocal cords	Ext. br. of sup. laryngeal n. (X)
Stylopharyngeus	Styloid process	Pharyngeal mm., thyroid cart.	Raises pharynx, expands sides of pharynx	Glossopharyngeal (IX)
Salpingopharyngeus	Auditory tube	Post. lat. pharynx	Raises upper and lat. pharynx	Pharyngeal plexus
Pharyngeal constrictors				
Superior	Med. pterygoid plate, hamulus, pterygomand. raphé, mandible	Pharyngeal tubercle of occipital bone, median raphé	Constricts upper pharynx	Pharyngeal plexus (X, XI?)
Middle	Great cornu of hyoid, lesser cornu	Median raphé	Constricts middle pharynx	Pharyngeal plexus (X, XI?)
Inferior	Sides of cricoid and thyroid cartilages	Median raphé	Constricts lower pharynx	Pharyngeal plexus (X, XI?)

II. Special features

A. PHARYNGOBASILAR FASCIA serves as the submucosa of the pharynx, located superiorly above the border of the superior constrictor, underlies all the pharyngeal muscles

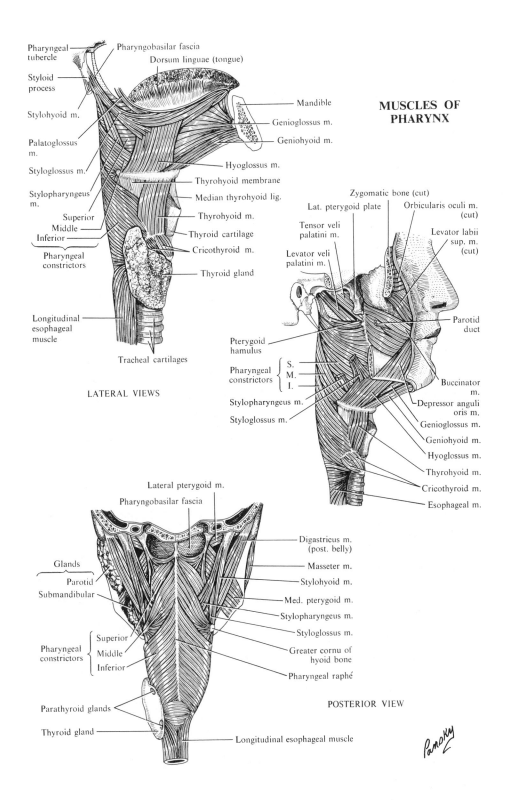

Pharyngeal tubercle

Pharyngobasilar fascia

Dorsum linguae (tongue)

Styloid process

Stylohyoid m.

Palatoglossus m.

Styloglossus m.

Stylopharyngeus m.

Superior
Middle
Inferior
Pharyngeal constrictors

Longitudinal esophageal muscle

Mandible

Genioglossus m.

Geniohyoid m.

Hyoglossus m.

Thyrohyoid membrane

Median thyrohyoid lig.

Thyrohyoid m.

Thyroid cartilage

Cricothyroid m.

Thyroid gland

Tracheal cartilages

LATERAL VIEWS

MUSCLES OF PHARYNX

Zygomatic bone (cut)

Lat. pterygoid plate

Orbicularis oculi m. (cut)

Tensor veli palatini m.

Levator labii sup. m. (cut)

Levator veli palatini m.

Parotid duct

Pterygoid hamulus

Pharyngeal constrictors
{ S.
M.
I. }

Stylopharyngeus m.

Styloglossus m.

Buccinator m.

Depressor anguli oris m.

Genioglossus m.

Geniohyoid m.

Hyoglossus m.

Thyrohyoid m.

Cricothyroid m.

Esophageal m.

Lateral pterygoid m.

Pharyngobasilar fascia

Glands
{ Parotid
Submandibular }

Pharyngeal constrictors
{ Superior
Middle
Inferior }

Parathyroid glands

Thyroid gland

Digastricus m. (post. belly)

Masseter m.

Stylohyoid m.

Med. pterygoid m.

Stylopharyngeus m.

Styloglossus m.

Greater cornu of hyoid bone

Pharyngeal raphé

POSTERIOR VIEW

Longitudinal esophageal muscle

Panoky

30. LARYNX: MUSCLES, VESSELS, AND NERVES

I. Muscles

Name	Origin	Insertion	Action
Cricothyroid			
Oblique	Arch of cricoid	Inf. cornu and lamina of	Draws up arch
Straight	Same	thyroid cartilage	Tenses vocal cords
Post. cricoarytenoid	Cricoid lamina	Muscular process of arytenoid	Separates vocal cords
Lat. cricoarytenoid	Arch of cricoid cartilage	Muscular process of arytenoid	Closes ant. part of rima glottidis
Arytenoid			
Oblique	Dorsal and lat. borders of arytenoid	Apex of opposite arytenoid	Closes rima glottidis, especially dorsal part
Transverse	Same	Dorsal surface of opposite arytenoid	Same
Thyroarytenoid (vocalis)	Angle of thyroid cartilage	Vent. base of arytenoid, vocal process	Shortens and relaxes vocal cords

II. Nerves

A. MOTOR
1. Ext. branch of sup. laryngeal (X) to cricothyroid; int. branch of sup. laryngeal (X) to part of transverse arytenoid
2. Recurrent laryngeal (X) to other intrinsic muscles of larynx

B. SENSORY: int. branch of sup. laryngeal to mucous membrane above vocal cords; cords to trachea innervated by recurrent laryngeal

III. Arteries

A. SUPERIOR LARYNGEAL ARTERY from the superior thyroid artery. Accompanies internal laryngeal branch of superior laryngeal nerve

B. INFERIOR LARYNGEAL ARTERY, a branch of the inf. thyroid a., ascends on back of larynx, under inf. constrictor of pharynx. Accompanied by recurrent nerve

IV. Veins follow same pattern as arteries

V. Lymphatic drainage: to infrahyoid, superior deep cervical, prelaryngeal, pretracheal, and paratracheal nodes

VI. General functions of larynx

A. IN RESPIRATION, framework is rigid to maintain passage. Even when vocal cords are closely approximated for extreme changes in pitch of voice, the glottis respiratoria, between the vocal processes of the arytenoids, remains open
1. To prevent substances from entering the trachea, both parts of the rima glottidis can be closed by the lateral cricoarytenoid and arytenoid muscles
2. In forced respiration, the rima glottidis can be widely opened by action of the posterior cricoarytenoid muscles

B. IN VOCALIZATION, pitch is changed through changes in tension on the vocal cords
1. Increased tension: through action of the cricothyroid muscles
2. Decreased tension: through action of the thyroarytenoid muscles

LARYNX

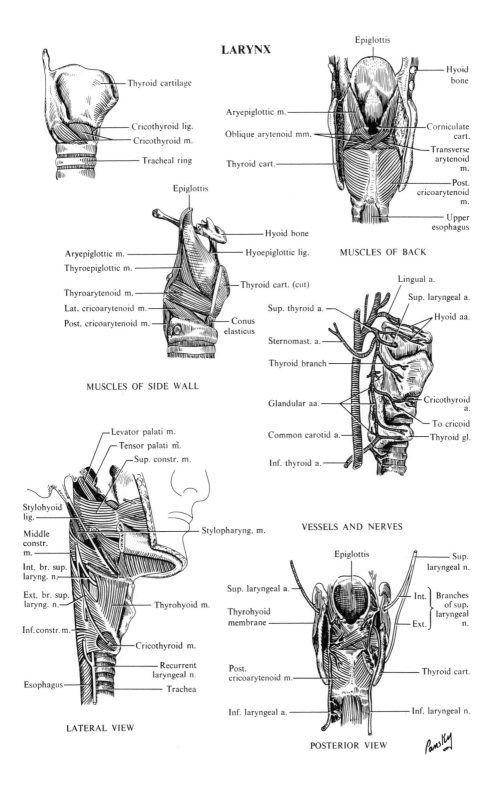

Thyroid cartilage

Cricothyroid lig.
Cricothyroid m.

Tracheal ring

Epiglottis

Hyoid bone

Aryepiglottic m.

Oblique arytenoid mm.

Thyroid cart.

Corniculate cart.

Transverse arytenoid m.

Post. cricoarytenoid m.

Upper esophagus

MUSCLES OF BACK

Epiglottis

Hyoid bone

Hyoepiglottic lig.

Aryepiglottic m.
Thyroepiglottic m.

Thyroarytenoid m.
Lat. cricoarytenoid m.
Post. cricoarytenoid m.

Thyroid cart. (cut)

Conus elasticus

MUSCLES OF SIDE WALL

Lingual a.

Sup. laryngeal a.

Sup. thyroid a.

Hyoid aa.

Sternomast. a.

Thyroid branch

Glandular aa.

Cricothyroid a.

To cricoid

Common carotid a.

Thyroid gl.

Inf. thyroid a.

VESSELS AND NERVES

Levator palati m.
Tensor palati m.
Sup. constr. m.

Stylohyoid lig.

Middle constr. m.

Stylopharyng. m.

Int. br. sup. laryng. n.

Ext. br. sup. laryng. n.

Thyrohyoid m.

Inf. constr. m.

Cricothyroid m.

Recurrent laryngeal n.

Esophagus

Trachea

LATERAL VIEW

Epiglottis

Sup. laryngeal n.

Sup. laryngeal a.

Int. } Branches of sup. laryngeal n.

Thyrohyoid membrane

Ext. }

Thyroid cart.

Post. cricoarytenoid m.

Inf. laryngeal a.

Inf. laryngeal n.

POSTERIOR VIEW

Pansky

31. LARYNX: FRAMEWORK AND STRUCTURE

I. Location: opposite third to sixth cervical vertebrae; shorter in women and children

II. Size: 44 mm long, 43 mm wide, and 36 mm deep

III. Cartilages: 9 in number (3 unpaired, 3 paired)

A. THYROID (largest and single): composed of *laminae,* one on either side, which meet at an acute angle in ventral midline of neck; *superior horn,* directed cephalically from posterior border of each lamina; *inferior horn,* directed caudally from posterior border of each lamina; *superior tubercle,* at root of superior cornu; *inferior tubercle,* on lower border; and *oblique line,* on lateral side of the lamina

B. CRICOID (single): composed of *lamina,* posteriorly, 2 to 3 cm long; *arch,* a narrow, convex ring of cartilage. *Articular facets: lateral,* at junction of lamina and arch; and *superior,* at border of lateral corners of lamina

C. EPIGLOTTIS: yellow elastic cartilage; single, leaf-shaped piece

D. ARYTENOIDS (paired): each consists of a pyramidal piece with *3 surfaces,* a *base,* and an *apex.* At caudal end of base is the *articular surface.* The *muscular process* projects backward and the *vocal process* anteriorly from the base

E. CORNICULATE AND CUNEIFORM: small paired pieces of yellow elastic cartilage rostral to the arytenoids and in the aryepiglottic folds

IV. Ligaments

A. EXTRINSIC: connect laryngeal cartilages and hyoid bone or trachea: *thyrohyoid membrane,* between thyroid cartilage and hyoid bone; *middle thyroid ligament,* thickened middle part of membrane; the *lateral thyrohyoid ligaments,* thickened, cordlike posterior edge of the membrane; *hyoepiglottic ligament,* from epiglottis to hyoid bone; and *cricotracheal ligament,* from cricoid cartilage to first tracheal ring

B. INTRINSIC: connect laryngeal cartilages together: *elastic membrane,* cephalic part between arytenoid and epiglottic cartilages; caudal part, *conus elasticus,* interconnecting thyroid, cricoid, and arytenoid cartilages. The middle, *cricothyroid ligament,* is the anterior part of the conus. The lateral part of the conus, the *thyroepiglottic ligament,* is continuous with the *vocal ligaments*

V. Articulations: cricothyroid, between inferior cornu of thyroid and lateral facet of cricoid; cricoarytenoid, between cricoid and base of arytenoid cartilage

VI. Interior of larynx extends from entrance to lower border of cricoid cartilage

A. AROUND THE ENTRANCE OF THE LARYNX

1. Three glossoepiglottic folds, 1 median and 2 lateral. Between these folds are the *valleculae*

2. Between the aryepiglottic folds and thyroid cartilage are the *piriform sinuses*

B. ENTRANCE: bounded anteriorly by epiglottis, posteriorly by apices of arytenoid cartilages and corniculate cartilages, and laterally by the aryepiglottic folds

C. LARYNGEAL CAVITY: divided by the *true vocal folds* into the *vestibule* above the folds, which contains the *vestibular* or *false vocal cords,* and the *ventricle.* Below the true vocal folds, the larynx is continuous with the trachea

VII. True vocal cords (folds): mucous membrane covering the vocal ligaments which extend from the vocal processes of the arytenoids to the thyroid cartilage

VIII. Rima glottidis: the fissure between the true vocal folds, including the vocal processes. Has two parts: intramembranous, ventral 3/5 between folds (glottis vocalis), and intercartilaginous, posterior 2/5 between arytenoid cartilages (glottis respiratoria)

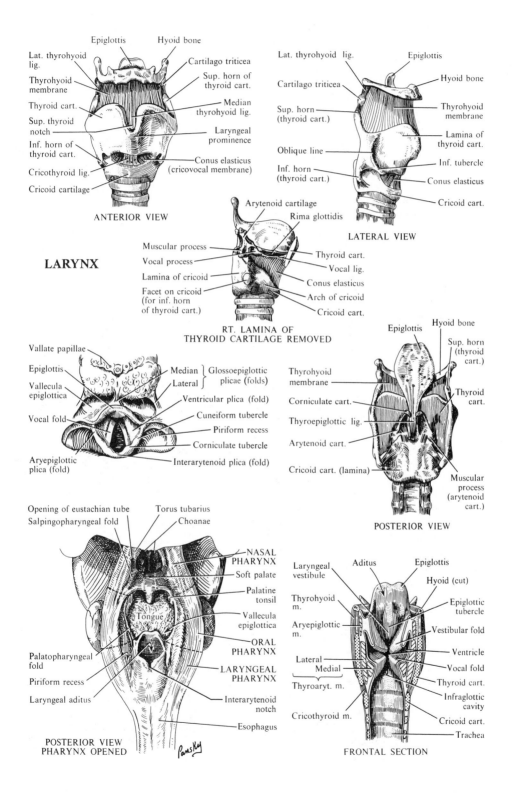

LARYNX

Epiglottis
Hyoid bone
Lat. thyrohyoid lig.
Cartilago triticea
Thyrohyoid membrane
Sup. horn of thyroid cart.
Thyroid cart.
Median thyrohyoid lig.
Sup. thyroid notch
Laryngeal prominence
Inf. horn of thyroid cart.
Conus elasticus (cricovocal membrane)
Cricothyroid lig.
Cricoid cartilage

ANTERIOR VIEW

Lat. thyrohyoid lig.
Epiglottis
Cartilago triticea
Hyoid bone
Sup. horn (thyroid cart.)
Thyrohyoid membrane
Lamina of thyroid cart.
Oblique line
Inf. tubercle
Inf. horn (thyroid cart.)
Conus elasticus
Cricoid cart.

LATERAL VIEW

Arytenoid cartilage
Rima glottidis
Muscular process
Thyroid cart.
Vocal process
Vocal lig.
Lamina of cricoid
Conus elasticus
Facet on cricoid (for inf. horn of thyroid cart.)
Arch of cricoid
Cricoid cart.

RT. LAMINA OF THYROID CARTILAGE REMOVED

Vallate papillae
Epiglottis
Vallecula epiglottica
Median ⎫ Glossoepiglottic
Lateral ⎬ plicae (folds)
Vocal fold
Ventricular plica (fold)
Cuneiform tubercle
Piriform recess
Corniculate tubercle
Aryepiglottic plica (fold)
Interarytenoid plica (fold)

Epiglottis
Hyoid bone
Sup. horn (thyroid cart.)
Thyrohyoid membrane
Corniculate cart.
Thyroid cart.
Thyroepiglottic lig.
Arytenoid cart.
Cricoid cart. (lamina)
Muscular process (arytenoid cart.)

POSTERIOR VIEW

Opening of eustachian tube
Torus tubarius
Salpingopharyngeal fold
Choanae
NASAL PHARYNX
Soft palate
Palatine tonsil
Tongue
Vallecula epiglottica
ORAL PHARYNX
Palatopharyngeal fold
LARYNGEAL PHARYNX
Piriform recess
Laryngeal aditus
Interarytenoid notch
Esophagus

POSTERIOR VIEW PHARYNX OPENED

Laryngeal vestibule
Aditus
Epiglottis
Hyoid (cut)
Thyrohyoid m.
Epiglottic tubercle
Aryepiglottic m.
Vestibular fold
Ventricle
Lateral
Medial
Vocal fold
Thyroaryt. m.
Thyroid cart.
Infraglottic cavity
Cricothyroid m.
Cricoid cart.
Trachea

FRONTAL SECTION

Pansky

32. CERVICAL NERVES, PLEXUS, AND CERVICAL SYMPATHETIC TRUNK

I. Cervical plexus: ventral primary divisions of cervical nerves 1–4, forming 3 loops
  A. LOCATION: opposite cervical vertebrae 1–3, ventral and lateral to levator scapulae and middle scalene muscles
  B. DISTRIBUTION
   1. To cranial nerves X and XII, mainly from C1 and C2; to XI, mainly from C2–4
   2. To skin (see p. 16)
   3. To muscles, directly: rectus capitis anterior and lateralis, longus capitis and cervicis, levator scapulae, and middle scalene
   4. To muscles, through communication of special branches: nerves to sternocleidomastoid from C2 and C3, and to trapezius from C3 and C4, may go with the spinal accessory nerve (XI)
     a. Communications with the hypoglossal from C1 and C2 run with the hypoglossal for a short distance, then leave this trunk as the *superior root of the ansa cervicalis.* This forms a loop with the *inferior root* from C2 and C3. This loop is the *ansa cervicalis,* which supplies the geniohyoid, thyrohoid, omohyoid, sternothyroid, and sternohyoid muscles

II. Cervical sympathetic system consists of 3 ganglia with intervening cords
  A. LOCATION: ventral to transverse processes of vertebrae, embedded in dorsal part of carotid sheath
  B. COMPOSITION: pre- and postganglionic autonomic fibers
  C. ORIGIN OF PREGANGLIONIC FIBERS: from the intermediolateral gray column of spinal cord segments T1 through T5. White rami communicantes leave the ventral roots of corresponding thoracic nerves and ascend in the trunk

III. Ganglia
  A. SUPERIOR: largest; lies on transverse process of C2 (and C3). Branches: internal carotid nerve to cranial nerves IX, X, and XII; gray rami to C1 through C4; pharyngeal to pharyngeal plexus; to internal carotid plexus; intercarotid to carotid body and sinus; superior cardiac nerve; to larynx and thyroid
  B. MIDDLE: smallest (sometimes absent); lies ventral to transverse process of C6 or C7. Branches: gray rami to C5 and C6; middle cardiac nerve; thyroid nerves to form plexus on inferior thyroid artery
  C. INFERIOR: lies between base of transverse process of C7 and neck of first rib, may be fused with first thoracic ganglion. Branches: gray rami to C7, C8, and T1; inferior cardiac nerve; and vertebral nerves along vertebral arteries

IV. Special features
  A. TRUNK between middle and inferior cervical ganglia splits to enclose the subclavian artery. The ventral part is longer and forms a loop, the *ansa subclavia*
  B. WHEN THE INFERIOR CERVICAL and first thoracic ganglia are fused, this mass is the cervicothoracic (stellate) ganglion

V. Clinical considerations
  A. INTERRUPTION OF CERVICAL SYMPATHETICS: when accomplished above C8 or T1, all sympathetic autonomic control for head, neck, and upper extremity is lost. This will result in Horner's syndrome, which shows narrowed palpebral fissure, constriction of pupil, flushing of skin, and lack of sweating. If bilateral, there is no acceleration of heart rate. Thus, when performed for the relief of Raynaud's disease, which is characterized by abnormal vasoconstriction of the upper limb, it is preferable to avulse the anterior roots of T2 and T3 or, still better, the white rami at those levels

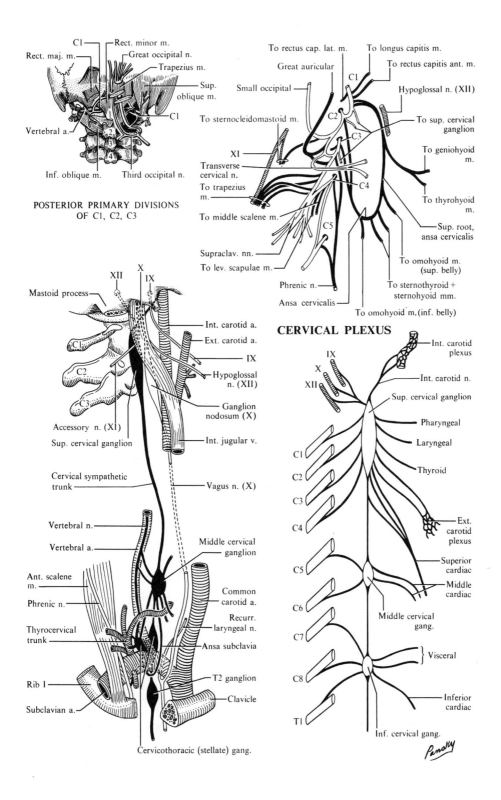

POSTERIOR PRIMARY DIVISIONS
OF C1, C2, C3

CERVICAL PLEXUS

33. THE FLOOR OF THE MOUTH, THE SUPRAHYOID, INFRAHYOID, AND ANTERIOR VERTEBRAL MUSCLES

I. **The floor of the mouth:** the mylohyoid and geniohyoid muscles, which form the floor of the mouth, are described elsewhere (see p. 52)

II. **The strap muscles,** attached to hyoid bone and thyroid cartilage (see p. 52)

III. **The anterior vertebral muscles**

Name	Origin	Insertion	Action	Nerve
Rectus capitis				
Anterior	Root of trans. proc. of atlas	Basilar part of occipital bone	Flexes head	C1,2
Lateralis	Upper trans. process of atlas	Lower jugular process of occipital bone	Bends head to side	C1,2
Longus capitis	Ant. tubercle, trans. proc. vertebrae C3–6	Basilar part of occipital bone	Flexes head	C1,2,3
Longus colli				
Sup. oblique	Ant. tubercle, trans. proc. vertebrae C3–5	Tubercle on ant. arch of atlas	Flexes neck, slight rotation of cervical part	C2–7
Inf. oblique	Bodies of vertebrae T1–3	Ant. tubercle, trans. proc. C5, C6	Slight rotation of cervical part	C2–7
Vertical	Bodies of vertebrae C5–7, T1–3	Bodies of C2–4	Flexes neck	C2–7
Anterior scalene	Ant. tubercle, trans. proc. vertebrae C3–6	Scalene tubercle & ridge on upper first rib	Raises first rib, bends neck	C5–8
Middle scalene	Post. tubercle, trans. proc. vertebrae C2–7	Upper first rib, behind subclav. groove	Raises first rib, bends neck	C5–8
Posterior scalene	Post. tubercle, trans. proc. vertebrae C5–7	Outer second rib	Raises second rib, bends neck	C6–8

IV. **Special features**
  A. RELATIONS OF ATTACHMENT of anterior scalene muscle to first rib and subclavian vessels
    1. Subclavian vein crosses first rib in front of insertion of anterior scalene muscle
    2. Subclavian artery crosses first rib in subclavian groove behind insertion of anterior scalene muscle
  B. THE FASCIA on the deep surface of the scalene mm. forms a conical fibrous dome called suprapleural membrane (Sibson's fascia) which arches over cupula of the pleura
  C. SPASM OR HYPERTROPHY of anterior scalene m. may lead to circulatory and neurologic symptoms of upper limb (scalenus anticus syndrome)

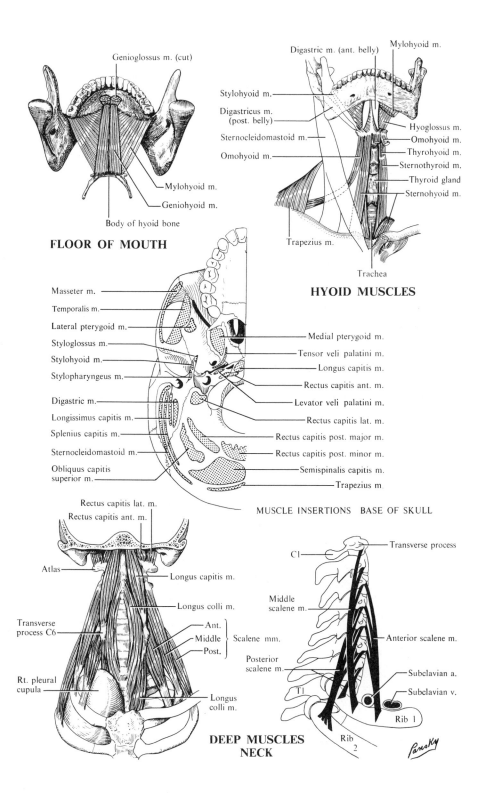

Genioglossus m. (cut)

Mylohyoid m.

Geniohyoid m.

Body of hyoid bone

FLOOR OF MOUTH

Digastric m. (ant. belly)

Mylohyoid m.

Stylohyoid m.

Digastricus m. (post. belly)

Sternocleidomastoid m.

Omohyoid m.

Hyoglossus m.

Omohyoid m.

Thyrohyoid m.

Sternothyroid m.

Thyroid gland

Sternohyoid m.

Trapezius m.

Trachea

HYOID MUSCLES

Masseter m.

Temporalis m.

Lateral pterygoid m.

Styloglossus m.

Stylohyoid m.

Stylopharyngeus m.

Digastric m.

Longissimus capitis m.

Splenius capitis m.

Sternocleidomastoid m.

Obliquus capitis superior m.

Medial pterygoid m.

Tensor veli palatini m.

Longus capitis m.

Rectus capitis ant. m.

Levator veli palatini m.

Rectus capitis lat. m.

Rectus capitis post. major m.

Rectus capitis post. minor m.

Semispinalis capitis m.

Trapezius m.

MUSCLE INSERTIONS BASE OF SKULL

Rectus capitis lat. m.

Rectus capitis ant. m.

Atlas

Longus capitis m.

Longus colli m.

Transverse process C6

Ant.

Middle

Post.

Scalene mm.

Rt. pleural cupula

Longus colli m.

DEEP MUSCLES NECK

C1

Transverse process

Middle scalene m.

Anterior scalene m.

Posterior scalene m.

T1

Subclavian a.

Subclavian v.

Rib 1

Rib 2

Pansky

34. SUBMANDIBULAR AND SUBLINGUAL GLANDS

I. Submandibular
A. SIZE: second largest of major salivary glands, size of walnut, about half the size of parotid
B. TYPE: compound, tubuloalveolar; mixed serous and mucous
C. LOCATION AND RELATIONS: mainly in digastric (submandibular) triangle (see p. 46)
   1. Superficial surface: cephalic part touches submandibular depression of mandible and internal pterygoid muscle; caudal part is covered by skin, superficial fascia, platysma, and deep cervical fascia
      a. Facial vein and small branches of facial nerve cross it; submandibular lymph nodes lie against it
   2. Deep surface: lies on mylohyoid, hyoglossus, styloglossus, stylohyoid, and posterior belly of digastric muscles
      a. Mylohyoid nerve and vessels are in contact with it
   3. Posterior border: facial artery
   4. Deep process: a projection from its deep surface is bounded below and laterally by mylohyoid muscle; medially, by hyoglossus and styloglossus muscles; above, by lingual nerve and submandibular ganglion; below by hypoglossal nerve
D. DUCT (WHARTON'S): formed on deep surface, runs rostrally between mylohyoid muscle laterally and hyoglossus and genioglossus muscles medially. Opens at caruncle at side of frenulum of tongue (on sublingual papilla). In part of course is between lingual and hypoglossal nerves; finally is crossed by lingual nerve
E. THE SUBMANDIBULAR LYMPH NODES lie on the gland and along the border of the mandible

II. Sublingual
A. SIZE: smallest of 3 major salivary glands and most deeply situated
B. TYPE: compound, tubuloalveolar; mixed serous and mucous
C. LOCATION AND RELATIONS: beneath mucous membrane of floor of mouth
   1. Above: floor of mouth
   2. Below: mylohyoid muscle
   3. Behind: deep part, submandibular gland
   4. Laterally: sublingual depression of mandible
   5. Medially: genioglossus muscle
      a. Lingual nerve and submandibular duct lie between gland and muscle
D. DUCTS
   1. Small (of Rivinus): some join submandibular duct, others (10–12) open separately in floor of mouth, forming a linear series along top of sublingual fold
   2. Large (Bartholin): 1 or 2 duct branches join submandibular duct

III. Innervation of both glands
A. PARASYMPATHETIC: preganglionic fibers, from superior salivatory nucleus through nervus intermedius of facial nerve: from facial via chorda tympani to lingual nerve, through parasympathetic root to submandibular ganglion, where they terminate in synapses. Postganglionic branches from ganglion go to both glands

IV. Vessels
A. ARTERIES: branches of the facial artery to submandibular gland; sublingual branch of lingual artery supplies sublingual
B. VEINS follow the arteries
C. LYMPHATIC DRAINAGE to submandibular nodes

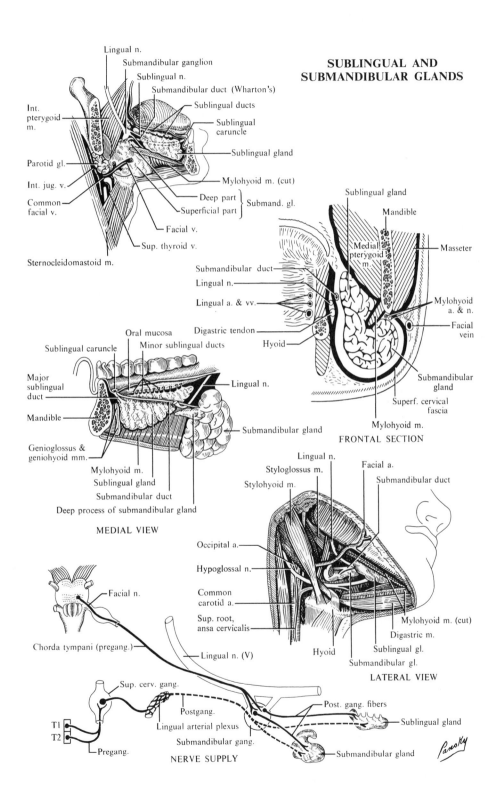

SUBLINGUAL AND SUBMANDIBULAR GLANDS

Lingual n.
Submandibular ganglion
Sublingual n.
Submandibular duct (Wharton's)
Sublingual ducts
Sublingual caruncle
Sublingual gland

Int. pterygoid m.

Parotid gl.
Int. jug. v.
Common facial v.

Mylohyoid m. (cut)
Deep part
Superficial part } Submand. gl.
Facial v.
Sup. thyroid v.

Sternocleidomastoid m.

Sublingual gland
Mandible
Medial pterygoid m.
Masseter

Submandibular duct
Lingual n.
Lingual a. & vv.
Digastric tendon
Hyoid

Mylohyoid a. & n.
Facial vein

Submandibular gland
Superf. cervical fascia
Mylohyoid m.

FRONTAL SECTION

Oral mucosa
Minor sublingual ducts
Sublingual caruncle
Major sublingual duct
Mandible
Lingual n.
Genioglossus & geniohyoid mm.
Mylohyoid m.
Sublingual gland
Submandibular duct
Deep process of submandibular gland
Submandibular gland

MEDIAL VIEW

Lingual n.
Styloglossus m.
Stylohyoid m.
Facial a.
Submandibular duct

Occipital a.
Hypoglossal n.
Common carotid a.
Sup. root, ansa cervicalis

Mylohyoid m. (cut)
Digastric m.
Sublingual gl.
Submandibular gl.
Hyoid

LATERAL VIEW

Facial n.
Chorda tympani (pregang.)
Lingual n. (V)

Sup. cerv. gang.
Postgang.
Lingual arterial plexus
Submandibular gang.

T1
T2
Pregang.

Post. gang. fibers
Sublingual gland
Submandibular gland

Pansky

NERVE SUPPLY

35. THE TONGUE

I. **Functions:** chief organ for taste; important in speech, mastication, and deglutition

II. **Parts:** root, apex, and body

III. **Surfaces**
   A. INFERIOR: mucous membrane continuous with that of gums and floor of mouth. *Frenulum linguae* in midline
   B. DORSAL: convex. Median sulcus divides tongue into lateral halves and ends posteriorly in a depression, the *foramen cecum*. The latter is the apex for 2 diverging grooves, the *terminal sulci*. Two thirds of the tongue lies rostral to the sulcus and is rough and covered with papillae, while the caudal third is more smooth and contains glands and crypts of the *lingual tonsil*

IV. **Papillae** are of 3 basic types: *vallate,* largest, 8 to 10 in number, lying in a row beginning just rostral to the foramen cecum and along the terminal sulci; *foliate* at sides and *fungiform* scattered over dorsum; and the *filiform,* smallest and most numerous, covering all of dorsum, arranged in rows parallel to sulcus terminalis

V. **Taste buds:** sensory organs of taste, scattered over mucous membrane of mouth and tongue and especially numerous on and around vallate papillae

VI. **Nerves of tongue**
   A. SENSORY
      1. General sensation: lingual branch of mandibular (V) for anterior two thirds; lingual branch of glossopharyngeal (IX) for caudal one third; and lingual branches from superior laryngeal of vagus (X) for root near epiglottis
      2. Special sensation (taste): taste buds on anterior two thirds, through lingual, chorda tympani, and nervus intermedius by way of facial (VII); taste buds on caudal one third, through lingual branch of glossopharyngeal (IX); and taste buds in epiglottic region by sup. laryngeal br. of vagus (X)
   B. MOTOR
      1. Hypoglossal nerve: from nucleus in medulla, nerve emerges through hypoglossal canal, becomes closely united with vagus nerve, passes between internal carotid artery and internal jugular vein, crosses superficial to both to hook around occipital artery, crosses lateral to external carotid and lingual arteries, curves upward above hyoid bone, deep to digastric and stylohyoid muscles, and then between mylohyoid and hyoglossus muscles to apex of tongue. Supplies both intrinsic and extrinsic muscles of tongue

VII. **Arteries of tongue**
   A. MAJOR: *lingual artery* from external carotid artery. Runs deep, along inferior tongue surface to end as *deep lingual artery*. The *dorsal lingual arteries,* 2 or 3 branches, ascend to dorsum of tongue
   B. MINOR: lingual branch of inferior alveolar artery, tonsillar branch of facial artery, and branches from ascending pharyngeal artery

VIII. **Veins**
   A. LINGUAL: from dorsum, sides, and undersurface to internal jugular vein
   B. VENA COMITANS OF HYPOGLOSSAL N. (RANINE): begins near apex and runs with hypoglossal nerve to terminate in lingual or common facial veins

IX. **Lymphatic drainage:** (see p. 78)

X. **Modalities of taste:** sweet at tip, bitter toward root, salty at sides and tip, and sour at sides

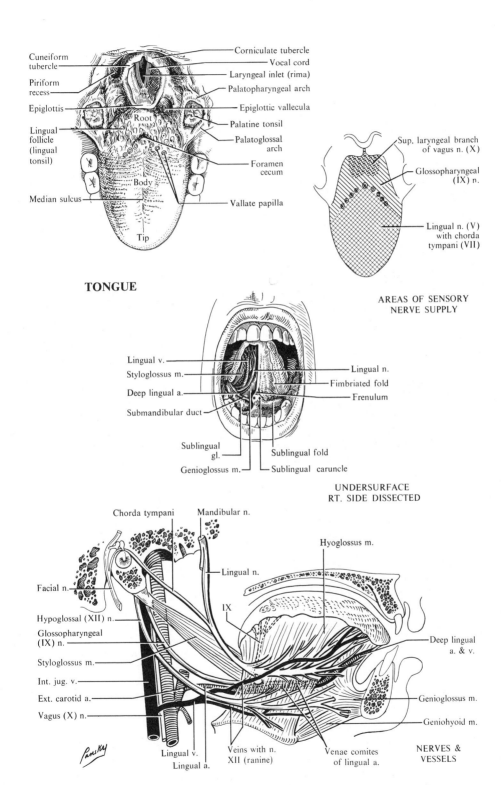

Cuneiform tubercle
Corniculate tubercle
Vocal cord
Piriform recess
Laryngeal inlet (rima)
Palatopharyngeal arch
Epiglottis
Epiglottic vallecula
Root
Palatine tonsil
Lingual follicle (lingual tonsil)
Palatoglossal arch
Foramen cecum
Body
Median sulcus
Vallate papilla
Tip

TONGUE

Sup. laryngeal branch of vagus n. (X)
Glossopharyngeal (IX) n.
Lingual n. (V) with chorda tympani (VII)

AREAS OF SENSORY NERVE SUPPLY

Lingual v.
Lingual n.
Styloglossus m.
Fimbriated fold
Deep lingual a.
Frenulum
Submandibular duct
Sublingual gl.
Sublingual fold
Genioglossus m.
Sublingual caruncle

UNDERSURFACE
RT. SIDE DISSECTED

Chorda tympani
Mandibular n.
Hyoglossus m.
Facial n.
Lingual n.
Hypoglossal (XII) n.
IX
Glossopharyngeal (IX) n.
Deep lingual a. & v.
Styloglossus m.
Int. jug. v.
Genioglossus m.
Ext. carotid a.
Geniohyoid m.
Vagus (X) n.
Lingual v.
Veins with n. XII (ranine)
Venae comites of lingual a.
Lingual a.

NERVES & VESSELS

– 75 –

36. MUSCLES OF THE TONGUE AND SWALLOWING

I. **Extrinsic:** (see p. 62)
A. GENIOGLOSSUS, HYOGLOSSUS, STYLOGLOSSUS, AND PALATOGLOSSUS MUSCLES

II. **Intrinsic:** all of which change the shape of the tongue
A. SUPERIOR LONGITUDINAL: on the dorsum of the tongue. Shortens tongue and raises tip and sides
B. INFERIOR LONGITUDINAL: on underside of tongue. Shortens tongue and depresses tip
C. TRANSVERSUS narrows and increases height of tongue
D. VERTICALIS makes tongue flat and broadens it

III. **Swallowing:** a complex of muscular action involving several activities, all in a proper sequence. This requires the propulsion of the bolus of food, enlargement of the alimentary canal in front of the bolus, closure of the tube behind the bolus, and closure of the other systems such as the nasopharynx and larynx

IV. **Actions involved in swallowing**
A. PROPULSION
   1. With the jaws and lips closed, the tongue is voluntarily raised against the hard palate, using the genioglossus, mylohyoid, and geniohyoid muscles, while the palatoglossus muscle relaxes. This moves the bolus to the level of the soft palate
   2. The soft palate is raised to enlarge the space by the levator and tensor veli palatini muscles
   3. At the same time, the base of the tongue is raised and drawn back, forcing the bolus farther down the tube. The styloglossus muscle is important in this function
   4. The pharynx, attached to the larynx, is raised and expanded to receive the bolus. With the jaw fixed, many muscles accomplish this action: the mylohyoid, geniohyoid, stylohyoid, posterior belly of the digastric, stylopharyngeus, salpingopharyngeus, and palatopharyngeus muscles. The stylopharyngeus also widens and enlarges the pharynx
   5. The bolus now lies in the region of the pharyngeal constrictors. When the superior contracts, the middle and inferiors relax. As the bolus passes to the next field, the middle constrictor then contracts, etc. After the inferior contracts, the bolus has reached the upper esophagus, where a peristaltic-like wave carries it down
B. WHEN THE BOLUS passes through the isthmus of the pharynx, the palatoglossus muscles contract. This closes off the mouth from the pharynx
C. CLOSURE OF THE OTHER SYSTEMS
   1. The nasopharynx is closed: by raising the soft palate through the levator muscle and by contraction of the 2 palatopharyngeus muscles, causing them to approach each other medially
   2. The larynx is closed:
      a. Superiorly, by the epiglottis being folded back under the retracted tongue as the larynx is raised; the epiglottis is thus folded over the laryngeal orifice
      b. Inferiorly, by the closure of the rima glottidis through the action of the arytenoid, lateral cricoarytenoid, and thyroarytenoid muscles

V. **Clinical considerations**
A. LESION OF THE HYPOGLOSSAL NERVE causes deviation of the protruded tongue toward the side of the lesion. This is due to the lack of function of the genioglossus muscle on the diseased side
B. IF GENIOGLOSSUS IS PARALYZED, tongue has tendency to fall back and obstruct oropharyngeal airway with risk of suffocation

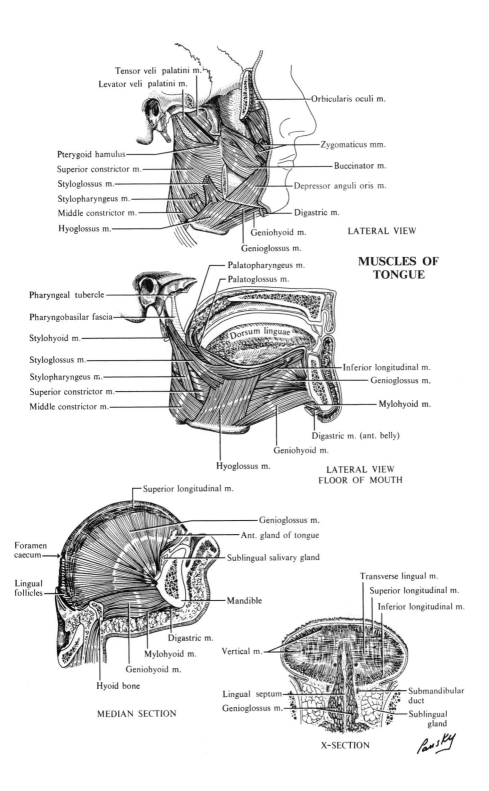

Tensor veli palatini m.

Levator veli palatini m.

Orbicularis oculi m.

Pterygoid hamulus

Superior constrictor m.

Styloglossus m.

Stylopharyngeus m.

Middle constrictor m.

Hyoglossus m.

Zygomaticus mm.

Buccinator m.

Depressor anguli oris m.

Digastric m.

Geniohyoid m.

Genioglossus m.

LATERAL VIEW

MUSCLES OF TONGUE

Palatopharyngeus m.

Palatoglossus m.

Pharyngeal tubercle

Pharyngobasilar fascia

Stylohyoid m.

Styloglossus m.

Stylopharyngeus m.

Superior constrictor m.

Middle constrictor m.

Dorsum linguae

Inferior longitudinal m.

Genioglossus m.

Mylohyoid m.

Digastric m. (ant. belly)

Geniohyoid m.

Hyoglossus m.

LATERAL VIEW
FLOOR OF MOUTH

Superior longitudinal m.

Genioglossus m.

Ant. gland of tongue

Sublingual salivary gland

Foramen caecum

Lingual follicles

Mandible

Transverse lingual m.

Superior longitudinal m.

Inferior longitudinal m.

Vertical m.

Digastric m.

Mylohyoid m.

Geniohyoid m.

Hyoid bone

Lingual septum

Genioglossus m.

Submandibular duct

Sublingual gland

MEDIAN SECTION

X-SECTION

Pansky

–77–

37. LYMPHATICS OF THE NOSE AND TONGUE

I. Lymphatic drainage of the nasal cavities and nasopharynx
A. NASAL CAVITY
1. Anterior part communicates with the vessels of the skin of the nose. These usually drain through the mandibular nodes to the submandibular and superior deep cervical nodes
2. Posterior part drains into the retropharyngeal nodes located in the fascia posterior to the pharynx. The efferents from these go to the deep cervical nodes. Part of this drainage may go directly to the superior deep cervical nodes. A few channels from this same area also may go to the subparotid nodes first and then to the superior deep cervical nodes
3. From floor: vessels may enter the parotid set and then follow the courses given above
4. From the nasopharynx: there is also drainage to both the retropharyngeal and parotid nodes
5. The paranasal sinuses drain partly into the retropharyngeal and partly by direct paths to the superior deep cervical nodes
6. The major part of the nasal cavity drains posteriorly to reach the upper deep cervical and retropharyngeal nodes. The most inferior part of the cavity drains into submandibular and superficial cervical nodes following the facial artery

II. Palatine tonsil
A. FROM THE TONSIL, the lymphatic channels drain directly into the most cephalic of the superior deep cervical nodes through 3 to 5 vessels

III. Tongue: it is usually divided into 4 main drainage areas
A. APEX drains into the suprahyoid and principal node* of the tongue
B. LATERAL (margins) drain to the submandibular and superior deep cervical nodes
C. BASAL (near vallate papillae) drain to the superior deep cervical nodes
D. MEDIAN drain to the submandibular and superior deep cervical nodes
E. CROSS DRAINAGE: channels from the opposite sides of the tongue cross the midline. Some of these enter the superior deep cervical nodes, and others enter the inferior deep cervical nodes
F. IN GENERAL, the anterior 2/3 of the tongue drains into the submandibular and superficial cervical nodes, and the posterior 1/3 drains into the upper deep cervical and retropharyngeal nodes

IV. Clinical considerations
A. JUGULODIGASTRIC NODE: because of the inaccessibility of the posterior third of the tongue and the tonsil, sometimes the first positive sign of carcinoma in these regions will be a swelling of this node, sometimes called the "main gland of the tonsil"
B. IN CONSEQUENCE OF THE LYMPHATIC DRAINAGE OF THE TONGUE, metastatic carcinoma may be widely disseminated through the submental and submandibular regions and along the internal jugular vein. The operation designed to remove such metastatic lesions is called "radical neck dissection" or "block dissection of the neck"
C. CARCINOMA OF THE TONGUE affects males chiefly, and the edges of the tongue are most often affected. The tumor is an epidermoid carcinoma and may be of any degree of malignancy. Metastasis is usually slow, involving the lymph nodes of the neck; rarely, there may be rapid, widespread metastases

---

*The "principal" node is a single constant node of the deep cervical group, lying at the bifurcation of the common carotid artery, which receives a large number of vessels from the tongue.

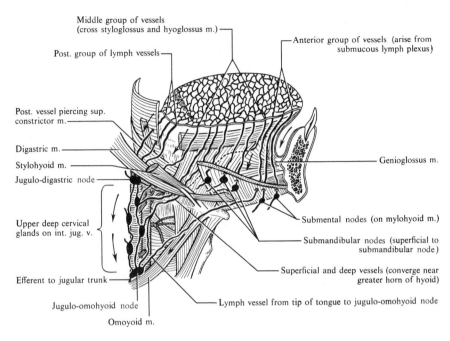

Middle group of vessels
(cross styloglossus and hyoglossus m.)

Anterior group of vessels (arise from
submucous lymph plexus)

Post. group of lymph vessels

Post. vessel piercing sup.
constrictor m.

Digastric m.

Stylohyoid m.

Jugulo-digastric node

Genioglossus m.

Upper deep cervical
glands on int. jug. v.

Submental nodes (on mylohyoid m.)

Submandibular nodes (superficial to
submandibular node)

Efferent to jugular trunk

Superficial and deep vessels (converge near
greater horn of hyoid)

Jugulo-omohyoid node

Omoyoid m.

Lymph vessel from tip of tongue to jugulo-omohyoid node

LYMPH VESSELS OF TONGUE
AND NASAL CAVITY

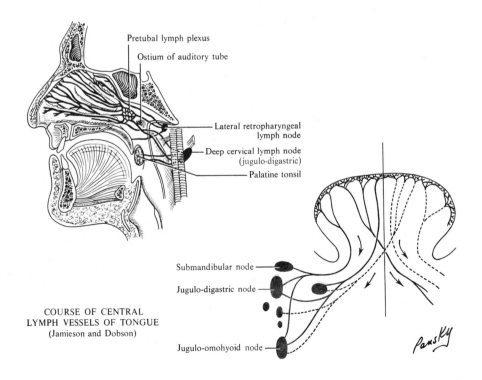

Pretubal lymph plexus

Ostium of auditory tube

Lateral retropharyngeal
lymph node

Deep cervical lymph node
(jugulo-digastric)

Palatine tonsil

COURSE OF CENTRAL
LYMPH VESSELS OF TONGUE
(Jamieson and Dobson)

Submandibular node

Jugulo-digastric node

Jugulo-omohyoid node

Pansky

38. TEETH AND PALATE

I. Teeth
A. VARIETIES: incisors, canines, premolars, and molars
B. SETS
1. Deciduous, 20: 2 incisors, 1 canine, 2 molars*
2. Permanent, 32: 2 incisors, 1 canine, 2 premolars, 3 molars*
C. GENERAL PARTS FOR ALL TEETH: *crown,* projecting above the gingiva (gum); *root,* embedded in an alveolus (socket) in the alveolar process (bone) of the jaw by a fibrous *periodontal membrane*; and *neck,* constriction between crown and root, also embedded in the gingiva
D. SURFACES: *labial* (or buccal), next to lip or cheek; *lingual,* toward tongue; and *contact edge* touching adjoining teeth
E. SPECIAL FEATURES
1. Premolars, or bicuspids: free surface of crown shows 2 eminences or cusps (a labial and lingual) separated by a groove. Root is usually single
2. Molars: have several cusps, separated by grooves
a. Upper, have 3 roots—2 buccal, 1 lingual; first molar has 4 cusps; second has 3 or 4 cusps; and third has 3 cusps
b. Lower, have 2 roots—anterior and posterior; first molar has 5 cusps; second and third have 4 or 5 cusps
F. STRUCTURE, from within, outward: *pulp cavity,* containing loose connective tissue, vessels, and nerves and continuous with periodontal tissue through the *root canal* and *apical foramen*; *dentin* (ivory), modified bone that forms bulk of tooth; *enamel,* covers exposed crown; and *cementum,* thin layer of modified bone covering the dentin of the root and neck of the tooth
G. AGE at time of eruption of permanent teeth

1st molar	6 years	2nd premolar	10 years
Central incisors	7 years	Canine	11–12 years
Lat. incisors	8 years	2nd molars	12–13 years
1st premolar	9 years	3rd molars	17–25 years

H. VESSELS: posterior, middle, and superior alveolar branches of maxillary artery for upper teeth; inferior alveolar for lower teeth. Veins run with arteries
I. NERVES: (see p. 38)

II. Palate
A. DIVISIONS
1. Hard: roof of mouth, separating nasal and oral cavities (see p. 12)
2. Soft: suspended from dorsal edge of hard palate. Consists of mucous membrane containing muscles (see p. 84), glands, etc. *Uvula* hangs from its caudal border
B. INNERVATION
1. Sensory: *greater (ant.) palatine nerve* via the pterygopalatine branches of the maxillary (V) nerve, mainly on hard palate and gums; *nasopalatine nerve* from posterior superior nasal branch of the pterygopalatine nerve to rostral hard palate; and *lesser (post.) palatine nerve,* from same source, to soft palate
2. Motor: to tensor veli palatini muscle via trigeminal (V) nerve and levator veli palatini via X and XI through the pharyngeal plexus
3. Autonomic. Parasympathetic: preganglionic from the sup. salivatory nucleus through the nervus intermedius, great petrosal, and nerve of pterygoid canal, to pterygopalatine ganglion. Postganglionics run with sensory fibers. Sympathetic: preganglionics from upper thoracic cord synapse in superior cervical ganglion. Postganglionics, through deep petrosal nerve, then nerve of pterygoid canal through pterygopalatine ganglion to be distributed as above

*Dental formula refers to number and variety of teeth in one half of each jaw, thus, in adult man: 2-1-2-3.

TEETH

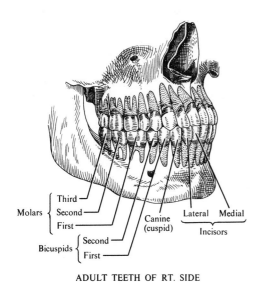

Molars { Third / Second / First }
Canine (cuspid)
Lateral   Medial — **Incisors**
Bicuspids { Second / First }

ADULT TEETH OF RT. SIDE

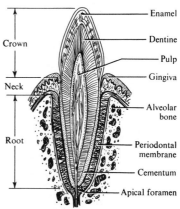

Enamel
Dentine
Pulp
Gingiva
Alveolar bone
Periodontal membrane
Cementum
Apical foramen

Crown
Neck
Root

VERTICAL SECTION
OF A TOOTH

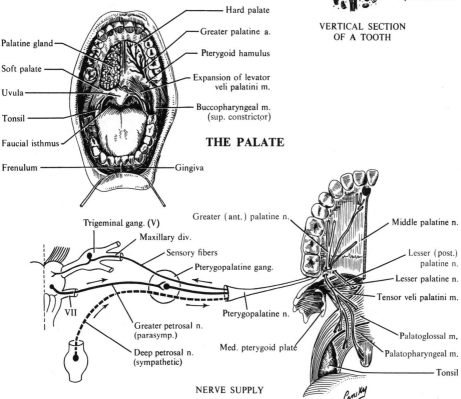

Hard palate
Greater palatine a.
Pterygoid hamulus
Expansion of levator veli palatini m.
Buccopharyngeal m. (sup. constrictor)

Palatine gland
Soft palate
Uvula
Tonsil
Faucial isthmus
Frenulum
Gingiva

THE PALATE

Trigeminal gang. (V)
Maxillary div.
Sensory fibers
Pterygopalatine gang.
Greater (ant.) palatine n.
Middle palatine n.
Lesser (post.) palatine n.
Lesser palatine n.
Tensor veli palatini m.
Palatoglossal m.
Palatopharyngeal m.
Tonsil
VII
Greater petrosal n. (parasymp.)
Deep petrosal n. (sympathetic)
Pterygopalatine n.
Med. pterygoid plate

NERVE SUPPLY

39. PALATINE TONSIL

I. Location: in triangular depression (tonsillar fossa), bounded rostrally by glossopalatine arch, caudally by dorsum of tongue, posteriorly by palatopharyngeal arch
A. SUPRATONSILLAR FOSSA: small area above cephalic pole of tonsil

II. Structure
A. SHAPE: oval
B. SIZE: variable, relatively and often actually larger in children
C. MEDIAL OR FREE SURFACE
  1. Covered by stratified squamous epithelium
  2. Surface epithelium extends inward at from 12 to 15 points. These orifices open into branching crypts
D. LYMPH NODULES lie beneath epithelium of surface and along crypts
E. LATERAL OR DEEP SURFACE is covered by a fibrous connective tissue capsule

III. Relations
A. MEDIAL SURFACE: free, faces pharynx
B. LATERAL SURFACE: separated from constrictor muscle by loose connective tissue. This muscle separates tonsil from facial artery and its tonsillar and ascending palatine branches. Internal carotid artery is about 1 in. behind and lateral
C. ROSTRAL BORDER: glossopalatine arch containing glossopalatine muscle
D. POSTERIOR BORDER: palatopharyngeal arch containing palatopharyngeal muscle

IV. Arterial supply: Tonsillar branches of
A. DORSAL LINGUAL ARTERY from lingual artery
B. ASCENDING PALATINE AND TONSILLAR ARTERIES from facial artery
C. ASCENDING PHARYNGEAL from external carotid artery
D. DESCENDING PALATINE from maxillary artery
E. ACCESSORY MENINGEAL from maxillary artery

V. Veins: from venous plexus lateral to tonsil, which enters pterygoid plexus of veins (see p. 20)

VI. Tonsillar ring (Waldeyer's) of lymphatic tissue, said to "guard" openings into respiratory and digestive systems. Ring composed of, caudally, the lingual tonsil on root of tongue, laterally, the palatine and tubal tonsils (at orifice of auditory tube), and dorsally, the pharyngeal tonsil (adenoid)

VII. Clinical considerations
A. THE TONSILS AND ADENOIDS tend to hypertrophy most commonly in childhood due to a general lymphatic hyperplasia
B. THE PALATINE TONSILS are subject to infection as a result of the numerous crypts on their surface
C. ADENOID HYPERTROPHY can block the nasopharynx, creating problems in breathing
D. BECAUSE OF THE TREMENDOUS VASCULAR SUPPLY of the palatine tonsils, extreme care must be used in tonsillectomy
E. BECAUSE OF THE THIN LATERAL WALL behind which lie many major vessels, excision must be done with great care
F. THE PHARYNGEAL TONSILS posteriorly and above, the palatine tonsils laterally, and the lingual tonsils anteriorly and below form an oblique ring of lymphoid tissue around the pharynx

PALATINE TONSIL

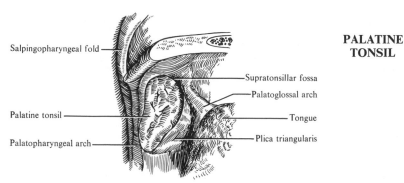

Salpingopharyngeal fold

Supratonsillar fossa

Palatoglossal arch

Palatine tonsil

Tongue

Palatopharyngeal arch

Plica triangularis

MEDIAL VIEW AND RELATED ARCHES

Palatopharyngeal m.

Pharyngeal aponeurosis

Sup. constrictor m.

Mucous membrane

Stylopharyngeus m.

Styloglossus m.

Med. pterygoid m.

Capsule

Palatoglossus m.

Tonsil

Ant. pillar

3rd molar

Mandibular n.

Masseter m.

Buccinator m.

Ramus of mandible

RELATIONS (HORIZONTAL SECTION)

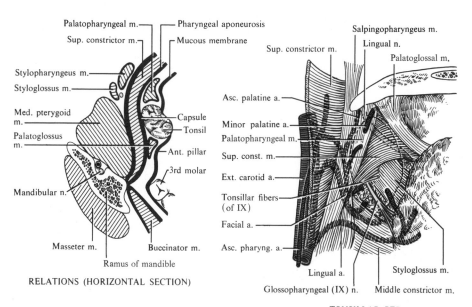

Sup. constrictor m.

Salpingopharyngeus m.

Lingual n.

Palatoglossal m.

Asc. palatine a.

Minor palatine a.

Palatopharyngeal m.

Sup. const. m.

Ext. carotid a.

Tonsillar fibers (of IX)

Facial a.

Asc. pharyng. a.

Lingual a.

Styloglossus m.

Glossopharyngeal (IX) n.

Middle constrictor m.

TONSILLAR BED

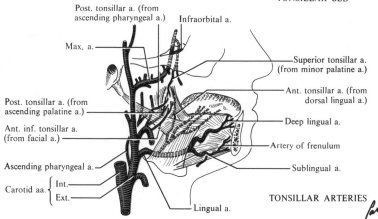

Post. tonsillar a. (from ascending pharyngeal a.)

Infraorbital a.

Max. a.

Superior tonsillar a. (from minor palatine a.)

Ant. tonsillar a. (from dorsal lingual a.)

Post. tonsillar a. (from ascending palatine a.)

Deep lingual a.

Ant. inf. tonsillar a. (from facial a.)

Artery of frenulum

Ascending pharyngeal a.

Sublingual a.

Carotid aa. { Int. / Ext. }

Lingual a.

TONSILLAR ARTERIES

-83-

40. MUSCLES OF THE PALATE

I.

Name	Origin	Insertion	Action	Nerve
Levator veli palatini	Petrous bone Medial auditory tube	Velum of palate	Raises soft palate	Pharyngeal plexus (X & XI)
Tensor veli palatini	Scaphoid fossa Angular spine Lateral auditory tube	Aponeurosis of palate	Tenses soft palate	Trigeminal (V)
Musculus uvulae	Post. nasal spine Palatine aponeurosis	Uvula	Raises uvula	Pharyngeal plexus (X & XI)
Palato-pharyngeus	Soft palate	Lat. post. pharynx Post. thyroid cart.	Raises pharynx, helps close nasopharynx	Pharyngeal plexus (X & XI)
Palatoglossus	Anterior soft palate	Dorsolateral tongue	Raises tongue	Pharyngeal plexus (X & XI)
Salpingo-pharyngeus (see p. 62)				
Pharyngeal constrictors (see p. 62)				

II. Special features

A. PALATINE APONEUROSIS: a fibrous sheet extending posteriorly from caudal edge of palatine bone into the soft palate to support and give attachment to muscles of that structure

B. MOUTH AND PHARYNX unite at palatoglossal arch.* The opening between the 2 arches is called the *isthmus of fauces*. The arch consists of mucous membrane over the palatoglossus muscle

C. PALATOPHARYNGEAL ARCH,† larger than above, produced by the membrane overlying the palatopharyngeus muscle, is separated from the palatoglossal arch by a triangular space containing the *palatine tonsil*

D. VERTICAL DESCENT of the fleshy part of the tensor veli palatini muscle is along the medial pterygoid plate and ends in a tendon that bends at right angles around hamulus to join the palatine aponeurosis

E. THERE IS A SEPARATION of the fibers of the palatopharyngeus muscle into anterior and posterior layers by the levator veli palatini muscle

F. CLEFT PALATE may or may not be associated with cleft lip and may involve only the uvula or involve entire palate

G. PARALYSIS OF SOFT PALATE: muscles do not rise in swallowing, and food and fluids may be forced into nasopharynx

---

*Also called anterior pillar of fauces.
†Also called posterior pillar of fauces.

MUSCLES OF PALATE

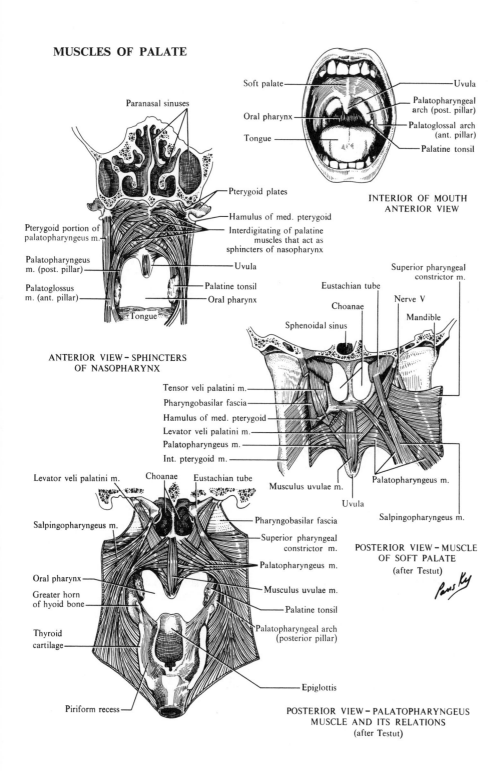

Paranasal sinuses

Soft palate

Oral pharynx

Tongue

Uvula

Palatopharyngeal
arch (post. pillar)

Palatoglossal arch
(ant. pillar)

Palatine tonsil

**INTERIOR OF MOUTH
ANTERIOR VIEW**

Pterygoid plates

Hamulus of med. pterygoid

Interdigitating of palatine
muscles that act as
sphincters of nasopharynx

Pterygoid portion of
palatopharyngeus m.

Palatopharyngeus
m. (post. pillar)

Palatoglossus
m. (ant. pillar)

Uvula

Palatine tonsil

Oral pharynx

Tongue

**ANTERIOR VIEW – SPHINCTERS
OF NASOPHARYNX**

Superior pharyngeal
constrictor m.

Eustachian tube

Choanae

Nerve V

Mandible

Sphenoidal sinus

Tensor veli palatini m.

Pharyngobasilar fascia

Hamulus of med. pterygoid

Levator veli palatini m.

Palatopharyngeus m.

Int. pterygoid m.

Musculus uvulae m.

Uvula

Palatopharyngeus m.

Salpingopharyngeus m.

**POSTERIOR VIEW – MUSCLE
OF SOFT PALATE**
(after Testut)

Levator veli palatini m.

Choanae

Eustachian tube

Salpingopharyngeus m.

Oral pharynx

Greater horn
of hyoid bone

Thyroid
cartilage

Piriform recess

Pharyngobasilar fascia

Superior pharyngeal
constrictor m.

Palatopharyngeus m.

Musculus uvulae m.

Palatine tonsil

Palatopharyngeal arch
(posterior pillar)

Epiglottis

**POSTERIOR VIEW – PALATOPHARYNGEUS
MUSCLE AND ITS RELATIONS**
(after Testut)

41. AUDITORY TUBE, NASOPHARYNX, AND ORAL PHARYNX

I. Auditory tube: (see p. 151)

A. EXTENT: 36 mm

B. COURSE: medially, rostrally, and caudally from middle ear to nasopharynx

C. PARTS
1. Osseous: proximal one third (between petrous temporal and greater wing of sphenoid)
2. Cartilaginous: distal two thirds

D. SPECIAL FEATURES
1. Torus tubarius (cushion): base of cartilage of the tube, forming an elevation dorsal to its orifice in the nasopharynx
2. Tubal tonsil: an accumulation of lymphoid tissue around the orifice of the auditory tube

II. Nasopharynx (entirely respiratory)

A. BOUNDARIES: rostrally, the choanae; posteriorly, the pharyngeal wall against the occipital bone, the anterior arch of vertebra C1 and body of C2; caudally, opens into the oral pharynx at border of soft palate

B. SPECIAL FEATURES
1. Orifice of auditory tube with its torus and tubal tonsil
2. Salpingopharyngeal fold: a fold of mucous membrane due to the muscle of the same name. The fold extends from the torus to the lateral pharyngeal wall
3. Salpingopalatine fold: a fold of mucous membrane overlying the salpingopalatine muscle from the torus to the soft palate
4. Pharyngeal recess: a recess behind the salpingopharyngeal fold
5. Pharyngeal tonsil (adenoid): in upper end of the posterior wall
6. Pharyngeal bursa: a small diverticulum lying just rostral to the pharyngeal tonsil, in the midline

III. Oral pharynx (respiratory and digestive)

A. BOUNDARIES: anteriorly, the dorsum of the tongue and the palatoglossal fold, which marks the limit of the oral cavity; caudally, the hyoid bone

B. SPECIAL FEATURES
1. Palatine tonsils: in the lateral wall (see p. 82)
2. Palatopharyngeal fold: due to the palatopharyngeal muscle

IV. Clinical considerations

A. ONE OF THE FUNCTIONS OF THE AUDITORY TUBE is to equalize air pressure on both sides of the tympanic membrane. Swallowing helps to open the tube (when changes in external pressure cause a disagreeable sensation) due to the action of the salpingopharyngeus muscle, which originates on the tube, and the dilator tubae, which arises on the cartilage of the tube and blends with the tensor veli palatini, both of which contract in swallowing

B. CARTILAGINOUS PART of tube remains closed except during swallowing or yawning

C. AUDITORY TUBE forms a route by which infections may pass from nasopharynx to tympanic cavity

D. TUBE IS EASILY BLOCKED by swelling of its mucous membrane because the walls of its cartilaginous part are normally apposed. When tube is blocked, residual air in tympanic cavity is absorbed, pressure is lowered, tympanic membrane is retracted, and hearing is affected

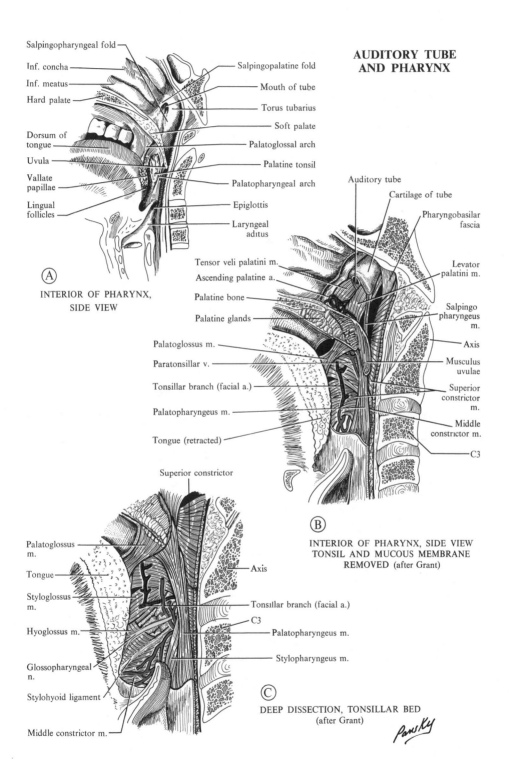

Salpingopharyngeal fold

Inf. concha

Inf. meatus

Hard palate

Dorsum of tongue

Uvula

Vallate papillae

Lingual follicles

Salpingopalatine fold

Mouth of tube

Torus tubarius

Soft palate

Palatoglossal arch

Palatine tonsil

Palatopharyngeal arch

Epiglottis

Laryngeal aditus

Ⓐ

INTERIOR OF PHARYNX,
SIDE VIEW

AUDITORY TUBE
AND PHARYNX

Auditory tube

Cartilage of tube

Pharyngobasilar fascia

Levator palatini m.

Salpingo pharyngeus m.

Axis

Musculus uvulae

Superior constrictor m.

Middle constrictor m.

C3

Tensor veli palatini m.

Ascending palatine a.

Palatine bone

Palatine glands

Palatoglossus m.

Paratonsillar v.

Tonsillar branch (facial a.)

Palatopharyngeus m.

Tongue (retracted)

Ⓑ

INTERIOR OF PHARYNX, SIDE VIEW
TONSIL AND MUCOUS MEMBRANE
REMOVED (after Grant)

Superior constrictor

Palatoglossus m.

Tongue

Styloglossus m.

Hyoglossus m.

Glossopharyngeal n.

Stylohyoid ligament

Middle constrictor m.

Axis

Tonsillar branch (facial a.)

C3

Palatopharyngeus m.

Stylopharyngeus m.

Ⓒ

DEEP DISSECTION, TONSILLAR BED
(after Grant)

Pansky

42. MENINGES OF THE BRAIN AND THE DURAL REFLECTIONS

I. Meninges, the fibrous coverings of the brain
A. PIA: innermost, thin, vascular, and intimately related to brain
B. ARACHNOID: middle, thin, cobweblike; forms outer wall of subarachnoid space
C. DURA: outermost layer, is very tough; is separated from arachnoid by slight, subdural space. Has 2 layers: the outer, *endosteal layer* is lightly adherent to bones, for which it serves as periosteum; inner, *meningeal layer,* is smooth with a mesothelial layer on its inner surface. Lines entire cranial cavity
D. THE DURA is occasionally called the *pachymenix* (Gk *pachys*, thick + *menix*, membrane). The pia and arachnoid together are *leptomeninges* (Gk *leptos*, slender + *meninges*, membrane)

II. Reflections or reduplication of the meningeal layer of dura
A. FALX CEREBRI extends caudally between cerebral hemispheres
   1. Attachments: to skull from crista galli to internal occipital protuberance, where it joins tentorium cerebelli. Its inferior border is free
B. TENTORIUM CEREBELLI extends horizontally between cerebellum and cerebrum
   1. Attachments: posterior and lateral to bone along transverse sinus; rostral and lateral to ridge of petrous bone and posterior clinoid process. Its free margin surrounds cerebral peduncles—this opening is the *incisura tentorii.*
C. FALX CEREBELLI projects between the cerebellar hemispheres
   1. Attachments: to internal occipital crest from protuberance to foramen magnum
D. DIAPHRAGMA SELLAE forms roof of sella turcica
   1. Attachments: to clinoid processes. Has opening for infundibulum

III. Dural venous sinuses: (see p. 90)

IV. Arteries of meninges

Name	Source	Area	Entry
Meningeal br.	Occipital	Post. fossa	Jugular foramen
Post. meningeal	Ascend. pharyngeal	Post. fossa	Jugular foramen
Meningeal br.	Ascend. pharyngeal	Post. fossa	Hypoglossal foramen
Meningeal br.	Vertebral	Post. fossa	Foramen magnum
Meningeal br.	Ascend. pharyngeal	Middle fossa	Foramen lacerum
Middle meningeal	Maxillary	Middle fossa	Foramen spinosum
Access. meningeal	Maxillary	Middle fossa	Foramen ovale
Recurrent br.	Lacrimal	Ant. fossa	Sup. orbital fissure
Meningeal br.	Ant. ethmoid	Ant. fossa	Ant. ethmoid canal
Meningeal br.	Post. ethmoid	Ant. fossa	Post. ethmoid canal

V. Nerves of meninges

Name	Nerve	Distribution
Meningeal br.	Semilunar ganglion (V)	Tentorium and middle fossa
Recurrent br.	Ophthalmic (V)	Middle fossa
Tentorial br.	Ophthalmic (V)	Tentorium
Middle meningeal	Maxillary (V)	Middle fossa and ant. fossa
Meningeal ramus	Mandibular (V)	Middle and post. fossa
Meningeal brs.	Hypoglossal (C1 & C2)	Post. fossa

VI. Meningeal veins: small, follow arteries, located between dura and skull. Communicate with emissary, diploic, and cerebral veins and end in various dural sinuses

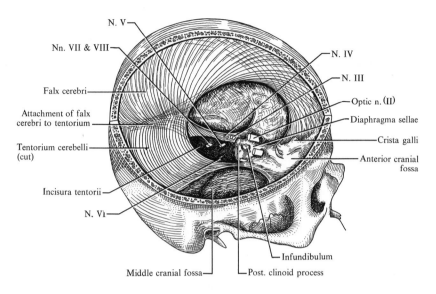

N. V

Nn. VII & VIII

Falx cerebri

Attachment of falx
cerebri to tentorium

Tentorium cerebelli
(cut)

Incisura tentorii

N. VI

N. IV

N. III

Optic n. (II)

Diaphragma sellae

Crista galli

Anterior cranial
fossa

Infundibulum

Middle cranial fossa

Post. clinoid process

CRANIAL DURAL PROCESSES

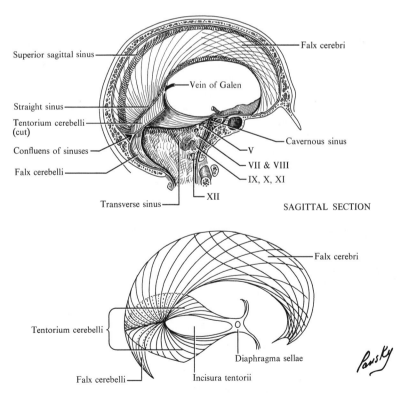

Superior sagittal sinus

Falx cerebri

Vein of Galen

Straight sinus

Tentorium cerebelli
(cut)

Confluens of sinuses

Falx cerebelli

Cavernous sinus

V

VII & VIII

IX, X, XI

XII

Transverse sinus

SAGITTAL SECTION

Falx cerebri

Tentorium cerebelli

Diaphragma sellae

Falx cerebelli

Incisura tentorii

Pansky

DIAGRAMMATIC ILLUSTRATION OF CRANIAL DURA

43. VENOUS SINUSES AND DRAINAGE OF HEAD AND FACE

I. **Diploic veins:** endothelial-lined channels between inner and outer tables of the flat bones of skull. Communicate with meningeal veins and dural sinuses internally, veins of the scalp externally. Found in frontal, ant., and post. temporal and occipital regions. Have no valves

II. **Emissary veins:** direct connections between dural sinuses and veins on outside of bone. Some major ones pass through foramen ovale, mastoid and parietal foramina, hypoglossal and condyloid canals

III. **Dural sinuses:** between endosteal and meningeal dura
   A. SUPERIOR SAGITTAL: in convexity of falx cerebri from foramen cecum to internal occipital protuberance, where it deviates to right transverse sinus
      1. Receives superior cerebral veins, venous lacunae, arachnoid granulations, diploic and dural veins, and parietal emissary veins
   B. INFERIOR SAGITTAL: in free edge of falx cerebri. Ends in straight sinus
      1. Receives veins from falx cerebri and medial side of hemispheres
   C. STRAIGHT: along line of attachment of falx cerebri to tentorium cerebelli. Terminates in left transverse sinus
      1. Receives inferior sagittal sinus, great cerebral vein (of Galen), left transverse sinus, and superior cerebellar veins
   D. TRANSVERSE: one on each side; begin at internal occipital protuberance; pass laterally in attached margin of tentorium to petrous bone, then bend caudally and medially as the *sigmoid sinus* to enter jugular foramina
      1. Receive superior sagittal, straight, and superior petrosal sinuses, mastoid and condyloid emissary veins, inferior cerebral and cerebellar veins, and diploic veins
   E. OCCIPITAL: in attached margin of falx cerebelli; begins at foramen magnum; ends at confluence of sinuses
   F. CONFLUENCE OF SINUSES: at internal occipital protuberance where superior sagittal, straight, occipital, and transverse sinuses meet
   G. CAVERNOUS lie on each side of body of sphenoid bone
      1. Receive superior ophthalmic and cerebral veins
      2. Communicate with transverse sinus via superior petrosal sinus, and internal jugular vein via inferior petrosal sinus. With the internal carotid venous plexus and pterygoid venous plexus through foramina of Vesalii, ovale, and lacerum. And with the angular vein, through superior ophthalmic vein
   H. INTERCAVERNOUS: anterior and posterior, connecting cavernous sinuses, rostral and dorsal, to pituitary
   I. SUPERIOR PETROSAL: along superior border of petrous bone in attached margin of tentorium; join cavernous and transverse sinuses
      1. Receive cavernous sinus; cerebellar, inferior cerebral, and tympanic veins
   J. INFERIOR PETROSAL: along suture between basilar portion of occipital and petrous temporal bones; join cavernous sinus and internal jugular vein
      1. Receive cavernous sinus, internal auditory vein, inferior cerebellar veins, and veins from medulla
   K. BASILAR PLEXUS: network on basilar portion of occipital bone; interconnects inferior petrosal sinuses
      1. Communicates with anterior vertebral plexus

IV. **Clinical considerations**
   A. IN LIFE, blood flows freely through the sinuses. They are a surgical hazard at operation; e.g., the superior petrosal sinus is located immediately superior to the sensory root of the 5th cranial nerve. Surgical procedures on that nerve in this region must take this sinus into account

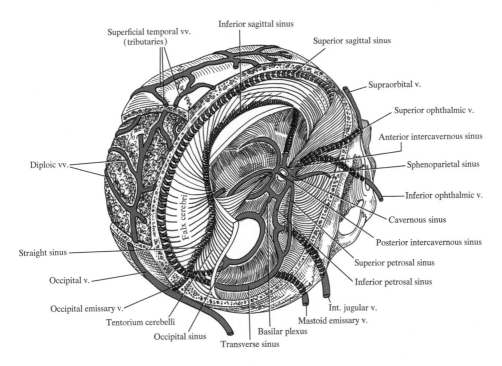

Superficial temporal vv. (tributaries)
Inferior sagittal sinus
Superior sagittal sinus
Supraorbital v.
Superior ophthalmic v.
Anterior intercavernous sinus
Sphenoparietal sinus
Diploic vv.
Inferior ophthalmic v.
Falx cerebri
Cavernous sinus
Posterior intercavernous sinus
Straight sinus
Superior petrosal sinus
Occipital v.
Inferior petrosal sinus
Occipital emissary v.
Int. jugular v.
Mastoid emissary v.
Tentorium cerebelli
Occipital sinus
Basilar plexus
Transverse sinus

VENOUS DRAINAGE OF HEAD

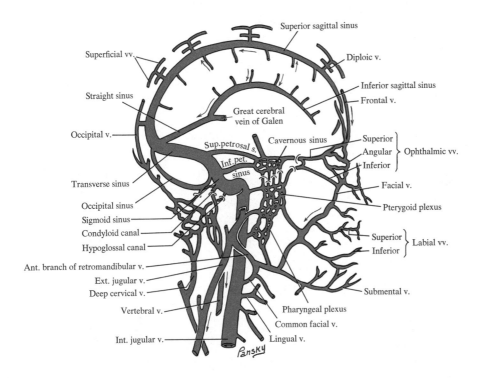

Superior sagittal sinus
Superficial vv.
Diploic v.
Inferior sagittal sinus
Straight sinus
Frontal v.
Great cerebral vein of Galen
Occipital v.
Sup. petrosal s.
Cavernous sinus
Superior
Angular
Ophthalmic vv.
Inf. pet. sinus
Inferior
Transverse sinus
Facial v.
Occipital sinus
Pterygoid plexus
Sigmoid sinus
Condyloid canal
Superior
Labial vv.
Hypoglossal canal
Inferior
Ant. branch of retromandibular v.
Ext. jugular v.
Deep cervical v.
Submental v.
Vertebral v.
Pharyngeal plexus
Int. jugular v.
Common facial v.
Lingual v.
Pansky

44. THE CAVERNOUS SINUSES

I. **Introduction:** the cavernous sinuses extend from the superior orbital fissures in front to the apices of the petrous temporals behind. They are found on each side of the sella turcica and the body of the sphenoid bone (and the frontal portions of the tentorium cerebelli) and are commonly made up of one or more venous channels (in the newborn, a venous plexus), lying in an osseodural compartment

II. **The cavernous sinuses** communicate with
A. THE SUPERIOR AND INFERIOR OPHTHALMIC VEINS via the superior orbital fissure
B. THE SUPERFICIAL MIDDLE CEREBRAL VEIN which runs along the lateral fissure of the cerebral hemisphere
C. THE CENTRAL VEIN OF THE RETINA usually opens into the sinus directly, but it can enter the superior ophthalmic vein
D. THE SPHENOPARIETAL SINUS which lies beneath the edge of the lesser wing of the sphenoid bone
E. EACH OTHER via the *intercavernous sinuses* that pass anterior and posterior to the hypophyseal stalk (infundibulum)
F. DRAIN POSTERIORLY AND INFERIORLY via the superior and inferior petrosal sinuses. The former join the transverse sinuses where they curve to form the sigmoid sinuses; the latter drain into the internal jugular veins just below the skull
G. THE FACIAL VEINS via the superior ophthalmic veins
H. THE PTERYGOID PLEXUS of veins via the emissary veins at the base of the skull that pass through the carotid canal, foramen ovale, and sphenoidal emissary foramina

III. **In addition to the major venous channel(s)** of the cavernous sinuses, the compartments contain
A. THE INTERNAL CAROTID ARTERY and its sympathetic plexus
B. THE ABDUCENT (VI) NERVE
C. IN THE COMPARTMENTS, but situated in the lateral walls of the sinuses are:
  1. The oculomotor (III) nerve
  2. The trochlear (IV) nerve
  3. The ophthalmic division of the trigeminal (V1) nerve
  4. The maxillary division of the trigeminal (V2) nerve is embedded in the dura mater lateral to the cavernous sinus
D. ALL OF THE ABOVE NERVES are separated from the blood in the sinus by endothelium

IV. **Clinical considerations**
A. CONNECTIONS OF THE FACIAL VEINS are important since they communicate with the ophthalmic veins and can carry infections from the face (fairly frequent) to the sinuses. Thus, thrombophlebitis of a facial vein can spread to the intracranial venous system (including cortical veins of the brain) since the facial veins have no valves
B. INFECTION can spread from one cavernous sinus to the other via the intercavernous sinuses
C. THROMBOSIS can extend from the sinus along the central retinal vein and produce thrombosis of retinal vein branches
D. SUPPURATION in the upper nasal cavities and in the paranasal sinuses can lead to thrombophlebitis of the cavernous sinus and subsequent meningitis
E. THROMBOPHLEBITIS of the sinuses themselves can result in poor drainage of blood posteriorly into the petrosal sinuses, resulting in an enlargement of the cavernous sinuses with inflammatory edema of their walls. This would involve cranial nerves III, IV, VI, and the ophthalmic and maxillary divisions of V, resulting in ocular signs
F. POOR OPHTHALMIC VENOUS DRAINAGE from the orbit as a result of pressure on the veins as they enter the cavernous sinus due to inflammatory edema may lead to blood stagnation resulting in exophthalmos and edema of the eyelids and conjunctiva

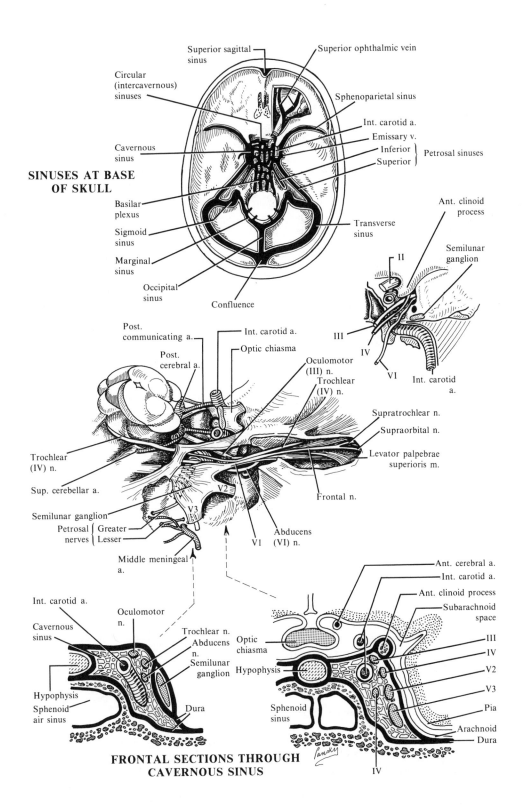

SINUSES AT BASE OF SKULL

Superior sagittal sinus

Superior ophthalmic vein

Circular (intercavernous) sinuses

Sphenoparietal sinus

Int. carotid a.

Emissary v.

Inferior ⎫
Superior ⎭ Petrosal sinuses

Cavernous sinus

Basilar plexus

Sigmoid sinus

Marginal sinus

Occipital sinus

Transverse sinus

Confluence

Ant. clinoid process

II

Semilunar ganglion

III

IV

VI

Int. carotid a.

Post. communicating a.

Int. carotid a.

Optic chiasma

Post. cerebral a.

Oculomotor (III) n.

Trochlear (IV) n.

Supratrochlear n.

Supraorbital n.

Levator palpebrae superioris m.

Trochlear (IV) n.

Sup. cerebellar a.

Semilunar ganglion

Petrosal ⎰ Greater
nerves ⎱ Lesser

Middle meningeal a.

V2

V3

VI

Frontal n.

Abducens (VI) n.

Int. carotid a.

Cavernous sinus

Oculomotor n.

Trochlear n.
Abducens n.
Semilunar ganglion

Optic chiasma

Hypophysis

Hypophysis

Sphenoid air sinus

Dura

Sphenoid sinus

Ant. cerebral a.

Int. carotid a.

Ant. clinoid process

Subarachnoid space

III

IV

V2

V3

Pia

Arachnoid

Dura

IV

FRONTAL SECTIONS THROUGH CAVERNOUS SINUS

45. VENOUS DRAINAGE OF THE BRAIN

I. **Superficial (external) cerebral veins:** thin-walled, devoid of valves, arise in brain substance, and first found in the pia mater. Crossing subarachnoid and subdural spaces, they pierce the arachnoid and meningeal dura and drain toward the superior and transverse sinuses. They enter the sinuses in the opposite direction to that of blood flow in the sinuses. Thus, the sup. cerebral vv. join the superior sagittal sinus in an anterior direction and transverse sinus in a posterior one

A. SUPERIOR CEREBRAL VV. may first empty into venous lakes (venous lacunae) on either side of the sup. sagittal sinus. The lacunae are invaginated by arachnoid villi or granulations (appear at age 7 and increase in size and number until adult life) which serve as a pathway for subarachnoid fluid to enter the venous circulation

B. MIDDLE CEREBRAL V. follows the lateral sulcus, sends superior (of Trolard) and inferior (of Labbe) anastomotic veins to the sup. and transverse sinuses, respectively, and ends in the cavernous and sphenoparietal sinuses

C. INFERIOR CEREBRAL VV. drain inferior aspect of hemispheres and join nearby sinuses

II. **Deep veins of brain** drain into the *great cerebral vein* (*of Galen*) which originates under splenium of corpus callosum from union of the 2 int. cerebral vv. The great vein joins the inf. sagittal sinus at a rt. angle and forms the *straight sinus*. The great vein receives 2 basal vv., the post. pericallosal, int. occipital, post. mesencephalic, precentral, sup. vermian, and sup. cerebellar vv.

A. INTERNAL CEREBRAL VEINS formed near the interventricular foramina by union of ant. septal and thalamostriate (terminal) vv. The cerebrals run posteriorly, enter the tela choroidea of the 3rd ventricle, pass through the velum interpositum to the level of the splenium of the corpus callosum, and join the 2 basal vv. to form great cerebral v. The int. cerebrals receive drainage from roof of the lateral ventricles, occipital horns, thalamic vv., and choroidal vv.

1. Thalamostriate (terminal) veins: drain caudate nucleus, int. capsule, and deep white matter of posterior frontal and anterior parietal lobes. They are formed by the ant. and post. terminal (caudate) vv.

2. Anterior septal veins: formed by union of many intermedullary vv. that drain the deep white matter of the anterior part of the frontal lobe. The ant. septal vv. join the ipsilateral thalamostriate vv. to form the internal cerebral v.

3. Superior choroidal veins: terminate in the anterior part of the int. cerebral or terminal vein and are the major drainage of the choroid plexus in this location

4. Direct lateral veins: are chief veins of the roof of the lateral ventricle. Drain the deep white matter of the ant. frontal and post. parietal lobes. They pass over the roof or floor of the ventricles to reach the int. cerebral vv.

5. Veins of posterior horn of lateral ventricle: drain post. temporal and occipital lobes, fimbria of fornix, and choroid plexus of atrium. Drain into cerebral veins

6. Veins of thalamus: drain into the internal cerebral veins

B. BASAL VEIN (OF ROSENTHAL) formed by union of the deep middle and ant. cerebral vv.; joins the int. cerebral vv. and basal v. of opposite side under splenium of corpus callosum to terminate in the great cerebral v. Tributaries drain optic chiasma and tract, hypothalamus, and medial part of cerebral peduncles. Major tributaries are:

1. Anterior cerebral vein(s): communicate with each other via ant. communicating v.; drain ant. 1/3 of corpus callosum, ant. medial surface of frontal lobe, and medial orbital frontal gyri

2. Deep cerebral vein: drains insula and inferior part of corpus striatum

3. Temporal and inferior ventricular veins: drain deep white matter of superior and lateral temporal lobe, hippocampus, dentate gyrus, and choroid plexus

4. Lateral mesencephalic vein: from mesencephalon

C. INTERNAL OCCIPITAL VEIN formed from veins on inferior and medial surface of occipital lobe. Drains to great cerebral vein or superior and transverse sinuses

D. POSTERIOR PERICALLOSAL VEIN(S) may be single or double. Enter great cerebral vein and drain posterosuperior surface of corpus callosum and splenium

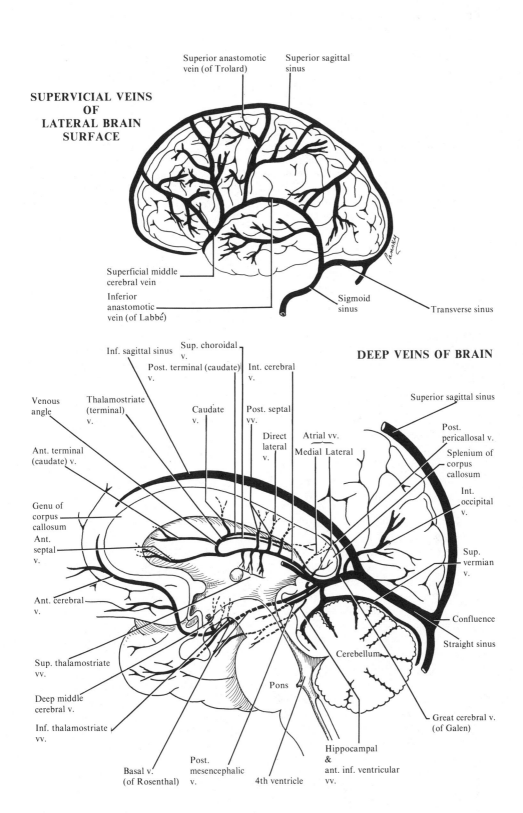

SUPERVICIAL VEINS OF LATERAL BRAIN SURFACE

Superior anastomotic vein (of Trolard)

Superior sagittal sinus

Superficial middle cerebral vein

Inferior anastomotic vein (of Labbé)

Sigmoid sinus

Transverse sinus

DEEP VEINS OF BRAIN

Inf. sagittal sinus

Sup. choroidal v.

Post. terminal (caudate) v.

Int. cerebral v.

Thalamostriate (terminal) v.

Caudate v.

Post. septal vv.

Venous angle

Direct lateral v.

Atrial vv.

Medial Lateral

Superior sagittal sinus

Ant. terminal (caudate) v.

Post. pericallosal v.

Splenium of corpus callosum

Genu of corpus callosum

Int. occipital v.

Ant. septal v.

Sup. vermian v.

Ant. cerebral v.

Confluence

Straight sinus

Sup. thalamostriate vv.

Cerebellum

Deep middle cerebral v.

Pons

Inf. thalamostriate vv.

Great cerebral v. (of Galen)

Basal v. (of Rosenthal)

Post. mesencephalic v.

4th ventricle

Hippocampal & ant. inf. ventricular vv.

46. VEINS OF THE POSTERIOR CRANIAL FOSSA

I. **Introduction:** veins of the posterior cranial fossa drain the cerebellum and the brain stem. According to their direction of drainage, they may be classified into three groups: superior or galenic group, anterior or petrosal group, and a posterior or tentorial group

II. **Superior (galenic) group of veins** drain the superior part of the cerebellum and upper portion of the brain stem. Major veins of the group include:
   A. PRECENTRAL VEIN: a major vessel that originates in the primary fissure of the cerebellum and runs parallel to the roof of the 4th ventricle. As it leaves the fissure to pass between the inferior colliculi and central lobule of the cerebellum, it angles above and behind the culmen. It leaves the culmen to enter the superior part of the quadrigeminal cistern and joins the great cerebral vein. It has numerous anastomotic connections with the basal and petrosal veins
   B. SUPERIOR VERMIAN VEIN is formed by preculminate, intraculminate, and supraculminate tributaries that drain the culmen of the cerebellar vermis. It joins the great cerebral vein by arching over the culmen
      1. It may unite with the precentral vein to form a common trunk called the *superior cerebellar vein,* which then drains into the great cerebral vein
   C. THE SUPERIOR CEREBELLAR HEMISPHERIC VEINS vary in number and location. They usually run rostral and medial to join one of the culminate tributaries, the stem of the superior vermian v., or the precentral v. Drainage is to the great cerebral v. or straight sinus. Others run laterally to the transverse or superior petrosal sinuses
   D. POSTERIOR MESENCEPHALIC VEIN may be single or double and originates in the interpeduncular fossa or on the lateral aspect of the mesencephalon. It passes backward to join the posterior portion of the great cerebral vein
   E. LATERAL MESENCEPHALIC VEIN is a constant anastomotic vein running in or near the lateral mesencephalic sulcus. It connects with the basal or posterior mesencephalic veins above or with the brachial tributary of the petrosal vein below. It may connect with the brachial tributaries of the precentral vein, occasionally
   F. ANTERIOR PONTOMESENCEPHALIC VEIN may drain to the superior group via the basal vein or to the petrosal group
   G. QUADRIGEMINAL VEINS drain the corpora quadrigemina and then pass superiorly and posteriorly to join the great cerebral vein

III. **Anterior or petrosal group** receives drainage from the anterior brain stem, superior and inferior surfaces of the cerebellar hemispheres, area of the cerebellomedullary fissure, and the lateral recess of the 4th ventricle. It drains into the superior petrosal sinus. Some of the group's major named tributaries are the *anterior medullary* (a direct continuation of the anterior spinal vv.); the *anterior* and *lateral pontomesencephalic; transverse* and *lateral pontine; anterolateral marginal; brachial* (pass on the lateral aspect of the brachium pontis); *vein of the great horizontal fissure* of the cerebellum; *superior* and *inferior cerebellar hemispheric* (vary in number and course); *retro-olivary* or *lateral medullary; vein of the restiform body* (superior continuation of the median posterior spinal v.); *medial tonsillars;* and the *vein of the lateral recess of the 4th ventricle* (receives tributaries from the dentate nucleus and nearby cerebellar white matter)

IV. **Posterior or tentorial group** drains the inferior part of the cerebellar vermis and the medial portion of the superior and inferior cerebellar hemispheres. Consists of the following major veins:
   A. INFERIOR VERMIAN VEIN formed in the region of the cupola pyramis by union of the superior and inferior retrotonsillar branches from the poles of the tonsil; drains into the straight sinus anterior to the confluence
   B. SUPERIOR CEREBELLAR HEMISPHERIC VEIN(S) drain the superomedial surface of the cerebellar hemisphere via a tentorial sinus into the lateral or straight sinus
   C. INFERIOR CEREBELLAR HEMISPHERIC VEIN(S) drain the inferomedial surface of the cerebellar hemisphere into the transverse sinus

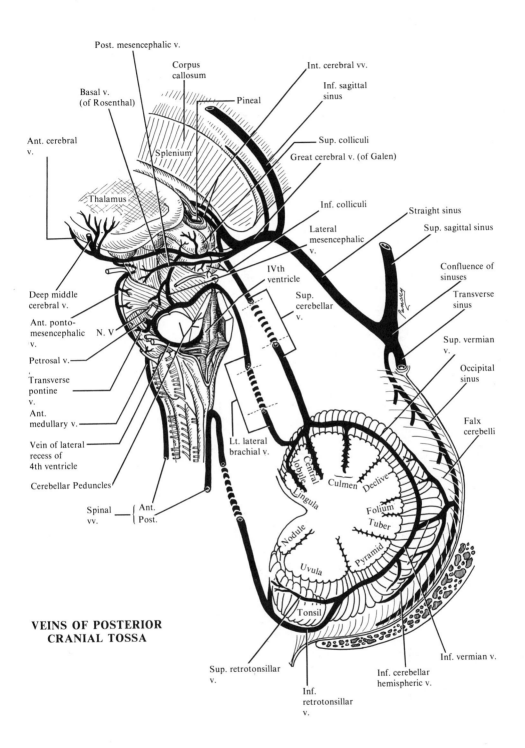

Post. mesencephalic v.

Corpus callosum

Int. cerebral vv.

Inf. sagittal sinus

Basal v. (of Rosenthal)

Pineal

Ant. cerebral v.

Splenium

Sup. colliculi

Great cerebral v. (of Galen)

Thalamus

Inf. colliculi

Straight sinus

Lateral mesencephalic v.

Sup. sagittal sinus

Confluence of sinuses

IVth ventricle

Transverse sinus

Deep middle cerebral v.

Sup. cerebellar v.

Ant. ponto-mesencephalic v.

N. V

Sup. vermian v.

Petrosal v.

Occipital sinus

Transverse pontine v.

Falx cerebelli

Ant. medullary v.

Vein of lateral recess of 4th ventricle

Lt. lateral brachial v.

Central lobule

Cerebellar Peduncles

Culmen

Declive

Lingula

Folium

Spinal vv.

Ant.

Post.

Tuber

Nodule

Pyramid

Uvula

Tonsil

Inf. vermian v.

VEINS OF POSTERIOR CRANIAL TOSSA

Sup. retrotonsillar v.

Inf. retrotonsillar v.

Inf. cerebellar hemispheric v.

47. SUBARACHNOID SPACES AND CISTERNS

I. **Subarachnoid space:** the space between arachnoid and pia, which receives cerebrospinal fluid from the 4th ventricle via the foramina of Magendie and Luschka. A subpia space does not exist. Above the layer of pial cells, a meshwork of arachnoid cells makes a filmy membrane with a depth of up to a few millimeters. The CSF is contained between the trabeculations of the arachnoid meshwork. Blood vessels of the brain enter and leave the brain through this space. Removal of this membrane will disrupt the vessels, producing a bloody CSF

II. **Subarachnoid cisterns:** in some areas at the base of the brain, the subarachnoid space is thicker than over the hemisphere and forms fluid spaces, or cisterns, which appear to have the function of supporting the brain or cushioning the base of the brain against the bony ledges of the fossae of the base of the skull. The major cisterns and their contents are the following:
A. DORSAL CISTERNS
   1. Superior (cerebellar): under the tentorium and over upper surface of cerebellum
   2. Pericallosal: above corpus callosum and between and below cingulate gyri
   3. Quadrigeminal: under splenium of corpus callosum and over quadrigeminal plate; in front of tentorial notch
      a. Contents: pineal gland, great cerebral vein, terminal parts of basal veins, and posterior cerebral artery
B. VENTRAL CISTERNS
   1. Medullary: between medulla and lower clivus of skull
      a. Contents: IX, X, XI, and XIIth nerves, vertebral and posterior inferior cerebellar aa.
   2. Interpeduncular: between cerebral peduncles, behind dorsum sellae
      a. Contents: cranial III; basilar, superior cerebellar, posterior cerebral, and posterior communicating arteries
   3. Pontine: between pons and upper clivus of skull
      a. Contents: cranial nerves V, VI; basilar and anterior inferior cerebellar arteries
   4. Cerebellopontine: in front of cerebellar hemisphere and cerebellar peduncle; behind petrous ridges
      a. Contents: cranial nerves VII, VIII, X, and XI; anterior inferior cerebellar artery
   5. Chiasmatic: anterior to optic chiasm, infundibulum, and anterior perforated substance; posterior to diaphragma sellae and anterior clinoid process
      a. Contents: internal carotid, anterior and middle cerebral, posterior communicating, and anterior choroidal arteries
   6. Crural: around the cerebral peduncle
      a. Contents: posterior cerebral and anterior choroidal arteries
C. INTERCOMMUNICATING CISTERNS
   1. Cerebellomedullary (great or magna): beneath cerebellar tonsils above and behind foramen magnum (between undersurface of cerebellum and dorsum of medulla)
      a. Contents: posterior inferior cerebellar arteries
   2. Ambient: lateral extensions of quadrigeminal cistern, continuous with crural and pontine cisterns; free edge of tentorial notch is lateral; between corpus callosum and thalamus
      a. Contents: cranial IV; posterior cerebral, superior cerebellar, and choroidal arteries; basal and internal cerebral veins
   3. Of lamina terminalis: in front of lamina terminalis; free edge of falx is anterior and corpus callosum forms its roof
      a. Contents: anterior cerebral and frontopolar arteries

III. **Clinical considerations**
A. CISTERNA MAGNA is used as a site for puncture to remove CSF samples
B. WHEN THE SUBARACHNOID SPACE IS OPENED, CSF leaks out because the fluid is normally under pressure. After a surgical procedure, it is not possible to close the subarachnoid space by suturing the arachnoid because it is flimsy and transparent. Leakage of CSF is prevented by carefully suturing the dura mater
C. THE CISTERNS CAN BE STUDIED by injecting air or radiopaque material into them

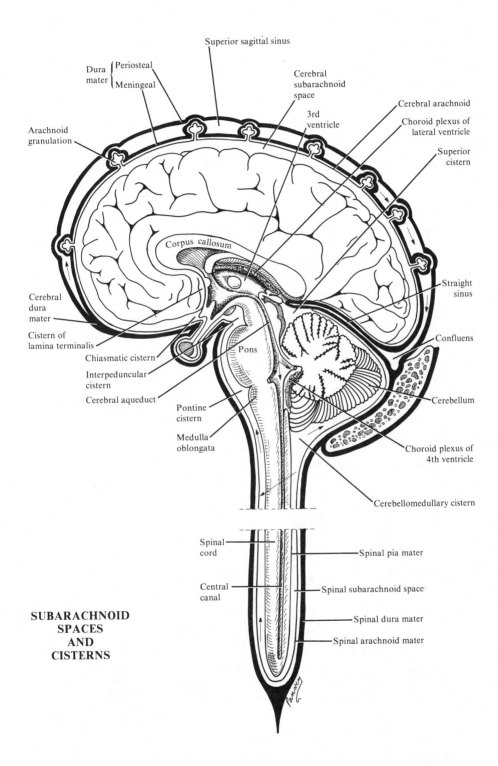

Superior sagittal sinus

Dura { Periosteal
mater { Meningeal

Cerebral
subarachnoid
space

Arachnoid
granulation

3rd
ventricle

Cerebral arachnoid

Choroid plexus of
lateral ventricle

Superior
cistern

Corpus callosum

Cerebral
dura
mater

Straight
sinus

Cistern of
lamina terminalis

Confluens

Chiasmatic cistern

Pons

Interpeduncular
cistern

Cerebral aqueduct

Cerebellum

Pontine
cistern

Medulla
oblongata

Choroid plexus of
4th ventricle

Cerebellomedullary cistern

Spinal
cord

Spinal pia mater

Central
canal

Spinal subarachnoid space

**SUBARACHNOID
SPACES
AND
CISTERNS**

Spinal dura mater

Spinal arachnoid mater

48. HEAD INJURIES AND INTRACRANIAL HEMORRHAGE (BLEEDING)

I. **Introduction:** injuries to the head are often associated with intracranial bleeding. Intracranial injuries occur with or without fracture of the skull. About 25% of fatal head injuries occur without a skull fracture. Failure of respiration is the usual cause of death when vital brain areas are damaged

II. **Epidural (extradural) hemorrhage** (about 30% mortality) is usually due to fracture, and bleeding occurs between dura and periosteum of the calvaria (outside dura). Bleeding is usually arterial in 90% of cases and needs immediate surgical treatment
   A. MOST COMMON FRACTURE is of greater wing of sphenoid, where ant. branch of middle meningeal a. courses through bone. The post. branch is also a source of bleeding
      1. Classically, the patient is knocked unconscious (brief concussion) at injury, has a lucid interval (of some hours) but may not, then becomes comatose as the hemorrhage increases
   B. A FEW CASES are due to bleeding from meningeal or dural venous sinuses or emissary veins. Onset of symptoms is often delayed for 1–3 weeks with venous bleeding
   C. AS FLUID MASS (HEMATOMA) INCREASES, brain compression occurs, requiring evacuation of blood and occlusion of bleeding vessel(s)
   D. THERE IS COMMONLY a 3rd nerve palsy and an ipsilateral dilated, fixed pupil due to pressure on the 3rd nerve as it passes through the superior orbital fissure
      1. Due to formation of a hematoma above the tentorium cerebelli resulting in a rise in supratentorial pressure, one may see herniation of the temporal lobe (usually uncus) through the tentorial incisura near the 3rd nerve origin

III. **Subdural hemorrhage** (bleeding between dura and arachnoid): seen in almost all head injuries of any severity as a result of stretching and/or tearing of dura and arachnoid. Small tears in arteries and veins lead to varying degrees of bleeding
   A. MOST COMMONLY RESULTS from tearing of cerebral veins where they enter the superior sagittal sinus, thus is venous bleeding
   B. MAY BE ACUTE OR CHRONIC, the latter being the end result of unrecognized acute bleeding
   C. MAY INITIALLY BE WIDESPREAD, but localizes in time, most often under parietal tuber
   D. BLEEDING IS RARELY of clinical significance being overshadowed by cerebral damage
   E. SUBDURAL BLEEDING IS COMMON in alcoholics due to their proneness to head injuries from falls, arachnoid edema, and a hemorrhagic tendency due to 2ndary liver damage
   F. CLINICAL SYMPTOMS are seen 3–6 weeks after injury when clot is organized: headache, confusion, somnolence, and coma
   G. CEREBROSPINAL FLUID is under pressure and is clear or xanthochromic

IV. **Subarachnoid hemorrhage:** most common type of intracranial bleeding due to trauma and is usually associated with cerebral or cerebellar contusion or laceration with bleeding into the subarachnoid space causing a bloody CSF
   A. TRAUMATIC SUBARACHNOID BLEEDING often occurs over occipital lobes and cerebellum
   B. NONTRAUMATIC BLEEDING is usually the result of a rupture of a berry (saccular) aneurysm of a branch of the circle of Willis, but may be "spontaneous" due to rupture of an apparent normal subarachnoid vessel
   C. BLOOD IN SUBARACHNOID SPACE acts as an irritant and results in a meningeal inflammatory reaction or irritation indicated by stiff neck, headache, and even loss of consciousness
   D. LARGE SPACE-OCCUPYING HEMATOMAS are seen in II and III but not evident here
   E. SUBARACHNOID BLEEDING is also seen in carbon monoxide poisoning, acute cerebral congestion due to status epilepticus, or in hemorrhagic diathesis (tendency to spontaneously bleed)

V. **Intracerebral hemorrhage:** bleeding into the brain substance (usually arterial)
   A. ARTERIES INVOLVED are often from the circle of Willis to the basal ganglia or internal capsule, resulting in a paralytic stroke due to interruption of pathways from the cortex to the brain stem and spinal cord

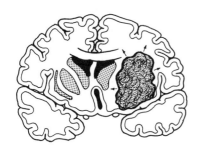

**SPONTANEOUS INTRACEREBRAL
HEMORRHAGE**

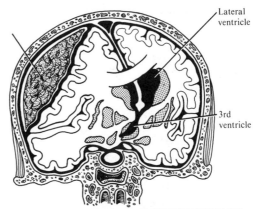

Lateral ventricle

3rd ventricle

**SUBDURAL HEMORRHAGE
(DEVIATION OF VENTRICLES)**

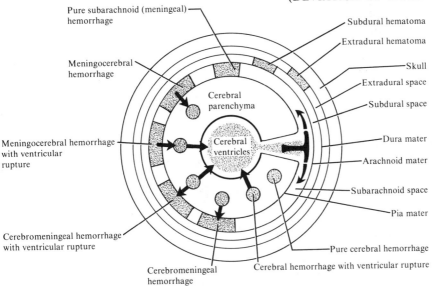

Pure subarachnoid (meningeal) hemorrhage

Meningocerebral hemorrhage

Cerebral parenchyma

Cerebral ventricles

Meningocerebral hemorrhage with ventricular rupture

Cerebromeningeal hemorrhage with ventricular rupture

Cerebromeningeal hemorrhage

Cerebral hemorrhage with ventricular rupture

Subdural hematoma

Extradural hematoma

Skull

Extradural space

Subdural space

Dura mater

Arachnoid mater

Subarachnoid space

Pia mater

Pure cerebral hemorrhage

**BLOOD DISTRIBUTION IN FORMS OF
INTRACRANIAL HEMORRHAGE
(AFTER ESCOUROLLE AND POIRIER)**

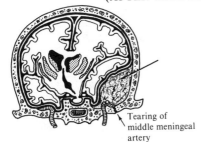

Tearing of middle meningeal artery

**EXTRADURAL HEMORRHAGE
(FRACTURE OF BASE OF SKULL)**

**CEPHALOHEMATOMA
(EXTRACRANIAL)**

49. BASAL VIEW OF THE BRAIN

I. Telencephalon hemispheres
A. FRONTAL LOBE
  1. Frontal pole
  2. Orbital gyri
B. TEMPORAL LOBE
  1. Temporal pole
  2. Temporal gyri
C. RHINENCEPHALON (OLFACTORY BRAIN)
  1. Olfactory bulb
  2. Olfactory tract
  3. Anterior perforated substance

II. Diencephalon
A. TUBER CINEREUM
B. MAMMILLARY BODIES
C. HYPOPHYSIS
D. OPTIC CHIASMA
E. OPTIC TRACT

III. Mesencephalon
A. CEREBRAL PEDUNCLES (basis pedunculi)
  1. Interpeduncular fossa
B. POSTERIOR PERFORATED SUBSTANCE

IV. Metencephalon
A. PONS
  1. Basilar sulcus
  2. Middle cerebellar peduncle (brachium pontis)
B. CEREBELLAR HEMISPHERES AND FLOCCULUS

V. Myelencephalon (medulla): continuous with spinal cord through foramen magnum
A. SULCI: anteromedian and anterolateral
B. PYRAMIDS
  1. Decussation of pyramids
  2. Inferior olive

VI. Nerves
A. OLFACTORY (I) attached to olfactory bulb
B. OPTIC (II)
C. OCULOMOTOR (III) emerging from the interpeduncular fossa
D. TROCHLEAR (IV) descends lateral to the cerebral peduncles
E. TRIGEMINAL (V) emerging from lateral side of pons
F. ABDUCENS (VI) emerging from the groove between pons and medulla
G. FACIAL (VII) AND VESTIBULOCOCHLEAR (VIII) join brain stem at the cerebellopontine angle
H. GLOSSOPHARYNGEAL (IX) AND VAGUS (X) emerge from lateral sulcus of medulla caudal to the facial (VII) nerve
I. SPINAL AND BULBAR PARTS OF ACCESSORY (XI) NERVE
  1. The bulbar part is associated with the vagus (X)
  2. The spinal part extends upward along the side of medulla
J. HYPOGLOSSAL (XII) emerges from sulcus between the inferior olive and the pyramid

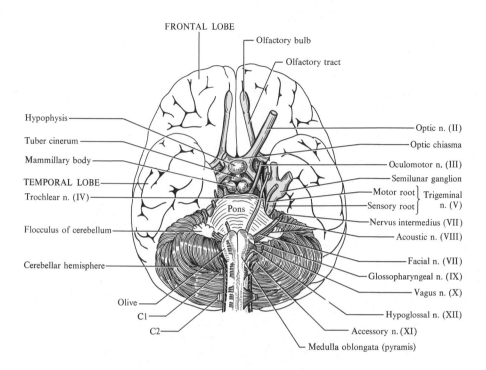

FRONTAL LOBE

Olfactory bulb

Olfactory tract

Hypophysis

Tuber cinerum

Mammillary body

TEMPORAL LOBE

Trochlear n. (IV)

Flocculus of cerebellum

Cerebellar hemisphere

Olive

C1

C2

Optic n. (II)

Optic chiasma

Oculomotor n. (III)

Semilunar ganglion

Motor root ⎱ Trigeminal
Sensory root ⎰ n. (V)

Nervus intermedius (VII)

Acoustic n. (VIII)

Facial n. (VII)

Glossopharyngeal n. (IX)

Vagus n. (X)

Hypoglossal n. (XII)

Accessory n. (XI)

Medulla oblongata (pyramis)

Pons

BRAIN - BASAL VIEW

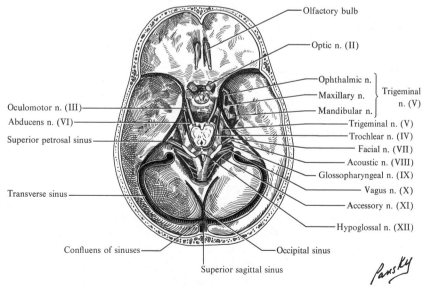

Olfactory bulb

Optic n. (II)

Ophthalmic n.

Maxillary n. ⎱ Trigeminal
n. (V)
Mandibular n. ⎰

Oculomotor n. (III)

Abducens n. (VI)

Superior petrosal sinus

Transverse sinus

Trigeminal n. (V)

Trochlear n. (IV)

Facial n. (VII)

Acoustic n. (VIII)

Glossopharyngeal n. (IX)

Vagus n. (X)

Accessory n. (XI)

Hypoglossal n. (XII)

Confluens of sinuses

Occipital sinus

Superior sagittal sinus

Pansky

BRAIN STEM AND CRANIAL NERVES
IN RELATION TO SKULL

50. LATERAL VIEW OF THE BRAIN

I. Poles: frontal, temporal, and occipital

II. Principal fissures or sulci
  A. LATERAL FISSURE OR SULCUS (sylvian): between temporal, frontal, and parietal lobes
    1. Anterior part, the stem
    2. Long posterior part, the posterior ramus (or limb)
      a. Near junction of 1 and 2; two short branches, anterior horizontal and anterior ascending, extend into frontal lobe
  B. CENTRAL SULCUS (of Rolando) runs obliquely downward, from superior margin of brain, almost to lateral fissure

III. Lobes
  A. FRONTAL LOBE, from frontal pole to central sulcus
    1. Precentral sulcus, rostral and parallel to central. Superior and inferior frontal sulci run rostrally toward frontal pole
    2. Precentral gyrus, between central and precentral sulci. Superior frontal gyrus lies above superior frontal sulcus; middle frontal gyrus, between superior and inferior sulci; inferior frontal gyrus, below inferior frontal sulcus. The anterior rami of lateral fissure usually divide this into triangular, opercular, and orbital portions
  B. PARIETAL LOBE extends from the central sulcus to parieto-occipital fissure, and a line extending from this to the preoccipital notch
    1. Postcentral sulcus lies parallel and dorsal to central sulcus. Intraparietal sulcus extends dorsally from middle of this to the transverse occipital sulcus
    2. Postcentral gyrus lies between central and posterior central sulci. Superior and inferior parietal lobules lie above and below intraparietal sulcus
      a. In the superior lobule, the parieto-occipital gyrus arches over the parieto-occipital sulcus
      b. In the inferior lobule, the *supramarginal gyrus* caps upper end of lateral fissure; *angular gyrus* arches over superior temporal sulcus
  C. OCCIPITAL LOBE, from occipital pole to parieto-occipital fissure and its extension to preoccipital notch
  D. TEMPORAL LOBE, from temporal pole to line extending from the parieto-occipital fissure to preoccipital notch
    1. Superior and middle temporal sulci divide lobe into superior, middle, and inferior temporal gyri
    2. On the upper surface of superior temporal gyrus are the transverse temporal gyri
  E. INSULA, at bottom of lateral fissure, covered by opercula, from frontal, parietal, and temporal lobes
    1. Encircled by circular sulcus
    2. Crossed by long and short gyri

IV. Special features
  A. THE PRIMARY VOLUNTARY MOTOR CORTEX lies along the posterior part of the precentral gyrus adjoining the central sulcus
  B. THE PRIMARY SENSORY CORTEX for the body (pain, temperature, touch) lies in the postcentral gyrus, some actually in the central sulcus
  C. THE PRIMARY AUDITORY AREA lies on the cephalic border of the superior temporal gyrus in the depths of the lateral fissure
  D. THE PRIMARY VISUAL CORTEX, as seen from the lateral side, lies at the occipital pole

BRAIN
LATERAL VIEW

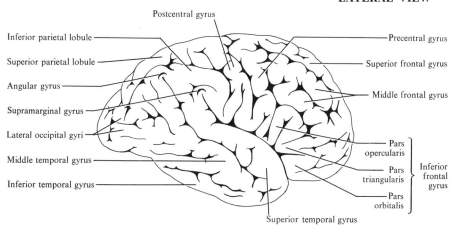

Postcentral gyrus

Inferior parietal lobule

Superior parietal lobule

Angular gyrus

Supramarginal gyrus

Lateral occipital gyri

Middle temporal gyrus

Inferior temporal gyrus

Precentral gyrus

Superior frontal gyrus

Middle frontal gyrus

Pars opercularis

Pars triangularis

Pars orbitalis

Inferior frontal gyrus

Superior temporal gyrus

PRINCIPAL GYRI

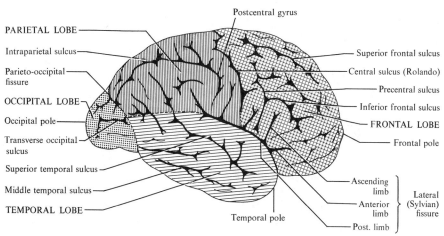

Postcentral gyrus

PARIETAL LOBE

Intraparietal sulcus

Parieto-occipital fissure

OCCIPITAL LOBE

Occipital pole

Transverse occipital sulcus

Superior temporal sulcus

Middle temporal sulcus

TEMPORAL LOBE

Superior frontal sulcus

Central sulcus (Rolando)

Precentral sulcus

Inferior frontal sulcus

FRONTAL LOBE

Frontal pole

Ascending limb

Anterior limb

Post. limb

Lateral (Sylvian) fissure

Temporal pole

PRINCIPAL SULCI AND LOBES

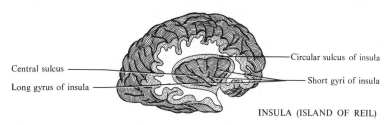

Central sulcus

Long gyrus of insula

Circular sulcus of insula

Short gyri of insula

INSULA (ISLAND OF REIL)

– 105 –

51. MEDIAL VIEW OF THE BRAIN

I. Telencephalon (hemisphere)
  A. OUTSTANDING LANDMARKS
1. Corpus callosum with its genu, body, and splenium
2. The fornix
3. The septum pellucidum, stretched between 1 and 2
4. Anterior commissure

  B. PRINCIPAL FISSURES AND SULCI
1. Calcarine fissure starts below splenium of corpus callosum and curves upward and posteriorly toward the occipital pole
2. Cingulate sulcus curves parallel to the body of the corpus callosum to a point dorsal to the central sulcus and then curves cephalically to its superior margin
3. Parieto-occipital fissure (sulcus) begins near the middle of the calcarine fissure and runs upward to its superior margin
4. Inferior temporal sulcus lies between the fusiform and inferior temporal gyri
5. Collateral fissure runs from near occipital pole, parallel to brain margin, almost to the temporal pole

  C. LOBES
1. Frontal bounded posteriorly by a line drawn obliquely downward and rostrally from the upper end of the central sulcus to the middle of the corpus callosum. Includes part of the cingulate gyrus and paracentral lobule
2. Parietal bounded rostrally by the posterior line of the frontal lobe and posteriorly by the parieto-occipital fissure
3. Occipital bounded rostrally by the parieto-occipital fissure and a line drawn caudally to the preoccipital notch
   a. The posterior part of the calcarine fissure divides its medial surface into the *cuneus* and *lingual* gyri
4. Temporal bounded dorsally by the rostral boundary of the occipital lobe
   a. Fusiform gyrus lies between the inferior temporal sulcus and the collateral sulcus
   b. Medial to the collateral fissure are the lingual gyrus and the hippocampus
   c. Parahippocampal gyrus lies medial to the collateral fissure and is continuous with the cingular gyrus through the isthmus, which lies below the splenium of the corpus callosum. Rostrally, it is curved around as the uncus
      i. The hippocampal fissure extends from the splenium of the corpus callosum to the medial side of the uncus

II. Diencephalon: the following structures should be noted, namely, the choroid plexus of the third ventricle, pineal body, posterior commissure, interthalamic adhesion, hypothalamic sulcus, optic recess, optic chiasma, infundibulum, thalamus, hypothalamus, and third ventricle

III. Mesencephalon: the following structures should be noted, namely, the corpora quadrigemina (superior and inferior colliculi), cerebral aqueduct, and cerebral peduncles

IV. Metencephalon: the following structures should be noted, namely, the superior medullary velum, cerebellum, fourth ventricle, and pons

V. Myelencephalon: the following structures should be noted, namely, the medulla and the inferior medullary velum

VI. Special feature: primary visual cortex lies on both sides of the entire length of calcarine fissure

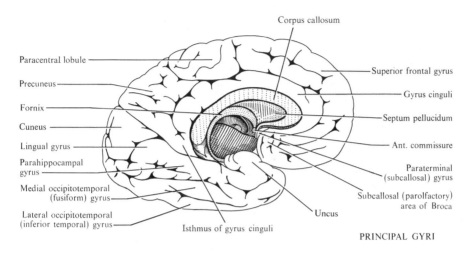

Corpus callosum

Paracentral lobule

Precuneus

Fornix

Cuneus

Lingual gyrus

Parahippocampal
gyrus

Medial occipitotemporal
(fusiform) gyrus

Lateral occipitotemporal
(inferior temporal) gyrus

Isthmus of gyrus cinguli

Superior frontal gyrus

Gyrus cinguli

Septum pellucidum

Ant. commissure

Paraterminal
(subcallosal) gyrus

Subcallosal (parolfactory)
area of Broca

Uncus

PRINCIPAL GYRI

BRAIN—MEDIAL VIEW

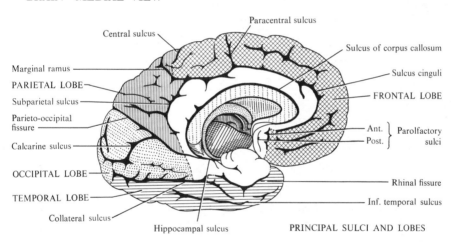

Paracentral sulcus

Central sulcus

Marginal ramus

PARIETAL LOBE

Subparietal sulcus

Parieto-occipital
fissure

Calcarine sulcus

OCCIPITAL LOBE

TEMPORAL LOBE

Collateral sulcus

Hippocampal sulcus

Sulcus of corpus callosum

Sulcus cinguli

FRONTAL LOBE

Ant. } Parolfactory
Post. } sulci

Rhinal fissure

Inf. temporal sulcus

PRINCIPAL SULCI AND LOBES

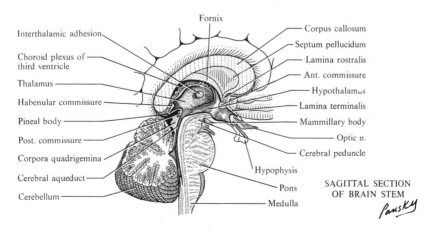

Fornix

Interthalamic adhesion

Choroid plexus of
third ventricle

Thalamus

Habenular commissure

Pineal body

Post. commissure

Corpora quadrigemina

Cerebral aqueduct

Cerebellum

Corpus callosum

Septum pellucidum

Lamina rostralis

Ant. commissure

Hypothalamus

Lamina terminalis

Mammillary body

Optic n.

Cerebral peduncle

Hypophysis

Pons

Medulla

SAGITTAL SECTION
OF BRAIN STEM

Pansky

52. PITUITARY GLAND (HYPOPHYSIS CEREBRI)

I. Position, size, and relations

A. LOCATION: in the sella turcica (hypophysial fossa) of sphenoid bone, attached to the hypothalamus by a stalk (infundibulum). A shelf formed by the meningeal layer of dura stretches between the clinoid processes to cover the sella (diaphragma sellae) except for a small hole for the stalk. Periosteal layer of dura lines the sella

B. SIZE: larger in females, increases in pregnancy, and is largest in multiparous women
   1. Length (rostrocaudal): 1.0 cm
   2. Width (transverse): 1.2–1.5 cm
   3. Thickness: 0.5 cm
   4. Weight: 0.5–0.6 g

C. RELATIONS
   1. Laterally: internal carotid arteries and structures in cavernous sinus
   2. Above: intercavernous (circular) sinuses in the diaphragma sellae and the base of the diencephalon
   3. In front and above: optic chiasma and optic tracts
   4. In front, below: sphenoid air sinus

II. Blood supply

A. ARTERIES
   1. Superior hypophysial arteries from internal carotid and posterior communicating arteries to stalk and adjacent anterior lobe
   2. Inferior hypophysial arteries from the internal carotid artery, passing through cavernous sinus, to posterior lobe
   3. Hypophysial portal veins carrying blood from stalk, lower hypothalamus, and pars tuberalis to most of anterior lobe

B. VEINS (LATERAL HYPOPHYSIAL) drain into cavernous and intercavernous sinuses

III. Parts

A. ADENOHYPOPHYSIS (anterior lobe): largest (75%), formed from oral ectoderm
   1. Developmentally, includes: pars distalis, pars infundibularis (tuberalis), and pars intermedia

B. NEUROHYPOPHYSIS (posterior lobe) formed from neural ectoderm of diencephalic floor
   1. Neurohypophysis includes: infundibular process (neural lobe), infundibular stem, and the median eminence of the hypothalamus

C. THE MEDIAN EMINENCE is often classified as a part of the tuber cinereum; the term "infundibulum" is used for the median eminence and the infundibular stem; the term "hypophysial stalk" refers to the pars infundibularis and the infundibulum

IV. Principal hormones and their effects

A. ADENOHYPOPHYSIS: somatotropic—promotes body growth; thyrotropic—stimulates thyroid secretion; adrenocorticotropic (ACTH)—stimulates secretion of adrenal steroid hormones; two gonadotropic hormones, FSH and LH, stimulate development of ovarian follicles and secretion of corpora lutea, respectively; prolactin—promotes secretion of mammary gland

B. NEUROHYPOPHYSIS: oxytocin—stimulates contraction of smooth muscle, especially of uterus; vasopressin—raises blood pressure and inhibits diuresis

V. Clinical considerations

A. GROSS ENLARGEMENT (TUMOR) can put pressure directly on internal carotid arteries, giving symptoms of occlusion; direct pressure on optic chiasma, causing bitemporal hemianopsia. Oversecretion of anterior lobe can cause gigantism, with hyperglycemia; overactivity of adrenals, leading to Cushing's syndrome; overactivity of thyroid, which may be a factor in Graves' disease

B. INSUFFICIENCY may lead to dwarfism; lower metabolism due to low thyroid activity; symptoms of adrenal insufficiency; and underdevelopment of the genital system

PITUITARY GLAND

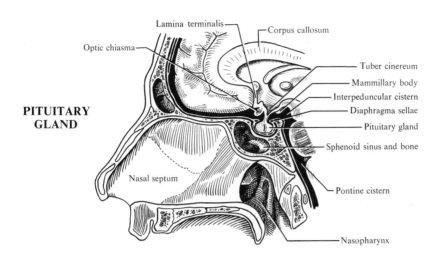

- Lamina terminalis
- Corpus callosum
- Optic chiasma
- Tuber cinereum
- Mammillary body
- Interpeduncular cistern
- Diaphragma sellae
- Pituitary gland
- Sphenoid sinus and bone
- Nasal septum
- Pontine cistern
- Nasopharynx

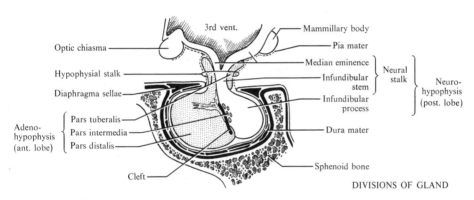

- 3rd vent.
- Mammillary body
- Optic chiasma
- Pia mater
- Median eminence
- Hypophysial stalk
- Infundibular stem
- Neural stalk
- Diaphragma sellae
- Infundibular process
- Neuro-hypophysis (post. lobe)
- Pars tuberalis
- Pars intermedia
- Pars distalis
- Adeno-hypophysis (ant. lobe)
- Dura mater
- Cleft
- Sphenoid bone

DIVISIONS OF GLAND

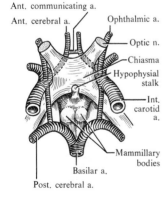

- Ant. communicating a.
- Ant. cerebral a.
- Ophthalmic a.
- Optic n.
- Chiasma
- Hypophysial stalk
- Int. carotid a.
- Mammillary bodies
- Basilar a.
- Post. cerebral a.

RELATIONSHIP TO
MAJOR VESSELS

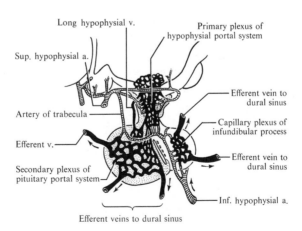

- Long hypophysial v.
- Primary plexus of hypophysial portal system
- Sup. hypophysial a.
- Efferent vein to dural sinus
- Artery of trabecula
- Capillary plexus of infundibular process
- Efferent v.
- Efferent vein to dural sinus
- Secondary plexus of pituitary portal system
- Inf. hypophysial a.
- Efferent veins to dural sinus

VESSELS OF PITUITARY

53. BRAIN STEM (BASAL GANGLIA TO MEDULLA)

I. Diencephalon
A. OBSERVE caudally directed pineal body, pulvinar (large bilateral projections of dorsal thalamus), and medial geniculate bodies beneath the pulvinar

II. Mesencephalon
A. OBSERVE the tectum (roof), composed of 2 pairs of swellings—the corpora quadrigemina (the cephalic pair are the superior colliculi, the caudal pair are the inferior colliculi); the brachium of the inferior colliculus; and the brachium conjunctivum

III. Metencephalon (with cerebellum cut away)
A. OBSERVE the three brachia of cerebellum, namely, the superior (conjunctivum), middle (pontis), and inferior (restiform body)
B. STRUCTURES IN THE FLOOR OF THE FOURTH VENTRICLE: median fissure, sulcus limitans, medial eminence, facial colliculus, locus ceruleus, medullary stria, and vestibular area

IV. Myelencephalon
A. STRUCTURES IN THE FLOOR OF THE FOURTH VENTRICLE: median fissure, sulcus limitans, vestibular area, hypoglossal trigone, and vagal trigone (ala cinerea)
B. OBSERVE the *taenia*, attachments of the ventricular roof to the brain stem with their horizontal part and caudally directed apex (the *calamus scriptorius*)
C. IN THE REGION CAUDAL TO THE TAENIA, OBSERVE
   1. Sulci: posterior median, intermediate, and lateral
   2. Fasciculi: gracilis, between median and intermediate sulci, ending rostrally in a swelling, the tubercle of nucleus gracilis; and the cuneatus, between the intermediate and lateral sulci, ending rostrally in an enlargement, the *cuneate tubercle* (tubercle of cuneate nucleus)

V. Nerves
A. TROCHLEAR (IV), emerging from roof, caudal to inferior colliculus
B. GLOSSOPHARYNGEAL (IX)
C. VAGUS (X)
D. ACCESSORY (XI)

VI. Cerebellum
A. LIES BELOW the posterior portions of the cerebral hemispheres and is separated from them by the tentorium cerebelli
B. CONSISTS OF paired lateral parts, the *cerebellar hemispheres* (consisting of an anterior and posterior lobe); and a smaller midline portion, the *vermis*
C. LIES IN the posterior cranial fossa
D. HAS A CORTEX, with folia and fissures, that covers a much larger center of white matter. It also has certain masses of gray matter, the *cerebellar nuclei*, embedded in it
E. NO CRANIAL NERVE is directly attached to the cerebellum

VII. Internal capsule: a large, fan-shaped band of fibers passing to and from the hemispheres. The thalamus lies to its medial side and the lenticular nucleus on its lateral side

VIII. Clinical considerations
A. TRAUMA OR DISEASE IN THE MYELENCEPHALON is often fatal because vital centers, such as those controlling circulation and respiration, are located here
B. A SINGLE LESION IN THE INTERNAL CAPSULE can result in complete unilateral motor and sensory loss
C. DAMAGE TO THE CEREBELLUM may result in disturbances of voluntary movement

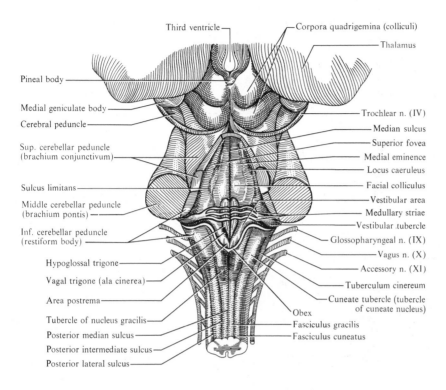

Third ventricle

Corpora quadrigemina (colliculi)

Thalamus

Pineal body

Medial geniculate body

Cerebral peduncle

Sup. cerebellar peduncle
(brachium conjunctivum)

Sulcus limitans

Middle cerebellar peduncle
(brachium pontis)

Inf. cerebellar peduncle
(restiform body)

Hypoglossal trigone

Vagal trigone (ala cinerea)

Area postrema

Tubercle of nucleus gracilis

Posterior median sulcus

Posterior intermediate sulcus

Posterior lateral sulcus

Trochlear n. (IV)

Median sulcus

Superior fovea

Medial eminence

Locus caeruleus

Facial colliculus

Vestibular area

Medullary striae

Vestibular tubercle

Glossopharyngeal n. (IX)

Vagus n. (X)

Accessory n. (XI)

Tuberculum cinereum

Cuneate tubercle (tubercle
of cuneate nucleus)

Obex

Fasciculus gracilis

Fasciculus cuneatus

DORSAL VIEW (CEREBELLUM REMOVED
TO SHOW RHOMBOID FOSSA)

BRAIN STEM

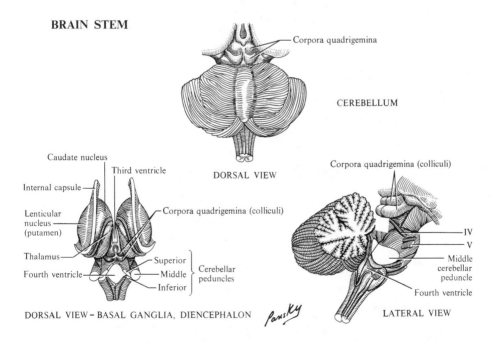

Corpora quadrigemina

CEREBELLUM

DORSAL VIEW

Caudate nucleus

Third ventricle

Internal capsule

Lenticular
nucleus
(putamen)

Thalamus

Fourth ventricle

Corpora quadrigemina (colliculi)

Superior

Middle

Inferior

Cerebellar
peduncles

Corpora quadrigemina (colliculi)

IV

V

Middle
cerebellar
peduncle

Fourth ventricle

DORSAL VIEW – BASAL GANGLIA, DIENCEPHALON

LATERAL VIEW

Pansky

54. ARTERIES OF THE BRAIN: ANTERIOR CIRCULATION

I. **Introduction:** the anterior part of the brain receives its blood from the *internal carotid aa.* (80% of total cerebral flow); the *vertebral-basilar system* (20% total blood flow) supplies the brain stem, cerebellum, and parts of the temporal and occipital lobes. Anatomic connections exist between vessels of the 2 systems at the circle of Willis, but functionally, the systems are almost completely separate. Cerebral aa. are considered to be end arteries since large anastomoses are not found between large branches of the circle of Willis. The capillary network of the cortex is dense, and neurons rarely are more than 50 µm from a capillary

II. **Circle of Willis:** a ring of vessels beneath the hypothalamus, enclosing the lamina terminalis, optic chiasm, infundibulum, tuber cinereum, mamillary bodies, and posterior perforated substance. It is composed of a single ant. communicating a. and the paired ant. cerebral, internal carotid, post. communicating, and post. cerebral arteries

III. **Anterior circulation**
  A. INTERNAL CAROTID enters skull through carotid canal in petrous temporal bone, passes along carotid groove, and exits anteriorly near apex of petrous bone to enter the posterior part of foramen lacerum in which it ascends to a juxtasellar location by piercing the dural layers of the cavernous sinus. It has an S-shaped course in the sinus, called the *carotid siphon*. As it pierces the dura, it gives off its 1st branch, the *ophthalmic a.*, then *post. communicating a.*, and then the *ant. choroidal a.* The artery then bifurcates into *ant.* and *middle cerebral aa.*
    1. Anterior cerebral: smaller of 2 terminal brs.; runs horizontal and anteromedial between optic n. and ant. perforated substance to enter interhemispheric (longitudinal) fissure at midline, anastomosing, via the *ant. communicating a.*, with its counterpart of the opposite side. Major branches are:
      a. Prior to ant. communicating a.: *inferior brs.* to optic n. and chiasm; *superior brs.* (via ant. perforating substance) to basal ganglia, int. capsule, and ant. hypothalamus. The largest branch is the *recurrent artery of Huebner* to the head of the caudate nucleus and ant. limb of the internal capsule
      b. Beyond the ant. communicating a. (interhemispheric): ant. carotid ascends in cistern of lamina terminalis and around genu of corpus callosum, giving rise to:
        i. Medial orbitofrontal a.: to orbital gyrus and olfactory tract and bulb
        ii. Frontopolar a.: to undersurface of frontal lobe and midline structures
        iii. Near genu of corpus callosum, ant. cerebral divides into a *callosomarginal a.* (runs in cingulate sulcus) and a *pericallosal a.* (runs above corpus callosum). They supply the medial and dorsal surfaces of frontal and parietal lobes, including the medial part of the motor and somatosensory cortex
    2. Middle cerebral: largest branch; runs along lateral (sylvian) fissure. Cortical areas supplied are insula, claustrum, lateral part of hemisphere (except for superior convexity from frontal to occipital pole), and inferior convexity from occipital to temporal pole. It gives off 2 groups of ganglionic branches (3–6 medial and 3–6 lateral *striate aa.* which enter brain through ant. perforated substance to int. capsule, basal ganglia, and thalamus); a *lateral orbitofrontal a.* (to lateral orbital and inf. frontal gyri); and *cortical brs.* (to lateral surface of frontal, parietal, and temporal lobes, including premotor, motor and somatosensory cortices of cerebral convexities). The largest striate branches are:
      a. Medial lenticulostriates: to outer segment of the globus pallidus
      b. Lateral lenticulostriates: to putamen, superior 1/2 of internal capsule and adjacent corona radiata, and much of the caudate nucleus
    3. Posterior communicating: runs posteriorly to join posterior cerebral a. of basilar
      a. Branches supply: optic tract and chiasm, hypothalamus, cerebral peduncle, thalamus, subthalamus, interpeduncular area, and hippocampal gyrus
    4. Anterior choroidal: passes along optic tract and around cerebral peduncle to enter temporal horn of lateral ventricle to end in glomus of choroid plexus of ventricle
      a. Branches supply: optic tract, hippocampus, caudate nucleus, internal capsule, cerebral peduncles, thalamus, and part of the midbrain

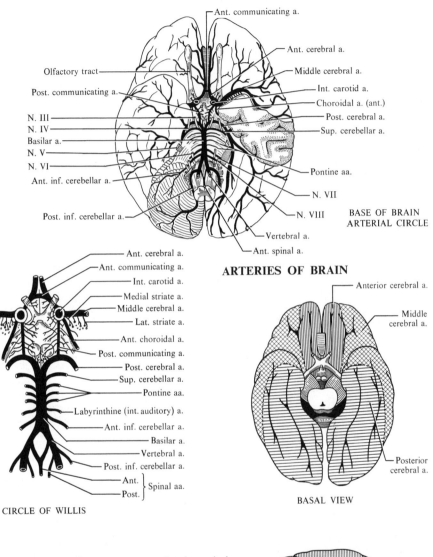

Ant. communicating a.

Ant. cerebral a.

Olfactory tract

Middle cerebral a.

Post. communicating a.

Int. carotid a.

Choroidal a. (ant.)

N. III

Post. cerebral a.

N. IV

Sup. cerebellar a.

Basilar a.

N. V

N. VI

Pontine aa.

Ant. inf. cerebellar a.

N. VII

Post. inf. cerebellar a.

N. VIII

BASE OF BRAIN
ARTERIAL CIRCLE

Vertebral a.

Ant. spinal a.

ARTERIES OF BRAIN

Ant. cerebral a.

Ant. communicating a.

Int. carotid a.

Medial striate a.

Middle cerebral a.

Lat. striate a.

Ant. choroidal a.

Post. communicating a.

Post. cerebral a.

Sup. cerebellar a.

Pontine aa.

Labyrinthine (int. auditory) a.

Ant. inf. cerebellar a.

Basilar a.

Vertebral a.

Post. inf. cerebellar a.

Ant. ⎱ Spinal aa.
Post. ⎰

CIRCLE OF WILLIS

Anterior cerebral a.

Middle cerebral a.

Posterior cerebral a.

BASAL VIEW

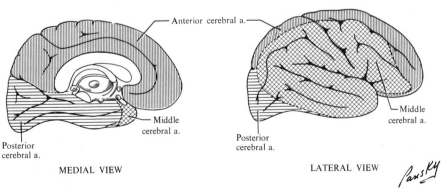

Anterior cerebral a.

Middle cerebral a.

Posterior cerebral a.

Middle cerebral a.

Posterior cerebral a.

MEDIAL VIEW

LATERAL VIEW

Pansky

55. ARTERIES OF THE BRAIN: POSTERIOR CIRCULATION

I. **Introduction of vertebral-basilar system:** the vertebral aa. pass up neck through foramina in transverse processes of cervical vertebrae 6 to 1. After looping between atlas and occipital bone, they penetrate the atlanto-occipital membrane and dura and enter skull through foramen magnum. The lt. vertebral is usually larger than the rt. The 2 vertebrals merge at the pontomedullary junction to form the *basilar a.* which passes in the midline ventral sulcus of the pons to end just posterior to the posterior clinoid process. This system supplies the posterior cerebrum, midbrain, pons, medulla, and cerebellum

II. **Vertebral arteries:** branches are
A. ANTERIOR SPINAL (single) descends on cord anteriorly in ant. median sulcus to supply ventral funiculi, ventral horns, and base of dorsal horns
B. POSTERIOR SPINAL (paired) descend on cord posteriorly, associated with dorsal nerve roots, and supply dorsal funiculi and dorsal horns
C. POSTERIOR INFERIOR CEREBELLAR (largest branch): to choroid plexus of 4th ventricle, lateral medulla, inferior cerebellar peduncle, and part of cerebellum

III. **Basilar artery** terminates in the interpeduncular (crural) cistern where it divides into the rt. and lt. cerebral aa. just superior to the oculomotor nerves. Branches:
A. PONTINE: numerous, small, penetrating to pons and ventrolateral cerebellar cortex and inferior part of the midbrain
B. INTERNAL AUDITORY (labyrinthine) passes with CN VII and VIII to internal auditory meatus and supplies dura of canal, cochlea, labyrinth, and facial nerve
C. ANTERIOR INFERIOR CEREBELLAR: to brain stem and superior and middle cerebellar peduncles and anterior part of undersurface of cerebellar hemisphere
D. SUPERIOR CEREBELLAR: medial branches to mesencephalon, pons, medial cerebellum, and deep cerebellar nuclei; lateral branches to anterolateral part of superior 1/2 of cerebellar cortex, superior and middle cerebellar peduncles, dentate n., and roof nuclei. In addition, the more medial *superior vermian br.* (anastomosis with inferior vermian br. of post. inf. cerebellar) supplies the inferior colliculi, superior cerebellar peduncles, and dentate n.
E. POSTERIOR CEREBRAL passes around midbrain in cisterna ambiens, then through incisura of tentorium, along medial surface of temporal and occipital lobes to end in the calcarine fissure to supply the visual cortex. Branches to:
1. Posteromedial ganglionic: post. hypothalamus, parts of thalamus and midbrain, and the *medial choroidal a.* to choroid plexus of the 3rd ventricle
2. Posterior choroidal branches: to posterior thalamus, subthalamus, internal capsule and choroid plexus of 3rd ventricle, and body of lateral ventricle
3. Cortical branches: to inferior surface of temporal and occipital lobes

IV. **Clinical considerations**
A. BLOOD-BRAIN BARRIER refers to a selective anatomic and physiologic complex that controls movement of substances from the general extracellular fluid of the body to extracellular fluid of the brain. Includes the arachnoid barrier layer and blood-cerebrospinal fluid barrier and a true blood-brain barrier of rows of tight junctions between adjacent endothelial cells of the cerebral capillaries
B. NERVOUS TISSUE is extremely sensitive to the lack of oxygen
C. ON SURFACE OF BRAIN, arterial anastomoses are numerous; in substance of CNS they are rare, and those present are small. Thus, occlusion or rupture of a vessel can lead to widespread and permanent destruction of nervous tissue since nerve cells do not reproduce
D. THE STRIATE ARTERIES are frequently involved in cerebrovascular accidents
E. "CONGENITAL" ANEURYSM (berry or miliary aneurysm) occurs usually on the circle of Willis, especially the anterior communicating a., the middle or anterior cerebral aa., or the basilar a., usually at or near a point of bifurcation. The media of the vessel are normally defective, and the internal elastic membrane is deficient. In 20% of cases, they are multiple. Rupture is found in about 75% of cases of cerebral arterial aneurysm

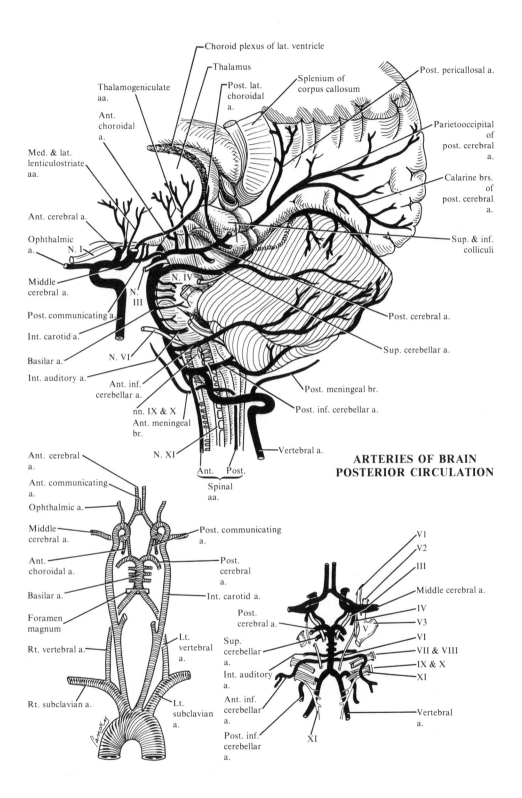

Choroid plexus of lat. ventricle

Thalamus

Post. lat. choroidal a.

Splenium of corpus callosum

Post. pericallosal a.

Thalamogeniculate aa.

Ant. choroidal a.

Parietooccipital of post. cerebral a.

Med. & lat. lenticulostriate aa.

Calarine brs. of post. cerebral a.

Ant. cerebral a.

Ophthalmic a.

N. I

Sup. & inf. colliculi

Middle cerebral a.

N. IV

N. III

Post. communicating a.

Int. carotid a.

N. VI

Basilar a.

Int. auditory a.

Post. cerebral a.

Sup. cerebellar a.

Ant. inf. cerebellar a.

nn. IX & X
Ant. meningeal br.

Post. meningeal br.

Post. inf. cerebellar a.

N. XI

Vertebral a.

Ant.   Post.

Spinal aa.

**ARTERIES OF BRAIN
POSTERIOR CIRCULATION**

Ant. cerebral a.

Ant. communicating a.

Ophthalmic a.

Post. communicating a.

Middle cerebral a.

Ant. choroidal a.

Post. cerebral a.

Int. carotid a.

Basilar a.

Foramen magnum

Post. cerebral a.

Sup. cerebellar a.

Int. auditory a.

Ant. inf. cerebellar a.

Post. inf. cerebellar a.

Rt. vertebral a.

Lt. vertebral a.

Rt. subclavian a.

Lt. subclavian a.

V1
V2
III

Middle cerebral a.

IV
V3
VI
VII & VIII
IX & X
XI

Vertebral a.

XI

56. OCCLUSION OF MAJOR ARTERIES TO BRAIN

Artery and Findings	*Area Involved in Lesion*
I. Anterior cerebral artery	
A. CONTRALATERAL MONOPLEGIA (LEG)	Paracentral lobule
B. CONTRALATERAL SENSORY LOSS	Thalamocortical radiations
C. IF ON DOMINANT SIDE	
1. Mental confusion	Frontal lobe
2. Apraxia, aphasia	Corpus callosum, cortical speech area
II. Anterior choroidal artery	
A. HOMONYMOUS HEMIANOPSIA	Geniculocalcarine tract (optic radiation)
B. CONTRALATERAL	
1. Hemiplegia	Corticospinal fibers of internal capsule
2. Hemianesthesia	Posterior limb of internal capsule
III. Middle cerebral artery	
A. HOMONYMOUS HEMIANOPSIA	Optic tract
B. CONTRALATERAL	
1. Hemiplegia and hemianesthesia	Ant. and post. limbs of internal capsule
C. IF ON DOMINANT SIDE	
1. Global aphasia	Motor and sensory speech areas
IV. Posterior cerebral artery	
A. HOMONYMOUS HEMIANOPSIA	Optic radiations
B. CONTRALATERAL	
1. Hemiplegia, ataxia	Internal capsule, spinocerebellar tract
2. Impaired sensation	Posterolateral ventral nucleus of thalamus
3. Burning pain	Dorsal nucleus of thalamus
4. Choreoathetoid movements	Red nucleus
V. Superior cerebellar artery	
A. HOMOLATERAL	
1. Cerebellar ataxia	Spinocerebellar tract
2. Choreiform movements	Red nucleus
3. Horner's syndrome	Reticular formation
B. CONTRALATERAL	
1. Loss of pain and temperature, face and body	Spinal nucleus of V and lemniscus system
2. Central facial weakness	Corticobulbar fibers to nucleus of VII
3. Partial deafness	Lateral lemniscus
VI. Anterior inferior cerebellar artery	
A. HOMOLATERAL	
1. Cerebellar ataxia, deafness	Spinocerebellar tract, cochlear nuclei
2. Loss of sensation, face	Spinal tract and nucleus of V
B. CONTRALATERAL	
1. Loss of pain and temp., body	Lemniscus system (spinothalamics)
VII. Posterior inferior cerebellar artery	
A. HOMOLATERAL	
1. Cerebellar ataxia, nystagmus	Spinocerebellar tract
2. Horner's syndrome	Reticular formation
3. Loss of sensation, face	Spinal tract and nucleus of V
4. Dysphagia and dysphonia	Nucleus ambiguus, to IX and to X
B. CONTRALATERAL	
1. Loss of pain and temp., body	Lateral spinothalamic tract

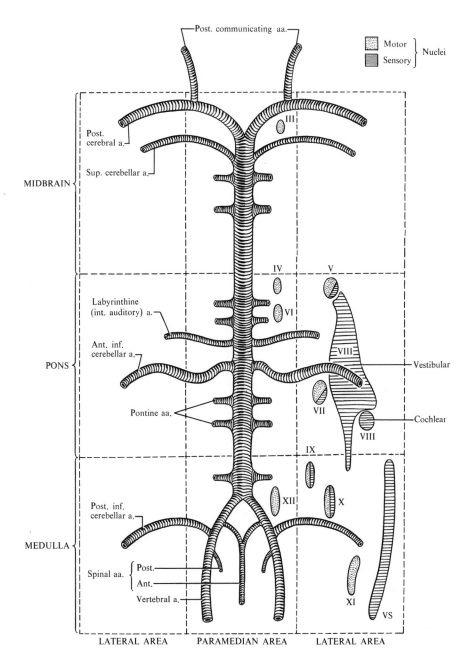

Post. communicating aa.

Motor ░░░ } Nuclei
Sensory ▦

III

Post.
cerebral a.

Sup. cerebellar a.

MIDBRAIN {

IV   V

Labyrinthine
(int. auditory) a.

VI

Ant. inf.
cerebellar a.

VIII — Vestibular

PONS {

VII

Pontine aa.

VIII — Cochlear

IX

Post. inf.
cerebellar a.

XII   X

MEDULLA {

Spinal aa. { Post.
            Ant.

XI

Vertebral a.

VS

LATERAL AREA   PARAMEDIAN AREA   LATERAL AREA

**SCHEME OF VASCULAR SUPPLY
AT BASE OF BRAIN WITH CRANIAL NUCLEI**
(AFTER HOLTZMAN, PANIN, AND EBEL)

Pansky

– 117 –

57. CEREBROVASCULAR CLINICAL CONSIDERATIONS

I. Cerebral angiography: femorocerebral angiography, with transcutaneous technique for arterial catheterization, is a common method for evaluating the brachiocephalic vessels. Using fluoroscopic x-ray control and television monitoring of the image, a preshaped, semirigid catheter is passed through a needle in the femoral a. and is guided up the iliac vessels and aorta to the aortic arch. The catheter can then be selectively maneuvered into the brachiocephalic, lt. common carotid, or lt. subclavian and then into the vertebral or internal carotid artery. With the catheter in place, an iodinated, water-soluble contrast medium is injected. Rapid serial x-rays (over 8–10 sec) are taken in the frontal and lateral projections, revealing, in sequence, the morphologic and physiologic status of the arterial, capillary, and venous phases of the cerebral circulation

II. Aneurysms of cerebral arteries often occur at division points in arteries at base of brain. Classified as saccular, fusiform, or tubular. Most often due to "congenital" weakening of wall. Media of vessel normally defective

A. UNRUPTURED ANEURYSMS: usually asymptomatic, but intermittent enlargements can cause throbbing headaches (due to meningeal stretching or subarachnoid bleeding) and neurologic signs: 3rd n. palsy if associated with post. communicating a.; anosmia (CN I compression); or unilateral hemianopia (CN II involvement)

B. SUDDEN RUPTURE produces marked dizziness and unbearable, sudden, severe headache due to gross bleeding into subarachnoid space with increased intracranial pressure, meningeal irritation, and bloody CSF (under pressure). Coma may follow severe hemorrhage

C. "CONGENITAL" (BERRY OR MILIARY) ANEURYSMS: most common, saccular, arise from circle of Willis and medium-size aa. at base of brain; 20% are multiple; vary from 0.1–3.0 cm; wall of thickened intima with complete absence of media and int. elastic lamina. May expand and rupture into subarachnoid space (subarachnoid hemorrhage) or into brain substance (intracerebral hemorrhage) or into subdural space (subdural hemorrhage)

D. ARTERIOSCLEROTIC ANEURYSM: at site of an atheroma in any cerebral artery; fusiform. Most common in vertebral-basilar and internal carotid–cerebral axes

E. ARTERIOVENOUS ANEURYSM: complex tortuous network of vessels due to congenital defect in capillary bed. Convulsions and/or motor and speech disturbances depend on location

F. MYCOTIC ANEURYSM: due to septic embolism (usually to middle or anterior cerebral artery).

III. Cerebrovascular accident (CVA or stroke): from either sudden hemorrhage into brain or interruption of blood supply leading to infarction and tissue death in area of brain supplied. Clinical picture relates to size and location of infarct or bleeding. "Soft" emboli that break up and pass through a vessel may produce neurologic deficits and *temporary strokes*

A. HEMORRHAGIC TYPE: usually due to rupture of an arteriosclerotic artery or aneurysm. Vary from tiny petechiae to massive hematomata. In the latter, brain bulges on affected side, gyri flatten, blood dissects through brain tissue and may rupture into ventricles. Common site is a lenticulostriate br. of a middle cerebral a. with hemorrhage into basal ganglia and capsule

B. THROMBOTIC TYPE results from thrombosis in an artery to brain or an embolus that is carried from a distant site via the vascular system. Emboli may be *blood clots* (usually from heart), *atheromatous material* (from ulcerated atheroma in medium aa.), *platelet aggregates* (from walls of medium aa.), or *gas bubbles* (from large veins opened traumatically or surgically)

IV. Stenosis or occlusion of vertebral a. may lead to dizziness, fainting, spots before eyes, and transient diplopia. Brain stem must then depend on the other vertebral a.

V. Occlusion of 1 of 3 major cerebral aa.: remaining vessels must take over circulation, but supply by anastomotic vessels is usually inadequate because there is little exchange of blood between cerebral aa. via the communicating aa. Infarcts usually develop, leading to vascular insufficiency. With cardiac arrest, the entire brain is affected, and unconsciousness is seen in 10 sec. With vascular insufficiency, one may begin to see irreversible neurologic damage in 5 minutes

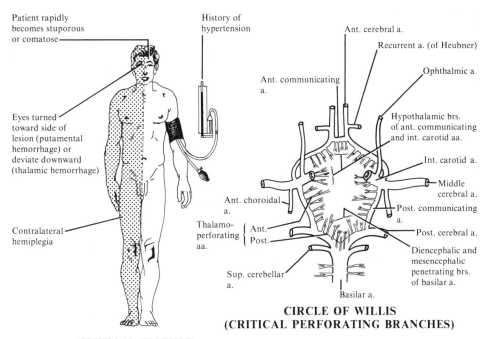

Patient rapidly
becomes stuporous
or comatose

History of
hypertension

Ant. cerebral a.

Recurrent a. (of Heubner)

Ant. communicating
a.

Ophthalmic a.

Eyes turned
toward side of
lesion (putamental
hemorrhage) or
deviate downward
(thalamic hemorrhage)

Hypothalamic brs.
of ant. communicating
and int. carotid aa.

Int. carotid a.

Middle
cerebral a.

Ant. choroidal
a.

Post. communicating
a.

Thalamo-
perforating
aa.

Ant.
Post.

Post. cerebral a.

Contralateral
hemiplegia

Diencephalic and
mesencephalic
penetrating brs.
of basilar a.

Sup. cerebellar
a.

Basilar a.

CIRCLE OF WILLIS
(CRITICAL PERFORATING BRANCHES)

CLINICAL PICTURE—
INTRACEREBRAL HEMORRHAGE

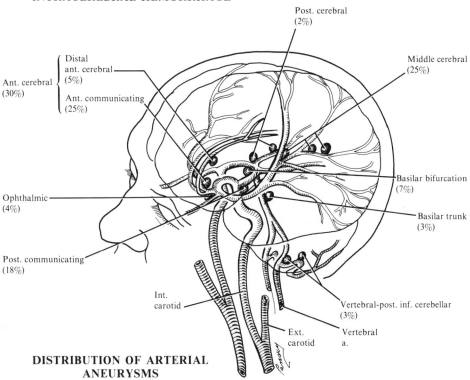

Post. cerebral
(2%)

Middle cerebral
(25%)

Ant. cerebral
(30%)

Distal
ant. cerebral
(5%)

Ant. communicating
(25%)

Basilar bifurcation
(7%)

Ophthalmic
(4%)

Basilar trunk
(3%)

Post. communicating
(18%)

Int.
carotid

Vertebral-post. inf. cerebellar
(3%)

Ext.
carotid

Vertebral
a.

DISTRIBUTION OF ARTERIAL
ANEURYSMS

58. THE VENTRICLES OF THE BRAIN

I. **Introduction:** the core of the embryonic neural tube forms a continuous fluid-filled system, the ventricles, ependymal-lined, and found in the cerebral hemispheres and brain stem. They are variable in size and contain about 35 ml of cerebrospinal fluid. Each lateral ventricle contains about 7–10 ml of fluid with the remainder in the 3rd and 4th ventricles. The system communicates with the subarachnoid space via the median foramen of Magendie and lateral foramina of Luschka. The central canal of the cord is probably not patent in adults

II. **Lateral ventricles:** C-shaped cavities in substance of each hemisphere. Have "horns" radiating from a center, the *collateral trigone (atrium)* of the ventricle, which lies under the parietotemporo-occipital junction. The *glomus of the choroid plexus* lies here. From the trigone, the horns radiate to an anterior (frontal) horn, an inferior (temporal) horn, and a posterior (occipital) horn. The interventricular foramina of Monro are at the junction of anterior horns and bodies and connect the lateral ventricles with the 3rd ventricle

  A. ANTERIOR HORN: rostral to interventricular foramen (of Monro). Floor and lateral wall are formed by head of caudate nucleus. Medial wall is the rostral part of the septum pellucidum. Roof is formed by the corpus callosum

  B. BODY (par centralis) runs posterior from interventricular foramen to the level of the splenium of corpus callosum. Roof is formed by midportion of body of corpus callosum. Floor is divided by terminal sulcus into a lateral part formed by the body of the caudate nucleus and a medial part formed by the thalamus. The medial wall is formed by the posterior part of the septum pellucidum above and the fornix below

  C. COLLATERAL TRIGONE (atrium): junction area of body and occipital and temporal horns

  D. POSTERIOR HORN extends into occipital lobe. Roof and lateral wall formed by the *tapetum of the corpus callosum.* In medial wall, two longitudinal elevations are seen: the *bulb of the posterior horn,* formed by the occipital portion of the radiation of the corpus callosum (*forceps major*), and the *calcar avis,* produced by the rostral part of the calcarine fissure

  E. INFERIOR HORN curves ventrally and then rostrally into temporal lobe. It lies in the medial part of lobe but not quite reaching temporal pole. The roof is formed by the white substance of the hemisphere, and along its medial border are the stria terminalis and tail of caudate nucleus. The amygdaloid nucleus bulges into the terminal end of the horn. The floor and medial wall are formed, from within out, by the fimbria, hippocampus, and collateral eminence. The choroid plexus is superimposed on the fimbria and hippocampus

  F. OPENINGS: interventricular foramina (of Monro) from lateral ventricles to 3rd ventricle

III. **Third ventricle:** small, narrow, midline vertical cleft of diencephalon

  A. BRIDGED BY INTERTHALAMIC ADHESION of thalamus. Lateral wall formed by thalamus and hypothalamus. Roof formed by choroid tela and choroid plexus. Floor contains optic chiasm, infundibulum, tuber cinereum, mamillary bodies, and subthalamus. Anterior wall contains ant. commissure, ant. pillars of fornix, and lamina terminalis. At post. end of roof is suprapineal recess. Below this, in post. wall, is the habenular commissure. The pineal body is attached below the latter. Beneath pineal body is post. commissure

  B. OPENINGS: 2 interventricular foramina laterally, cerebral aqueduct posteriorly

IV. **Fourth ventricle:** lozenge-shaped cavity of rhombencephalon, lying between pons and medulla oblongata ventrally and cerebellum dorsally. It is continuous with the central canal of the closed portion of the medulla posteriorly

  A. THE FLOOR (rhomboid fossa) is formed by the dorsal surfaces of the pons and open part of the medulla. The lateral boundaries of the floor are formed by sup. and inf. cerebellar peduncles, cuneate tubercles, and clavae. (For structures in floor, see p. 111.) The roof is formed by sup. and inf. medullary vela, cerebellum, and choroid tela

  B. OPENINGS: rostrally, the cerebral aqueduct; caudally, closed part of medulla; foramen of Magendie, a median opening at level of obex; foramina of Luschka, bilateral openings at lateral angles at level of 8th cranial nerve. The foramen of Magendie and foramina of Luschka open into the subarachnoid space

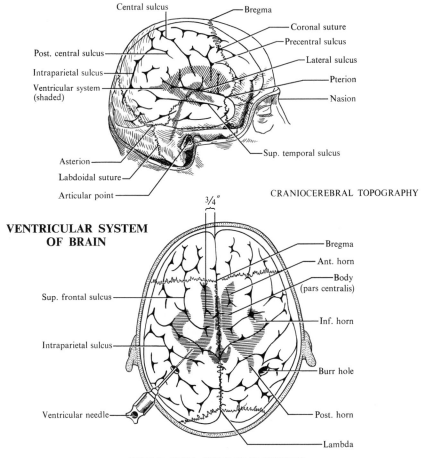

Central sulcus — Bregma
Coronal suture
Post. central sulcus — Precentral sulcus
Intraparietal sulcus — Lateral sulcus
Ventricular system (shaded) — Pterion
Nasion
Asterion
Labdoidal suture
Articular point — Sup. temporal sulcus

CRANIOCEREBRAL TOPOGRAPHY

VENTRICULAR SYSTEM
OF BRAIN

3/4″

Bregma
Ant. horn
Body (pars centralis)
Sup. frontal sulcus
Inf. horn
Intraparietal sulcus
Burr hole
Ventricular needle
Post. horn
Lambda

DORSAL VIEW – SURFACE PROJECTION
OF VENTRICULAR SYSTEM

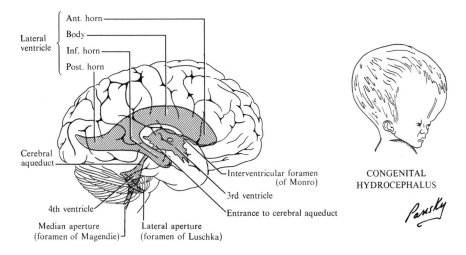

Ant. horn
Body
Lateral ventricle
Inf. horn
Post. horn
Cerebral aqueduct
Interventricular foramen (of Monro)
3rd ventricle
Entrance to cerebral aqueduct
4th ventricle
Median aperture (foramen of Magendie)
Lateral aperture (foramen of Luschka)

CONGENITAL
HYDROCEPHALUS

Pansky

59. CHOROID PLEXUS AND CEREBROSPINAL FLUID (CSF)

I. Choroid plexus: the ventricles are covered by ependymal epithelium and pia mater (together called the *tela choroidea*). In some areas, a rich capillary plexus develops and invaginates the ependymal layer into the ventricular system. These invaginations are called the choroid plexuses. The line of attachment of a plexus is called a *taenia*

A. LOCATIONS: found in the roof of the 3rd and 4th ventricles and on the floor of the bodies and inferior horns of the lateral ventricles. None is seen in the frontal or occipital horns. The plexus in the lateral ventricle is the largest and most important and is continuous with that of the 3rd ventricle via the interventricular foramina. The lateral openings of the 4th ventricle (foramina of Luschka) also contain choroid plexus which protrudes through them and secretes CSF into the subarachnoid space

B. THE PLEXUS IS ENLARGED in the region of the collateral trigone, and here it is called the *glomus* (L *glomus,* ball of thread)

II. Cerebrospinal fluid (CSF)

A. COMPOSITION: colorless, crystal clear; contains Na (148 mEq/L), K (2.88 mEq/L), Cl (120–130 mEq/L), glucose (50–75 mg/100 ml), HCO_3 (22.9 mEq/L), and has a pH of 7.3. Differs from blood in having virtually no cells (0–5 WBC per mm^3—usually lymphocytes) or protein (15–45 mg/100 ml—80% albumin, 6–10% gamma globulin)

1. Modified, after secretion, by choroid plexuses, during its circulation
2. Since it does not resemble an ultrafiltrate of blood, suggests active secretion and absorption of ions rather than it being a simple diffusion across membranes
3. Ventricular fluid and CSF are not the same. The former is more dilute and has a different protein content than that of CSF of the subarachnoid space

B. FORMATION: in part by choroid plexuses and perhaps by vessels in the subarachnoid space. CSF is formed at the rate of 0.3–0.4 ml/min, with the total volume being replaced about every 6 hours. Formation is independent of ventricular, subarachnoid, or systemic blood pressures (rate does decrease at high back pressures)

1. Each lateral ventricle has about 7 ml of fluid; entire ventricular system has about 35 ml; and subarachnoid space and spinal spaces about 100–125 ml of CSF

C. DRAINAGE OF CSF: as a result of hydrostatic pressure in the highly convoluted choroid plexuses, CSF passes from the lateral to the 3rd ventricle via the interventricular foramina (of Monro) and from the 3rd to the 4th ventricle via the midbrain aqueduct (of Sylvius). From the 4th ventricle, CSF passes through the median foramen of Magendie and the lateral foramina of Luschka into the subarachnoid cisterns (cerebellomedullary and pontine) of the posterior cranial fossa

1. From the cisterns, some CSF passes inferiorly around the spinal cord and postero-superiorly over the cerebellum, but most flows up through the tentorial incisure in the subarachnoid space around the midbrain (interpeduncular cistern, rt. and lt. cisterna ambiens, and superior cistern). From these, CSF spreads up through the sulci and fissures on the median and superolateral surfaces of the cerebral hemispheres (probably aided by cerebral arterial pulsations). CSF also passes into extensions of the subarachnoid spaces around the cranial nerves

D. ABSORPTION: rate is directly related to CSF pressure. Major sites of absorption are into venous blood, especially the sup. sagittal sinus and adjacent lateral lacunae, via protrusion of arachnoid into the dural sinuses (the so-called *arachnoid villi* or *pacchionian granulations*) with the villi acting as one-way "valves." CSF is also absorbed by the ventricular ependymal lining, in the spinal subarachnoid space, through the walls of capillaries in the pia, and into lymphatics adjacent to the subarachnoid space around the cerebrospinal nerves

E. FUNCTION: protects the brain by providing a cushion against blows to the head and also against vibration. It helps separate the upper part of brain from the calvaria, and, at its base, separates brain and bones where only meninges intervene

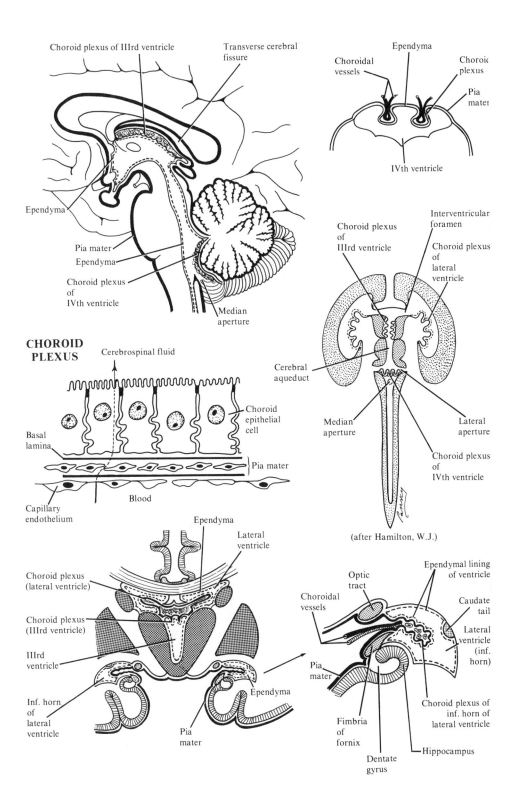

Choroid plexus of IIIrd ventricle

Transverse cerebral fissure

Ependyma

Choroidal vessels

Choroid plexus

Pia mater

Ependyma

IVth ventricle

Pia mater
Ependyma

Choroid plexus of IVth ventricle

Median aperture

Interventricular foramen

Choroid plexus of IIIrd ventricle

Choroid plexus of lateral ventricle

CHOROID PLEXUS

Cerebrospinal fluid

Choroid epithelial cell

Basal lamina

Pia mater

Blood

Capillary endothelium

Cerebral aqueduct

Median aperture

Lateral aperture

Choroid plexus of IVth ventricle

(after Hamilton, W.J.)

Ependyma

Lateral ventricle

Choroid plexus (lateral ventricle)

Choroid plexus (IIIrd ventricle)

IIIrd ventricle

Inf. horn of lateral ventricle

Ependyma

Pia mater

Ependymal lining of ventricle

Optic tract

Choroidal vessels

Caudate tail

Lateral ventricle (inf. horn)

Pia mater

Choroid plexus of inf. horn of lateral ventricle

Fimbria of fornix

Dentate gyrus

Hippocampus

– 123 –

60. CLINICAL CONSIDERATIONS: VENTRICULAR SYSTEM AND CSF

I. Introduction: ventriculography and encephalography are means by which one replaces the fluid in the subarachnoid space and ventricular system with a contrast medium of differing density to x-rays. Air injected into the spaces outlines and fills them and, being less opaque on the x-ray film, the cavities appear more intensely exposed and darker on radiographs. Sometimes radiopaque oil is injected

II. Fractional pneumoencephalography (PEG): about 35 cc of air, in 10– to 15–cc increments, is slowly injected into the lumbar subarachnoid space with the patient sitting upright

A. CEREBROSPINAL FLUID PRESSURE is monitored carefully, and only if it is not abnormally high is air injected and CSF removed alternately in small amounts

B. THE PATIENT'S HEAD is flexed and extended in such a way as to fill the different parts of the subarachnoid cisterns and ventricular system sequentially

C. THE AIR BUBBLE can be kept in different parts of the ventricular system by maneuvering the head, thus the system can be outlined

III. Ventriculography: if the intracranial pressure is elevated, it is safer to inject air (or a positive contrast medium) into one lateral ventricle, following trephination (perforating skull with surgical instrument) in the posterior part of the frontal or in a parietal bone. The air is exchanged for ventricular fluid

A. THIS METHOD is of value in diagnosing the size of the ventricular system and possible obstructions or encroachments on the ventricular system

IV. Radionuclide encephalography: injection of a radioactive isotope into the lumbar subarachnoid space and scanning the brain at intervals over 24–48 hours

A. IF CSF IS NORMAL, most of the isotope passes into the basilar cisterns and over the convexities of the hemispheres

B. IF OBSTRUCTION OCCURS, the isotope refluxes into the ventricles

C. USEFUL METHOD for locating an atrophic or space-occupying lesion

V. Substances can be added to the CSF via puncture for anesthesia

VI. Fluid can be withdrawn from the subarachnoid space or cisterns for diagnostic purposes

VII. Hydrocephalus (Gk *hydōr*, water + *kephalē*, head): overproduction or failure of CSF to drain properly due to mechanical obstruction of ventricular outflow (either in the ventricular system or in the basilar cisterns) or interference with its absorption

A. RESULTS IN A RAISED CSF pressure and progressive dilation of the ventricular system and subsequent enlargement of the ventricles, as well as the developing head

B. NONCOMMUNICATING OR OBSTRUCTIVE TYPE: operational definition. A condition in which dye placed in the lateral ventricle does not appear in the lumbar subarachnoid space
   1. Caused by an obstruction in the ventricular system or in the 4th ventricular outflow

C. COMMUNICATING TYPE (of unknown mechanisms): dye does appear in the lumbar subarachnoid space
   1. Due to blockage of absorption of CSF either in the basilar cisterns or at the pacchionian granulations

D. HYDROCEPHALUS IS MORE COMMON in infants than in adults and, in most cases, no definitive cause can be found (of the communicating type)

E. THE ADULT TYPE of hydrocephalus is generally due to some mechanical obstruction in the circulation of CSF

VIII. Meningitis (inflammation of the meninges) may produce "pus" in the subarachnoid spaces over the brain and in the cisterns, resulting in obstruction of CSF circulation

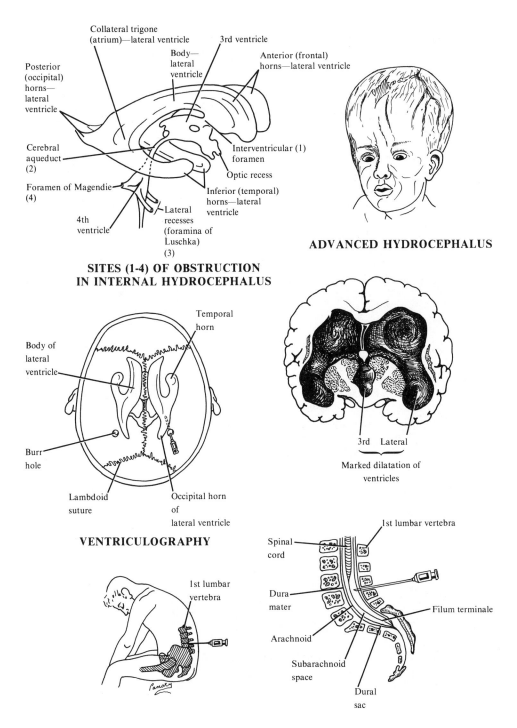

Collateral trigone
(atrium)—lateral ventricle

3rd ventricle

Body—
lateral
ventricle

Anterior (frontal)
horns—lateral ventricle

Posterior
(occipital)
horns—
lateral
ventricle

Cerebral
aqueduct
(2)

Interventricular (1)
foramen

Optic recess

Foramen of Magendie
(4)

Inferior (temporal)
horns—lateral
ventricle

4th
ventricle

Lateral
recesses
(foramina of
Luschka)
(3)

**SITES (1-4) OF OBSTRUCTION
IN INTERNAL HYDROCEPHALUS**

ADVANCED HYDROCEPHALUS

Temporal
horn

Body of
lateral
ventricle

Burr
hole

Lambdoid
suture

Occipital horn
of
lateral ventricle

3rd   Lateral

Marked dilatation of
ventricles

VENTRICULOGRAPHY

1st lumbar
vertebra

1st lumbar vertebra

Spinal
cord

Dura
mater

Arachnoid

Subarachnoid
space

Dural
sac

Filum terminale

PNEUMOENCEPHALOGRAPHY

– 125 –

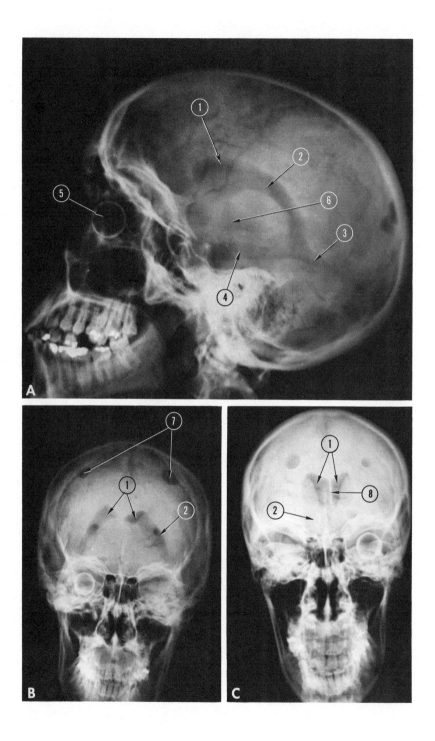

FIGURE 5. **Pneumoencephalogram.** **A, Lateral view; B, anterior view; C, posterior view.** *1*, Anterior horn of lateral ventricle; *2*, body of lateral ventricle; *3*, posterior horn of lateral ventricle; *4*, inferior horn of lateral ventricle; *5*, glass eye; *6*, third ventricle; *7*, burr holes; *8*, septum pellucidum.

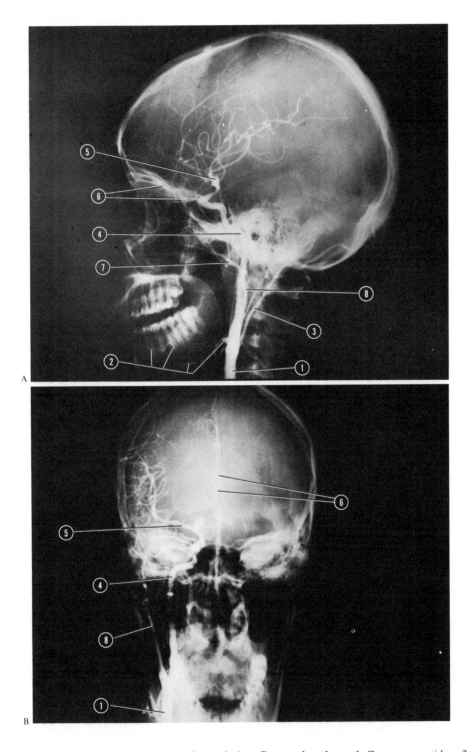

FIGURE 6. **Carotid arteriogram. A, Lateral view; B, anterior view.** *1*, Common carotid a.; *2*, facial a.; *3*, occipital a.; *4*, internal carotid a.; *5*, middle cerebral a.; *6*, anterior cerebral a.; *7*, maxillary a.; *8*, external carotid a.

61. SURFACE ANATOMY OF HEAD AND NECK

I. Vessels
A. ARTERIES
1. Common carotid: on a line from upper border of sternal end of clavicle to a point midway between apex of mastoid and angle of mandible
2. Subclavian: indicated by an arch, the medial end at the sternoclavicular articulation and the lateral end at middle of clavicle
3. Facial (on face): on a line from the facial groove on the inferior border of the mandible, 1 in. from the angle, to the medial corner of the eye, the line passing 1/2 in. lateral to the angle of the mouth
4. Superficial temporal: runs upward just in front of the ear, crossing posterior root of the zygomatic process
5. Middle meningeal and its anterior branch: by a line beginning at midpoint of the zygomatic arch, curving slightly forward, then back through the pterion,* then upward and backward toward the vertex

B. VEINS
1. Internal jugular: follows same line as internal carotid artery
2. Superior sagittal sinus: from nasion to external occipital protuberance, running in the midsagittal plane
3. Transverse sinus: horizontal part—on a line from the external occipital protuberance laterally, approximating the superior nuchal line, to a point just posterosuperior to the external auditory meatus: sigmoid part—begins with the line above and is continued vertically downward to the level of lower border of external auditory meatus

II. Nerves
A. VAGUS: same line as internal carotid artery
B. ACCESSORY: passes under anterior border of sternocleidomastoid, 3.75 cm (1.5 in.) below tip of mastoid; emerges from posterior border of that muscle at junction of upper and middle thirds; passes obliquely downward and backward across posterior triangle to pass under anterior border of trapezius, 5 cm (2 in.) above clavicle
C. PHRENIC: begins at level of middle of lamina of thyroid cartilage, and its caudal course is indicated by a line down the middle of the sternocleidomastoid, parallel to the direction of the muscle

III. Viscera and Sinuses
A. BRAIN
1. Lateral fissure begins at pterion* with the posterior ramus extending upward and backward to the parietal eminence
2. Central sulcus: a line extending from a point halfway between nasion and external occipital protuberance downward and forward to a point 1 in. behind the pterion*
3. Base of cerebrum: this lies one fingerbreadth above Reid's line, a line drawn between lower border of the orbit and the auricular point (center of external auditory meatus)

B. THYROID GLAND: upper pole at junction of middle and caudal thirds of lamina of thyroid cartilage, lateral and caudal to prominence; caudal pole at level of 5th or 6th tracheal ring; isthmus covering tracheal rings 2–4 across the midline

C. AIR SINUSES OR PARANASAL SINUSES: vary greatly in size, shape, and position. The *frontal sinus* occupies the area in the bone deep to the medial part of the superciliary ridge. The *maxillary sinus* occupies the body of the maxilla, the area between the orbit, nasal cavity, and upper teeth

---

*Pterion—located 35 mm behind and 12 mm above the frontozygomatic suture.

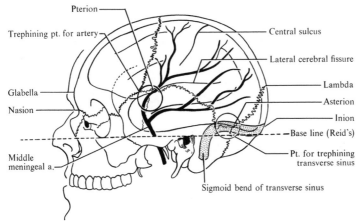

Pterion

Trephining pt. for artery

Central sulcus

Lateral cerebral fissure

Glabella

Lambda

Nasion

Asterion

Inion

Base line (Reid's)

Middle
meningeal a.

Pt. for trephining
transverse sinus

Sigmoid bend of transverse sinus

SURFACE RELATIONS

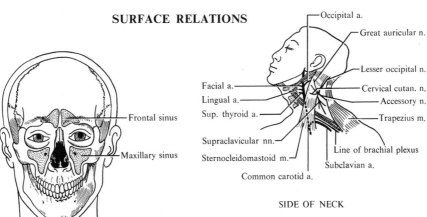

Frontal sinus

Maxillary sinus

Occipital a.

Great auricular n.

Lesser occipital n.

Facial a.

Cervical cutan. n.

Lingual a.

Accessory n.

Sup. thyroid a.

Trapezius m.

Supraclavicular nn.

Sternocleidomastoid m.

Line of brachial plexus

Subclavian a.

Common carotid a.

SIDE OF NECK

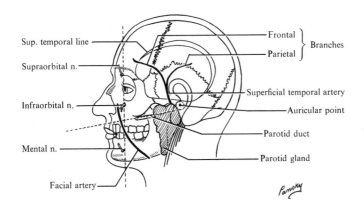

Sup. temporal line

Frontal

Parietal

} Branches

Supraorbital n.

Infraorbital n.

Superficial temporal artery

Auricular point

Mental n.

Parotid duct

Parotid gland

Facial artery

Panaky

62. THE EYELID

I. Eye, as seen from in front

A. PARTS OF EYEBALL: pupil and iris are seen through the transparent cornea, which is continuous with the white sclera or outer fibrous coat of the eye

B. EYELIDS with the *palpebral fissure* (*rima*) between them, *cilia or eyelashes, medial* and *lateral commissures* (*canthi*) at respective corners; medial is prolonged toward nose, leaving triangular space between the lids, the *lacus lacrimalis* (lacrimal lake). On each lid is an elevation, the *lacrimal papilla,* surmounted with a small opening, the *lacrimal punctum*

II. Structure of lid protects eyes from injury and excessive light, keeps cornea moist

A. GROSS: eyelashes (cilia) are short hairs set in double or triple rows; sebaceous glands associated with hairs are *ciliary glands* (of Zeiss); between hair follicles are apocrine type *sweat glands* (glands of Moll). Along border of lid, behind cilia and glands, is a single row of openings for *tarsal* (meibomian) *glands* which are embedded in the tarsal plates. Their fatty secretion lubricates edges of eyelids to prevent sticking and help seal lids in closing

   1. In medial angle (canthus or corner) between lids, the *lacrimal caruncle* (small, fleshy hillock) is seen within the lacrimal lake

   2. Lateral to caruncle is a vertical curved fold of conjunctiva, the *semilunar fold* (plica semilunaris), a remnant of the nictitating membrane seen in some animals

B. MICROSCOPIC, from outside to inside

   1. Skin: very thin, continuous with conjunctiva at margins, no hairs, little fat

   2. Loose connective tissue: without fat

   3. Orbicularis oculi skeletal (voluntary) muscle: a sphincter lying in lids and around them as well (see pp. 26, 27)

   4. Tarsi (tarsal plates) with orbital septum

      a. Tarsi: thin plates of fibrous connective tissue in each lid, upper being larger; attached to palpebral ligaments laterally and medially and to orbital septum

      b. Orbital septum (palpebral fascia): fibrous membrane attached to margins of bony orbit where it is continuous with periorbita (periosteum). In upper lid, joins tendon of levator palpebrae muscle and attaches to superior tarsus. In lower lid, also attaches to tarsus. Pierced by vessels and nerves from orbit to face. Forms barrier between inside and outside contents of orbit

   5. Tarsal (meibomian) glands: embedded in tarsal plates. Seen as yellow streaks through the conjunctiva. Secretion decreases evaporation of lacrimal fluid from conjunctival surface

   6. Conjunctiva: mucous membrane deep to tarsi and orbital septum and lines inside of eyelids. Folds back from lids onto anterior surface of eyeball

C. LEVATOR PALPEBRAE SUPERIORIS M. originates from bone above optic foramen to insert via a broad aponeurosis onto ant. surface of superior tarsus with some fibers penetrating orbicularis oculi m. to attach to skin of upper lid and other fibers attaching to conjunctiva. Tendon penetrates orbital septum. Aponeurosis attaches to bony orbit laterally, checking movement

   1. Posterior part contains a layer of smooth m. attached to superior edge of tarsus and innervated by sympathetic nerves via sup. cervical ganglion. Interruption of sympathetic system induces ptosis (drooping) of eyelids, pupil constriction, flushing of face, and anhidrosis (lack of sweating) and is known as *Horner's syndrome*

III. Motor nerves: sympathetic (see above); zygomatic br. of facial to orbicularis oculi; oculomotor to levator palpebrae superioris

IV. Sensory nerves: palpebral branches of supraorbital, supratrochlear, infratrochlear and lacrimal brs. of V1 (trigeminal) and palpebral br. of infraorbital of V2

V. Arteries: palpebral brs. of lacrimal and ophthalmic aa. Veins similar to arteries

VI. Lymphatics: from upper lid drain to superficial parotid and superficial cervical nodes; from lower lid tends to drain to submandibular nodes

EYELID AND
LACRIMAL GLAND

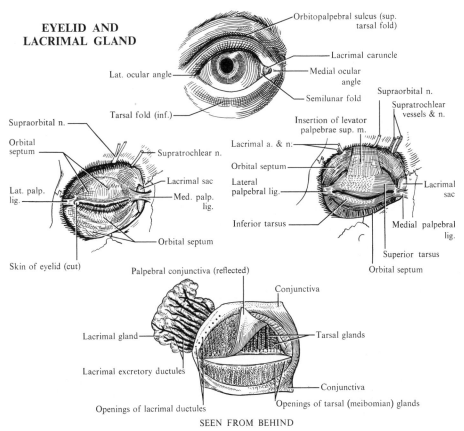

Orbitopalpebral sulcus (sup. tarsal fold)

Lacrimal caruncle

Lat. ocular angle

Medial ocular angle

Semilunar fold

Tarsal fold (inf.)

Insertion of levator palpebrae sup. m.

Supraorbital n.

Supratrochlear vessels & n.

Supraorbital n.

Orbital septum

Supratrochlear n.

Lacrimal a. & n.

Lacrimal sac

Orbital septum

Lat. palp. lig.

Lateral palpebral lig.

Lacrimal sac

Med. palp. lig.

Inferior tarsus

Medial palpebral lig.

Orbital septum

Superior tarsus

Skin of eyelid (cut)

Orbital septum

Palpebral conjunctiva (reflected)

Conjunctiva

Lacrimal gland

Tarsal glands

Lacrimal excretory ductules

Conjunctiva

Openings of lacrimal ductules

Openings of tarsal (meibomian) glands

SEEN FROM BEHIND

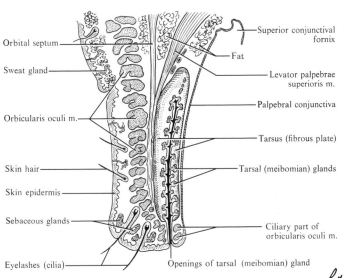

Orbital septum

Superior conjunctival fornix

Sweat gland

Fat

Levator palpebrae superioris m.

Orbicularis oculi m.

Palpebral conjunctiva

Skin hair

Tarsus (fibrous plate)

Skin epidermis

Tarsal (meibomian) glands

Sebaceous glands

Ciliary part of orbicularis oculi m.

Eyelashes (cilia)

Openings of tarsal (meibomian) gland

Pansky

VERTICAL SECTION (after Netter)

63. THE EYELID: CLINICAL CONSIDERATIONS

I. Anomalies of the eyelids include
A. ABLEPHARON: the eyelids are absent, and the eyes are exposed
B. ANKYLOBLEPHARON: the eyelids are fused
C. CRYPTOPHTHALMOS: the eyes are hidden by the overlying skin, and there is no indication of any lid formation present
D. COLOBOMA (a vertical fissure) of one or both upper lids
E. ECTROPION (eversion) OR ENTROPION (inversion) of the lids: quite rare
F. AN EPICANTHUS: a fold of skin covering the inner or, rarely, the outer canthus
   1. In the Mongolian race is normal at the inner canthus but occurs abnormally in mongolian (Down's) syndrome and bilateral renal agenesis
G. DISTICHIASIS: presence of an accessory row of eyelashes

II. Inflammations
A. DERMATITIS OF EYELIDS: common, usually allergic in origin and may be caused by medications used or chemicals present in eyelash dyes and cosmetics, and poison ivy
B. BLEPHARITIS: inflammation of lid margin due to infection of sebaceous glands
   1. Usually chronic with acute exacerbations
   2. There are photophobia, excessive lacrimation, itching, loss of lashes, lid margin is red and scaly. May see edema, congestion, perifollicular abscesses, and lymphocytic infiltration
C. HORDEOLUM (STY): an acute, circumscribed, suppurative inflammation of a gland of Zeis or Moll (external hordeolum) or a meibomian gland (acute chalazion or internal hordeolum)
   1. Usually caused by *Staphylococcus aureus*
   2. Common in children
   3. Reddish swelling near the eyelash roots which can develop into an abscess which not uncommonly may rupture spontaneously
D. CHALAZION: a chronic granuloma of the eyelid due to infection and obstruction of a meibomian gland or its duct
   1. An oval or round mass of a few millimeters develops in the lid and causes no symptoms except for a feeling of pressure
E. CONJUNCTIVITIS
   1. Acute type (pinkeye): inflammation of the bulbar and palpebral conjunctiva and may be caused by several organisms. Children are affected more often than adults
   2. Chronic type may be caused by bacteria, irritating fumes, or allergies. See a moderate conjunctival thickening, reddening, and itching in allergic cases

III. Mechanical injuries
A. SUBCUTANEOUS HEMORRHAGE into eyelids (black eye, shiner): caused by blunt force injury. Hemorrhage is slowly resorbed
B. MECHANICAL INJURY is the most common form of eye disease and may be produced by contusion, concussion, penetrating wound, or perforating wound of eye

IV. Neoplasms: skin of eyelids is a common site for epithelial neoplasms. Squamous papilloma develops at the lid margin and on the palpebral conjunctiva
A. SEBORRHEIC KERATOSIS (not a true neoplasm) appears frequently on lid skin
B. BASAL CELL CARCINOMA is the most common malignant neoplasm of eyelids with 80% developing at inner canthus or on lower lid margin
C. NEOPLASMS ARISING in the eyelid adnexa are uncommon

V. Drooping of eyelid (ptosis) is usually due to an involvement of the oculomotor (III) nerve

VI. Lesions of the facial (VII) nerve eliminate the blink reflex of the eye

**BLEPHARITIS
(EYELID INFLAMMATION)**

**CONJUNCTIVAL
INJECTION**

PERIORBITAL EDEMA

ECTROPION

ENTROPION

CLINICAL EYELID

**DACRYOCYSTITIS
(INFLAMMATION OF LACRIMAL
SAC)**

CHALAZION

**ACUTE HORDEOLUM
(STY)**

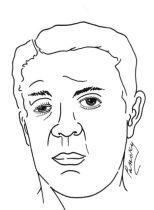

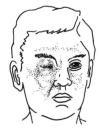

**RT. UPPER EYELID
PTOSIS**

**BASAL CELL
CARCINOMA**

**LACRIMAL GLAND
ENLARGEMENT**

CONTACT DERMATITIS

64. THE LACRIMAL APPARATUS

I. Conjunctiva: the mucous membrane of eye
A. PALPEBRAL PORTION: thick, red, very vascular, lines inside of eyelids
1. At margin of lid, continuous with skin; at medial angle, forms a crescentic fold, plica semilunaris
2. Above and below leaves lid to read eyeball—these are the superior and inferior conjunctival fornices
B. BULBAR (OCULAR) PORTION: thin, transparent, slightly vascular over sclera; on the cornea only the epithelial part is present
C. CONJUNCTIVAL SAC: when lids are closed, a space lined with conjunctiva lies in front of eye

II. Lacrimal gland
A. LOCATION: lies in lacrimal fossa on superior lateral aspect of roof of orbit. Separated from eyeball by levator palpebrae superioris and lateral rectus mm. The levator indents the gland to divide it into an orbital and palpebral portion. In structure, it is like a serous salivary gland. Its 3 to 9 excretory ducts open into the superior fornix of the conjunctival sac
B. BLOOD SUPPLY: lacrimal branches of the ophthalmic artery
C. NERVE SUPPLY: lacrimal br. of trigeminal (sensory); secretomotor via facial (VII) n.

III. Lacrimal canals (ducts)
A. LOCATION: 1 at the medial end of each lid
B. ORIGIN: at puncta lacrimalia, which open on summit of lacrimal papilla, at lateral end of lacus lacrimalis
C. COURSE
1. Superior: first upward, then medially and downward
2. Inferior: first descends, then directly medially
D. TERMINATION: both in lacrimal sac

IV. Lacrimal sac: cephalic, dilated end of nasolacrimal duct
A. LOCATION: lacrimal groove formed by lacrimal bone and frontal process of maxillary
1. Covered anteriorly by expansion of medial palpebral ligament and posteriorly by fibers of orbicularis oculi muscle
B. TERMINATION: caudally into nasolacrimal duct

V. Nasolacrimal duct
A. LOCATION AND COURSE: lies in bony canal formed by maxilla, lacrimal bone, and inferior nasal concha
B. TERMINATION: inferior meatus of nose

VI. Function
A. PURPOSES: (1) to prevent drying of eyeball, (2) to wash out foreign bodies that might damage eyeball
B. BLINKING OF EYE by contraction of levator palpebrae and orbicularis oculi muscles helps (1) to spread secretion throughout conjunctival sac, (2) action of orbicularis oculi, because of relations to lacrimal ducts and sac, "pumps" tears through duct system

VII. Clinical considerations
A. DESTRUCTION OF THE SENSORY ROOT OF THE TRIGEMINAL NERVE or its ophthalmic division may lead to ulcerations of the cornea of eye because of the loss of the afferent limb of the tearing reflex. The conjunctiva becomes dry and is constantly irritated by foreign substances which abrade the surface as the eye and lid move

LACRIMAL APPARATUS

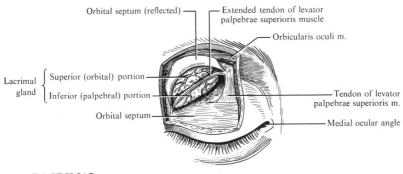

Orbital septum (reflected) — — Extended tendon of levator palpebrae superioris muscle

— Orbicularis oculi m.

Lacrimal gland { Superior (orbital) portion / Inferior (palpebral) portion }

Orbital septum —

— Tendon of levator palpebrae superioris m.

— Medial ocular angle

LACRIMAL GLAND

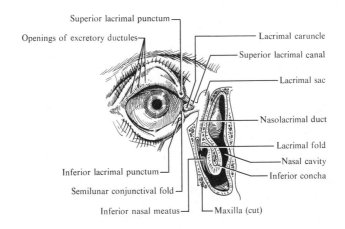

Superior lacrimal punctum —

Openings of excretory ductules —

— Lacrimal caruncle

— Superior lacrimal canal

— Lacrimal sac

— Nasolacrimal duct

— Lacrimal fold

— Nasal cavity

— Inferior concha

Inferior lacrimal punctum —

Semilunar conjunctival fold —

Inferior nasal meatus — — Maxilla (cut)

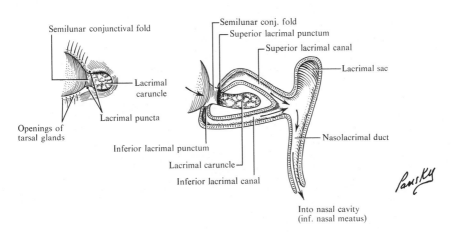

Semilunar conjunctival fold

— Lacrimal caruncle

— Lacrimal puncta

Openings of tarsal glands

Inferior lacrimal punctum

— Semilunar conj. fold
— Superior lacrimal punctum
— Superior lacrimal canal

— Lacrimal sac

— Nasolacrimal duct

Lacrimal caruncle —

Inferior lacrimal canal

Into nasal cavity (inf. nasal meatus)

65. THE ORBIT AND FASCIA BULBI

I. **Bony orbit:** a pyramidal cavity, with base, apex, roof, floor, medial and lateral walls. Apex points dorsomedially; base directed rostrolaterally

A. APEX: optic foramen for optic nerve

B. FLOOR: orbital part of maxilla, orbital process of zygomatic bone, orbital process of palatine bone. Medially, opening of nasolacrimal duct; laterally, depression for origin of inferior oblique muscle; in floor, groove for infraorbital nerve

C. ROOF: orbital plate of frontal bone and small wing of sphenoid. Medially, trochlea for superior oblique muscle; laterally, lacrimal fossa

D. MEDIAL: frontal process of maxilla, lacrimal bone, orbital lamina (lamina papyracea) of ethmoid, body of sphenoid bone, lacrimal groove with posterior lacrimal crest near rostral end of wall, and near roof, anterior and posterior ethmoidal foramina

E. LATERAL: orbital process of zygomatic bone and orbital part of great wing of sphenoid bone. Orifices for branches of zygomatic nerve, superior orbital fissure, inferior orbital fissure

F. BASE (ORBITAL MARGIN): above, supraorbital arch of frontal bone with its supraorbital notch (foramen); laterally, zygomatic and zygomatic process of frontal bone; below, zygomatic and maxillary bones; medially, frontal bone and frontal process of maxillary bone

II. **Orbital septum:** continuous with the periorbita (the periosteum of the bony orbit) at margins of base of orbit; in upper lid joins tendon of levator palpebrae muscle; in both lids attaches to tarsal plates

III. **Vagina bulbi (fascia bulbi, Tenon's capsule)**

A. THIN MEMBRANE to envelop eyeball from optic nerve to level of ciliary muscle. Surrounded by periorbital fat

B. SEPARATED FROM SCLERA by episcleral space—continuous with subdural and subarachnoid spaces

C. FUSES WITH SHEATH OF OPTIC NERVE and with sclera at entrance of this nerve

D. FUSES WITH BULBAR CONJUNCTIVA

E. AT TENDONS OF EXTRINSIC MUSCLES is reflected back, along muscles as sheath
   1. Extension of this, from superior rectus muscle, unites with tendon of levator palpebrae muscle. Prevents eye from being raised too far superiorly
   2. Extension from inferior rectus muscle joins inferior tarsus
   3. From medial and lateral rectus muscles, fascia extends to lacrimal and zygomatic bones (these are strong and known as *medial* and *lateral check ligaments* [*lacertus of medial* and *lateral recti mm.*], for they are thought to check action of respective muscles)
   4. Fascia coverings of inf. oblique and inf. rectus mm. form a sling, the *suspensory ligament of the eyeball*

IV. **Special features**

A. THE ORBIT IS LINED by periosteum that is given the special name *periorbita*. It is continuous over the rim of the orbit and through the inferior orbital fissure with the periosteum of the outer surface of the skull (pericranium), and through the superior orbital fissure and the optic canal with the periosteal or outer layer of the dura mater (endosteal dura or endocranium). The periorbita forms a funnel-shaped sheath to enclose orbital contents, is tough, and may be easily detached from the roof and medial wall of orbit

B. THE ORBITS are related to the frontal sinus above, the maxillary sinus below, and the ethmoid and sphenoid sinuses medially

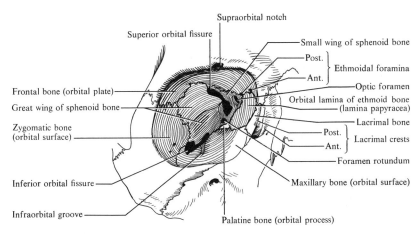

Supraorbital notch

Superior orbital fissure

Small wing of sphenoid bone

Post.
Ant. } Ethmoidal foramina

Frontal bone (orbital plate)

Optic foramen

Great wing of sphenoid bone

Orbital lamina of ethmoid bone
(lamina papyracea)

Zygomatic bone
(orbital surface)

Lacrimal bone

Post.
Ant. } Lacrimal crests

Inferior orbital fissure

Foramen rotundum

Maxillary bone (orbital surface)

Infraorbital groove

Palatine bone (orbital process)

BONY ORBIT

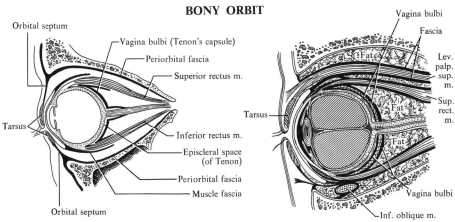

Orbital septum

Vagina bulbi (Tenon's capsule)

Vagina bulbi

Periorbital fascia

Fascia

Superior rectus m.

Fat

Lev.
palp.
sup.
m.

Tarsus

Sup.
rect.
m.

Inferior rectus m.

Fat

Episcleral space
(of Tenon)

Periorbital fascia

Fat

Muscle fascia

Vagina bulbi

Orbital septum

Inf. oblique m.

FASCIA BULBI - TENON'S FASCIA

SAGITTAL
SECTION

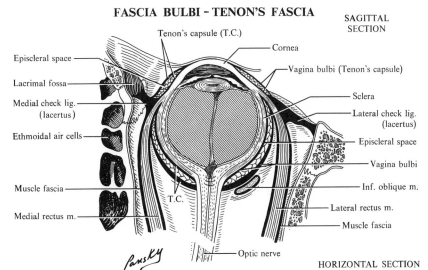

Tenon's capsule (T.C.)

Cornea

Episcleral space

Vagina bulbi (Tenon's capsule)

Lacrimal fossa

Sclera

Medial check lig.
(lacertus)

Lateral check lig.
(lacertus)

Ethmoidal air cells

Episcleral space

Vagina bulbi

Muscle fascia

Inf. oblique m.

T.C.

Lateral rectus m.

Medial rectus m.

Muscle fascia

Optic nerve

HORIZONTAL SECTION

66. THE EXTRINSIC MUSCLES OF THE EYE

I. **Extrinsic muscles:** 7 in orbit; 6 move eyeball, 1 moves lid. The first 6 listed below have their origins from bone or a ring of fibrous tissue (common tendinous ring) around the optic nerve. The seventh, inferior oblique muscle, has a different origin

A. SUPERIOR RECTUS inserts into sclera between equator and corneal margin. Turns eye up and slightly medially; slight medial rotation. Nerve: oculomotor (III)

B. INFERIOR RECTUS inserts in sclera between equator and corneal margin. Turns eye down and medially; slight lateral rotation. Nerve: oculomotor (III)

C. MEDIAL RECTUS inserts in sclera, between equator and corneal margin. Turns eye medially. Nerve: oculomotor (III)

D. LATERAL RECTUS has 2 heads of origin, from upper and from lower parts of common ring. Inserts in sclera between equator and corneal margin. Turns eye laterally. Nerve: abducens (VI)

E. SUPERIOR OBLIQUE, after its tendon bends through the trochlea (pulley) at angle of 50°, inserts in sclera behind equator. Turns eye down and laterally, with medial rotation. Nerve: trochlear (IV)

F. LEVATOR PALPEBRAE arises as A–E, above; inserts in upper eyelid. Raises lid. Nerve: oculomotor (III)

G. INFERIOR OBLIQUE arises from medial edge of orbital floor and inserts in sclera behind equator posterolaterally. Turns eye up and laterally with lateral rotation. Nerve: oculomotor (III)

II. **Nerves**

A. MOTOR: to extrinsic muscles
   1. Oculomotor divides into 2 branches, superior and inferior, before entering orbit through superior orbital fissure; passes between heads of lateral rectus
      a. Superior branch, above optic nerve to superior rectus and levator palpebrae muscles
      b. Inferior branch, below optic nerve, to inferior and medial recti and inferior oblique muscles
   2. Trochlear (IV) enters orbit through superior orbital fissure above lateral rectus muscle, rises to roof of orbit, and then passes medially to superior oblique muscle
   3. Abducens (VI) enters orbit through superior orbital fissure between heads of the lateral rectus muscle, but below inferior branch of the oculomotor nerve, and runs along the inner surface of the lateral rectus muscle, which it supplies

B. SENSORY (see p. 146)

III. **Clinical considerations**

A. DESTRUCTION OF A NERVE or any nerve involvement results in abnormal deviations of the eye and faulty eye movements. For example, if the oculomotor nerve (III) is destroyed while the trochlear (IV) and abducens (VI), which supply the superior oblique and lateral rectus muscles, respectively, are functional, the eye will look lateralward and downward because these muscles are now unopposed by those normally supplied by the oculomotor nerve

B. IT SHOULD BE POINTED OUT that under normal conditions both eyes work together (*conjugate movement*)

C. JUST BEFORE THE FASCIAL SHEATHS of the 4 rectus muscles blend with the bulbar sheath, they expand laterally to fuse with each other, thus forming what is called the *intermuscular membrane*. Tumors or other masses lying internal to this membrane and the rectus muscles may not be visible unless the space among the muscles is explored, therefore, the orbit is frequently described as being subdivided into 2 spaces, 1 within and 1 outside the muscle cone

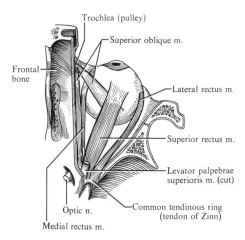

Trochlea (pulley)

Superior oblique m.

Frontal
bone

Lateral rectus m.

Superior rectus m.

Levator palpebrae
superioris m. (cut)

Optic n.

Common tendinous ring
(tendon of Zinn)

Medial rectus m.

SUPERIOR VIEW

EXTRINSIC EYE MUSCLES

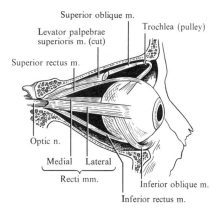

Superior oblique m.

Levator palpebrae
superioris m. (cut)

Trochlea (pulley)

Superior rectus m.

Optic n.

Medial   Lateral

Recti mm.

Inferior oblique m.

Inferior rectus m.

LATERAL VIEW

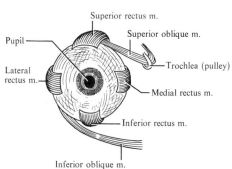

Superior rectus m.

Superior oblique m.

Pupil

Trochlea (pulley)

Lateral
rectus m.

Medial rectus m.

Inferior rectus m.

Inferior oblique m.

ANTERIOR VIEW

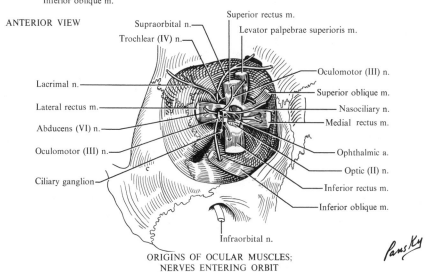

Supraorbital n.

Superior rectus m.

Trochlear (IV) n.

Levator palpebrae superioris m.

Oculomotor (III) n.

Lacrimal n.

Superior oblique m.

Lateral rectus m.

Nasociliary n.

Abducens (VI) n.

Medial rectus m.

Oculomotor (III) n.

Ophthalmic a.

Ciliary ganglion

Optic (II) n.

Inferior rectus m.

Inferior oblique m.

Infraorbital n.

ORIGINS OF OCULAR MUSCLES;
NERVES ENTERING ORBIT

Pansky

67. STRUCTURE OF THE EYE

I. The eyeball (bulbus oculi)
A. OUTER OR FIBROUS COAT consists of a white, opaque posterior 5/6, the *sclera,* and a transparent anterior 1/6, the *cornea*
   1. Sclera: made up of densely packed collagenous fibers; loosely attached to choroid layer; pierced by optic nerve, ciliary nerves, and blood vessels
      a. Lamina cribrosa sclerae: weakest point of sclera. Circular posterior area of sclera which is perforated by fibers of the optic nerve
   2. Corneoscleral junction: where cornea and sclera meet; contains a small canal, the *sinus venosus sclerae,* which encircles the eye
B. MIDDLE OR VASCULAR COAT consists of the *choroid,* the *ciliary body,* and the *iris*
   1. The choroid: dark-brown membrane between retina and sclera; posterior 2/3 of middle coat which ends anteriorly in the ciliary body
      a. Contains venous plexuses and capillaries for nutrition of the retina
      b. Is firmly attached to retina, but is easily stripped from the sclera
   2. Ciliary body connects choroid with circumference of iris, continuous with choroid
      a. Folds on its inner surface, the *ciliary processes,* secrete the *aqueous humor*
      b. Externally, the ciliary body contains *ciliary muscle,* which, when it contracts, allows the lens to bulge by relaxing its suspensory ligament
   3. Iris: heavily pigmented colored part of eye; contractile diaphragm in front of lens with central opening, the *pupil.* Found between cornea and lens, attached radially to ciliary body and cornea by short *pectinate ligaments*
      a. Contains 2 sets of muscles: radial group for enlargement of pupil (*dilator pupillae*) and a circular group set to decrease pupil size on contraction (*sphincter pupillae*). Muscles regulate amount of light entering eye
      b. Eye color depends on pigment distribution in iris: in blue eyes, pigment is limited to posterior surface of iris; in brown eyes, pigment is scattered throughout the iris
C. INTERNAL OR RETINAL COAT is divisible into a *pars optica* (visual part) and a *pars caeca* (nonvisual part); consists of two primary layers, an *inner neural layer* and an *outer pigmented layer.* The *ora serrata,* near edge of ciliary body, is junction point of retina where pars optica transforms to pars caeca
   1. Pars optica: light sensitive; posteriorly at optic fundus, it contains a circular, depressed white area, the *optic papilla* or *optic disk* where the optic fibers leave the eyeball and retinal vessels also enter and leave
      a. The optic disk contains nerve fibers but no photoreceptors and is insensitive to light, thus is called the *blind spot*
      b. Lateral to the disk is a small, oval, yellow area, the *macula lutea* which has a central depressed area, the *fovea centralis.* This area is the area of most acute vision
   2. Pars caeca: nonvisual part lying anterior to ora serrata. Consists of
      a. Pars ciliaris: part of retina covering ciliary body posteriorly
      b. Pars iridica: retinal pigmented cells covering posterior surface of iris

II. Divisions of the eyeball
A. OCULAR CHAMBER lies anterior to lens and suspensory ligament. Further divided into an *anterior* and *posterior chamber* by the iris
B. VITREOUS CHAMBER contains vitreous body, lies behind lens (between lens and retina)

III. The refractive media
A. CORNEA: the transparent, strongly curved, rostral portion of the sclera
B. AQUEOUS HUMOR: clear, watery fluid of the anterior and posterior chambers
C. CRYSTALLINE BICONVEX LENS: composed of elastic capsule containing lens fibers and is suspended from ciliary processes by the *ciliary zonule (suspensory ligament)*
D. VITREOUS BODY: colorless, transparent jelly mass in space between lens and retina; contained in a thin, *hyaloid (vitreous) membrane.* Thickens to form ciliary zonule. Contains *hyaloid canal:* embryonic channel from optic n. to lens for hyaloid a.

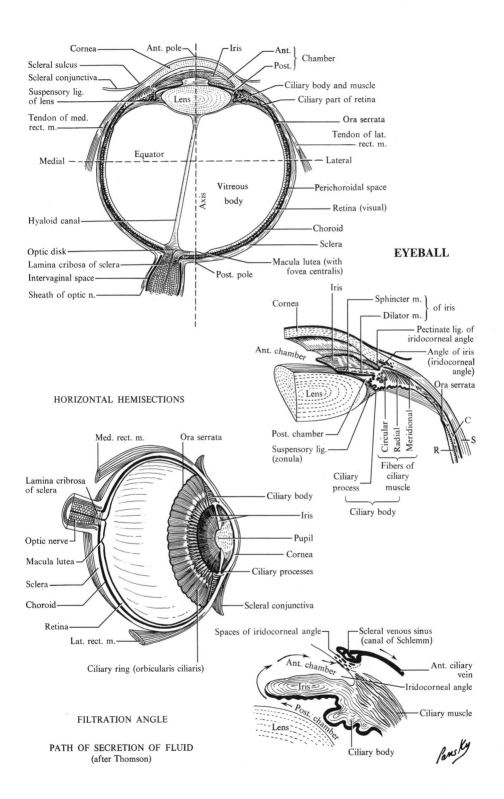

Cornea — Ant. pole — Iris — Ant. } Chamber
Post. }

Scleral sulcus
Scleral conjunctiva
Suspensory lig. of lens

Ciliary body and muscle
Ciliary part of retina

Lens

Tendon of med. rect. m.

Ora serrata
Tendon of lat. rect. m.

Medial — Equator — Lateral

Vitreous body

Axis

Perichoroidal space

Retina (visual)

Hyaloid canal

Optic disk
Lamina cribosa of sclera
Intervaginal space
Sheath of optic n.

Post. pole

Perichoroidal space
Retina (visual)
Choroid
Sclera
Macula lutea (with fovea centralis)

EYEBALL

Iris
Cornea
Sphincter m. } of iris
Dilator m. }

Ant. chamber

Pectinate lig. of iridocorneal angle
Angle of iris (iridocorneal angle)
Ora serrata

Lens

Post. chamber
Suspensory lig. (zonula)

Circular
Radial
Meridional

C
S
R

Ciliary process
Fibers of ciliary muscle

Ciliary body

HORIZONTAL HEMISECTIONS

Med. rect. m. — Ora serrata

Lamina cribosa of sclera

Optic nerve

Macula lutea

Sclera

Choroid

Retina

Lat. rect. m.

Ciliary body
Iris
Pupil
Cornea
Ciliary processes

Scleral conjunctiva

Ciliary ring (orbicularis ciliaris)

FILTRATION ANGLE

PATH OF SECRETION OF FLUID
(after Thomson)

Spaces of iridocorneal angle

Ant. chamber
Iris
Post. chamber
Lens

Scleral venous sinus (canal of Schlemm)
Ant. ciliary vein
Iridocorneal angle
Ciliary muscle
Ciliary body

Pansky

68. CLINICAL CONSIDERATIONS OF VISUAL SYSTEM

I. **Formation and circulation of aqueous humor:** fluid leaves capillary net in the ciliary processes of posterior chamber, flows medially to edge of pupil, and enters the anterior chamber. Here it flows laterally to the iridocorneal angle to enter a meshwork of spaces (Fontana) of the angle. From these, fluid enters the scleral venous sinus (canal of Schlemm) and drains via aqueous veins into scleral plexuses

II. **Accommodation (for near vision)** involves *convergence of the eyes* (effected by a voluntary act of contracting the muscles that move the eyes medially: medial, superior, and inferior recti), an *increase in thickness of the lens* (accomplished by the ciliary muscle which, on contraction, decreases tension on the suspensory ligament and the naturally elastic lens thickens—*note,* when the eye is at rest, the lens is flattened by the natural tension on the suspensory ligament, and since the eye at rest is adjusted for distant vision, accommodation is done for near vision), and *constriction of the pupil* (accomplished by the sphincter pupillae muscle innervated by the parasympathetic system)

III. **Visual field:** area within which objects are distinctly seen by the eye in a fixed position

IV. **Glaucoma:** a complex of disease entities characterized by an increase in intraocular pressure sufficient to cause degeneration. Almost always due to tissue changes which decrease outflow of aqueous humor from eyes. 90% are ''open-angle'' type

V. **Nystagmus:** involuntary, rhythmically repeated oscillations of one or both eyes in any or all fields of gaze. Movements are pendular with undulatory movements of equal speed, amplitude, and duration in each direction, or jerky with slower movements in one direction (slow component) followed by a rapid return (rapid component)

VI. **Strabismus:** a squinting or constant lack of parallelism of visual axes

VII. **Other abnormalities**
  A. ASTIGMATISM: distorted vision caused by a variation in refractive power along different meridians of eye. Most due to irregularities in corneal shape
  B. EMMETROPIA (SIGHT IN PROPER MEASURE): normal refraction in eye where parallel rays of light are focused on the retina without use of accommodation
  C. MYOPIA (NEARSIGHTEDNESS): axis of eyeball too long and focus is in front of retina
  D. HYPEROPIA (FARSIGHTEDNESS): axis of eyeball too short and focus is behind retina
  E. PRESBYOPIA (OLD SIGHT): lessening of effective powers of accommodation due to greater difficulty in shaping lens with increasing age
  F. DIPLOPIA: double vision. Paralysis of one or more eye muscles due to nerve injury
  G. EXOPHTHALMOS (PROPTOSIS): abnormal protrusion of eyeball (seen in Graves' disease)
  H. CONJUNCTIVITIS (PINKEYE): inflammation of bulbar and palpebral conjunctiva
  I. COLOBOMA: any defect, congenital, pathologic, or artificial, especially of eye
  J. PHOTOPHOBIA: abnormal sensitivity of eyes to light
  K. OPAQUE CORNEA: corneal fibers lose their transparency
  L. RETINAL SEPARATION: separation of sensory (neural) from pigment layer of retina
  M. KAYSER-FLEISCHER RING: pigmented ring of variable color (usually green or brown) just inside limbus. Sign of hepatolenticular degeneration (Wilson's disease)
  N. ARCUS SENILIS: benign peripheral corneal degeneration
  O. PTERYGIUM: winglike plaque usually lying across nasal half of conjunctiva and cornea due to repeated exposure to wind and dust and hot, dry climates
  P. BLINDNESS: visual acuity less than 20/200 with limitation of visual field to an angle of 20°
  Q. KERATITIS: inflammation of the cornea (may be ulcerative)
  R. DIABETIC RETINOPATHY: small, fatty retinal deposits and petechiae are seen. Common in patients who have had diabetes for over 15 years.

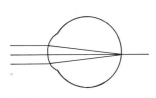

EMMETROPIA (NORMAL)

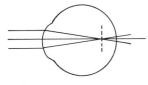

MYOPIA (NEARSIGHTED)

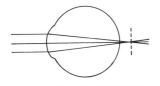

HYPEROPIA (FARSIGHTED)

Lt. visual field

Rt. visual field

Nasal

Temporal

Temporal

Left eye

Rt. eye

Optic nerve

Optic tract

Optic
chiasma

Lateral geniculate
body

Optic radiation

Visual area (brain)

VISUAL PATHWAYS

CLINICAL VISUAL SYSTEM

LINEAR TYPE KERATITIS

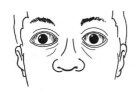

BILATERAL EXOPHTHALMOS
(THYROTOXIC)

PTERYGIUM

ACUTE GLAUCOMA

ARCUS SENILIS

CATARACT

69. BLOOD VESSELS OF THE ORBIT AND EYE

I. Ophthalmic artery
A. ORIGIN: from internal carotid artery at end of cavernous sinus
B. COURSE: through optic foramen, below and lateral to optic nerve; in orbit, crosses above optic nerve to medial wall of orbit, then passes rostrally to divide into supratrochlear and dorsal nasal arteries
C. BRANCHES
1. In orbit, for surrounding parts
   a. Lacrimal near optic foramen, along border of lateral rectus muscle to lacrimal gland, eyelids, and conjunctiva
   b. Supraorbital passes rostrally, along medial border of superior rectus muscle, through supraorbital notch to forehead
   c. Anterior and posterior ethmoidal to sinuses and nasal cavity
   d. Medial palpebral arteries: 1 to each lid
   e. Supratrochlear leaves orbit with supratrochlear nerve to forehead
   f. Dorsal nasal to outer surface of nose
   g. Meningeal turns back through superior orbital fissure to meninges
   h. Muscular come off at intervals to supply orbital muscles
2. In orbit, for eyeball
   a. Central artery of retina leaves main channel as it crosses optic nerve, pierces the latter, and runs in its center to spread over retina
   b. Ciliary arteries arranged in 3 groups
      i. Short posterior ciliary: 6–12, pierce sclera around entrance of optic nerve; to choroid and ciliary processes
      ii. Long posterior ciliary: 2 enter sclera on either side of the optic nerve; run between choroid and sclera to ciliary body, where their branches form anterior major arterial circle
         (1) Send branches inward, which form anterior minor circle
      iii. Anterior ciliary artery: from muscular branches running with tendons of recti muscles to form vascular zone under conjunctiva; pierce sclera to join major circle

II. Distribution of central artery to retina
A. IN EYEBALL, immediately bifurcates giving upper and lower branches
B. BOTH UPPER AND LOWER BRANCHES divide into medial and lateral branches
C. AT FIRST, ARTERIES are between hyaloid membrane and nervous layer of retina but soon pierce latter, forming fine capillary net. Do not go deeper than inner nuclear layer
D. MACULA gets 2 small twigs from temporal branch and a few directly from central artery

III. Venous drainage of eyeball
A. RETINA DRAINED BY VEINS that accompany branches and trunk of central artery
B. OUTER COATS DRAINED BY VORTICOSE VEINS in outer layer of choroid. These converge into 4 or 5 trunks, pierce sclera between optic nerve and corneoscleral junction to drain into superior ophthalmic vein

IV. Clinical considerations
A. EYEGROUNDS (fundus of the eye as seen with the ophthalmoscope): a great deal can be learned from the appearance of the retinal arteries as observed by ophthalmoscopic examination
B. BLIND SPOT: the area of retina where the optic nerve fibers leave eyeball and retinal vessels enter
C. CHOKED DISK: a swelling of the optic disk due to increased intracranial pressure

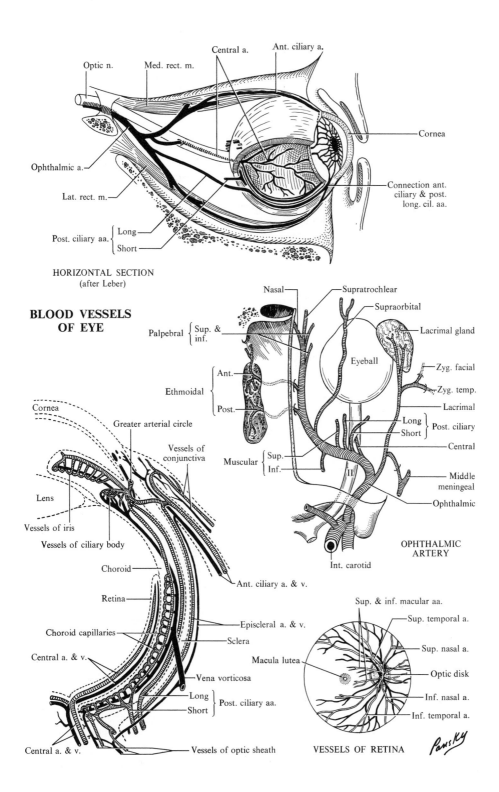

Central a.

Ant. ciliary a.

Optic n.　Med. rect. m.

Cornea

Ophthalmic a.

Connection ant.
ciliary & post.
long. cil. aa.

Lat. rect. m.

Post. ciliary aa. { Long
Short

HORIZONTAL SECTION
(after Leber)

**BLOOD VESSELS
OF EYE**

Nasal　Supratrochlear

Supraorbital

Palpebral { Sup. &
inf.

Lacrimal gland

Eyeball

Zyg. facial

Ant.

Zyg. temp.

Ethmoidal {

Lacrimal

Post.

Long
Short } Post. ciliary

Cornea

Greater arterial circle

Muscular { Sup.
Inf.

Central

Vessels of
conjunctiva

II

Middle
meningeal

Lens

Ophthalmic

Vessels of iris

Vessels of ciliary body

Int. carotid

OPHTHALMIC
ARTERY

Choroid

Retina

Ant. ciliary a. & v.

Sup. & inf. macular aa.

Sup. temporal a.

Episcleral a. & v.

Choroid capillaries

Sclera

Sup. nasal a.

Central a. & v.

Macula lutea

Optic disk

Vena vorticosa

Inf. nasal a.

Long
Short } Post. ciliary aa.

Inf. temporal a.

Central a. & v.

Vessels of optic sheath

VESSELS OF RETINA

Pansky

70. NERVES OF THE ORBIT

I. Sensory
A. SPECIAL SENSE (see also p. 170)
   1. Optic: actually not a true nerve but a tract of central nervous system; enters orbit through optic foramen and joins eyeball just medial to posterior pole
B. GENERAL SENSE
   1. Ophthalmic nerve
      a. Origin: cell bodies in trigeminal (semilunar) ganglion
      b. Course and distribution: along lateral wall of cavernous sinus, below oculomotor and trochlear nerves, enters orbit through superior orbital fissure. Gives off branch to meninges (tentorium cerebelli and falx cerebri) and then branches just before entering orbit:
         i. Lacrimal (smallest branch): along upper border of lateral rectus muscle to lacrimal gland, then rostral to conjunctiva and skin of upper lid
         ii. Frontal (largest branch) runs rostrally, above levator palpebrae muscle to divide into *supraorbital,* which leaves orbit through supraorbital notch, supplying upper lid, forehead, and scalp; *supratrochlear* passes over trochlea of superior oblique muscle to conjunctiva of upper lid and forehead
         iii. Nasociliary crosses optic nerve, passes obliquely below superior rectus and superior oblique muscles to medial wall of orbit. Branches:
            (1) Communication to ciliary ganglion
            (2) Long ciliary to eyeball, especially iris and cornea
            (3) Infratrochlear, to medial angle of eye for conjunctiva, lacrimal sac, skin of lid, and side of nose
            (4) Anterior and posterior ethmoidals to sinuses (frontal, ethmoid, sphenoid) and nasal cavity

II. Motor (for visceral motor, see p. 148)
A. GENERAL MOTOR
   1. Oculomotor (III). For divisions and distribution, see page 138
      a. Origin: nucleus in midbrain tegmentum. The nerve appears in interpeduncular fossa, lies in most cephalic, lateral wall of cavernous sinus to enter orbit through superior orbital fissure
   2. Trochlear (IV). For distribution, see page 138
      a. Origin: nucleus in midbrain tegmentum. Fibers leave central nervous system through anterior medullary velum dorsally, cross to opposite side, pass rostrally and caudally, run in lateral wall of cavernous sinus between III and ophthalmic V, cross III, and pass through superior orbital fissure above other nerves
   3. Abducens (VI). For distribution, see page 138
      a. Origin: nucleus in tegmentum of pons. Fibers leave central nervous system in ventral groove between medulla and pons, pass through cavernous sinus between internal carotid artery and ophthalmic (V) to enter orbit through superior orbital fissure

III. Clinical considerations
A. PATHOLOGY IN THE CAVERNOUS SINUS: tumors or aneurysms of the carotid artery in this location may encroach upon all the nerves passing through the sinus, such as III, IV, and VI, thus causing complete paralysis of all the ocular muscles
B. THE LARGEST NERVE in the orbit, and the central structure in the muscle cone, is the *optic nerve.* This leaves the back of the eyeball and in the orbit is surrounded by a heavy sheath, the *external sheath* of the nerve, which is continuous with the dura and arachnoid in the skull; on the surface of the nerve is a very thin layer of pia mater, the *internal sheath.* Between the 2 layers is the *intervaginal space,* which is a continuation of the subarachnoid space

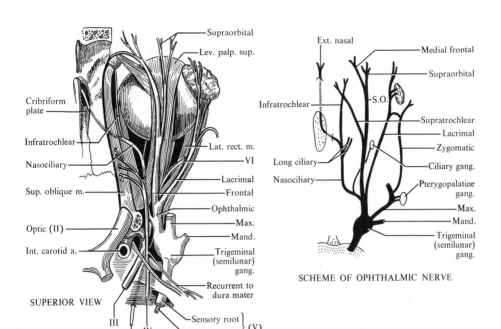

Supraorbital
Lev. palp. sup.
Cribriform plate
Infratrochlear
Nasociliary
Sup. oblique m.
Optic (II)
Int. carotid a.
Lat. rect. m.
VI
Lacrimal
Frontal
Ophthalmic
Max.
Mand.
Trigeminal (semilunar) gang.
Recurrent to dura mater
SUPERIOR VIEW
III
VI
IV
Sensory root
Motor root } (V)

Ext. nasal
Infratrochlear
Long ciliary
Nasociliary
Medial frontal
Supraorbital
S.O.
Supratrochlear
Lacrimal
Zygomatic
Ciliary gang.
Pterygopalatine gang.
Max.
Mand.
Trigeminal (semilunar) gang.

SCHEME OF OPHTHALMIC NERVE

NERVES OF ORBIT

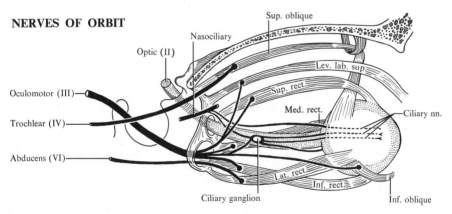

Sup. oblique
Nasociliary
Optic (II)
Lev. lab. sup.
Sup. rect.
Oculomotor (III)
Trochlear (IV)
Med. rect.
Ciliary nn.
Abducens (VI)
Lat. rect.
Inf. rect.
Ciliary ganglion
Inf. oblique

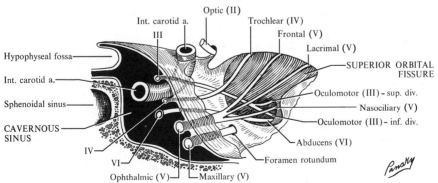

Optic (II)
Int. carotid a.
III
Trochlear (IV)
Frontal (V)
Lacrimal (V)
Hypophyseal fossa
Int. carotid a.
Sphenoidal sinus
CAVERNOUS SINUS
IV
VI
Ophthalmic (V)
Maxillary (V)
SUPERIOR ORBITAL FISSURE
Oculomotor (III) - sup. div.
Nasociliary (V)
Oculomotor (III) - inf. div.
Abducens (VI)
Foramen rotundum
Pansky

71. AUTONOMIC NERVES OF THE ORBIT

I. Lacrimal gland
A. PARASYMPATHETIC (CRANIOSACRAL)—CRANIAL NERVE VII (FACIAL)
   1. Preganglionic fibers arise from cells in superior salivatory nucleus to run in nervus intermedius (with facial) then to greater petrosal nerve, which becomes part of nerve of pterygoid canal, to enter and synapse in pterygopalatine ganglion
   2. Postganglionic fibers arise from cells of pterygopalatine ganglion, pass through pterygopalatine nerves to maxillary nerve, and then through the zygomatic branch of the latter. These fibers enter the zygomaticotemporal branch of the zygomatic nerve in the orbit and communicate with the lacrimal branch of ophthalmic nerve to reach the gland
B. SYMPATHETIC (THORACOLUMBAR)
   1. Preganglionics arise in intermediolateral gray column of upper thoracic cord, enter the sympathetic trunk, ascend in cervical sympathetic trunk to end by synapses in superior cervical ganglion
   2. Postganglionics arise in cells of superior cervical ganglion, pass through carotid plexus, continue rostrally in the deep petrosal nerve, which becomes part of the nerve of the pterygoid canal, pass through pterygopalatine ganglion, without synapse, and are distributed as above (A2)

II. Muscles of iris (sphincter and dilator pupillae)
A. PARASYMPATHETICS TO SPHINCTER MUSCLE (CONSTRICTS PUPIL)
   1. Preganglionics arise in Edinger-Westphal nucleus of midbrain, travel through oculomotor (III) nerve, run into its inferior division and through short root to end by synapse in ciliary ganglion
   2. Postganglionics arise from cells of ciliary ganglion, leave by short ciliary nerves, which pierce sclera posteriorly and run to iris
B. SYMPATHETIC TO DILATOR PUPILLAE MUSCLE (DILATES PUPIL)
   1. Preganglionics: as IB
   2. Postganglionics arise in cells of superior cervical ganglion, ascend through carotid and cavernous plexuses. Some fibers join ophthalmic (V) nerve to continue in its nasociliary branch and are carried to the eye with the long ciliary branches of this nerve; others from cavernous plexus enter sympathetic "root" of ciliary ganglion, pass through without synapse, and run with short ciliary nerves

III. Muscles of ciliary body:
ciliary muscle composed of circular and radiating fibers of smooth muscle. The circularly arranged muscle fibers act as a sphincter and move the ciliary body and suspensory ligaments toward the central axis of the eyeball. The same movement is accomplished when the radially arranged portion is contracted. Thus, the ciliary zonule is moved toward the central axis, as is the suspensory ligament. This decreases tension on the ligament and the lens thickens as in accommodation to near vision
A. PARASYMPATHETICS: same as IIA
B. SYMPATHETICS: none

IV. Clinical considerations
A. HORNER'S SYNDROME: characterized by constriction of pupil, partial drooping of upper eyelid, sinking in of eyeball, dilation of vessels of face and conjunctiva, and lack of sweating. This can result from any lesion in the sympathetic pathways described above
B. ARGYLL ROBERTSON PUPIL: in this condition, the pupil does not constrict for light but will constrict for near vision. Usually caused by a lesion in the pretectal zone of the midbrain, the center for light reflexes

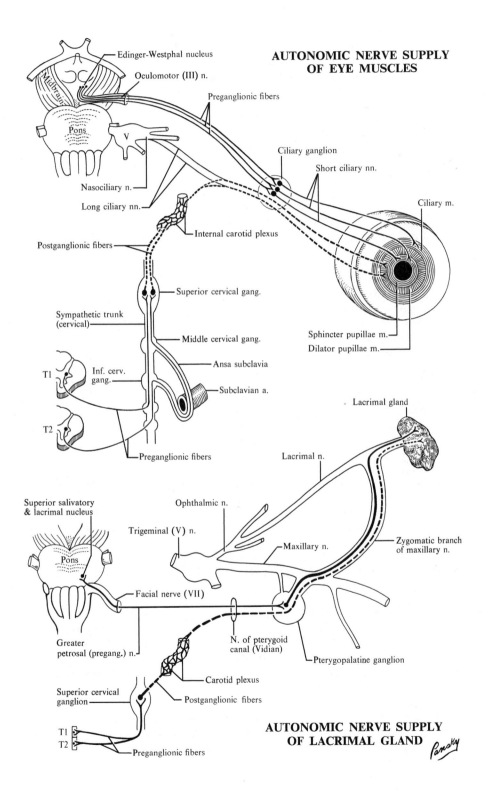

AUTONOMIC NERVE SUPPLY
OF EYE MUSCLES

Edinger-Westphal nucleus

Oculomotor (III) n.

Preganglionic fibers

Ciliary ganglion

Short ciliary nn.

Nasociliary n.

Long ciliary nn.

Internal carotid plexus

Ciliary m.

Postganglionic fibers

Superior cervical gang.

Sympathetic trunk
(cervical)

Middle cervical gang.

Ansa subclavia

Sphincter pupillae m.

Dilator pupillae m.

T1

Inf. cerv.
gang.

Subclavian a.

T2

Preganglionic fibers

Lacrimal gland

Lacrimal n.

Superior salivatory
& lacrimal nucleus

Ophthalmic n.

Trigeminal (V) n.

Maxillary n.

Zygomatic branch
of maxillary n.

Pons

Facial nerve (VII)

Greater
petrosal (pregang.) n.

N. of pterygoid
canal (Vidian)

Pterygopalatine ganglion

Carotid plexus

Superior cervical
ganglion

Postganglionic fibers

T1
T2

Preganglionic fibers

AUTONOMIC NERVE SUPPLY
OF LACRIMAL GLAND

Pansky

72. THE EXTERNAL EAR

I. Ear consists of three parts: external, middle, internal

II. Structure of the external ear: 2 parts

A. AURICLE (PINNA) projects from side of head

    1. Lateral surface: irregular and concave with crests and grooves

        a. Helix (rim): with small auricular tubercle

        b. Anthelix: elevation rostral and parallel to helix

            i. Splits into 2 crura, which bound triangular fossa

        c. Scapha: groove between a and b above

        d. Concha: deep concavity rostral to and bordered by anthelix

        e. Tragus: a dorsal projection, rostral to concha and over meatus

        f. Antitragus: a dorsal tubercle opposite tragus

        g. Intertragic notch lies between e and f, above

        h. Lobule: caudal part of pinna, below antitragus and notch

    2. Medial (cranial) surface shows 2 eminences, corresponding to depressions of lateral surface

    3. Internal structure: basic framework of elastic cartilage and dense fibrous tissue to which thin skin is firmly attached

    4. Auricular muscles (see p. 26): innervated by facial (VII) nerve

B. EXTERNAL AUDITORY MEATUS

    1. Extent: from bottom of concha to tympanic membrane, about 2.4 cm

    2. Course: S-shaped, at first being convex upward and convex backward

    3. Form: cylindrical, with 2 constrictions, one near outer end, the other, the *isthmus,* located deeper

    4. Tympanic membrane, at medial end, set obliquely, so that floor and rostral wall are longer than roof and dorsal wall

    5. Parts

        a. Cartilaginous: about 8.0 mm long and continuous with cartilage of pinna. Is firmly attached to auricular process of temporal bone

        b. Osseous: about 16 mm long, narrower than cartilaginous part, inner end narrower than outer

            i. Tympanic sulcus: a narrow groove to which periphery of tympanic membrane is attached

    6. Canal is lined with skin, epidermis continuing onto tympanic membrane. There are hairs and modified sweat glands that give rise to cerumen or earwax (ceruminous glands)

III. Relations of external ear

A. ROSTRALLY: condyle of mandible and parotid gland

B. DORSALLY: mastoid air cells

IV. Arterial supply: branches of posterior auricular of external carotid artery, anterior auricular from superficial temporal artery, and deep auricular branch from maxillary artery

V. Nerves: auricular branch of vagus and auriculotemporal of mandibular (V) nerve. In addition, auricle supplied by great auricular and lesser occipital nn. of cervical plexus

VI. Clinical considerations

A. LESIONS IN EXTERNAL AUDITORY CANAL: infections or boils in canal may cause nausea and vomiting because the general somatic afferent fibers (pain, etc.) from this area are carried in the vagus nerve (X), which also carries all the parasympathetic fibers to the upper gastrointestinal tract

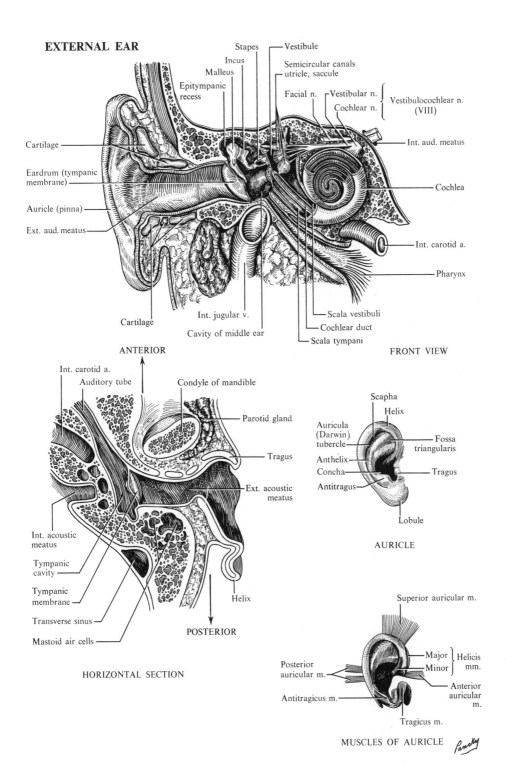

EXTERNAL EAR

Stapes

Incus

Malleus

Epitympanic recess

Vestibule

Semicircular canals utricle, saccule

Facial n.

Vestibular n.

Cochlear n.

Vestibulocochlear n. (VIII)

Cartilage

Int. aud. meatus

Eardrum (tympanic membrane)

Cochlea

Auricle (pinna)

Ext. aud. meatus

Int. carotid a.

Pharynx

Cartilage

Int. jugular v.

Cavity of middle ear

Scala vestibuli

Cochlear duct

Scala tympani

FRONT VIEW

ANTERIOR

Int. carotid a.

Auditory tube

Condyle of mandible

Parotid gland

Tragus

Ext. acoustic meatus

Int. acoustic meatus

Tympanic cavity

Tympanic membrane

Transverse sinus

Mastoid air cells

Helix

POSTERIOR

HORIZONTAL SECTION

Scapha

Helix

Auricula (Darwin) tubercle

Anthelix

Concha

Antitragus

Fossa triangularis

Tragus

Lobule

AURICLE

Superior auricular m.

Major

Minor

Helicis mm.

Posterior auricular m.

Anterior auricular m.

Antitragicus m.

Tragicus m.

MUSCLES OF AURICLE

Pansky

73. THE MIDDLE EAR

I. Middle ear: 2 parts
A. TYMPANIC CAVITY PROPER: opposite tympanic membrane
B. EPITYMPANIC RECESS: above level of membrane

II. Form: a box with a roof, floor, lateral, medial, posterior, and anterior walls
A. ROOF (TEGMENTAL WALL): bony plate, between cranial and tympanic cavities
B. FLOOR (JUGULAR WALL): plate of bone between cavity and jugular fossa
C. LATERAL (MEMBRANOUS WALL): formed by tympanic membrane
    1. Structure and relations
        a. Set obliquely, forming angle of 55° with floor of external meatus
        b. Periphery, where attached to bone in tympanic sulcus, is fibrocartilaginous (anulus)
        c. Where tympanic sulcus is lacking at *tympanic notch (of Rivinus)* the 2 bands, *anterior* and *posterior malleolar folds,* extend from notch to lateral process of malleus
            i. Triangular area between (and superior to) folds is *pars flaccida* of membrane
        d. Manubrium of malleus is firmly fixed to membrane at its middle
            i. Umbo: caudal end of manubrium, where membrane is concave
        e. Chorda tympani nerve runs between handle of malleus and incus
    2. Special features
        a. Notch of Rivinus: deficiency in bony rim
        b. Iter chordae anterius and posterius
        c. Petrotympanic fissure
D. MEDIAL (LABYRINTHIC WALL): bone between middle and inner ear
    1. Vestibular window (oval) with long axis horizontal
    2. Cochlear window (round) contains secondary tympanic membrane
    3. Promontory: rounded eminence between 1 and 2
    4. Facial canal: above oval window, then curves downward along posterior wall
    5. Prominence of lateral semicircular canal
E. POSTERIOR (MASTOID WALL)
    1. Entrance (aditus) to *tympanic antrum,* actually from epitympanic recess to mastoid
    2. Pyramidal eminence
    3. Fossa of incus
F. ANTERIOR (CAROTID WALL): bone separating cavity from carotid artery
    1. Perforated by tympanic branch of internal carotid artery and sympathetic nerves
    2. Auditory tube and canal for tensor tympani

III. Auditory ossicles, which transfer sound waves across middle ear
A. MALLEUS with head, neck, anterior process, and lateral process
B. INCUS with body, short crus, and long crus
C. STAPES with head, neck, 2 crura, and foot plate

IV. Intrinsic muscles
A. TENSOR TYMPANI arises from cartilaginous part of auditory tube and great wing of sphenoid. Inserts on malleus. Action: tenses membrane. Nerve: mandibular (V)
B. STAPEDIUS arises from inner wall of pyramidal eminence. Inserts on neck of stapes. Nerve: branch of facial (VII). Action: draws stapes back

V. Clinical considerations
A. MIDDLE EAR INFECTIONS: quite prevalent and may become extensive due to connection to both the mastoid cells and the nasopharynx by way of the eustachian tube
B. OTOSCOPIC EXAMINATION: when reflected light is used, the cone of light seen has its apex at the umbo and expands downward and forward. This cone changes in disease

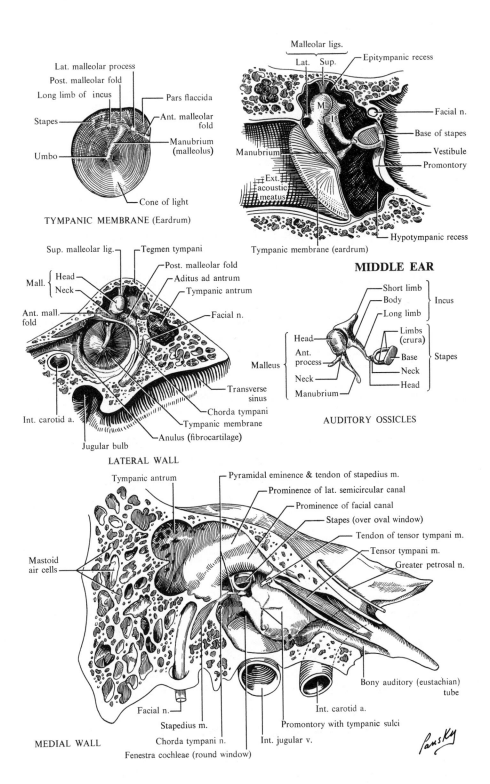

Lat. malleolar process
Post. malleolar fold
Long limb of incus
Stapes
Umbo
Pars flaccida
Ant. malleolar fold
Manubrium (malleolus)
Cone of light

TYMPANIC MEMBRANE (Eardrum)

Malleolar ligs.
Lat. Sup.
Epitympanic recess
M
Facial n.
Base of stapes
Vestibule
Promontory
Manubrium
Ext. acoustic meatus
Hypotympanic recess
Tympanic membrane (eardrum)

MIDDLE EAR

Sup. malleolar lig.
Tegmen tympani
Post. malleolar fold
Mall. { Head Neck
Aditus ad antrum
Tympanic antrum
Ant. mall. fold
Facial n.
Int. carotid a.
Transverse sinus
Chorda tympani
Tympanic membrane
Anulus (fibrocartilage)
Jugular bulb

LATERAL WALL

Short limb
Body
Long limb
} Incus
Limbs (crura)
Head
Ant. process
Neck
Manubrium
} Malleus
Base
Neck
Head
} Stapes

AUDITORY OSSICLES

Tympanic antrum
Pyramidal eminence & tendon of stapedius m.
Prominence of lat. semicircular canal
Prominence of facial canal
Stapes (over oval window)
Tendon of tensor tympani m.
Tensor tympani m.
Greater petrosal n.
Mastoid air cells
Bony auditory (eustachian) tube
Facial n.
Stapedius m.
Chorda tympani n.
Int. jugular v.
Int. carotid a.
Promontory with tympanic sulci
Fenestra cochleae (round window)

MEDIAL WALL

Pansky

74. THE INNER EAR

I. Inner ear (labyrinth) consists of 2 parts
  A. OSSEOUS: cavities in the petrous bone
  B. MEMBRANOUS: made up of interconnected sacs and tubes (ducts)

II. Osseous labyrinth has 3 parts, lined with periosteum. Filled with *perilymph*
  A. VESTIBULE: central part of labyrinth
    1. In lateral wall: oval window
    2. In medial wall: perforations for nerves and opening for endolymphatic duct (vestibular aqueduct)
    3. Dorsally: 5 openings for semicanals
    4. Rostrally: opening for cochlea
  B. SEMICIRCULAR CANALS: 3 in number
    1. Anterior (superior): lateral end has a dilation, the *ampulla,* which opens into vestibule; medial end joins posterior canal, forming the *common crus*
    2. Posterior: caudal end has an *ampulla,* which opens into vestibule
    3. Lateral has *ampulla,* which opens into vestibule at lateral end; other end opens into dorsal part of vestibule
  C. COCHLEA (SNAIL SHELL) has apex and base. Apex directed rostrally and laterally; base faces dorsomedially and lies at bottom of internal acoustic meatus
    1. Modiolus: the bony, conical axis
    2. Bony canal makes 2½ turns around modiolus; lower end diverging from modiolus has openings for *round window.* At apex, canal called *helicotrema*
    3. Osseous spiral lamina: a bony shelf projecting from modiolus. With basilar membrane, it divides an upper scala vestibuli and lower scala tympani

III. Membranous labyrinth: connective tissue and epithelium, filled with *endolymph*
  A. UTRICLE: in bony vestibule. Has 5 openings for semicircular canals and *ductus utriculosaccularis* (becomes endolymphatic duct and sac)
  B. SACCULE: in vestibule. Has opening into endolymphatic duct and cochlear duct (*ductus reuniens*)
  C. SEMICIRCULAR CANALS open into utricle, conform to bony labyrinth
  D. COCHLEAR DUCT: that part of labyrinth separated from scala tympani by the basilar membrane and from scala vestibuli by the vestibular membrane

IV. Sense organs
  A. FOR EQUILIBRIUM: macula utriculi, macula sacculi, and cristae ampullaris
  B. FOR HEARING: the *spiral organ* (*of Corti*) in floor of cochlear duct

V. Nerves
  A. VESTIBULAR PORTION OF VIII for equilibrium
    1. Ganglion: vestibular, located in acoustic meatus
    2. Branches of nerve: *superior,* to macula utriculi and cristae of superior and lateral canals; *inferior,* to macula sacculi; and *posterior,* to crista of posterior canal
  B. AUDITORY PORTION OF VIII for hearing
    1. Ganglion; spiral, in spiral canal at center of bony modiolus
    2. Distribution: fibers pass through osseous spiral lamina to the organ of Corti

VI. Clinical considerations
  A. DEAFNESS CAN RESULT from damage to the organ of Corti, cochlear duct, or acoustic (VIII) nerve and its ganglion
  B. MÉNIÈRE'S DISEASE is characterized by a loss of balance and ringing in the ears, which are caused by edema of the labyrinth or inflammation of the vestibular nerve

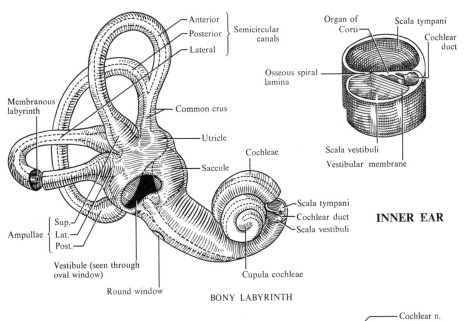

Anterior ⎫
Posterior ⎬ Semicircular canals
Lateral ⎭

Organ of Corti
Scala tympani
Cochlear duct
Osseous spiral lamina
Scala vestibuli
Vestibular membrane

Membranous labyrinth
Common crus
Utricle
Saccule
Cochleae
Scala tympani
Cochlear duct
Scala vestibuli

INNER EAR

Ampullae ⎰ Sup.
⎱ Lat.
Post.

Vestibule (seen through oval window)
Round window
Cupula cochleae

BONY LABYRINTH

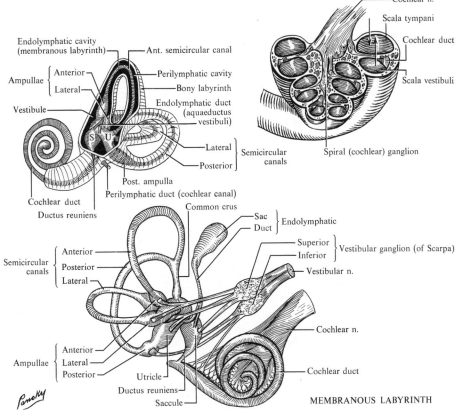

Cochlear n.
Scala tympani
Cochlear duct
Scala vestibuli

Endolymphatic cavity (membranous labyrinth)
Ant. semicircular canal
Perilymphatic cavity
Bony labyrinth
Endolymphatic duct (aquaeductus vestibuli)

Ampullae ⎰ Anterior
⎱ Lateral

Vestibule

Lateral ⎫
Posterior ⎬ Semicircular canals

Spiral (cochlear) ganglion

Post. ampulla
Perilymphatic duct (cochlear canal)
Cochlear duct
Ductus reuniens

Common crus
Sac ⎱ Endolymphatic
Duct ⎰

Superior ⎱ Vestibular ganglion (of Scarpa)
Inferior ⎰
Vestibular n.

Semicircular canals ⎰ Anterior
⎱ Posterior
Lateral

Cochlear n.

Ampullae ⎰ Anterior
⎱ Lateral
Posterior

Utricle
Ductus reuniens
Saccule
Cochlear duct

MEMBRANOUS LABYRINTH

Pansky

75. THE SPIRAL ORGAN OF CORTI AND FACIAL NERVE RELATIONS TO EAR

I. **The spiral organ (of Corti)** is the receptor of auditory stimuli and is situated on the basilar membrane and extends from the base to the apex of the cochlea

A. IT PROJECTS INTO the cochlear duct and is thus bathed by endolymph

B. IN TRANSVERSE SECTION, it is seen to consist of a row of *inner hair cells* and a row of *outer hair cells* which have hairlike projections into the endolymph
   1. These cells are accompanied by various types of supporting cells
   2. The nerves are a direct continuation of the hair cells

C. A MEMBRANE, the *tectorial membrane,* is seen to project into the endolymph from the osseous spiral lamina and is suspended over the hair cells
   1. When a hair cell touches the membrane, an impulse is carried over the cochlear nerve to the brain and is recognized as sound

D. THE BASE OF THE COCHLEA is involved with high sounds, while the low sounds involve the area near the helicotrema (due to fact that basilar membrane is longer near the apex of the coil and shorter toward the base)

II. **Mechanism of hearing sound waves:** air waves enter external auditory meatus and cause tympanic membrane to vibrate. Vibrations are carried through the malleus, incus, and stapes to oval window in vestibule of osseous labyrinth. If too loud, reflex stimulation of tensor tympani and stapedius mm. decreases the magnitude of the vibrations at the oval window. Vibrations caused by the stapes induce vibrations in the perilymph of the scala vestibuli (and scala tympani across the helicotrema). The membrane of the round window bulges out, as that of the oval window bulges in. The vibrations pass through the vestibular membrane into the cochlear duct. A theory of hearing states that a specific portion of the basilar membrane responds to any given frequency of vibration in the perilymph and endolymph, and the stimulated hair cells send impulses to the brain for interpretation

III. **The facial (VII) nerve** as related to the ear structures

A. THE FACIAL NERVE ACCOMPANIES the VIIIth cranial nerve in the internal acoustic meatus until it reaches the *geniculate ganglion* (a sensory ganglion). Here the nerve makes a sharp posterior bend in the medial wall of the middle ear beneath the lateral semicircular canal (creating a prominence) and then turns laterally and inferiorly in the posterior wall of the middle ear to finally exit from the *stylomastoid foramen.* It gives off 7 branches
   1. Greater petrosal: 1st branch of the facial nerve, in the region of the ganglion, turns anterior and medial and enters the middle cranial fossa through the *facial hiatus* on the anterior surface of the petrous temporal. The nerve crosses the foramen lacerum and is joined by the *deep petrosal nerve* (sympathetic, postganglionic). The two become the *nerve of the pterygoid canal* (vidian). The latter enters the pterygopalatine fossa, and its facial parasympathetic fibers synapse in the pterygopalatine ganglion. From here, both postganglionic fibers are distributed to the mucous membranes and vessels of the oral cavity, nasal cavity, pharynx, and other regions (including the lacrimal gland)
   2. Communication with lesser petrosal: the latter is a contribution from the tympanic plexus on the promontory and is parasympathetic IX. Function of fibers of VIIth nerve in lesser petrosal is uncertain
   3. Nerve to stapedius muscle
   4. Chorda tympani arises from facial while it is in facial canal posterior to middle ear. It courses through the middle ear behind the tympanic membrane, exits from the petrotympanic fissure to the infratemporal fossa to join lingual V. Contains taste fibers from ant. 2/3 of tongue and secretory fibers to submandibular and sublingual glands
   5, 6, and 7. Emerging from the skull via the stylomastoid foramen, it gives off its *posterior auricular br., branch to posterior belly of digastric m.,* and *branch to stylohyoid muscle* before dividing into terminal branches on face (temporal, zygomatic, buccal, mandibular, and cervical)

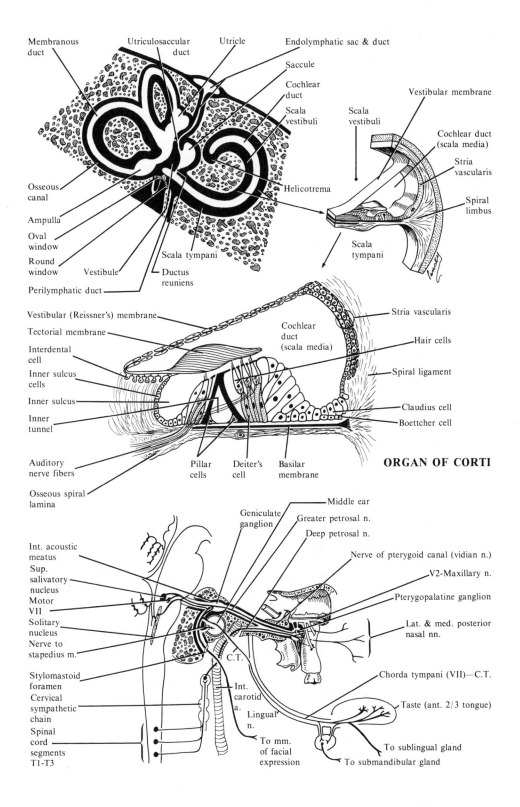

Membranous duct
Utriculosaccular duct
Utricle
Endolymphatic sac & duct
Saccule
Cochlear duct
Scala vestibuli
Scala vestibuli
Vestibular membrane
Cochlear duct (scala media)
Stria vascularis
Spiral limbus
Osseous canal
Helicotrema
Ampulla
Oval window
Round window
Vestibule
Ductus reuniens
Scala tympani
Scala tympani
Perilymphatic duct

Vestibular (Reissner's) membrane
Tectorial membrane
Interdental cell
Inner sulcus cells
Inner sulcus
Inner tunnel
Cochlear duct (scala media)
Stria vascularis
Hair cells
Spiral ligament
Claudius cell
Boettcher cell
Auditory nerve fibers
Pillar cells
Deiter's cell
Basilar membrane
Osseous spiral lamina

ORGAN OF CORTI

Int. acoustic meatus
Sup. salivatory nucleus
Motor VII
Solitary nucleus
Nerve to stapedius m.
Stylomastoid foramen
Cervical sympathetic chain
Spinal cord segments T1-T3
Geniculate ganglion
Middle ear
Greater petrosal n.
Deep petrosal n.
Nerve of pterygoid canal (vidian n.)
V2-Maxillary n.
Pterygopalatine ganglion
Lat. & med. posterior nasal nn.
Chorda tympani (VII)—C.T.
Taste (ant. 2/3 tongue)
C.T.
Int. carotid a.
Lingual n.
To mm. of facial expression
To sublingual gland
To submandibular gland

– 157 –

76. THE NOSE—PART I

I. External nose
A. PARTS: root, dorsum or bridge (from root to apex), tip, alae, base, with nares (nostrils) separated by a nasal septum
B. NERVES (see also p. 168)
   1. Motor: facial (VII) to muscles (see p. 26)
   2. Sensory: infratrochlear from nasociliary branches of ophthalmic (V), to skin of root, alae, and nostrils; infraorbital branch of maxillary (V) to sides of nose
C. VESSELS (see p. 42)
   1. Arteries: facial to sides, alae, and septum; dorsal nasal branch of ophthalmic to root and dorsum; infraorbital branch of maxillary to sides
   2. Veins drain into the ophthalmic and anterior facial veins
D. FRAMEWORK: hyaline cartilage and bone
   1. Cartilage
      a. Septal: partition, between right and left nasal cavities
         i. Attachments: perpendicular plate of ethmoid, vomer, maxilla, nasal, septal process of lower nasal cartilage
      b. Lateral nasal (upper lateral)
         i. Attachments: nasal bone, frontal process of maxilla, lower nasal cartilage becomes continuous with septal cartilage
      c. Greater alar (lower lateral, alar)
         i. U-shaped, open posteriorly; has medial and lateral crura
         ii. Medial crura attach with each other and upper nasal cartilage
      d. Lesser alar
   2. Bones: 2 nasal bones and frontal process of maxillae

II. Nasal cavities: right and left separated by nasal septum
A. ANTERIOR APERTURE OR NARES (NOSTRILS) open into:
   1. Vestibule, a dilated area bounded by ala and crus of nasal cartilage
B. LATERAL WALL: divisions created by the nasal conchae
   1. Sphenoethmoidal recess: above superior concha
   2. Superior meatus: below superior concha, above middle concha
   3. Middle meatus: below and lateral to middle concha, above inferior concha
   4. Inferior meatus: below and lateral to inferior concha, above palate
C. OPENINGS INTO THE MEATUSES OR RECESSES
   1. Sphenoethmoidal recess: sphenoidal sinus
   2. Superior meatus: posterior ethmoidal air cells
   3. Middle meatus: frontonasal duct anteriorly (50% of cases), accessory ostium for maxillary sinus below posterior end of middle concha, and semilunar hiatus which receives:
      a. Anteriorly: frontonasal duct (50%), anterior ethmoid air cells
      b. Posteriorly: ostium from maxillary sinus
      c. On or above bulla: middle ethmoidal air cells
   4. Inferior meatus: nasolacrimal duct
D. MEDIAL WALL: recesses, spaces, or hiatuses
   1. Nasopalatine recess: depression in septum over incisive canal
   2. Vomeronasal organ (see p. 161)
E. POSTERIOR APERTURE: 2 *choanae* opening into nasopharynx

III. Clinical considerations
A. FRACTURES OF THE NOSE are common and are usually transverse. If caused by a direct blow, the horizontal plate of the ethmoid is often fractured

NASAL CAVITY

Nasal bone

Septal cartilage

Great alar (lower) cartilage

Lateral (upper) nasal cartilage

Lesser alar cartilages

Fibro-fatty tissue

CARTILAGES OF NOSE

Great alar (lower) cartilage

Septal cartilage

Fibro-fatty tissue

CARTILAGES OF NOSE
FROM BELOW

Frontal sinus

Crista galli

Superior concha

Sphenoethmoidal recess

Middle concha

Atrium

Vestibule

Hard palate

Inferior meatus

Inferior concha

Middle meatus

Soft palate

Semilunar hiatus

Bulla ethmoidalis

Maxillary sinus opening

Opening of nasolacrimal duct

Inferior nasal concha (cut away)

LATERAL WALL

Frontal sinus & opening

Middle ethmoidal air cells (sinus) opening

Sphenoidal sinus & opening

Posterior ethmoidal air cells (sinus) opening

Superior concha

Middle concha (cut away)

Auditory (eustachian) tube opening

Infundibular opening (frontal)

Infundibulum

Hiatus semilunaris

Bulla ethmoidalis

DETAILS OF
MIDDLE MEATUS

Middle ethmoidal air cells (sinus) opening

Superior concha

Middle concha (cut)

Anterior ethmoidal sinus opening

Uncinate process (cut)

Maxillary sinus opening

Superior concha

Middle concha (cut)

Pansky

– 159 –

77. THE NOSE—PART II

I. Nasal cavity

A. ANTERIOR APERTURE: pear-shaped, bounded by nasal bone and anterior border of maxillae
B. FRAMEWORK
  1. Medial (septal) wall
    a. Septal cartilage
    b. Perpendicular plate of ethmoid
    c. Vomer
    d. Projection of other bones which join those listed above
      i. Palatine
      ii. Maxillary
      iii. Frontal
      iv. Nasal
      v. Sphenoid
  2. Lateral wall
    a. Superior and middle conchae (turbinates) of ethmoid bone
    b. Inferior concha (turbinate)
    c. Nasal bone
    d. Frontal process and nasal surface of maxilla
    e. Lacrimal bone
    f. Perpendicular plate of the palatine
    g. Medial pterygoid plate and body of sphenoid
    h. The inferior and middle conchae project medially and inferiorly, creating air passageways beneath them called the inferior and middle *meatuses*. The short superior concha conceals the superior meatus. The space posterosuperior to the superior concha into which the sphenoid sinus opens is called the *sphenoethmoidal recess*
  3. Roof
    a. Upper nasal cartilage
    b. Nasal bone
    c. Spine of frontal bone
    d. Cribriform plate of ethmoid bone
    e. Body of sphenoid bone
  4. Floor
    a. Palatine process of maxilla
    b. Horizontal plate of palatine bone
C. POSTERIOR APERTURE (CHOANA) opens into the nasopharynx

II. Clinical considerations

A. DEVIATED SEPTUM: a deflection of the septum from the midline can block or partly block the nasal passageways on the side toward which the deviation occurs
B. THE CLOSE RELATIONSHIP between the roof of the nasal cavity and the anterior cranial cavity, as well as the orbit, should be recalled when dealing with trauma, disease, or surgery of the nose
C. RHINITIS is an inflammation of the nasal mucous membrane (nasal catarrh)
D. THE NASAL POLYP is a focal submucosal thickening due to edema, which is pinkish-gray and edematous and may attain a remarkably large size
E. RHINORRHEA: a discharge from the nasal mucous membrane
  1. CSF rhinorrhea: owing to a fracture of the cribriform plate and tearing of the meninges

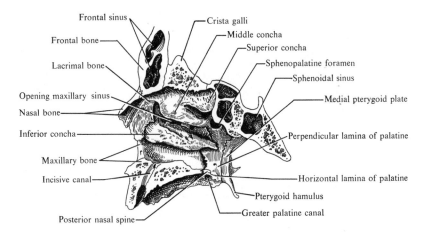

Frontal sinus — Crista galli
Frontal bone — Middle concha
Lacrimal bone — Superior concha
— Sphenopalatine foramen
Opening maxillary sinus — Sphenoidal sinus
Nasal bone — Medial pterygoid plate
Inferior concha —
— Perpendicular lamina of palatine
Maxillary bone —
Incisive canal — Horizontal lamina of palatine
— Pterygoid hamulus
— Greater palatine canal
Posterior nasal spine —

BONES OF LATERAL NASAL WALL

NASAL CAVITY

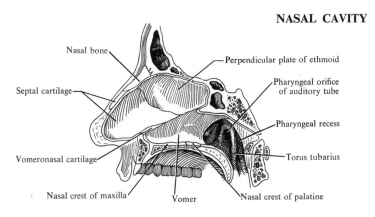

Nasal bone — Perpendicular plate of ethmoid
— Pharyngeal orifice of auditory tube
Septal cartilage —
— Pharyngeal recess
Vomeronasal cartilage —
— Torus tubarius
Nasal crest of maxilla — Nasal crest of palatine
Vomer

NASAL SEPTUM

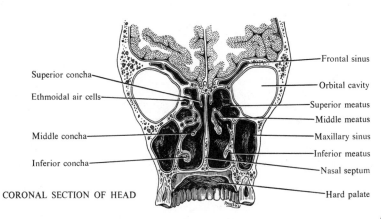

Superior concha — Frontal sinus
Ethmoidal air cells — Orbital cavity
— Superior meatus
— Middle meatus
Middle concha — Maxillary sinus
— Inferior meatus
Inferior concha — Nasal septum
CORONAL SECTION OF HEAD — Hard palate

78. MUCOUS MEMBRANE OF THE NOSE AND SINUSES

I. Mucous membrane
A. DIVISIONS: vestibule, respiratory area, olfactory area
B. STRUCTURE
1. Vestibule: skin turned in at nares with coarse hairs and sebaceous glands
2. Olfactory: over superior concha, roof, and upper third of septum; thick epithelial layer with supporting and olfactory cells (proximal ends are part of olfactory n.)
3. Respiratory: covers remainder of nasal cavity, continues into sinuses. Epithelium, pseudostratified columnar ciliated, with goblet cells

II. Paranasal sinuses (see also p. 161): air-filled extensions of nasal cavities lined with mucous membrane
A. FRONTAL
1. Location: behind supraciliary ridges of frontal bone, often extends dorsally into orbital roof, usually divided by a septum
2. Drainage: frontonasal duct, through rostral ethmoidal cells to middle meatus
B. ETHMOIDAL
1. Location: aggregations of thin-walled spaces (cells) in ethmoidal labyrinth between orbit and nasal cavities, arranged in 3 sets
2. Drainage
a. Anterior, into infundibulum of middle meatus
b. Middle, on or above ethmoid bulla (formed by middle ethmoidal cells) in middle meatus
c. Posterior, into superior meatus, with interconnections to sphenoidal sinus
C. SPHENOIDAL
1. Location: in body of sphenoid, usually asymmetrically divided by a septum
2. Drainage: into sphenoethmoidal recess
D. MAXILLARY (largest)
1. Location: in body of maxilla; roofed by orbit, wall of nasal cavity is medial; the alveolar process is lateral, and the sinus extends into the zygoma
2. Drainage: into semilunar hiatus of middle meatus; accessory opening frequently found

III. Blood and nerve supply of paranasal sinuses
A. SUPRAORBITAL ARTERY and nerve are major suppliers of frontal sinus
B. ANTERIOR AND POSTERIOR ETHMOIDAL vessels and nerves and orbital branches of pterygopalatine ganglion supply ethmoidal sinuses
C. POSTERIOR ETHMOIDAL VESSELS and nerves and orbital branches of pterygopalatine ganglion supply the sphenoidal sinus
D. BRANCHES of the ant., middle, and post. superior alveolar and infraorbital nn. supply the maxillary sinus. Vessels are the facial, infraorbital, and greater palatine

IV. Clinical considerations
A. STRUCTURE OF MUCOUS MEMBRANE, with its motile cilia, glands, and rich blood supply is adapted to purifying, moistening, and warming air in order to protect lungs
B. SINUSES EFFECTIVE IN MAKING HEAD LIGHTER but frequently cause difficulty because:
1. Numerous connections to nasal cavities lead to easy infection
2. When sinuses are infected, swelling of nasal mucosa around orifices slows drainage
3. In maxillary and sphenoid, the lowest portions of the sinus lie below opening into nose, thus making complete drainage difficult
4. Because of proximity of roots of upper teeth to maxillary sinus, it is sometimes difficult to differentiate between toothache and sinusitis
5. Due to the thin bone between many of the sinuses and meninges, there is a chance of infecting the latter
C. SINUSITIS is an inflammation of the mucosa of one or more accessory nasal sinuses and may be acute or chronic

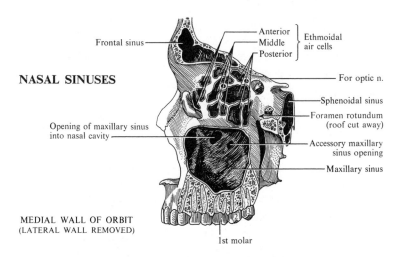

NASAL SINUSES

Frontal sinus

Anterior
Middle
Posterior
} Ethmoidal
air cells

For optic n.

Sphenoidal sinus

Foramen rotundum
(roof cut away)

Opening of maxillary sinus
into nasal cavity

Accessory maxillary
sinus opening

Maxillary sinus

MEDIAL WALL OF ORBIT
(LATERAL WALL REMOVED)

1st molar

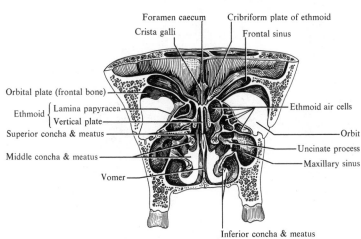

Foramen caecum

Crista galli

Cribriform plate of ethmoid

Frontal sinus

Orbital plate (frontal bone)

Ethmoid { Lamina papyracea
Vertical plate

Superior concha & meatus

Middle concha & meatus

Vomer

Ethmoid air cells

Orbit

Uncinate process

Maxillary sinus

Inferior concha & meatus

FRONTAL SECTION THRU NASAL CAVITY

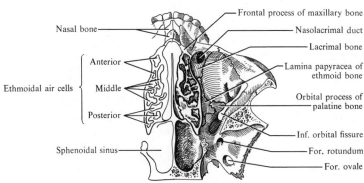

Nasal bone

Ethmoidal air cells {
Anterior
Middle
Posterior

Sphenoidal sinus

Frontal process of maxillary bone

Nasolacrimal duct

Lacrimal bone

Lamina papyracea of
ethmoid bone

Orbital process of
palatine bone

Inf. orbital fissure

For. rotundum

For. ovale

NASAL SINUSES SEEN FROM ABOVE (ALSO FLOOR OF ORBIT)

Panolty

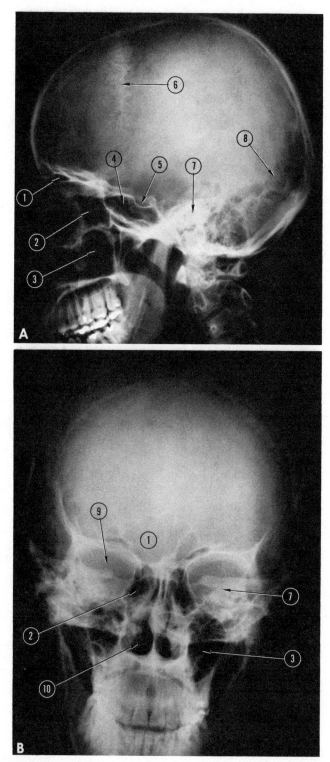

FIGURE 7. **Normal skull. A, Lateral view; B, anterior view.** *1*, Frontal sinus; *2*, ethmoid sinus; *3*, maxillary sinus; *4*, sphenoid sinus; *5*, sella turcica; *6*, coronal suture; *7*, petrous temporal bone; *8*, lambdoidal suture; *9*, orbit; *10*, nasal cavity.

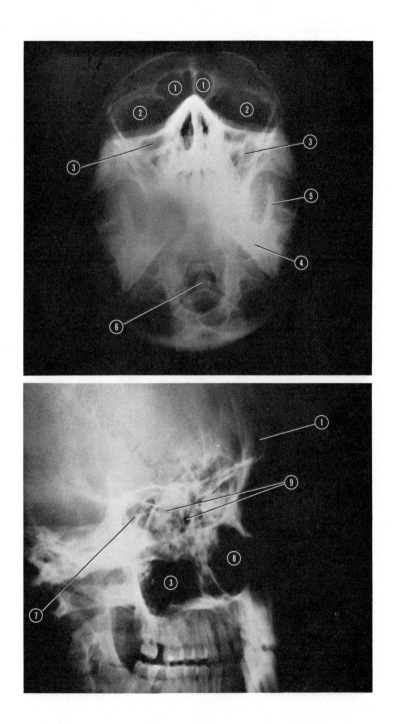

FIGURE 8. **Paranasal sinuses.** *1*, Frontal sinus; *2*, orbits; *3*, maxillary sinuses; *4*, mandible; *5*, coronoid process of mandible; *6*, dens of axis; *7*, sphenoid sinus; *8*, nasal cavity; *9*, ethmoidal sinuses.

79. ARTERIES OF THE NOSE

I. Anterior ethmoidal
A. ORIGIN: ophthalmic
B. COURSE: from orbit through anterior ethmoid canal with nasal branches that pass through groove near crista galli
C. DISTRIBUTION: to both lateral wall and septum; external branch to dorsolateral external surface of nose (external nasal artery); ant. meningeal br.; also ant. and middle ethmoidal sinuses

II. Posterior ethmoidal
A. ORIGIN: ophthalmic
B. COURSE: from orbit through posterior ethmoid canal with nasal branches through cribriform plate of ethmoid
C. DISTRIBUTION: anastomoses with branches of sphenopalatine artery and distributed with them—post. ethmoidal sinuses and lateral wall and septum of nose

III. Sphenopalatine
A. ORIGIN: maxillary, third part
B. COURSE: leaves pterygopalatine fossa and enters nose through sphenopalatine foramen
C. DISTRIBUTION
1. Posterior lateral nasal branches: conchae, meatuses
2. Posterior septal branch: to nasal septum

IV. Greater palatine
A. ORIGIN: maxillary, third part
B. COURSE: leaves pterygopalatine fossa through pterygopalatine canal, runs in roof of mouth forward to incisive canal where terminal branch ascends to septum
C. DISTRIBUTION: to septum with septal branch of III, above

V. Septal and alar branches
A. ORIGIN: superior labial branch of facial
B. COURSE: cephalically from upper lip
C. DISTRIBUTION: to rostral inferior septum and vestibule
1. Source of nosebleeds, particularly in children

VI. Clinical considerations
A. VASCULAR BED OF NASAL MUCOSA: vascularity is extremely great. Thus:
1. Injuries to the membrane lead to profuse bleeding, especially where the arteries are large as they enter the nose. When severe, as at the posterior end of the middle concha, ligation of the external carotid may be necessary. When in the roof of the nose, intraorbital ligation of the ethmoidal arteries may be needed
2. Irritating substances, either infectious or allergenic agents, which could lead to either engorgement or increased capillary permeability, will cause great swelling and occlusion of the air passageways
B. THE ARTERIES OF THE NOSE run, in general, with the larger nerves
C. IN THE ANTERIOR LOWER PORTION of the nasal septum there are broad anastomoses between the major arteries of the nose, and this area often is involved when there is a nosebleed (epistaxis)
D. BLEEDING close to the back end of the middle concha is usually from the sphenopalatine artery and can often be checked by packs placed here (if severe, the artery is ligated); bleeding from the roof of the nasal cavity originates from one of the ethmoidal arteries

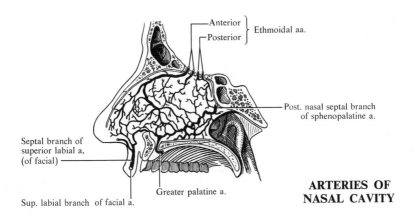

Anterior ⎫
Posterior ⎬ Ethmoidal aa.
⎭

Post. nasal septal branch
of sphenopalatine a.

Septal branch of
superior labial a.
(of facial)

Greater palatine a.

Sup. labial branch of facial a.

**ARTERIES OF
NASAL CAVITY**

ARTERIES OF NASAL SEPTUM

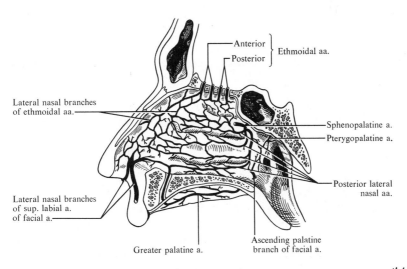

Anterior ⎫
Posterior ⎬ Ethmoidal aa.
⎭

Lateral nasal branches
of ethmoidal aa.

Sphenopalatine a.

Pterygopalatine a.

Posterior lateral
nasal aa.

Lateral nasal branches
of sup. labial a.
of facial a.

Greater palatine a.

Ascending palatine
branch of facial a.

ARTERIES OF LATERAL NASAL WALL

Pansky

80. NERVES OF THE NASAL CAVITY

I. Special sensory

A. OLFACTORY (I): nerve of smell
   1. Receptors: the neuroepithelial cells of the olfactory mucosa lie on the upper third of the nasal septum and on the superior nasal concha. Are a part of nerve
   2. Course: the fibers are gathered into bundles that pass through the cribriform plate of the ethmoid to enter the olfactory bulb

II. General sensory

A. OPHTHALMIC DIVISION OF THE TRIGEMINAL (V) NERVE, with its cell bodies in the trigeminal (semilunar) ganglion
   1. Nasociliary branch gives off an anterior ethmoid nerve, which leaves the orbit through the anterior ethmoid canal and sends nasal branches through the cleft near the crista galli to the anterior part of the nasal septum and lateral wall; also gives cutaneous innervation to the tip of the nose (external nasal branch)

B. MAXILLARY DIVISION OF THE TRIGEMINAL (V) NERVE, with its cell bodies in the trigeminal (semilunar) ganglion
   1. Pterygopalatine nerves
      a. Greater palatine nerve gives off posterior inferior nasal branches as this nerve lies in the pterygopalatine canal. Distributed to inferior concha, inferior and middle meatuses
      b. Posterior superior nasal branches enter nose through the sphenopalatine foramen to the superior and middle conchae
         i. Nasopalatine nerve runs over septum to the incisive canal
   2. Anterior superior alveolar nerve runs to the rostral part of the inferior meatus and the floor of the nasal cavity

III. Motor: autonomic system distributed to the glands and blood vessels of the mucous membrane

A. CRANIOSACRAL DIVISION (PARASYMPATHETIC)
   1. Preganglionic fibers, arising in the superior salivatory nucleus of the pons via the nervus intermedius and greater petrosal nerve to the pterygopalatine ganglion where they terminate
   2. Postganglionic fibers arise from cells in the pterygopalatine ganglion and are distributed with the pterygopalatine nerves

B. THORACOLUMBAR (SYMPATHETIC)
   1. Preganglionic fibers arise in the intermediolateral gray column of the upper thoracic cord and ascend in the cervical sympathetic trunk to terminate in the superior cervical ganglion
   2. Postganglionic fibers arise from cells in the superior cervical ganglion, ascend to the carotid plexus, and form the deep petrosal nerve, which joins the greater petrosal nerve to form the nerve of the pterygoid canal (vidian). The sympathetic fibers of this nerve pass through the pterygopalatine ganglion without synapse and are distributed as given above under IIIA, 2. It is thought that the effect of activity of this system is due to vasoconstriction of the arteries

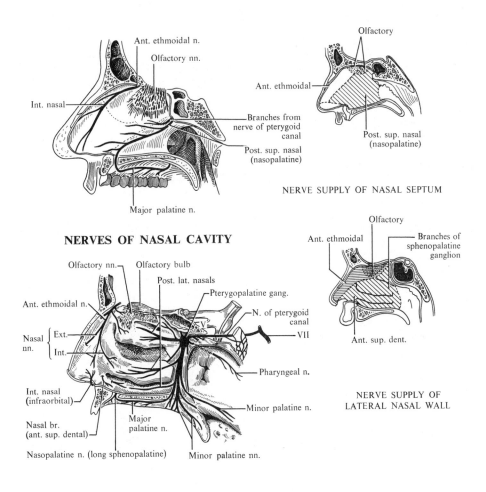

Ant. ethmoidal n.

Olfactory nn.

Int. nasal

Branches from nerve of pterygoid canal

Post. sup. nasal (nasopalatine)

Major palatine n.

Olfactory

Ant. ethmoidal

Post. sup. nasal (nasopalatine)

NERVE SUPPLY OF NASAL SEPTUM

NERVES OF NASAL CAVITY

Olfactory nn.

Olfactory bulb

Post. lat. nasals

Pterygopalatine gang.

Ant. ethmoidal n.

N. of pterygoid canal

Nasal nn. { Ext. Int.

V

VII

Pharyngeal n.

Int. nasal (infraorbital)

Minor palatine n.

Nasal br. (ant. sup. dental)

Major palatine n.

Nasopalatine n. (long sphenopalatine)

Minor palatine nn.

Olfactory

Ant. ethmoidal

Branches of sphenopalatine ganglion

Ant. sup. dent.

NERVE SUPPLY OF LATERAL NASAL WALL

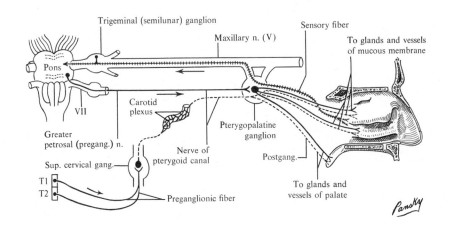

Trigeminal (semilunar) ganglion

Maxillary n. (V)

Sensory fiber

To glands and vessels of mucous membrane

Pons

Carotid plexus

Pterygopalatine ganglion

VII

Greater petrosal (pregang.) n.

Nerve of pterygoid canal

Postgang.

Sup. cervical gang.

T1
T2

Preganglionic fiber

To glands and vessels of palate

Pansky

81. SUMMARY OF THE CRANIAL NERVES

Name	Nuclei of Origin and Termination	Distribution	Function
I. Olfactory (sensory)	Central or deep process of olfactory bulb	Nasal mucous membranes	Sense of smell
II. Optic (sensory)	Ganglionic cells of retina	Retina of eye	Sense of sight
III. Oculomotor (motor)	Nucleus in floor of cerebral aqueduct	Sup., inf., and med. recti; inf. oblique; ciliaris; sphincter pupillae mm.	Motion
IV. Trochlear (motor)	Nucleus in floor of cerebral aqueduct	Superior oblique of eye	Motion
V. Trigeminal (mixed)	Fibers of sensory root arise from semilunar ganglion	1. *Ophthalmic* to cornea; ciliary body; iris; lacrimal gl.; conjunct.; mucous membrane of nasal cavity; skin of forehead, eyelid, eyebrow, and nose	Sensation
		2. *Maxillary* to dura, forehead, lower eyelid, lat. angles of orbit, upper lip, gums and teeth of upper jaw, mucous membrane, and skin of cheek and nose	Sensation
	Fibers of motor root arise from nucleus in pons	3. *Mandibular* to temple, auricle of ear, lower lip, lower part of face, teeth and gums of mandible; muscles of mastication; to mucous membrane of ant. part of tongue	Sensation and motion (some of VII reach tongue via lingual for taste)
VI. Abducens (motor)	Nucleus beneath floor of fourth ventricle	Lateral rectus muscle of eye	Motion

Name	Nuclei of Origin and Termination	Distribution	Function
VII. Facial (mixed)	Sensory from geniculate ganglion on nerve	Ant. 2/3 tongue (taste) via *chorda tympani* Middle ear	Taste
	Motor from nucleus in lower part of pons Sup. salivatory nucleus	To muscles of face, scalp, auricle, and superfic. neck. Stimulatory or excitatory to subman. and sublingual gls.	Gen. sense Secretion
VIII. Acoustic (sensory): 2 sets	Cochlear from bipolar cells in spiral gang. of cochlea	To organ of Corti	Sense of hearing
	Vestibular from bipolar cells in vestibular gang.	To semicircular canals	Sense of equilib.
IX. Glossopharyngeal (mixed)	Sensory fibers from sup. and inf. gang. on trunk of nerve Nucleus ambiguus and inf. salivatory nucleus	To mucous memb. of fauces, tonsils, pharynx, and post. 1/3 tongue To mm. of pharynx and secretory fibers to parotid gland	Sense of taste and gen. sense Motion Secretion
X. Vagus (mixed)	Sensory fibers from sup. gang. and inf. gang. (nodosum) on nerve trunk	To mucous memb. of larynx, trachea, lungs, esophagus, stomach, intestines, and gallbl.	Sensation
	Motor fibers from nucleus ambiguus in medulla and dorsal motor nucleus	To larynx, esophagus, stomach, sm. intestine, and part of lg. intest. Excitatory fibers to gastric and pancreatic glands	Motion Secretion
XI. Accessory (2 parts: cranial, spinal)	Cranial: from n. ambiguus	To pharyngeal and laryngeal brs. of vagus to pharynx and larynx	Motion
	Spinal: from spinal cord as low as C5	To sternocleidomastoic and trapezius mm.	Motion
XII. Hypoglossal (motor)	From hypoglossal nucleus in medulla	To muscles of tongue	Motion

82. HEAD, VISCERAL ARCHES, AND DETAILED NECK ANATOMY

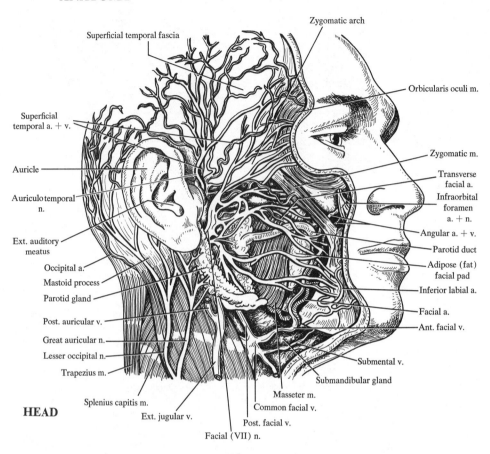

Zygomatic arch

Superficial temporal fascia

Orbicularis oculi m.

Superficial temporal a. + v.

Zygomatic m.

Transverse facial a.

Infraorbital foramen a. + n.

Auricle

Auriculotemporal n.

Angular a. + v.

Parotid duct

Adipose (fat) facial pad

Ext. auditory meatus

Inferior labial a.

Occipital a.

Mastoid process

Parotid gland

Facial a.

Ant. facial v.

Post. auricular v.

Great auricular n.

Lesser occipital n.

Trapezius m.

Submental v.

Submandibular gland

Masseter m.

Splenius capitis m.

Common facial v.

Ext. jugular v.

Post. facial v.

Facial (VII) n.

HEAD

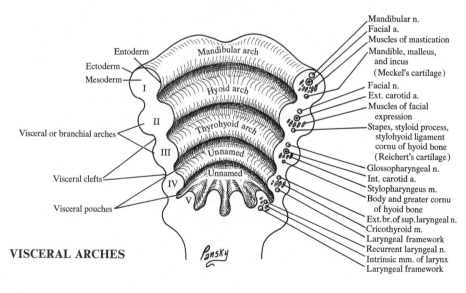

Mandibular n.
Facial a.
Muscles of mastication
Mandible, malleus, and incus (Meckel's cartilage)

Entoderm

Mandibular arch

Ectoderm

Mesoderm

I

Facial n.
Ext. carotid a.
Muscles of facial expression
Stapes, styloid process, stylohyoid ligament cornu of hyoid bone (Reichert's cartilage)

Hyoid arch

Visceral or branchial arches

II

Thyrohyoid arch

III

Unnamed

Glossopharyngeal n.
Int. carotid a.
Stylopharyngeus m.
Body and greater cornu of hyoid bone
Ext. br. of sup. laryngeal n.
Cricothyroid m.
Laryngeal framework
Recurrent laryngeal n.
Intrinsic mm. of larynx
Laryngeal framework

Visceral clefts

Unnamed

IV

V

Visceral pouches

VISCERAL ARCHES

Pansky

-172-

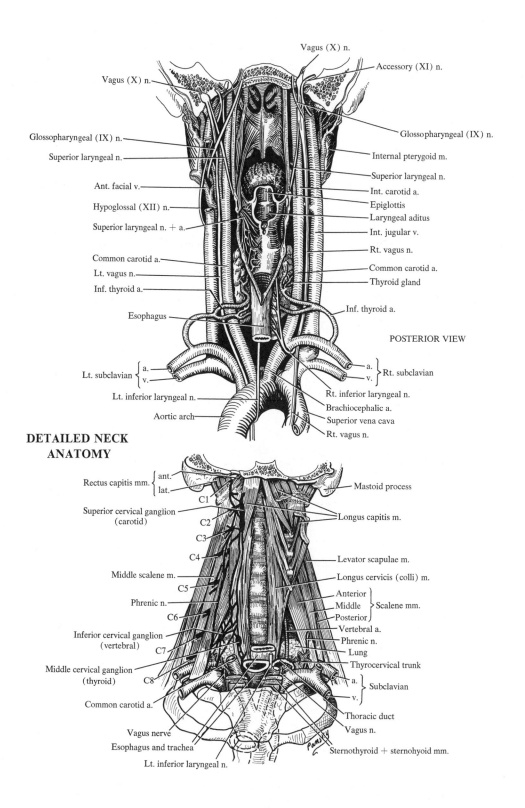

Vagus (X) n.

Accessory (XI) n.

Vagus (X) n.

Glossopharyngeal (IX) n.

Superior laryngeal n.

Ant. facial v.

Hypoglossal (XII) n.

Superior laryngeal n. + a.

Common carotid a.

Lt. vagus n.

Inf. thyroid a.

Esophagus

Glossopharyngeal (IX) n.

Internal pterygoid m.

Superior laryngeal n.

Int. carotid a.

Epiglottis

Laryngeal aditus

Int. jugular v.

Rt. vagus n.

Common carotid a.

Thyroid gland

Inf. thyroid a.

POSTERIOR VIEW

Lt. subclavian { a. v.

Lt. inferior laryngeal n.

Aortic arch

a. v. } Rt. subclavian

Rt. inferior laryngeal n.

Brachiocephalic a.

Superior vena cava

Rt. vagus n.

**DETAILED NECK
ANATOMY**

Rectus capitis mm. { ant. lat.

Superior cervical ganglion
(carotid)

Middle scalene m.

Phrenic n.

Inferior cervical ganglion
(vertebral)

Middle cervical ganglion
(thyroid)

Common carotid a.

Vagus nerve

Esophagus and trachea

Lt. inferior laryngeal n.

C1

C2

C3

C4

C5

C6

C7

C8

Mastoid process

Longus capitis m.

Levator scapulae m.

Longus cervicis (colli) m.

Anterior

Middle } Scalene mm.

Posterior

Vertebral a.

Phrenic n.

Lung

Thyrocervical trunk

a. v. } Subclavian

Thoracic duct

Vagus n.

Sternothyroid + sternohyoid mm.

– 173 –

83. COMPUTERIZED TOMOGRAPHIC SCANNING (CT SCAN)

I. Definition:* computerized tomographic (CT) scanning is a noninvasive means for imaging sections of the body

  A. A THIN X-RAY BEAM is projected through the area to be scanned, and the emerging rays are picked up by sensitive detectors which are constructed of solid crystal or gases

    1. This is unlike conventional x-rays which use photographic film

  B. THE X-RAYS ARE CONVERTED into digits by a computer, which "reconstructs" images of the body sections

    1. The images are displayed on a cathode-ray tube (or another display system) and are photographed

II. Radiation: the amount picked up by the detector varies with the absorption or attenuation by structures in the body

  A. VERY DENSE MATERIALS, such as bone (containing calcium), absorb a great deal

  B. AIR AND WATER absorb barely any radiation

  C. THE STANDARD NUMERICAL scale for recording absorption (CT attenuation values) uses water as a zero point

    1. Extreme positive or negative absorption is expressed as $+1000$ or -1000 Hounsfield units

  D. RADIATION DOSAGE from a complete series of skull CT scans is about 4 rads (equal to 6 routine skull series radiographs)

III. Common indicators for CT scanning: dysphasia, diplopia, facial pain (atypical spasm), motor or sensory deficits of nonspinal origin, ataxia (incoordination, new tremor, etc.), neurosensory deafness or unexplained vertigo, visual loss (transient, permanent, field defects), epilepsy (of any kind), blackout (drop attack, seizure, etc.), headache with atypical migraine or focal features, and dementia of unknown cause

IV. General indications for CT scanning

  A. ANY FOCAL NEUROLOGIC SIGNS, head trauma with unconsciousness, focal signs or persistent headache, suspected cerebral metastases, or brain abscesses

V. Abnormalities are detected by observing

  A. DISTORTION AND DISPLACEMENT of normal structures

  B. DISTENTION OF THE VENTRICLES

  C. DESTRUCTION OF NORMAL TISSUE

  D. DIFFERENCES IN DENSITY between normal and abnormal tissue (present before or after intravenous injection of iodine contrast material)

  E. INCREASED DENSITY may be due to abnormal vascularity or breakdown of the blood-brain barrier (normally prevents intravascular material from entering the brain substance)

---

*For additional references, See Appendix IV, p. 565.

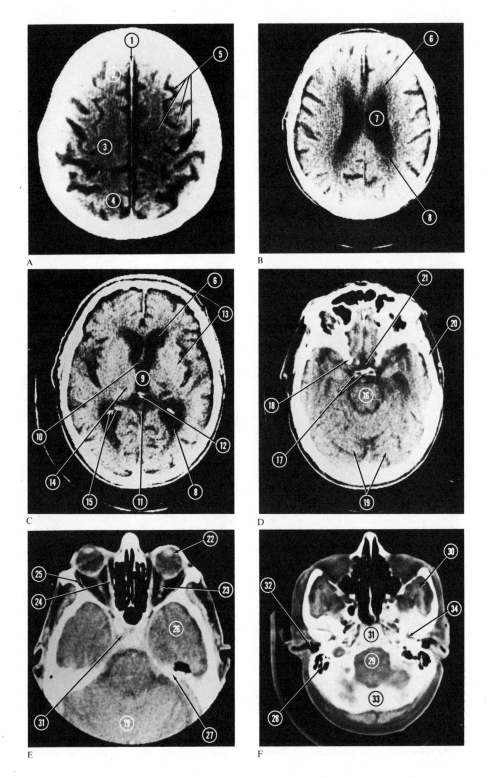

CT scans through head (A–F, top to bottom). *1*, Falx cerebri; *2*, frontal lobe; *3*, parietal lobe; *4*, occipital lobe; *5*, gyri and sulci; *6, 7,* and *8*, anterior, body, and posterior horns of lateral ventricle; *9*, 3rd ventricle; *10*, interventricular foramen; *11*, 4th ventricle; *12*, pineal; *13*, insula; *14*, thalamus; *15*, midbrain; *16*, posterior clinoid processes and dorsum sellae; *17*, anterior clinoid process; *18*, cerebellum; *19*, zygomatic arch; *20*, hypophyseal fossa; *21*, eyeball; *22*, optic nerve; *23*, medial rectus m.; *24*, lateral rectus m.; *25*, temporal lobe; *26*, petrous temporal; *27*, middle ear; *28*, foramen magnum; *29*, maxillary sinus; *30*, body of sphenoid; *31*, ext. acoustic meatus; and *32*, occipital bone

84. NUCLEAR MAGNETIC RESONANCE (NMR)

I. **Characteristics:** the image is basically anatomic, comparable to the CT scan, but can discriminate more sensitively between normal and pathologic tissue. X-ray absorption properties, on the other hand, are similar to surrounding tissues and are undetectable unless large differences are present. NMR spatial relationships are not often as fine as CT images, but contrast between gray and white matter of the brain is greater

A. NMRs DO NOT CARRY the risk of physiologic harm as do x-rays, which expose the patient to ionizing radiation

B. NMRs CAN DISCRIMINATE AMONG overlapping structures (deficiency of radiography)

C. NMR SPECTROSCOPY IS BASED on the ability to elucidate the conformation of organic molecules and present it in a pictorial form by generating tomographic images of body structures, providing a new diagnostic tool

II. **Basic fundamentals of how it functions**

A. ATOMIC NUCLEI have an angular momentum arising from their inherent property of rotation or spin

1. The spin, being electrically charged, corresponds to a current flowing on a spin axis which generates a small magnetic field

2. Only nuclei with an odd number of protons or neutrons have a net spin and lend themselves to NMR spectroscopy

3. The magnetic moments or dipoles of the nuclei with spin are pointed in random directions, but orient themselves, in a magnetic field, with the field's lines of induction or lines of force

   a. The magnetic behavior of the entire population of nuclei is predictable and can be controlled in a magnetic field

B. A ROTATING MAGNETIC FIELD is applied by surrounding the "sample" or "patient" with a coil connected to a source of radiofrequency power

1. The frequency of the applied electromagnetic radiation must match the natural precessional frequency of the nuclei, thus, the term NMR

   a. The frequencies used are in the radiofrequency band of the electromagnetic spectrum, far below those of x-rays, or even visible light, and do not disrupt the molecules of living systems

C. THE PATIENT IS INSERTED in a body-size chamber, in a large magnet, which also contains radiofrequency coils

1. He is immersed in a magnetic field of up to 3000 gauss, which causes a magnetic alignment of the H^+ nuclei in the body, as well as other nuclei with an odd number of protons or neutrons (thus, a net spin)

2. When subjected to energy of the correct phase and radiofrequency, the spins of these nuclei change direction and absorb energy in the process, depending on the specific type of nucleus and magnetic field strength

3. The energy is subsequently reemitted at a characteristic radiofrequency, and the signal is picked up by a receiver

   a. Signal characteristics depend on the strength of the field and density of the nuclei in the plane of the body and correspond to the particular field (number of H^+ or other nuclei present)

D. DATA FROM THE RECEIVER are fed into a computer that plots the distribution of the nuclei responsible for the emission and shows the abundance of nuclei, as well as the molecular environment in which they reside

1. Resolution is achieved in the 2–4-mm range, improving on standard CT images

2. NMR is inherently a 3-dimensional phenomenon, and signals are obtained from the total volume of material enclosed in the transmitter and receiver coils

3. NMR imaging systems can be designed to receive data from a single point, from a line, from a plane, and from a complete 3-dimensional volume all at once, which can then be converted into a composite picture

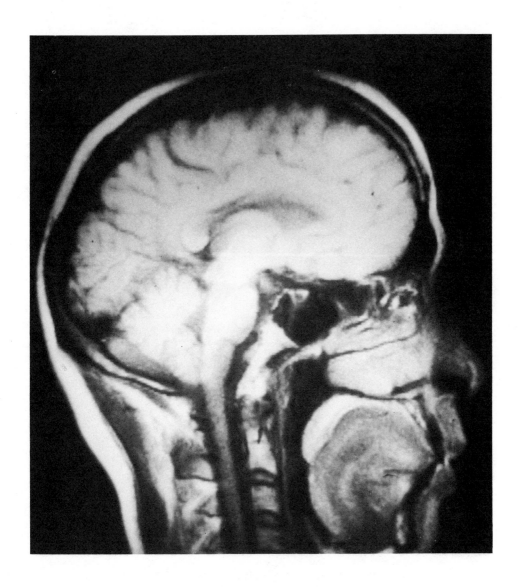

Nuclear magnetic resonance (NMR). Sagittal slice of cranial anatomy acquired using a saturation recovery pulse technique at a 5.0 kilogauss magnetic field strength. (NMR image courtesy of Technicare Corporation, Cleveland, Ohio.)

UNIT TWO

Back

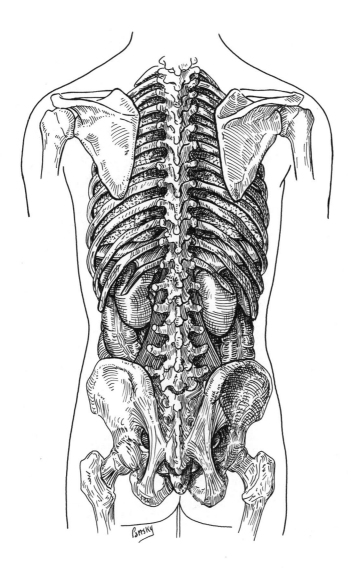

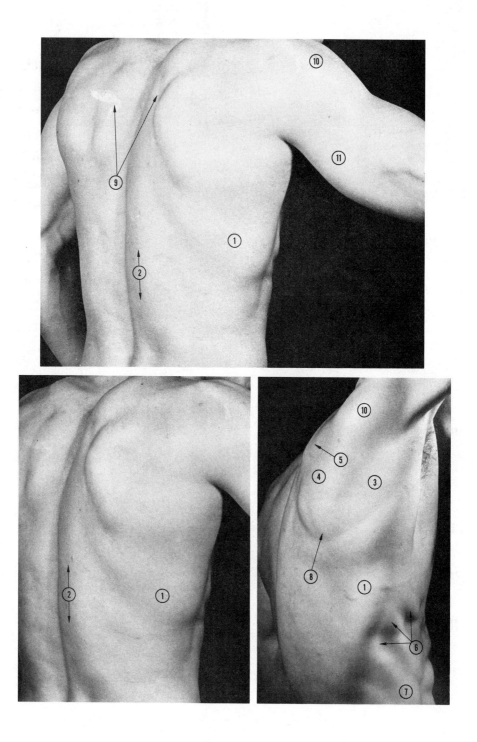

FIGURE 9. **Surface anatomy of back.** *1*, Latissimus dorsi; *2*, erector spinae; *3*, teres major m.; *4*, infraspinatus m.; *5*, supraspinatus m.; *6*, serratus anterior m.; *7*, external oblique m.; *8*, inferior angle of scapula; *9*, trapezius m.; *10*, deltoid m.; *11*, triceps brachii m.

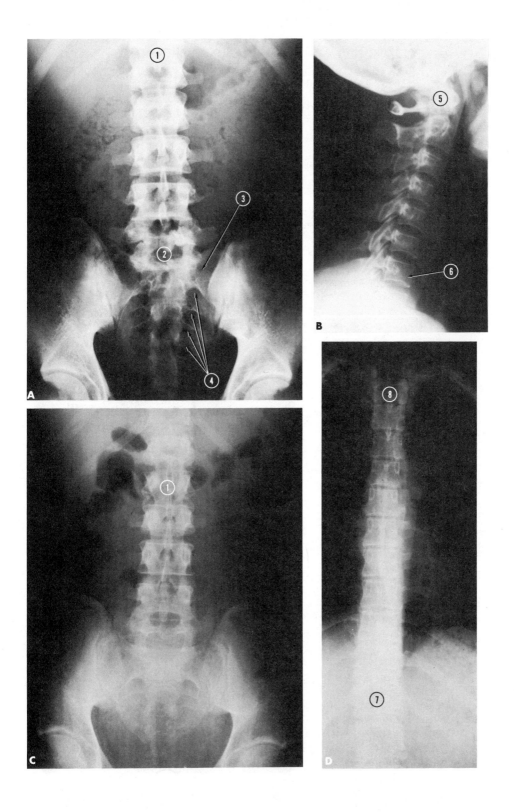

FIGURE 10. **Vertebral column. A, Lumbar flat plate; B, cervical spine; C, lumbar projection, upright film; D, thoracic spine.** *1*, First lumbar vertebra; *2*, 5th lumbar vertebra; *3*, sacrum; *4*, anterior sacral foramina; *5*, atlas; *6*, 7th cervical vertebra; *7*, 12th thoracic vertebra; *8*, 2nd thoracic vertebra.

85. SPINAL NERVES AND CUTANEOUS NERVES OF THE BACK

I. Spinal nerves: arranged in 31 pairs grouped regionally: 8 cervical, 12 thoracic, 5 lumbar, 5 sacral, and 1 coccygeal

A. ATTACHMENTS TO SPINAL CORD
 1. Posterior root: numerous rootlets along posterolateral sulcus
  a. Has an oval swelling composed of nerve cell bodies—*posterior root ganglion*
 2. Anterior root: rootlets from anterolateral sulcus
  a. White ramus: preganglionic sympathetic fibers from root to sympathetic trunk
B. BEYOND GANGLION, roots join to form nerve trunk
C. BRANCHES OF SPINAL NERVE TRUNK
 1. Gray ramus: on all spinal nerves; contains postganglionic sympathetic fibers
 2. Terminal branches: posterior and anterior primary divisions

II. Distribution of primary divisions

A. POSTERIOR PRIMARY DIVISION: all, except C1, S4, S5, and coccygeal have lateral and medial branches
 1. C1: entire dorsal division supplies suboccipital muscles
 2. C2: medial branch large, *great occipital nerve:* lateral branch, muscular
 3. C3: medial branch to lower part of back of head; lateral branch, muscular
 4. C4–C8: medial branches of C4 and 5 muscular and cutaneous; lateral branch is muscular; below C5, both divisions muscular
 5. T1–T12: medial branches of upper 6 nerves to skin; lateral branches muscular. Medial branches of lower 6 nerves muscular; lateral branches cutaneous
 6. L1–L5: medial branches, all muscular; lateral branches, L1–L3 to buttock as superior cluneal nerves; other lateral branches muscular
 7. S1–S3: medial branches all muscular; lateral branches to skin on buttock
 8. S4, S5, and coccygeal: posterior division to skin over coccyx
B. ANTERIOR PRIMARY DIVISIONS
 1. C1–C4, C5–T1, L1–L4, and L4–S3 form plexuses
 2. Thoracic nerves: T1–T11 *intercostal,* T12 *subcostal*
  a. T1 has larger and smaller parts; larger to brachial plexus; small, first intercostal nerve
  b. T2–T6: in intercostal spaces are thoracic intercostal nerves. Have muscular and cutaneous branches
   i. Lateral cutaneous: to lateral part of chest and back. From T2: to arm, as *intercostobrachial nerve*
   ii. Anterior cutaneous: termination of intercostal nerves. Supply medial chest
  c. T6–T11 continue to abdomen as *thoracoabdominal nerves.* Have same arrangement as T2–T6
  d. T12: below twelfth rib, the subcostal nerve
   i. Lateral branch to anterolateral gluteal region

III. Special features

A. NOT ALL THE DORSAL RAMI OF SPINAL NERVES have cutaneous branches, and there is some variation in distribution
 1. The first cervical nerve usually has no cutaneous branch
 2. The cutaneous branches of the 2nd and 3rd cervical nerves are distributed upward to the scalp
 3. The medial branches of the 4th and 5th cervical nerves typically reach skin on the back of the neck, while the next 3 typically have no cutaneous branches
 4. All the posterior rami of the thoracic nerves typically have cutaneous branches

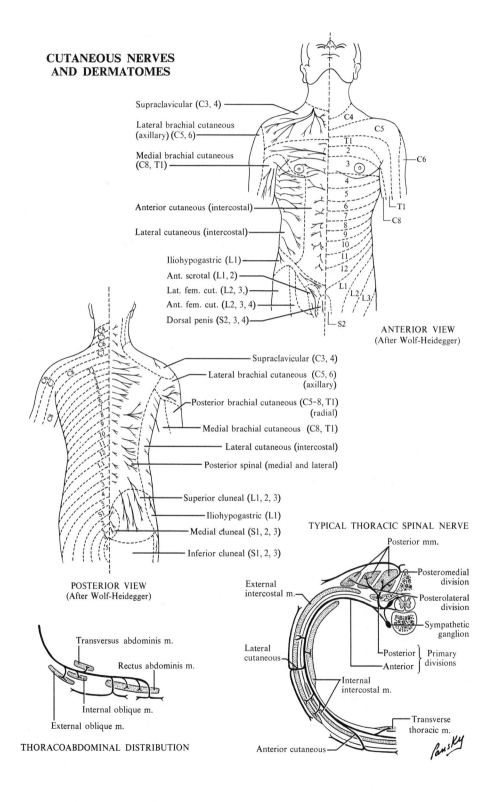

CUTANEOUS NERVES
AND DERMATOMES

Supraclavicular (C3, 4)

Lateral brachial cutaneous
(axillary) (C5, 6)

Medial brachial cutaneous
(C8, T1)

Anterior cutaneous (intercostal)

Lateral cutaneous (intercostal)

Iliohypogastric (L1)

Ant. scrotal (L1, 2)

Lat. fem. cut. (L2, 3,)

Ant. fem. cut. (L2, 3, 4)

Dorsal penis (S2, 3, 4)

C4

C5

T1

2

3

4

5

6

7

8

9

10

11

12

L1

L2

L3

C6

T1

C8

S2

ANTERIOR VIEW
(After Wolf-Heidegger)

Supraclavicular (C3, 4)

Lateral brachial cutaneous (C5, 6)
(axillary)

Posterior brachial cutaneous (C5-8, T1)
(radial)

Medial brachial cutaneous (C8, T1)

Lateral cutaneous (intercostal)

Posterior spinal (medial and lateral)

Superior cluneal (L1, 2, 3)

Iliohypogastric (L1)

Medial cluneal (S1, 2, 3)

Inferior cluneal (S1, 2, 3)

POSTERIOR VIEW
(After Wolf-Heidegger)

Transversus abdominis m.

Rectus abdominis m.

Internal oblique m.

External oblique m.

THORACOABDOMINAL DISTRIBUTION

TYPICAL THORACIC SPINAL NERVE

Posterior mm.

Posteromedial
division

Posterolateral
division

Sympathetic
ganglion

External
intercostal m.

Lateral
cutaneous

Internal
intercostal m.

Transverse
thoracic m.

Anterior cutaneous

Posterior
Anterior

Primary
divisions

Pansky

86. THE VERTEBRAL COLUMN

I. Composition: 32–34 vertebrae; 7 cervical, 12 thoracic, 5 lumbar, 5 sacral, 3–5 coccygeal

II. Length
- A. IN MALE: 71 cm
- B. IN FEMALE: 61 cm

III. Curvatures
- A. CERVICAL: CONVEX ANTERIORLY, from apex of odontoid process to T2
- B. THORACIC: CONCAVE ANTERIORLY, from middle T2 to middle T12
- C. LUMBAR: CONVEX ANTERIORLY, from middle T12 to sacrovertebral articulation; convexity greatest at lower 3 segments
- D. SACRAL: CONCAVE ANTEROCAUDALLY
- E. AT BIRTH, only the thoracic and sacral curves (primary) are present. The secondary curves develop after birth: cervical from elevation and extension of head in infancy; lumbar from assumption of erect position when child begins to walk

IV. Surfaces
- A. ANTERIOR: in general, width gradually broadens with widest point at base of sacrum, then tapers sharply to apex at tip of coccyx
- B. POSTERIOR: spinous processes in midline
  1. Spinous processes
     a. Cervical, except for 2 and 7, are short and horizontal, have bifid extremity
     b. Thoracic: long, upper spines directed obliquely caudally with some separation; middle spines, almost vertical in direction with overlapping; lower spines, shorter and directed almost posteriorly with separation between them
     c. Lumbar: short, thick, and separated by interval
  2. Vertebral groove, on either side of spinous processes
     a. Formed by lamina and transverse processes; shallow in cervical and lumbar areas; deep in thoracic area
  3. The articular processes are lateral to the groove
  4. The transverse processes are the most lateral
- C. LATERAL: separated from posterior surface by articular processes in cervical and lumbar areas, transverse processes in thoracic region
  1. Anteriorly: sides of bodies of vertebrae show articular facets for ribs in thorax
  2. Intervertebral foramina formed by the opposition of notches in adjoining vertebrae
     a. Increase in size from above downward

V. Vertebral canal consists of the foramina of adjoining vertebrae

VI. Special features
- A. IN THE MIDTHORACIC REGION, the tips of the spines are below the level of the bodies of the corresponding vertebrae
- B. THE TRANSVERSE PROCESSES are in front of the articular processes in the cervical region and in line with the intervertebral foramina, in the thoracic region they are behind both, and in the lumbar region they are behind the foramina but in front of the articular processes

VII. Clinical considerations
- A. ABNORMAL CURVATURE
  1. Kyphosis: increased backward curvature of spine (mostly in thorax)
  2. Lordosis: an exaggerated forward curvature of the lumbar segment
  3. Scoliosis: commonest. Lateral deviation most frequent in thorax

VERTEBRAL COLUMN

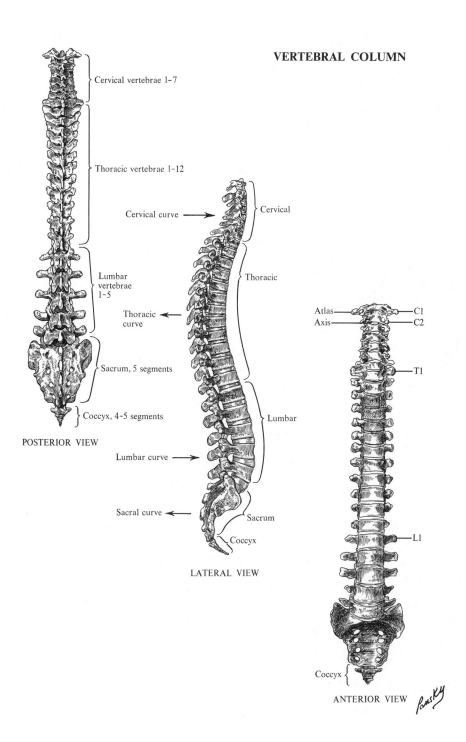

Cervical vertebrae 1-7

Thoracic vertebrae 1-12

Lumbar vertebrae 1-5

Sacrum, 5 segments

Coccyx, 4-5 segments

POSTERIOR VIEW

Cervical curve →

Cervical

Thoracic

Thoracic curve ←

Lumbar

Lumbar curve →

Sacral curve ←

Sacrum

Coccyx

LATERAL VIEW

Atlas — C1
Axis — C2

T1

L1

Coccyx

ANTERIOR VIEW

87. TYPICAL AND CERVICAL VERTEBRAE

I. **Vertebrae made up of 2 parts:** a body and vertebral (neural) arch. The *vertebral foramen* is surrounded by parts of both

A. BODY is largest and heaviest part

B. VERTEBRAL ARCH consists of 2 *pedicles* and 2 *laminae* and has 7 *processes:* 1 *spinous,* 4 *articular,* and 2 *transverse*

    1. Pedicle joins arch to posterolateral body

        a. Concavities in upper and lower surfaces: *vertebral notches*

    2. Laminae: plates extending posteriorly and medially from pedicle

    3. Spinous processes directed posteriorly and caudally from union of laminae

    4. Articular processes extend upward and downward from point where pedicles and laminae join

        a. 2 superior, articular surfaces face posteriorly

        b. 2 inferior, articular surfaces face anteriorly

    5. Transverse processes project laterally between the superior and inferior articular processes

II. **Typical cervical vertebrae:** smallest of true vertebrae

A. BODY, small

B. VERTEBRAL FORAMEN, large and triangular

C. SPINOUS PROCESS, short and bifid

D. TRANSVERSE PROCESS contains a foramen—*transverse foramen*

    1. Anterior and posterior tubercles on process

III. **Atypical cervical vertebrae** (only points of difference will be given)

A. FIRST—THE ATLAS

    1. Has no body, is more or less circular

    2. Anterior arch has an anterior tubercle, posterior to which is an oval articular facet

    3. Posterior arch has a posterior tubercle

    4. Superior articular facets are very large concave ovals, facing upward

    5. Inferior articular facets are circular

    6. Transverse processes are large, anterior and posterior tubercles fused

B. SECOND—THE AXIS (or epistropheus)

    1. Body has a long, pointed projection directed cranially—the *dens* (odontoid process)

        a. Process has an oval articular facet on anterior surface

    2. Pedicles are strong, are fused with sides of body and *dens* (odontoid process), and their upper surface forms the superior articular facet

    3. Transverse processes are small, end in single tubercle

        a. Foramen transversarium set obliquely

C. SEVENTH—VERTEBRA PROMINENS

    1. Spinous process: thick, directed almost straight posteriorly; is not bifid, but ends in tubercle

    2. Transverse processes are large, tubercles not clear

IV. **Special features**

A. THE CERVICAL VERTEBRAE are distinguished by their small size

B. THE SPINOUS PROCESSES are typically short, but those of the 6th and 7th are much longer; the 3rd through the 6th are usually bifid in white persons, but not in black persons; and there is no spinous process on the 1st cervical vertebra

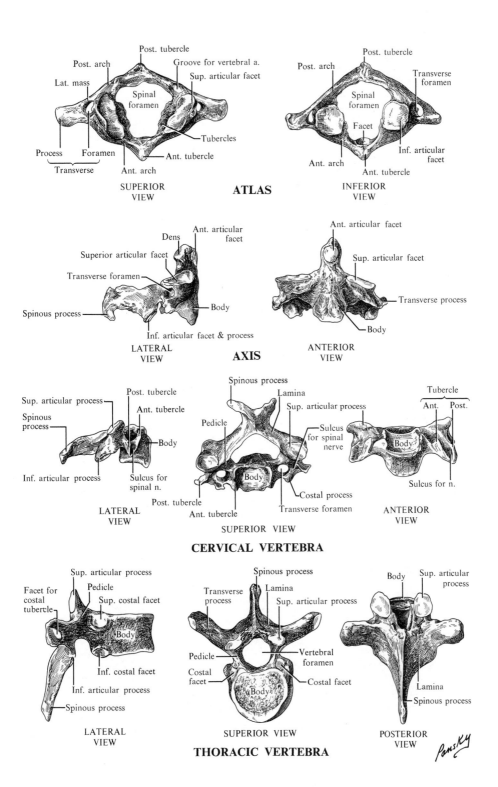

ATLAS

SUPERIOR VIEW

Post. tubercle

Groove for vertebral a.

Sup. articular facet

Post. arch

Lat. mass

Spinal foramen

Process — Foramen — Transverse

Ant. tubercle

Ant. arch

Tubercles

INFERIOR VIEW

Post. tubercle

Post. arch

Transverse foramen

Spinal foramen

Facet

Ant. arch

Ant. tubercle

Inf. articular facet

AXIS

LATERAL VIEW

Dens

Ant. articular facet

Superior articular facet

Transverse foramen

Spinous process

Inf. articular facet & process

Body

ANTERIOR VIEW

Ant. articular facet

Sup. articular facet

Transverse process

Body

CERVICAL VERTEBRA

LATERAL VIEW

Sup. articular process

Spinous process

Post. tubercle

Ant. tubercle

Body

Inf. articular process

Sulcus for spinal n.

Post. tubercle

SUPERIOR VIEW

Spinous process

Lamina

Sup. articular process

Pedicle

Sulcus for spinal nerve

Body

Costal process

Transverse foramen

Ant. tubercle

ANTERIOR VIEW

Tubercle
Ant. Post.

Body

Sulcus for n.

THORACIC VERTEBRA

LATERAL VIEW

Sup. articular process

Facet for costal tubercle

Pedicle

Sup. costal facet

Body

Inf. costal facet

Inf. articular process

Spinous process

SUPERIOR VIEW

Spinous process

Transverse process

Lamina

Sup. articular process

Pedicle

Vertebral foramen

Costal facet

Body

Costal facet

POSTERIOR VIEW

Body

Sup. articular process

Lamina

Spinous process

Pansky

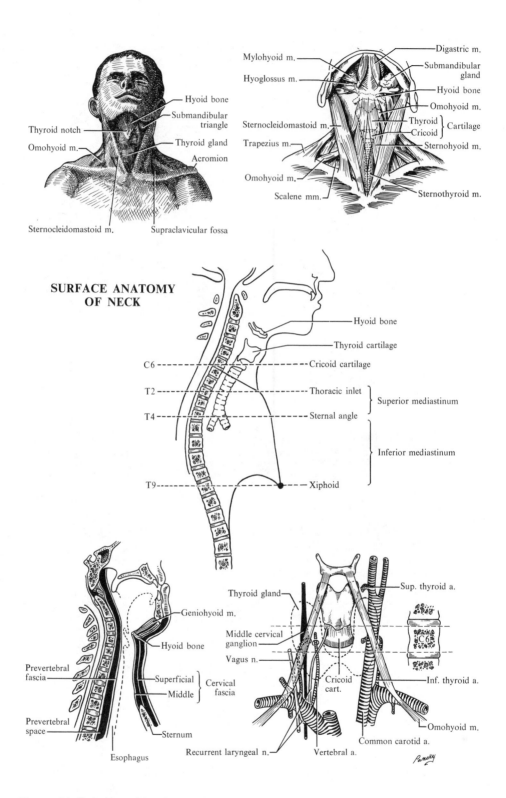

**SURFACE ANATOMY
OF NECK**

FIGURE 11. **Relations of cervical region.**

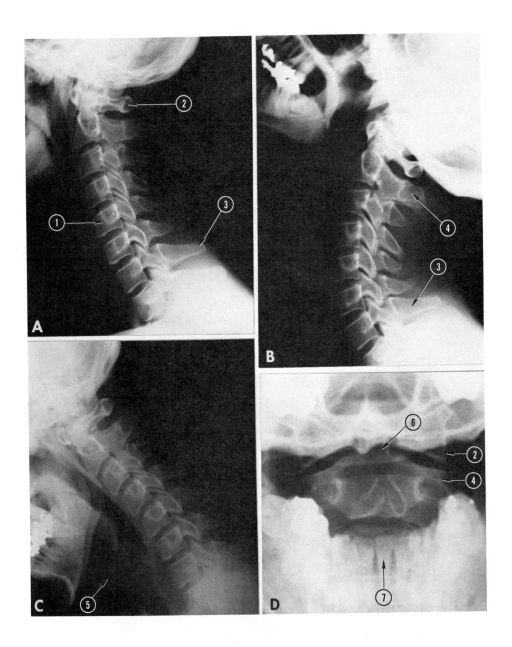

FIGURE 12. **Cervical vertebrae.   A, Head erect; B, head extended; C, head flexed; D, view through open mouth showing second cervical vertebra.**   *1,* Body, C5; *2,* atlas, C1; *3,* spine C7; *4,* axis, C2; *5,* hyoid bone; *6,* dens of axis; *7,* teeth of lower jaw.

88. THE THORACIC AND LUMBAR VERTEBRAE— SACRUM AND COCCYX

I. Thoracic vertebrae (see p. 187)

A. GENERAL CHARACTERISTICS: body increases in size from above downward, has facets or demifacets for rib articulation; laminae are broad and thick; spinous processes are long and directed obliquely caudally; the superior articular processes are thin with facets directed posteriorly; the inferior articular processes are short, and their facets are directed anteriorly; the transverse processes are thick, strong, and have articular facets for rib tubercles

B. SPECIAL FEATURES
1. T1 has one entire facet on each side of body for first rib
2. T10 has a single articular facet on each side
3. T11: large body, large articular facet; spinous process is short and almost horizontal; transverse process is short with no articular facet
4. T12 resembles both T11 and L1; inferior articular facet is directed laterally; transverse process has superior, inferior, and lateral tubercles

II. Lumbar vertebrae (largest)

A. GENERAL CHARACTERISTICS: body is large, wide, and thick; pedicles are strong and directed posteriorly; laminae are broad and strong; spinous processes are thick, broad, and directed posteriorly, superior articular processes are directed medially and posteriorly; inferior articular processes are directed anteriorly and laterally; transverse processes are long, slender, and have upper tubercle at junction with superior articular process called *mammillary process* and inferior tubercle at base of process called *accessory process*

B. SPECIAL FEATURE: L5 has a heavy body, small spinous process, and thick transverse process

III. Sacrum: a fusion of 5 segments, triangular in shape

A. PELVIC SURFACE: concave, crossed by 4 transverse ridges; *pelvic (anterior) sacral foramina* are seen at ends of the ridges

B. POSTERIOR SURFACE: convex, *median sacral crest* at midline, *sacral groove* on either side of median crest, *sacral articular (intermediate) crest* lateral to groove, which terminate as the *sacral cornua*, row of *posterior sacral foramina* lateral to articular crest; *lateral crests* lie lateral to foramina

C. LATERAL SURFACE: upper half, *auricular surface* with *sacral tuberosity* just behind this; inferolateral angle is at lower end of this surface

D. BASE: directed upward with a large, oval articular surface in middle of body just behind which is the *sacral canal*

E. SUPERIOR SURFACE exhibits a projecting anterior border—the *sacral promontory; superior articular processes* are supported by short, heavy pedicles and laminae which enclose sacral canal

F. ALA: on either side of body of sacrum, formed of costal and transverse processes

G. APEX: directed caudally, has oval articular facet at end

IV. Coccyx: formed by fusion of 3–5 segments

A. ANTERIOR SURFACE: slightly convex with transverse ridges

B. POSTERIOR SURFACE: convex, with transverse ridges; has articular crest (as sacrum), the cephalic end of which projects upward as *coccygeal cornua*

C. BASE: oval articular facet

D. APEX: caudally directed, rounded, but may be bifid

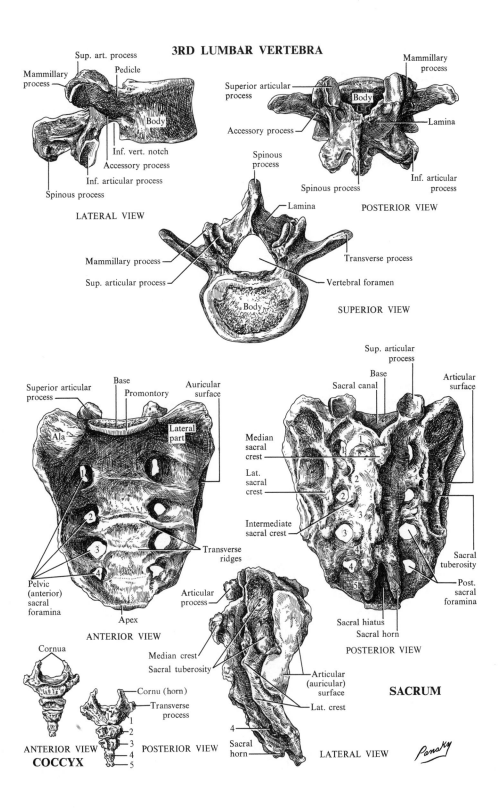

3RD LUMBAR VERTEBRA

LATERAL VIEW

Mammillary process
Sup. art. process
Pedicle
Body
Inf. vert. notch
Accessory process
Inf. articular process
Spinous process

POSTERIOR VIEW

Mammillary process
Superior articular process
Body
Lamina
Accessory process
Spinous process
Spinous process
Inf. articular process

SUPERIOR VIEW

Lamina
Mammillary process
Transverse process
Sup. articular process
Vertebral foramen
Body

ANTERIOR VIEW

Superior articular process
Base
Promontory
Auricular surface
Ala
Lateral part
Pelvic (anterior) sacral foramina
Transverse ridges
Apex

POSTERIOR VIEW

Sup. articular process
Base
Sacral canal
Articular surface
Median sacral crest
Lat. sacral crest
Intermediate sacral crest
Sacral tuberosity
Post. sacral foramina
Sacral hiatus
Sacral horn

SACRUM

COCCYX

Cornua

ANTERIOR VIEW

POSTERIOR VIEW

Cornu (horn)
Transverse process
1
2
3
4
5

Median crest
Sacral tuberosity
Articular process
Articular (auricular) surface
Lat. crest
4
Sacral horn

LATERAL VIEW

Pansky

89. SUPERFICIAL BACK MUSCLES

I.

Name	Origin	Insertion	Action	Nerve
Trapezius	External occipital protuberance Medial part superior nuchal line of occipital bone Ligamentum nuchae Spinous proc. C7, T1–T12 vertebrae Supraspinous ligament	Post. border lat. clavicle Med. margin acromion Post. border spine of scapula Tubercle on spine of scapula	Rotates scapula to raise point of shoulder Adducts scapula Upper part raises scapula Lower part lowers and pulls scapula down Upper part draws head to same side and turns face to opposite side Two sides together draw head back	Spinal accessory (XI) C3 C4
Latissimus dorsi	Through lumbar aponeurosis to spines T6–12 vertebrae Spines of lumbar and sacral vertebrae and supra-spinous lig. Post. iliac crest Directly from iliac crest Lower 4 ribs	Bottom of intertubercular groove of humerus	Extends, adducts, and rotates arm medially Draws shoulder downward and backward Helps in climbing	Thoraco-dorsal (long sub-scapular)

II. Special features

A. THE APONEUROSIS in the lower fibers of the trapezius as they converge near scapula glides over a smooth area at the medial end of scapular spine

B. THERE ARE INTERDIGITATIONS of the costal origin of the latissimus dorsi with the external abdominal oblique muscle

C. THERE IS A TWISTING of the fibers of the latissimus dorsi as they converge toward insertion on humerus

D. THE UPPER PART OF THE TRAPEZIUS, in rotating the scapula, helps the serratus anterior (see p. 238) in making possible abduction of more than 90°

E. WHEN THE HANDS ARE FIXED by gripping an object above the head, the latissimus dorsi helps the pectoralis major (see p. 238) in drawing the body upward

F. TRIANGLE OF AUSCULTATION: medially, the trapezius; laterally, the scapula; below, the latissimus dorsi. The triangle can be enlarged by flexing the trunk with the arms across the chest

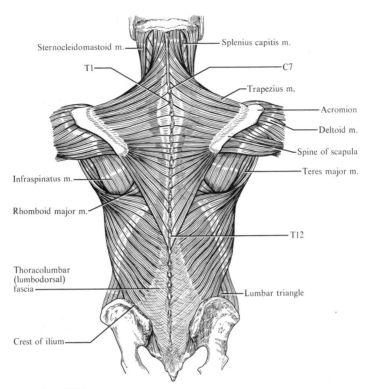

Sternocleidomastoid m.

Splenius capitis m.

T1

C7

Trapezius m.

Acromion

Deltoid m.

Spine of scapula

Teres major m.

Infraspinatus m.

Rhomboid major m.

T12

Thoracolumbar
(lumbodorsal)
fascia

Lumbar triangle

Crest of ilium

SUPERFICIAL
POSTERIOR (DORSAL) MUSCLES

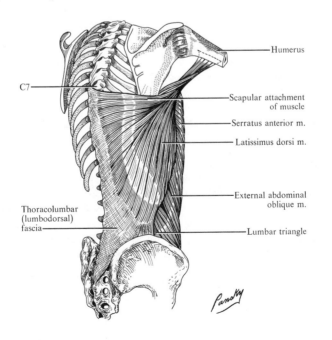

Humerus

C7

Scapular attachment
of muscle

Serratus anterior m.

Latissimus dorsi m.

External abdominal
oblique m.

Thoracolumbar
(lumbodorsal)
fascia

Lumbar triangle

90. MUSCLES OF THE BACK, THE RHOMBOID LAYER, AND THE THORACOLUMBAR (LUMBODORSAL) FASCIA

I. Muscles

Name	Origin	Insertion	Action	Nerve
Levator scapulae	Posterior tubercles, transverse process C1–4	Medial border, scapula above spine	Elevates scapula Rotates scapula Scapula fixed, extends & lat. bends neck	C3, C4; C5, through dorsal scapular
Rhomboid minor	Lig. nuchae Spines C7 through T1	Medial border, scapula at root of spine	Draws scapula medially Holds scapula to chest Depresses shoulder	Dorsal scapular
Rhomboid major	Spines, T2 through T5 Supraspinous lig.	Medial border, scapula below spine	As above	Dorsal scapular

II. **Nuchal and thoracolumbar (lumbodorsal) fascia:** muscles of back and dorsum of neck are enclosed by fascia. This fascial sheath attaches medially to the ligamentum nuchae, tips of spinous processes, and supraspinous ligaments of entire column and to the medial crest of sacrum. In the cervical and lumbar regions, it is attached to the transverse processes of the vertebrae. In the thoracic region, it joins the angles of the ribs (lateral to the iliocostalis muscle and to the intercostal fascia)

A. IN THE THORACIC REGION, it is thin and transparent; in the lumbar region, it is dense and very strong

B. IN THE LUMBAR REGION, it extends lateral to the transverse processes, investing the sacrospinalis muscle to become continuous with the aponeurosis of origin of the transversus abdominis muscle. Thus, it has 2 layers
1. Posterior layer (lumbar aponeurosis) passes over the sacrospinalis muscle
2. Anterior layer lies deep to the sacrospinalis. Medially it is attached to the transverse processes of the lumbar vertebrae
   a. The posterior lumbocostal ligament is a thickening of this layer between twelfth rib and the transverse process of L1

C. BELOW THE LUMBAR REGION (see p. 193), it attaches to the iliac crest and lateral crest of sacrum

III. Clinical considerations

A. NODOSE LUMBAGO (RHEUMATISM): a form of rheumatism characterized by nodule formation, often in areas where the posterior layer of the thoracolumbar fascia attaches to the bones (especially the iliac and sacral crests). This may lead to severe and incapacitating pain

B. RHEUMATOID SPONDYLITIS (Marie-Strümpell disease) is characterized by inflammation of the cartilaginous joints between the vertebral bodies and of the gliding joints between the vertebral arches of the spine. It begins in the lumbar area and extends upward

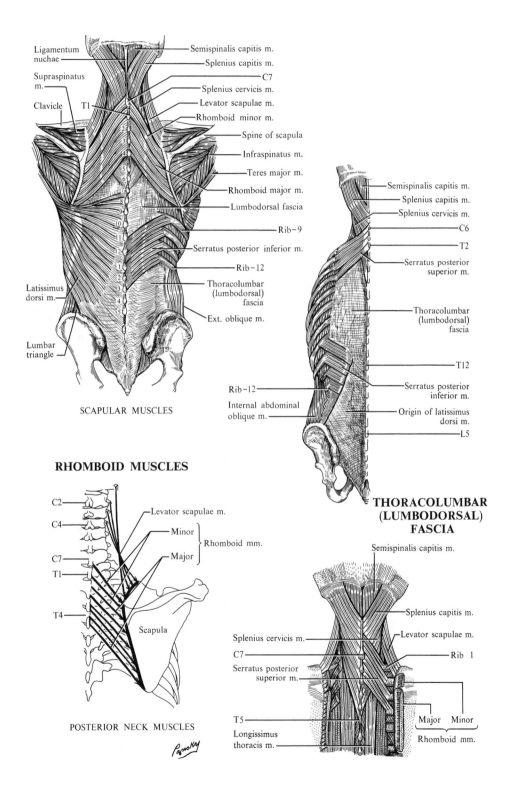

Ligamentum nuchae
Semispinalis capitis m.
Splenius capitis m.
Supraspinatus m.
C7
Splenius cervicis m.
Clavicle
T1
Levator scapulae m.
Rhomboid minor m.
Spine of scapula
Infraspinatus m.
Teres major m.
Rhomboid major m.
Lumbodorsal fascia
Rib–9
Serratus posterior inferior m.
Rib–12
Thoracolumbar (lumbodorsal) fascia
Ext. oblique m.
Latissimus dorsi m.
Lumbar triangle

SCAPULAR MUSCLES

RHOMBOID MUSCLES

C2
C4
Levator scapulae m.
Minor
Rhomboid mm.
Major
C7
T1
T4
Scapula

POSTERIOR NECK MUSCLES

Semispinalis capitis m.
Splenius capitis m.
Splenius cervicis m.
C6
T2
Serratus posterior superior m.
Thoracolumbar (lumbodorsal) fascia
T12
Serratus posterior inferior m.
Rib–12
Origin of latissimus dorsi m.
Internal abdominal oblique m.
L5

THORACOLUMBAR (LUMBODORSAL) FASCIA

Semispinalis capitis m.
Splenius capitis m.
Levator scapulae m.
Splenius cervicis m.
C7
Rib 1
Serratus posterior superior m.
T5
Longissimus thoracis m.
Major Minor
Rhomboid mm.

91. DEEP MUSCLES OF THE BACK—PART I

I. Superficial layer, transversocostal group (see p. 199)

Name	Origin	Insertion
Splenius capitis	Lig. nuchae Spines C7, T1–3	Lat. part of occipital bone Mastoid of temporal bone
Splenius cervicis	Spines T3–6	Trans. proc. C1–3
Erector spinae Iliocostalis lumborum	Mid. crest, sacrum Spines T11–L5 Post. iliac crest Lat. crest of sacrum	Angles, lower 6 ribs
Iliocostalis thoracis	Upper borders of angles of lower 6 ribs	Upper borders, ribs 1–6 Trans. proc. C7
Iliocostalis cervicis	Angles, ribs 3–6	Trans. proc. C4–6
Longissimus thoracis	See Erector spinae above	Trans. proc. lumbar and thoracic vertebrae Lower 10 ribs
Longissimus cervicis	Trans. proc. T1–5	Trans. proc. C2–6
Longissimus capitis	Trans. proc. T1–5 Artic. proc. C5–7	Post. margin, mastoid proc.
Spinalis thoracis	Spines T11–L2	Spines of T1–8
Spinalis cervicis	Spine C7 (T1–2)	Spine C2 (C2–4)

II. Deep layer, transversospinal group (see p. 199)

Name	Origin	Insertion
Semispinalis thoracis	Trans. proc. T6–10	Spines C6–T4
Semispinalis cervicis	Trans. proc. T1–6	Spines C2–5
Semispinalis capitis	Trans. proc. C7–T7 Artic. proc. C4–6	Planum nuchale of occip. bone
Spinalis capitis	With semispinalis capitis	
Multifidus Sacral part	Post. sacrum Aponeurosis of sacrospinalis Post. sup. iliac spine Post. sacroiliac lig.	Each part of the muscle crosses 1 to 4 vertebrae to reach spines of vertebrae from C2–L5
Lumbar part	From all mammillary proc.	
Thoracic part	From all transverse proc.	
Cervical part	Artic. proc. C4–7	
Rotatores Longi Breves	Trans. proc. of 1 vertebra Trans. proc. of 1 vertebra	Spine, 2 vertebrae above Spine, next vertebrae above

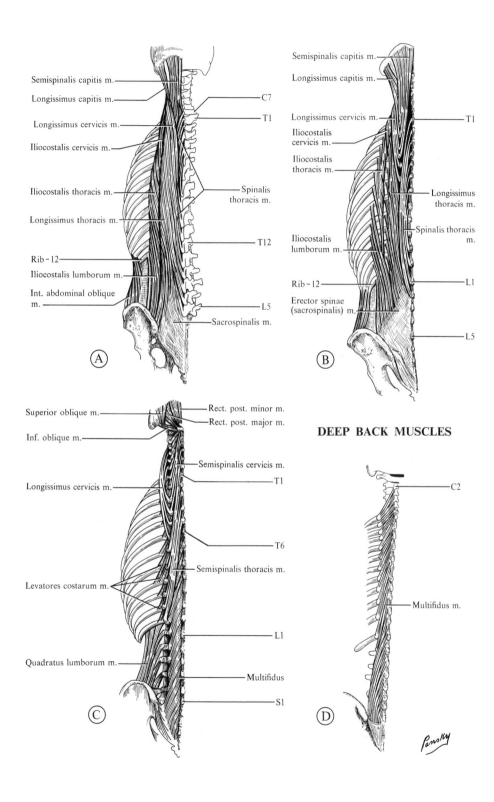

Semispinalis capitis m.
Longissimus capitis m.
Longissimus cervicis m.
Iliocostalis cervicis m.
Iliocostalis thoracis m.
Longissimus thoracis m.
Rib - 12
Iliocostalis lumborum m.
Int. abdominal oblique m.

C7
T1
Spinalis thoracis m.
T12
L5
Sacrospinalis m.

(A)

Semispinalis capitis m.
Longissimus capitis m.
Longissimus cervicis m.
Iliocostalis cervicis m.
Iliocostalis thoracis m.
Iliocostalis lumborum m.
Rib - 12
Erector spinae (sacrospinalis) m.

T1
Longissimus thoracis m.
Spinalis thoracis m.
L1
L5

(B)

Superior oblique m.
Inf. oblique m.
Longissimus cervicis m.
Levatores costarum m.
Quadratus lumborum m.

Rect. post. minor m.
Rect. post. major m.
Semispinalis cervicis m.
T1
T6
Semispinalis thoracis m.
L1
Multifidus
S1

(C)

DEEP BACK MUSCLES

C2
Multifidus m.

(D)

Pansky

92. MUSCLES OF THE BACK—PART II

II. Deep layer (continued)

Name	Location and Attachments
Interspinalis	Connect the apices of spinous processes of adjoining vertebrae from C2 to C3, C7 to T2, T11 to T12, L1 to L5
Intertrans-verse	
Anterior	Interconnect the ant. tubercles of trans. processes from C1 to T1, T10 to L1
Lateral	Between adjoining trans. processes of lumbar vertebrae
Medial	Between accessory process and mammillary processes of adjoining vertebrae of lumbar region
Posterior	Interconnect post. tubercles of trans. processes of lumbar vertebrae

III. Nerve supply: all by posterior primary divisions of spinal nerves

IV. Action: in general, to extend vertebral column
  A. SPECIFIC, actions in addition to the above:
    1. Splenius draws head back, bends head laterally, and rotates face to the same side
    2. Iliocostalis bend vertebral column to side; the lumborum group depress ribs
    3. Longissimus thoracis and cervicis bend column to one side, depress ribs
    4. Longissimus capitis extends head, bends head to side, rotates face to same side
    5. Semispinalis thoracis and cervicis rotate column to opposite side
    6. Semispinalis capitis extends head, rotates head to opposite side
    7. Multifidis rotates column to opposite side
    8. Rotatores rotate column to opposite side
    9. Intertransverse bends column to same side

V. Suboccipital muscles

A. Name	Origin	Insertion	Action
Rectus capitis post. major	Spinous process, axis	Inf. nuchal line and bone below	Extends head Rotates to same side
Rectus capitis post. minor	Tubercle on post. arch of atlas	Inf. nuchal line and bone medial to it	Extends head
Obliquus capit. inferior	Apex, spine of axis	Trans. proc. atlas	Turns head, same side
Obliquus capit. superior	Trans. proc. of atlas	Occip. bone between sup. and inf. nuchal lines	Extends head and bends it to same side

  B. SPECIAL FEATURES
    1. Suboccipital triangle (see p. 221): medially and above, by rectus capitis posterior major; above and laterally, by obliquus capitis superior; below and laterally, obliquus capitis inferior; roof, fascia under semispinalis capitis; floor, atlanto-occipital membrane and posterior arch of atlas. The vertebral artery and the first cervical nerve lie on the upper surface of the arch of atlas

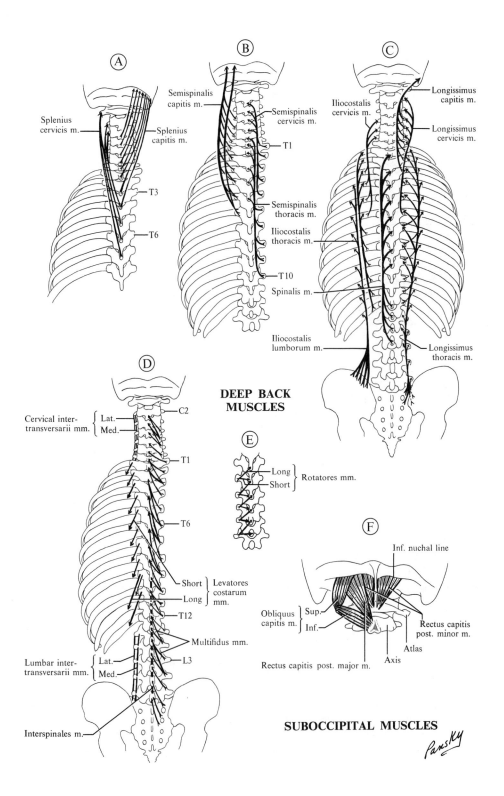

A

Splenius
cervicis m.

Splenius
capitis m.

T3

T6

B

Semispinalis
capitis m.

Semispinalis
cervicis m.

T1

Semispinalis
thoracis m.

Iliocostalis
thoracis m.

T10

Spinalis m.

Iliocostalis
lumborum m.

C

Iliocostalis
cervicis m.

Longissimus
capitis m.

Longissimus
cervicis m.

Longissimus
thoracis m.

**DEEP BACK
MUSCLES**

D

Cervical inter-
transversarii mm. { Lat.
Med.

C2

T1

T6

Short } Levatores
Long } costarum
mm.

T12

Multifidus mm.

Lumbar inter-
transversarii mm. { Lat.
Med.

L3

Interspinales m.

E

Long } Rotatores mm.
Short }

F

Inf. nuchal line

Obliquus
capitis m. { Sup.
Inf.

Rectus capitis
post. minor m.

Atlas

Axis

Rectus capitis post. major m.

SUBOCCIPITAL MUSCLES

Pansky

93. ARTERIAL BLOOD SUPPLY OF THE SPINAL CORD

I. **Posterior spinal arteries:** 1 on each side
A. ORIGIN: from vertebral artery, lateral to medulla oblongata or post. inferior cerebellar artery
B. COURSE: descends through foramen magnum to spinal cord, where it runs downward anterior to dorsal roots of spinal nerves. Extends to lower end of cord and onto the cauda equina. Supplies posterior 1/3 of spinal cord
C. BRANCHES: form free anastomoses around dorsal roots with communications to:
1. Opposite side
2. Posterior funiculus
3. Posterior gray columns
4. Lateral funiculus (in part)

II. **Anterior spinal arteries**
A. ORIGIN: medial branch from each vertebral artery
B. COURSE: 2 arteries course downward and fuse into a single channel, anterior to medulla at level of the foramen magnum. Single *anterior spinal artery* runs caudally along anterior side of spinal cord, just in front of the anterior median fissure, and extends as a fine vessel along filum terminale. Supplies anterior 2/3 of cord
C. DISTRIBUTION
1. Through anterior median fissure to gray commissure
a. To medial side of anterior gray column
2. To anterior roots and lateral side of anterior gray columns
3. To anterior funiculus

III. **Reinforcing system:** both vessels in I and II, above, are small. The blood supplied by the ant. and post. spinal arteries is sufficient only for the upper cervical segment of the cord. The rest of the cord receives most of its blood from the many radicular arteries which join the spinal aa. They are joined in their caudal extent by small vessels, entering the vertebral column through intervertebral or anterior sacral foramina, which arise from sources indicated below. Each of these reinforcing radicular vessels divides and follows the posterior and anterior spinal arteries
A. SOURCES
1. Vertebral, deep cervical, ascending cervical, and inferior thyroid in cervical region
2. Posterior intercostals in thoracic region
3. Lumbar in abdomen
4. Lateral sacral in pelvis
B. DISPOSITION
1. The reinforcing vessels have ascending and descending branches, thus continuing the longitudinal course of the anterior and posterior spinal arteries
2. The main channels have medial and lateral branches. These anastomose posteriorly, anteriorly, and laterally, thus forming a circular arterial complex around periphery of cord

IV. **Clinical considerations**
A. THERE ARE SAID TO BE ARTERIAL WEAKNESSES anteriorly at the level of T4 and L1. Posteriorly, the weakness is said to be between T1 and T3. This is due to few and small radicular aa. here. This area is called the *watershed area* of cord (midthoracic) and is junctional between well-supplied superior and inferior segments of cord
B. BOTH THE ARTERIES AND VEINS of the cord are primarily longitudinally running vessels that communicate above with cranial vessels, but are reinforced at irregular intervals by segmental radicular vessels that come in along the nerve roots. These are important for the cord since the anterior and posterior spinal arteries are small even at their origins and can supply by themselves only a short upper portion of the cord

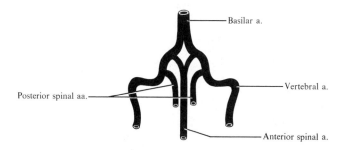

ORIGIN OF SPINAL ARTERIES (SCHEMATIC)

**ARTERIES OF
SPINAL CORD**

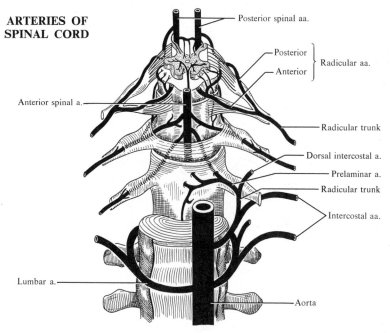

SOURCE, COURSE, AND DISTRIBUTION

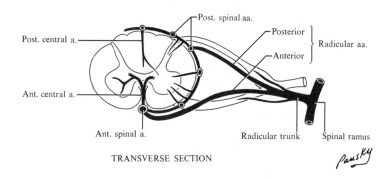

TRANSVERSE SECTION

94. VEINS OF CORD AND VERTEBRAL COLUMN

I. **The veins of the vertebral column** are arranged in a plexiform manner along the length of the vertebral column, both inside and outside the vertebral canal
  A. EXTERNAL VENOUS PLEXUS: composed of 2 freely anastomosing parts
    1. Anterior in front of vertebrae
      a. Communicate with *basivertebral* and *intervertebral* veins
      b. Receive tributaries from bodies of vertebrae
    2. Posterior on posterior surfaces of vertebral arches and processes
      a. Anastomose with vertebral, occipital, and deep cervical veins
  B. INTERNAL VENOUS PLEXUS lies in vertebral canal, between dura and bone. Tends to run longitudinally, receiving drainage from bone and cord. Four main channels:
    1. Anterior (2) lie on posterior surface of bodies of the vertebrae, 1 on each side of posterior longitudinal ligament, interconnected by transverse branches that receive basivertebral veins
    2. Posterior (2): on either side of midline, anterior to vertebral arches and ligamenta flava
      a. Communications
        i. Anastomose with posterior external plexus by veins piercing ligament
        ii. Venous rings at each vertebral level connect with anterior internal plexus
      b. Terminations
        i. Vertebral veins
        ii. Occipital sinus
        iii. Basilar plexus
        iv. Condyloid emissary veins
  C. BASIVERTEBRAL VEINS lie in large network of channels in bodies of vertebrae
    1. Communications: anteriorly with anterior external plexus; posteriorly converge into 1 or 2 large channels which open into transverse veins between anterior internal veins
  D. INTERVERTEBRAL VEINS run through intervertebral foramina
    1. Drain spinal cord and both internal and external plexuses
    2. Terminate in vertebral, intercostal, lumbar, and lateral sacral veins

II. **Veins of cord** lie in fine plexuses in the pia, with 6 longitudinal channels: posterior and anterior median, related to the posterior median sulcus and anterior median fissure; a vein just posterior to each of the posterior roots; and a vein just posterior to each of the anterior roots
  A. LATERALLY, send communications to the intervertebral veins
  B. CEPHALICALLY, coalesce into 2 or 3 vessels that communicate with the vertebral veins and terminate in the inferior petrosal sinus

III. **Clinical considerations**
  A. SINCE VEINS FROM ALL PARTS OF THE BODY interconnect with those around the spinal cord, these channels act as collateral venous pathways. Thus, metastases from cancer in the pelvis could involve the spinal cord

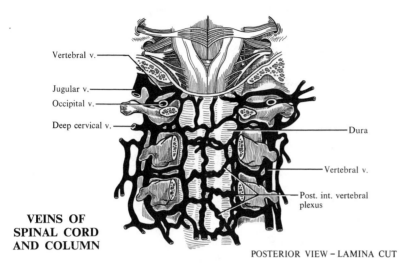

Vertebral v.

Jugular v.

Occipital v.

Deep cervical v.

Dura

Vertebral v.

Post. int. vertebral plexus

**VEINS OF
SPINAL CORD
AND COLUMN**

POSTERIOR VIEW – LAMINA CUT

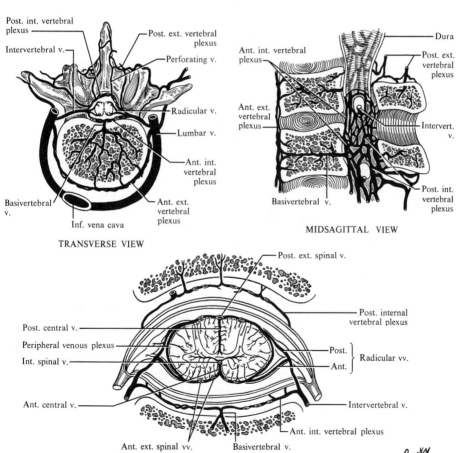

Post. int. vertebral plexus

Post. ext. vertebral plexus

Intervertebral v.

Perforating v.

Radicular v.

Lumbar v.

Ant. int. vertebral plexus

Basivertebral v.

Ant. ext. vertebral plexus

Inf. vena cava

TRANSVERSE VIEW

Dura

Ant. int. vertebral plexus

Post. ext. vertebral plexus

Ant. ext. vertebral plexus

Intervert. v.

Basivertebral v.

Post. int. vertebral plexus

MIDSAGITTAL VIEW

Post. ext. spinal v.

Post. internal vertebral plexus

Post. central v.

Peripheral venous plexus

Int. spinal v.

Post.
Ant. } Radicular vv.

Ant. central v.

Intervertebral v.

Ant. int. vertebral plexus

Ant. ext. spinal vv.

Basivertebral v.

TRANSVERSE SECTION

Panasky

95. THE ATLANTO-OCCIPITAL, ATLANTOEPISTROPHIC, AND INTERVERTEBRAL ARTICULATIONS

I. **Articulation between atlas and axis** consists of a medial and 2 lateral joints. Since the lateral are somewhat similar to those between arches of other vertebrae, these will not be included here
  A. TYPE: trochoid (pivot)
  B. MOVEMENTS: rotation
  C. BONES: dens (odontoid process) of axis with anterior arch and transverse ligament of atlas
  D. LIGAMENTS
    1. Anterior atlantoaxial: from body of axis to border of anterior arch of atlas
    2. Posterior atlantoaxial: lamina of axis to posterior arch of atlas
    3. Transverse: across arch of atlas to hold dens against anterior arch of atlas
      a. Has an upward prolongation to occipital bone and a caudal extension to the posterior surface of body of axis, forming *cruciform ligament of atlas*

II. **Articulation between atlas and occipital bone**
  A. TYPE: condyloid
  B. MOVEMENTS: flexion, extension, and lateral motion (abduction)
  C. BONES: condyles of occipital bone and superior facets of atlas
  D. LIGAMENTS OR MEMBRANES
    1. Articular capsules surround condyles and superior articular processes of atlas
    2. Anterior atlanto-occipital: membrane between foramen magnum and anterior arch of atlas
    3. Posterior atlanto-occipital: membrane between margin of foramen magnum and posterior arch of atlas
    4. Lateral: transverse processes of atlas to jugular processes of occipital

III. **Union of axis with occipital bone**
  A. TECTORIAL MEMBRANE: inside vertebral canal and is a cephalic extension of posterior longitudinal ligament extending from body of axis to occipital bone
  B. ALAR (2): from sides of dens to condyles of occipital bone
  C. APICAL ODONTOID: from apex of dens to foramen magnum

IV. **Muscles acting between atlas and axis**
  A. ROTATION OF HEAD
    1. To opposite side: sternocleidomastoid muscle
    2. To same side: semispinalis capitis, longus capitis, splenius capitis, longissimus capitis, rectus capitis posterior major, oblique capitis superior and inferior muscles

V. **Muscles acting on the atlanto-occipital joint**

Flexion	Extension	Lateral Bending
Longus capitis	Rectus capitis post. major	Trapezius of same side
Rectus capitis anterior	Rectus capitis post. minor	Splenius capitis of same side
	Superior oblique	Sternocleidomastoid of same side
	Semispinalis capitis	
	Splenius capitis	
	Sternocleidomastoid	
	Upper trapezius	

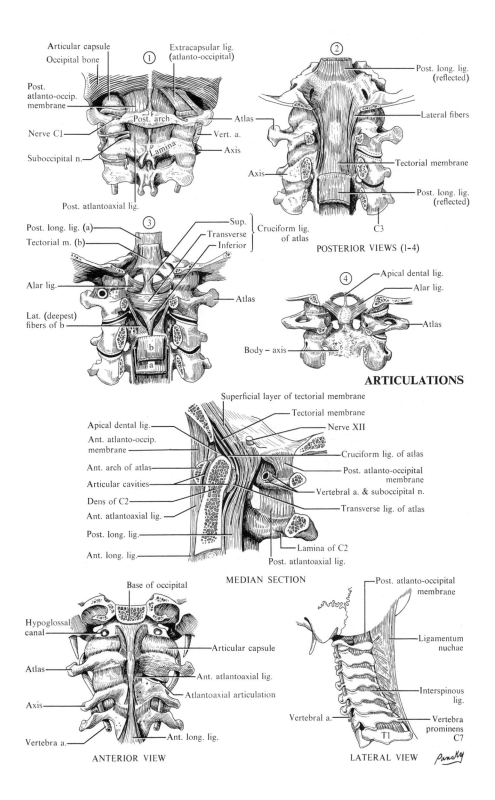

Articular capsule
Occipital bone
① Extracapsular lig.
(atlanto-occipital)

Post.
atlanto-occip.
membrane

Post. arch
Atlas
Nerve C1
Vert. a.
Lamina
Axis
Suboccipital n.

Post. atlantoaxial lig.

Post. long. lig. (a)
③ Sup.
Tectorial m. (b)
Transverse
Inferior
Cruciform lig.
of atlas

Alar lig.

Atlas

Lat. (deepest)
fibers of b

b
a

② Post. long. lig.
(reflected)

Lateral fibers

Tectorial membrane

Axis

Post. long. lig.
(reflected)

C3

POSTERIOR VIEWS (1-4)

④ Apical dental lig.

Alar lig.

Atlas

Body – axis

ARTICULATIONS

Superficial layer of tectorial membrane

Apical dental lig.
Tectorial membrane
Ant. atlanto-occip.
membrane
Nerve XII

Ant. arch of atlas
Cruciform lig. of atlas

Articular cavities
Post. atlanto-occipital
membrane

Dens of C2
Vertebral a. & suboccipital n.

Ant. atlantoaxial lig.
Transverse lig. of atlas

Post. long. lig.

Ant. long. lig.
Lamina of C2
Post. atlantoaxial lig.

MEDIAN SECTION

Base of occipital
Post. atlanto-occipital
membrane

Hypoglossal
canal
Ligamentum
nuchae

Articular capsule

Atlas
Ant. atlantoaxial lig.

Axis
Atlantoaxial articulation

Vertebral a.

Interspinous
lig.

Vertebra a.
Ant. long. lig.
Vertebra
prominens
C7

ANTERIOR VIEW
T1
LATERAL VIEW
Pansky

96. INTERVERTEBRAL AND COSTOVERTEBRAL ARTICULATIONS—PART I

I. Intervertebral

A. ARTICULATION BETWEEN BODIES OF VERTEBRAE
1. Type: amphiarthrodial
2. Movement: slight
3. Ligaments
   a. Anterior longitudinal covers ventral surfaces of vertebrae from C2 to sacrum
      i. Closely attached to disks and margins of vertebral bodies
   b. Posterior longitudinal in vertebral canal over posterior surface of bodies, extending from C2 to sacrum
      i. Closely adherent to disks and margins of bodies; more loosely attached over concavities and permits transverse veins to cross
   c. Intervertebral disks form chief connections between bodies from C2 to sacrum. Variable in size, with size of adjacent bodies
      i. In cervical and lumbar regions, it is thicker in front than behind, thus helping to form the convex curvatures in these regions
      ii. One-fourth length of column due to disks
      iii. Where disk adjoins bones, cartilage is hyaline; in center is fibrocartilage
      iv. Nucleus pulposus: soft, pulpy, yellowish elastic material lying in center of disk (embryonic notochord)
      v. Fibrous ring (anulus): concentric ring of fibrous tissue and fibrocartilage
      vi. Not only do disks join bones, but they absorb shock

B. ARTICULATIONS BETWEEN ARCHES OF VERTEBRAE
1. Type: arthrodial
2. Bones: superior articular process with inferior articular process
3. Movements: gliding
4. Ligaments
   a. Articular capsule attached around articular facets of articular processes
   b. Ligamenta flava interconnect laminae of vertebrae
   c. Supraspinal ligament joins tips of vertebrae from C7 to sacrum
   d. Ligamentum nuchae corresponds to B4c, above, in cervical region. Extends longitudinally from external occipital protuberance and median nuchal line to spine of C7. Anteriorly, it is attached to the posterior tubercle of atlas and spines of C2–C6
   e. Interspinal interconnects spines of vertebrae, extending from root to apex of spinous process
   f. Intertransverse joins transverse processes of vertebrae

II. Movements of vertebral column as a whole

A. FLEXION: bending forward. Anterior parts of disks flattened; posterior parts expanded; ligamenta flava, posterior longitudinal, supraspinous, and interspinal ligaments all stretch and help to check action. Strongest check is tension of extensor muscles
B. EXTENSION: bending backward. Here anterior longitudinal ligament is stretched and helps check action. Contact of spinous processes also checks
C. LATERAL FLEXION: bending to either side. Stretches opposite ligamentum flavum and part of anterior longitudinal ligament and intertransverse ligaments, which help to check action. The opposing extensor muscles also check
D. ROTATION at any one articulation is slight, but total rotation is a summation of movement at several joints
E. FOR MUSCLES PRODUCING MOVEMENTS at *all* intervertebral joints (see p. 218)

ARTICULATIONS

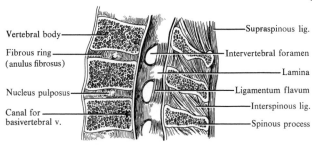

Vertebral body

Fibrous ring
(anulus fibrosus)

Nucleus pulposus

Canal for
basivertebral v.

Supraspinous lig.

Intervertebral foramen

Lamina

Ligamentum flavum

Interspinous lig.

Spinous process

MEDIAN SECTION - LUMBAR REGION

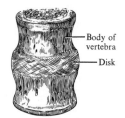

Body of
vertebra

Disk

INTERVERTEBRAL DISK - ANTERIOR

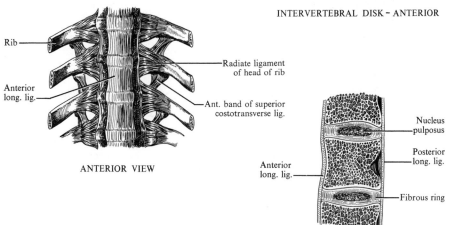

Rib

Anterior
long. lig.

Radiate ligament
of head of rib

Ant. band of superior
costotransverse lig.

ANTERIOR VIEW

Nucleus
pulposus

Posterior
long. lig.

Anterior
long. lig.

Fibrous ring

SAGITTAL SECTION - LUMBAR REGION

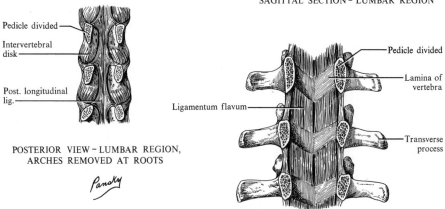

Pedicle divided

Intervertebral
disk

Post. longitudinal
lig.

POSTERIOR VIEW - LUMBAR REGION,
ARCHES REMOVED AT ROOTS

Panaky

Pedicle divided

Lamina of
vertebra

Ligamentum flavum

Transverse
process

FRONT VIEW - BODIES OF VERTEBRAE REMOVED

97. INTERVERTEBRAL AND COSTOVERTEBRAL ARTICULATIONS—PART II

III. Articulations between ribs and vertebrae
A. COSTOVERTEBRAL at head of rib
1. Type: arthrodial
2. Bones: head of rib and facet on side of vertebral body
3. Movements: gliding
4. Ligaments
   a. Articular capsule surrounds joint
   b. Radiate: from anterior head of rib to bodies and intervertebral disk
   c. Intra-articular: from interarticular crest of rib to fibrocartilage. Located within joint and divides it into 2 parts
B. COSTOTRANSVERSE between rib and transverse process
1. Type: arthrodial
2. Bones: tubercle (and neck) of rib with transverse process
3. Movements: gliding
4. Ligaments
   a. Articular capsule attached to edges of articular facets
   b. Superior costotransverse from cephalic border of neck of rib to transverse process above
   c. Posterior costotransverse from neck of rib to transverse process and inferior articular process of vertebra above
   d. Ligament of neck of rib (costotransverse ligament) from back of neck of rib to ventral surface of adjoining transverse process
   e. Ligament of tubercle of rib (lateral costotransverse ligament) from tubercle of rib to apex of transverse process

IV. Movement of ribs: total effect is a rotation of head of rib in its own axis so that ribs are raised or lowered
A. MUSCLES INVOLVED IN PROCESS are those used in respiration (see p. 370)

V. Clinical considerations
A. HERNIATED (SLIPPED) DISK: after unusual strain to the vertebral column, the disk itself or its center, the nucleus pulposus, may be extruded beyond its normal limits and fail to return to normal position. Slight posterior protrusion may cause nerve root pain because of pressure on spinal nerve roots. The protrusion may occur at any level, but the most common location is low in the lumbar region, and the most frequent symptom is sciatic pain
B. MOST DISK RUPTURES occur in the third and fourth decades, which represent the most active period of adult life
C. DISK HERNIATION THAT DOES NOT INVOLVE A NERVE ROOT may result in back pain without sciatica, but from a clinical standpoint, diagnosis of ruptured disk is made only when nerve root symptoms are present
D. HERNIATION OF NUCLEAR MATERIAL usually occurs through a small tear or rent in the anulus in a posterolateral direction, lateral to the posterior longitudinal ligament
E. THE NERVE THAT EXITS THROUGH THE FORAMEN BELOW the ruptured disk is usually the one affected
F. THE DISKS contribute about 25% of the length of the vertebral column above the sacrum. Their high water content means they are subject to dehydration, accounting for as much as 1.875 cm (0.75 in.) in loss of height by a man, and 1.25 cm (0.5 in.) by a woman, during the course of a day. This dehydration is usually made up by reabsorption of water while we lie down, but over a period of years reabsorption does not equal water loss, and the disks gradually thin. This accounts for part of the loss in height between young adulthood and old age

COSTOVERTEBRAL ARTICULATIONS

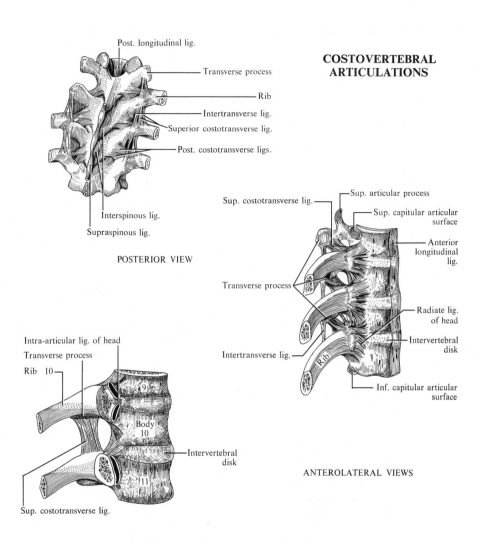

Post. longitudinal lig.

Transverse process

Rib

Intertransverse lig.

Superior costotransverse lig.

Post. costotransverse ligs.

Interspinous lig.

Supraspinous lig.

POSTERIOR VIEW

Sup. costotransverse lig.

Sup. articular process

Sup. capitular articular surface

Anterior longitudinal lig.

Transverse process

Radiate lig. of head

Intervertebral disk

Intertransverse lig.

Rib

Inf. capitular articular surface

Intra-articular lig. of head

Transverse process

Rib 10

9

Body 10

11

Intervertebral disk

Sup. costotransverse lig.

ANTEROLATERAL VIEWS

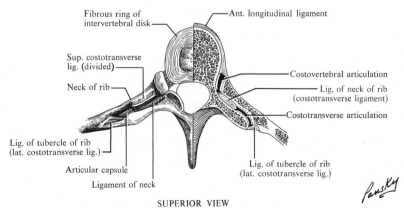

Fibrous ring of intervertebral disk

Ant. longitudinal ligament

Sup. costotransverse lig. (divided)

Neck of rib

Costovertebral articulation

Lig. of neck of rib (costotransverse ligament)

Costotransverse articulation

Lig. of tubercle of rib (lat. costotransverse lig.)

Articular capsule

Ligament of neck

Lig. of tubercle of rib (lat. costotransverse lig.)

SUPERIOR VIEW

98. CLINICAL CONSIDERATIONS: VERTEBRAL COLUMN

I. **Quadriplegia:** with rupture of the transverse ligament in which the dens of the axis rotates, the dens can be driven posteriorly into the cervical region of the spinal cord, resulting in paralysis of the upper and lower limbs. Sudden death may occur if it is driven into the lower end of the medulla

II. **Fractures of the vertebral column:** neck and back injuries are dangerous because they may involve fractures of the vertebrae and injuries to the cord and/or cauda equina
   A. THE CERVICAL REGION is very vulnerable to injury which may result in compression or transection of the cord, leading to a loss of sensation and voluntary movement below the lesion or to sudden death, depending on the level of injury
   B. FRACTURES, DISLOCATIONS, and combinations of both of the vertebral column are often the result of forceful flexion or a violent blow to the back of the head
      1. The common fracture is a *crush* or *compression fracture* of the body of one or more vertebrae, often in the middle or lower cervical region or near the junction of the thoracic and lumbar regions
      2. In severe flexion injuries, the posterior longitudinal and interspinous ligaments may be torn, and the vertebral arches may be dislocated and/or fractured, along with crush fractures of the vertebral bodies. In these cases, the spinal cord is usually injured
   C. INJURIES CAN ALSO be caused by extension of the vertebral column. Extension fractures and/or dislocations vary from region to region, but, invariably, the posterior parts of the column are injured
      1. Hyperextension injuries are rarely seen in the thoracic region due to the support given by the ribs. This is less true in the lower 2 thoracic vertebrae which are freer to move

III. **Rotatory dislocation** or fracture-dislocation is most common in the cervical region. Rotatory forces often cause fractures of the articular processes
   A. ARTICULAR PROCESS FRACTURES also can be seen in the lower thoracic and lumbar regions. The fractures are unstable and require spinal fusion for stability of the column

IV. **Dislocation of vertebrae** without fracture is rare, except in the cervical region, because of the interlocking of thoracic and lumbar articular processes

V. **Displacement of the vertebral column**
   A. DISLOCATION is the complete and persistent displacement of the articular surface of one of the bones of a joint from that of its fellow
   B. SUBLUXATION is a partial displacement, in which the normal relation of the articular surface is disturbed, although the surfaces may remain partly in contact
      1. Sprain is a temporary subluxation, in which the articular surfaces return to their normal position with some damage to the ligaments, tendons, and muscles about the joint
   C. SPONDYLOLISTHESIS is an anterior displacement where there is a forward movement of the body of one of the lower lumbar vertebrae on the vertebra below it, or upon the sacrum
   D. SPONDYLOLYSIS is a breaking down or dissolution of the body of a vertebra and does not include a forward movement of a vertebra

VI. **Spondylomalacia** is a softening of vertebrae

VII. **Whiplash (necklash):** injury usually occurring in an automobile passenger when the car is struck sharply from behind. The impact accelerates the automobile momentarily, and the passenger's torso, padded against the seat, moves forward. The unprotected neck and head are thrown violently backward because of their inertia. The neck muscles react to right the head, usually overcompensating and snapping it forward. This latter action probably causes the most severe damage in the neck. The extent of the cervical injuries depends on the degree of violence of the whiplash effect. They may include muscle strain, cervical root damage, subluxation of vertebrae, compression fractures, and herniation of an intervertebral disk

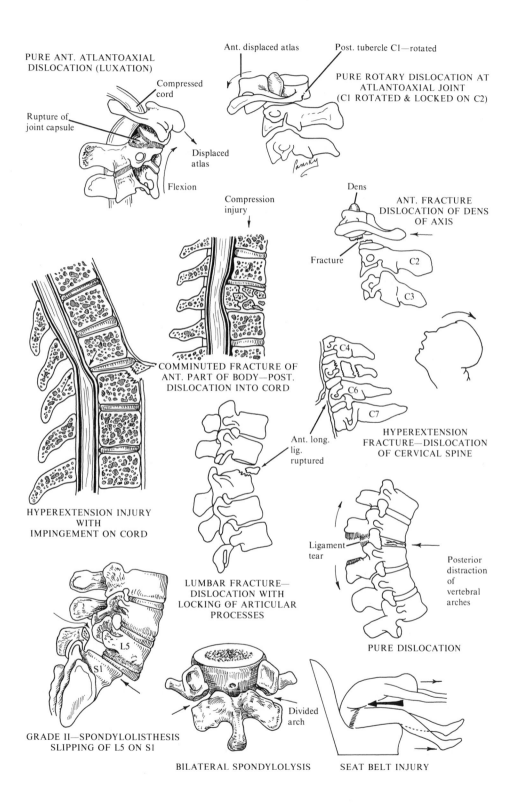

PURE ANT. ATLANTOAXIAL
DISLOCATION (LUXATION)

Compressed cord

Rupture of joint capsule

Displaced atlas

Flexion

Ant. displaced atlas

Post. tubercle C1—rotated

PURE ROTARY DISLOCATION AT
ATLANTOAXIAL JOINT
(C1 ROTATED & LOCKED ON C2)

Compression injury

Dens

ANT. FRACTURE
DISLOCATION OF DENS
OF AXIS

Fracture

C2

C3

COMMINUTED FRACTURE OF
ANT. PART OF BODY—POST.
DISLOCATION INTO CORD

C4

C6

C7

HYPEREXTENSION
FRACTURE—DISLOCATION
OF CERVICAL SPINE

Ant. long.
lig.
ruptured

HYPEREXTENSION INJURY
WITH
IMPINGEMENT ON CORD

LUMBAR FRACTURE—
DISLOCATION WITH
LOCKING OF ARTICULAR
PROCESSES

Ligament
tear

Posterior
distraction
of
vertebral
arches

PURE DISLOCATION

L5

S1

GRADE II—SPONDYLOLISTHESIS
SLIPPING OF L5 ON S1

Divided
arch

BILATERAL SPONDYLOLYSIS

SEAT BELT INJURY

–211–

99. SPINAL MENINGES

I. Dura, outer covering: dense and tough fibrous membrane

A. CORRESPONDS WITH MENINGEAL LAYER of cranial dura; vertebrae have their own periosteal layer

    1. The spinal dura mater (dural sac) is free within the vertebral canal but is adherent to the margin of the foramen magnum of the skull above and to the coccyx below

B. EPIDURAL SPACE filled with fat, loose connective tissue, and veins that lie between dura and bony canal

C. EXTENT

    1. Cephalic end: foramen magnum to which it is attached (also attached to vertebrae C2 and C3)

    2. Extends downward as sheath larger than cord. Narrows down and fuses with filum terminale at level of second sacral segment

    3. From second sacral segment continues to be attached to back of coccyx

    4. Continues with spinal nerves through intervertebral foramina to fuse with their sheaths (epineurium)

II. Arachnoid, middle covering: thin and delicate, filamentous, avascular

A. CONTINUOUS WITH CRANIAL ARACHNOID

B. OUTER SURFACE covered with a mesothelium, separated from dura by *subdural space*

C. INNER SURFACE connected across the subarachnoid space by delicate connective tissue, the *arachnoid trabeculae*

D. SUBARACHNOID SPACE continuous with that of cranium and extends to second sacral segment and outward to intervertebral foramina

    1. Longitudinal subarachnoid septum (septum posticum) incompletely subdivides space. Septum joins pia to arachnoid along post. median sulcus of cord

III. Pia, inner covering: delicate connective tissue, highly vascular, closely applied to cord

A. COMPOSED OF 2 LAYERS

    1. Outer layer consists of longitudinal connective tissue fibers

        a. Forms a median, longitudinal, anterior band, the *linea splendens*

        b. Along sides of cord, longitudinal fibers are concentrated as *denticulate* ligaments

            i. Extend full length of spinal cord, lying between attachments of posterior and anterior roots

            ii. At 21 points, beginning at foramen magnum and ending at conus medullaris, are attached (on each side) to dura

    2. Inner layer, intimately adherent to cord, sends a septum into anterior median fissure

B. AT APEX OF CAUDAL END OF CORD the pia continues along filum terminale and, with addition of dura below S2, forms *central ligament of spinal cord,* which is attached to posterior of coccyx to help anchor the cord

IV. Clinical considerations

A. INFECTION, HEMORRHAGE, AND TUMOR FORMATION: the meninges may be the seat of any of these conditions. Infection of the meninges is called meningitis; it may involve the dura (pachymeningitis) or the pia-arachnoid (leptomeningitis). Headache; neck stiffness; flexion at ankle, knee, and hip when the neck is bent; inability to extend the leg completely when in the sitting position or when lying with the thigh flexed upon the abdomen; fever; mental confusion—all are indications of meningeal infection

B. ANESTHETIC AGENTS introduced into the subarachnoid space mix with cerebrospinal fluid and bathe the spinal cord and nerve roots, producing spinal anesthesia

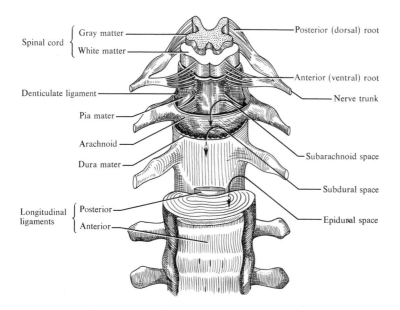

Spinal cord { Gray matter — White matter —

Denticulate ligament —

Pia mater —

Arachnoid —

Dura mater —

Longitudinal { Posterior — ligaments { Anterior —

Posterior (dorsal) root

Anterior (ventral) root

Nerve trunk

Subarachnoid space

Subdural space

Epidural space

SPINAL MENINGES AND THEIR SPACES

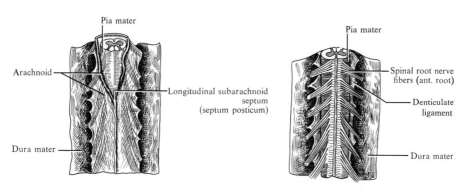

Pia mater

Arachnoid —

Longitudinal subarachnoid septum (septum posticum)

Dura mater —

CERVICAL CORD - POSTERIOR VIEW

Pia mater

Spinal root nerve fibers (ant. root)

Denticulate ligament

Dura mater

ANTERIOR VIEW - ARACHNOID REMOVED

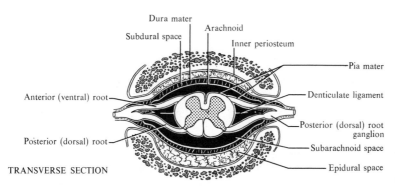

Dura mater

Subdural space

Arachnoid

Inner periosteum

Pia mater

Anterior (ventral) root —

Denticulate ligament

Posterior (dorsal) root ganglion

Posterior (dorsal) root —

Subarachnoid space

Epidural space

TRANSVERSE SECTION

100. SPINAL CORD

I. Extent and size
A. BEGINS: cephalic border of atlas; ends in filum terminale at lower first or upper second lumbar vertebrae
B. LENGTH: 42 to 45 cm
C. DIAMETER varies at different levels

II. External appearance
A. ENLARGEMENTS
   1. Cervical: from C3 to T2 vertebral levels
   2. Lumbar, beginning at end of ninth thoracic, reaching largest diameter at twelfth thoracic
      a. Tapers down rapidly as conus medullaris, ending as the filum terminale
B. FISSURES AND SULCI
   1. Posterior median sulcus and anterior median fissure divide cord into halves
   2. Posterior lateral and anterior lateral sulci extend longitudinally along length of cord, represent line for attachment of posterior and anterior roots
   3. Posterior intermediate sulcus between posteromedial and posterolateral sulci in upper thoracic and cervical regions

III. Relationship between vertebrae and pairs of spinal nerves
A. BECAUSE OF GROWTH INEQUALITIES between vertebral column and spinal cord, the nerves, especially at lower levels, do not emerge from vertebral canal at level of attachment to cord
   1. In third fetal month, cord fills canal; at birth, end of cord lies at the lower border of third lumbar vertebra
   2. Due to a more accelerated growth of the bony structures, the spinal cord ends at the lower end of the 1st lumbar or the upper end of the 2nd lumbar vertebrae
   3. There is also a discrepancy in the number of cervical vertebrae and cervical spinal nerves: 8 cervical spinal nerves, 7 cervical vertebrae
   4. Relationship in adult (see following table)

Spinal Nerve and Cord Segment	Vertebral Segment	Exit
C1	C1	Above C1
C8	C7	Between C7–T1
T6	T5	Between T6–T7
T12	T8	Between T12–L1
L2	T10	Between L1–L2
L5	T11	Below L5
S3	T12	Third sacral foramen

IV. Internal structure
A. COMPOSITION
   1. Centrally located, H-shaped gray matter containing cell bodies of efferent and internuncial neurons
      a. Two, narrow posterior projections—posterior columns (horns)
      b. Two, usually thick anterior projections—anterior columns (horns)
      c. Cross bar—the gray commissure (intermediate substance)
   2. Peripherally located white matter made up of nerve fibers and glia
      a. The posterior lateral and anterior lateral sulci divide cord into posterior, lateral, and anterior funiculi
         i. The posterior funiculus of cervical cord is subdivided into fasciculus gracilis and fasciculus cuneatus by the posterior intermediate sulcus

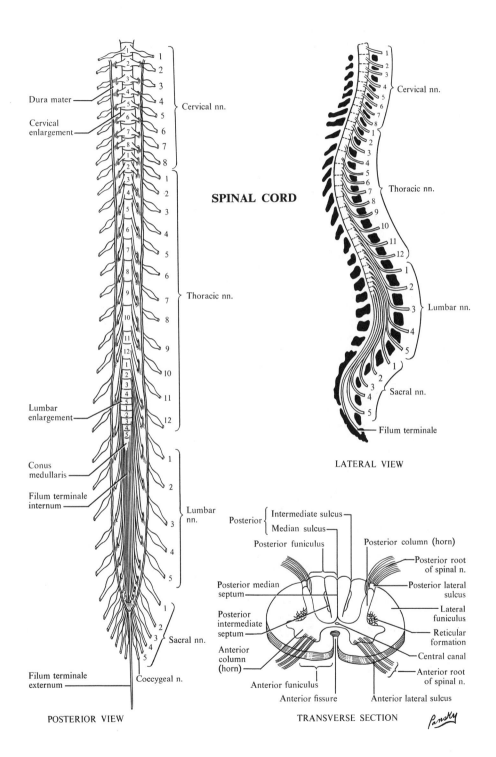

SPINAL CORD

Posterior View labels:

Dura mater

Cervical enlargement

Lumbar enlargement

Conus medullaris

Filum terminale internum

Filum terminale externum

Cervical nn. 1 2 3 4 5 6 7 8

Thoracic nn. 1 2 3 4 5 6 7 8 9 10 11 12

Lumbar nn. 1 2 3 4 5

Sacral nn. 1 2 3 4 5

Coccygeal n.

POSTERIOR VIEW

Lateral View labels:

Cervical nn.

Thoracic nn.

Lumbar nn.

Sacral nn.

Filum terminale

LATERAL VIEW

Transverse Section labels:

Posterior { Intermediate sulcus / Median sulcus }

Posterior funiculus

Posterior median septum

Posterior intermediate septum

Anterior column (horn)

Anterior funiculus

Anterior fissure

Posterior column (horn)

Posterior root of spinal n.

Posterior lateral sulcus

Lateral funiculus

Reticular formation

Central canal

Anterior root of spinal n.

Anterior lateral sulcus

TRANSVERSE SECTION

Pansky

101. CAUDAL END OF SPINAL CORD AND ITS MENINGES

I. Caudal extent of spinal cord
A. IN ADULT, tip of conus medullaris lies at lower border of first lumbar or upper border of second lumbar vertebrae (see p. 215)

II. Filum terminale: a filamentous continuation of caudal tip of cord (conus medullaris) with a covering of pia
A. LIES IN CENTER OF THE DURAL SHEATH, which continues at same size to level of second sacral segment where dura moves down to join filum, thus forming the *central ligament* of the cord
B. WITHIN DURAL SHEATH, filum is surrounded by cauda equina, which is composed of the lumbar and sacral nerves arising from cord to descend to the appropriate intervertebral or sacral foramen for exit from the vertebral canal

III. Clinical considerations
A. SPINAL PUNCTURE: for withdrawal of fluid for diagnostic purposes; treatment, as antibiotics in meningitis; anesthesia, for operations below the thorax
   1. Site: below L2, since spinal cord ends at or above this level. Most usually at L4, since it is easily located by a line drawn across level of highest point on iliac crests. This line passes through spinous process of L4
   2. In the newborn, the spinal cord terminates at the lower border of L3
B. EXTRADURAL INJECTIONS: to produce anesthesia over a more limited area and without the dangers inherent in spinal puncture
   1. Caudal, usually for obstetric, gynecologic, and urologic surgery
      a. Site: *sacral hiatus,* located between sacral and coccygeal cornua
   2. Transsacral: to inject anesthetic material by way of posterior sacral foramina
      a. Site: using posterior superior iliac spine as landmark: 1.0 cm directly medial and slightly above locates first sacral foramen. Line drawn 0.5 cm downward and 1.0 cm medially from first foramen should locate second sacral foramen; one fingerbreadth down and slightly medial is point for the third foramen, while the fourth foramen is a fingerbreadth below and medial to the third. It usually is not necessary to locate the fifth, but, when present, is a small fingerbreadth below and slightly medial to the fourth
C. PILONIDAL SINUS is variously explained as a persistence of the neurenteric canal in the coccygeal region, an epidermal inclusion, or a failure of fusion of the skin of the 2 halves of the body in this region. The sinus opens onto the skin over the coccyx
D. SPINA BIFIDA is a failure of development of the dorsal arches of 1 or more vertebrae, usually in the lumbosacral region. A *tight filum terminale* may be associated with spina bifida occulta (no protrusion above skin). If there is protrusion, a sac is formed (spina bifida aperta). Its contents may be of 3 types: (1) *meningocele,* a leptomeningeal protrusion through a bony and dural defect in the skull or vertebral column (usually sacral); (2) *meningomyelocele,* protrusion of both leptomeninges and cord through a bony and dural defect in the vertebrae; or (3) *syringomyelocele,* a rare defect in which the central canal of the cord in myelocele is distended with CSF with resultant pressure atrophy of the spinal cord

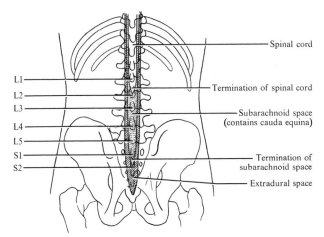

Spinal cord

L1 — Termination of spinal cord

L2

L3 — Subarachnoid space
(contains cauda equina)

L4

L5

S1 — Termination of
subarachnoid space

S2 — Extradural space

VERTEBRAL SPINE LEVELS OF CORD AND MENINGES

SPINAL CORD

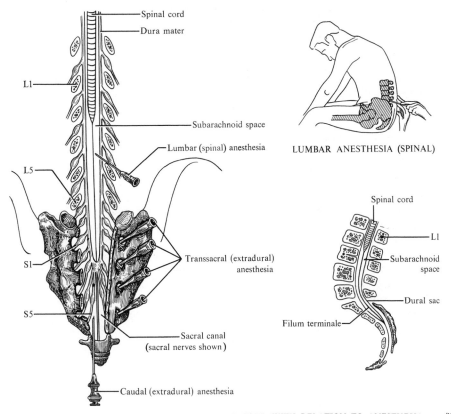

Spinal cord

Dura mater

L1

Subarachnoid space

Lumbar (spinal) anesthesia

L5

S1 — Transsacral (extradural)
anesthesia

S5

Sacral canal
(sacral nerves shown)

Caudal (extradural) anesthesia

LUMBAR ANESTHESIA (SPINAL)

Spinal cord

L1

Subarachnoid
space

Dural sac

Filum terminale

SPINAL CORD WITH RELATION TO ANESTHESIA   Pansky

102. MUSCLES PRODUCING MOVEMENTS ON THE VERTEBRAL COLUMN

Flexion*	Extension*	Lateral Bending†	Rotation†
Sternocleidomastoid	Erector spinae	Same muscles	All the muscles
Longus capitis	Semispinalis	listed under	listed under flexion
Rectus capitis ant.	Splenius	flexion and	and extension rotate
Longus coli	Multifidus	extension plus	column to the same
Scaleni	Rotatores	rectus capitis	side except
Psoas major	Interspinous	lateralis	sternocleidomastoid,
Ext. abdominal		(except for the	ext. abd. oblique,
oblique		interspinous)	semispinalis cervicis
Int. abdominal	Intertransverse		and capitis, multi-
oblique			fidus, and the
Rectus abdominis	All posterior sub-		rotatores, which
	occipital muscles		rotate to opposite
			side (the longus
			coli and scalenes
			give only a slight
			rotation)

*Flexion and extension are accomplished when the bilateral muscles act together.
†Lateral bending and rotation are accomplished when the muscles on one side contract unilaterally.

I. General function

A. THE POSTURE of the vertebral column is maintained by the intrinsic back muscles, but these muscles do not act simultaneously. They come into action to maintain balance but are quiescent most of the time. They are most completely relaxed in extreme flexion since the ligaments of the column then take over the load

B. IN WALKING, the pelvis is prevented from falling to the side where the foot is raised by the contraction of the vertebral muscles on that same side

C. IN GENERAL, all the extensor muscles on the dorsum of the back can be considered to be antigravity muscles enabling us to maintain the erect position

II. Special features

A. THE FLEXIBILITY OF THE COLUMN is controlled by groups of muscles that are vertically disposed (therefore, they are parallel to the column), as well as by obliquely disposed muscles swathed around the torso

B. THE MUSCLES OF THE BACK are the principal extensors; the two rectus abdominis muscles are powerful flexors; the oblique fibers of the abdominal muscles act as rotators, and their vertical fibers form part of the side flexor group, along with the appropriate rectus abdominis, quadratus lumborum, and erector spinae

C. GRAVITY must be taken into account, thus, the muscle groups mentioned above have the specific actions when they act against a resistance, which may, indeed, be gravity

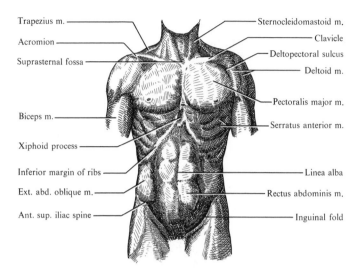

Trapezius m. — — Sternocleidomastoid m.

Acromion — — Clavicle

Suprasternal fossa — — Deltopectoral sulcus

— Deltoid m.

— Pectoralis major m.

Biceps m. — — Serratus anterior m.

Xiphoid process —

Inferior margin of ribs — — Linea alba

Ext. abd. oblique m. — — Rectus abdominis m.

Ant. sup. iliac spine — — Inguinal fold

SURFACE ANATOMY – THORAX AND DORSUM

POSTERIOR VIEW

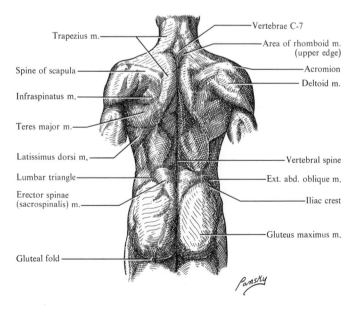

Trapezius m. — — Vertebrae C-7

— Area of rhomboid m.
(upper edge)

Spine of scapula — — Acromion

— Deltoid m.

Infraspinatus m. —

Teres major m. —

Latissimus dorsi m. — — Vertebral spine

Lumbar triangle — — Ext. abd. oblique m.

Erector spinae
(sacrospinalis) m. — — Iliac crest

— Gluteus maximus m.

Gluteal fold —

Pansky

103. SYMPATHETIC SYSTEM, SUBOCCIPITAL AND BACK DETAIL

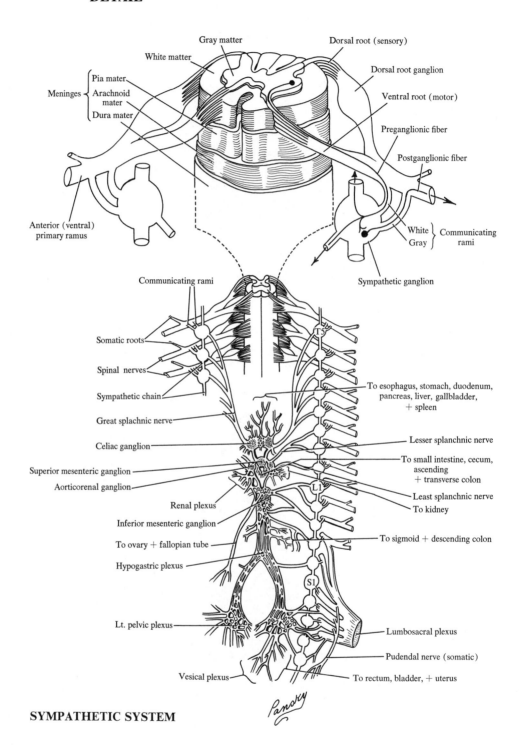

SYMPATHETIC SYSTEM

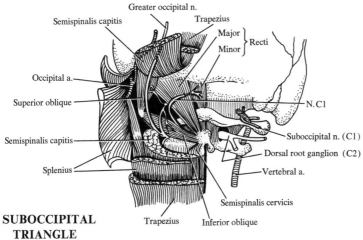

Greater occipital n.

Semispinalis capitis

Trapezius

Major

Minor

} Recti

Occipital a.

Superior oblique

N.C1

Suboccipital n. (C1)

Semispinalis capitis

Dorsal root ganglion (C2)

Splenius

Vertebral a.

Semispinalis cervicis

**SUBOCCIPITAL
TRIANGLE**

Trapezius

Inferior oblique

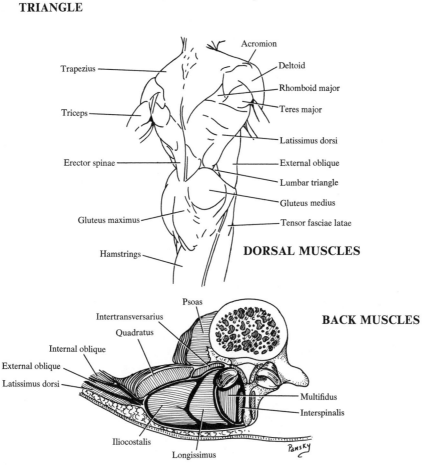

Acromion

Trapezius

Deltoid

Rhomboid major

Triceps

Teres major

Latissimus dorsi

Erector spinae

External oblique

Lumbar triangle

Gluteus medius

Gluteus maximus

Tensor fasciae latae

DORSAL MUSCLES

Hamstrings

Psoas

BACK MUSCLES

Intertransversarius

Quadratus

Internal oblique

External oblique

Latissimus dorsi

Multifidus

Interspinalis

Iliocostalis

Longissimus

Pansky

Upper
Extremity

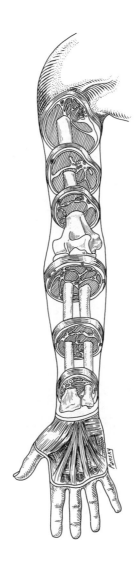

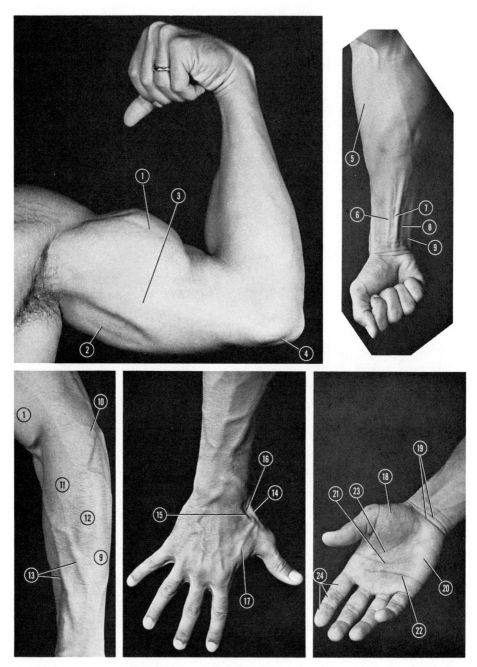

Figure 13. **Surface anatomy of upper extremity.** *1*, Biceps m.; *2*, long head of triceps m.; *3*, medial head of triceps m.; *4*, medial epicondyle; *5*, brachioradialis m.; *6*, flexor carpi radialis m.; *7*, palmaris longus tendon; *8*, flexor digitorum superficialis m.; *9*, flexor carpi ulnaris m.; *10*, epicondyle, *11*, extensor digitorum m.; *12*, extensor carpi ulnaris m.; *13*, superficial veins; *14*, extensor pollicis brevis m.; *15*, extensor pollicis longus m.; *16*, anatomical snuffbox; *17*, first dorsal interosseous m.; *18*, thenar eminence; *19*, wrist creases; *20*, hypothenar eminence; *21*, proximal transverse crease; *22*, distal transverse crease; *23*, transverse crease; *24*, phalanges.

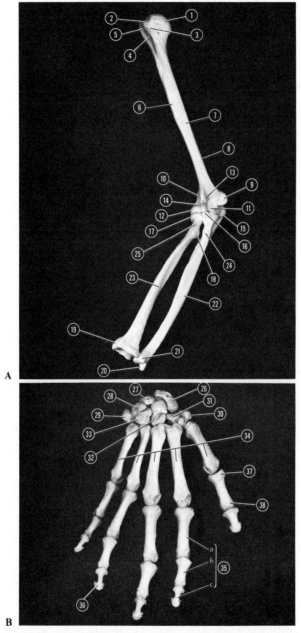

FIGURE 14. **A, Right humerus, ulna, and radius; B, Posterior view of hand.** *1*, Head of humerus; *2*, anatomical neck; *3*, lesser tubercle; *4*, intertubercular groove; *5*, greater tubercle; *6*, deltoid tuberosity; *7*, shaft of humerus; *8*, medial supracondylar ridge; *9*, medial epicondyle; *10*, lateral supracondylar ridge; *11*, trochlea; *12*, lateral epicondyle; *13*, coronoid fossa; *14*, radial fossa; *15*, olecranon; *16*, coronoid process; *17*, head of radius; *18*, tuberosity of radius; *19*, styloid process of radius; *20*, styloid process of ulna; *21*, head of ulna; *22*, shaft of ulna; *23*, shaft of radius; *24*, tuberosity of ulna; *25*, neck of radius; *26*, scaphoid; *27*, lunate; *28*, triquetrum; *29*, pisiform; *30*, trapezium; *31*, trapezoid; *32*, capitate; *33*, hamate; *34*, metacarpal bones; *35*, phalanges (*a*, proximal; *b*, middle; *c*, distal); *36*, tuberosity; *37*, base of phalanx; *38*, head of phalanx.

104. CUTANEOUS NERVES AND DERMATOMES OF THE UPPER EXTREMITY

I. Sources of cutaneous nerves: cervical plexus, brachial plexus, posterior primary divisions of upper thoracic nerves, and lateral cutaneous branches of second and third intercostals

II. Distribution

A. CERVICAL PLEXUS (C3 and C4) as medial, intermediate, and lateral *supraclavicular nerves* to upper pectoral, deltoid, and outer trapezius areas

B. POSTERIOR DIVISIONS of upper 3 thoracic nerves over trapezius area to spine of scapula

C. BRACHIAL PLEXUS (see p. 240)

1. Superior lateral brachial cutaneous nerve and other cutaneous twigs from axillary over deltoid and upper arm
2. Medial brachial cutaneous to medial side of arm as far as elbow
3. Posterior brachial cutaneous of radial over the posterior arm, nearly to the olecranon
4. Posterior antebrachial cutaneous from radial to lateral side of volar surface of arm, most of the posterior side of the lower arm, and middle posterior of the forearm
5. Lateral antebrachial cutaneous from musculocutaneous to lateral third of posterior of forearm and lateral half of anterior side of forearm
6. Medial antebrachial cutaneous from medial cord to middle third of anterior side of arm, medial half of anterior forearm, and medial third of posterior forearm
7. Palmar branch of median nerve to midpalm and thenar eminence
8. First, second, and third common palmar digitals to palmar surfaces of thumb, first and second fingers, and lateral side of ring finger, and posterior surface of terminal phalanges of these same fingers
9. Palmar cutaneous branch of ulnar to medial palm
10. Dorsal digital and metacarpal nerves from dorsal branch of ulnar to medial side of dorsum of hand, both sides of little finger, and medial side of ring finger to the tip
11. Lateral and medial branches of the superficial radial nerve to lateral part of dorsum of hand, dorsum of thumb, first 2 fingers, and lateral side of ring finger as far as terminal phalanx

D. LATERAL CUTANEOUS BRANCHES

1. Intercostobrachial from second intercostal to medial arm, usually with medial brachial cutaneous (another branch from the third sometimes occurs)

III. The dermatome is a specific skin area supplied by a specific spinal nerve, regardless of the cutaneous nerve within which the spinal segmental fibers lie. Knowledge of the dermatomes is helpful in localizing certain lesions of the nervous system. However, it is not enough to know merely the cord segment involved. When surgery is contemplated, it is also necessary to know at which level of the vertebral column a particular cord segment lies. For example, a patient showing numbness on the fifth digit and the medial side of the hand has a lesion in the eighth spinal cord segment, and this lies at the level of the seventh cervical vertebra

IV. Special features

A. THE 3RD AND 4TH CERVICAL NERVES supply a limited area of skin over the upper parts of the pectoral region and shoulder, and the 2nd thoracic nerve usually sends a branch to skin of the medial and upper part of the arm; otherwise, the skin of the upper limb is supplied by branches from the brachial plexus

B. IN THE LIMBS, the overlap between peripheral nerves is less extensive than in the trunk, so that complete interruption of a single peripheral nerve typically produces changes in sensation that are appreciated by the patient

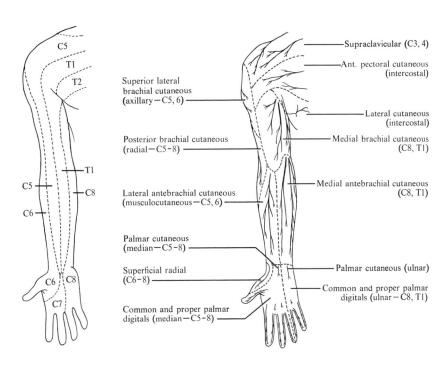

Supraclavicular (C3, 4)

Ant. pectoral cutaneous
(intercostal)

Superior lateral
brachial cutaneous
(axillary—C5, 6)

Lateral cutaneous
(intercostal)

Medial brachial cutaneous
(C8, T1)

Posterior brachial cutaneous
(radial—C5-8)

Medial antebrachial cutaneous
(C8, T1)

Lateral antebrachial cutaneous
(musculocutaneous—C5, 6)

Palmar cutaneous
(median—C5-8)

Superficial radial
(C6-8)

Palmar cutaneous (ulnar)

Common and proper palmar
digitals (ulnar—C8, T1)

Common and proper palmar
digitals (median—C5-8)

DERMATOMES

CUTANEOUS NERVES

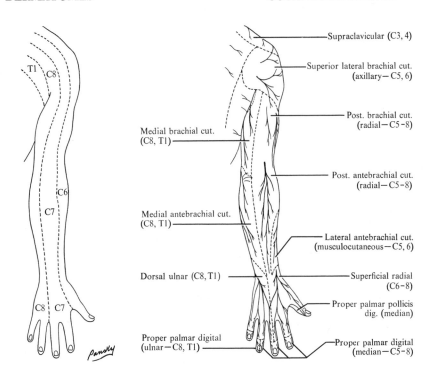

Supraclavicular (C3, 4)

Superior lateral brachial cut.
(axillary—C5, 6)

Medial brachial cut.
(C8, T1)

Post. brachial cut.
(radial—C5-8)

Post. antebrachial cut.
(radial—C5-8)

Medial antebrachial cut.
(C8, T1)

Lateral antebrachial cut.
(musculocutaneous—C5, 6)

Dorsal ulnar (C8, T1)

Superficial radial
(C6-8)

Proper palmar pollicis
dig. (median)

Proper palmar digital
(ulnar—C8, T1)

Proper palmar digital
(median—C5-8)

Pansky

– 227 –

105. SUPERFICIAL VEINS OF THE UPPER EXTREMITY

I. General: lie between 2 layers of superficial fascia

II. Specific veins

A. CEPHALIC VEIN
   1. Origin: lateral side of dorsal venous rete (network, arch) on hand
   2. Course: from dorsum curves around to lateral anterior side of forearm, ascends in front of elbow in groove between biceps and brachialis, then along lateral side of biceps to lie in deltopectoral triangle (groove), and pierces clavipectoral fascia to reach axillary vein
   3. Termination: axillary vein, below clavicle
   4. Tributaries from:
      a. Lateral side of palmar venous plexus
      b. Posterior and anterior veins on radial side of forearm
      c. Lower lateral arm
      d. Entire upper arm and shoulder region
   5. Communications
      a. Median cubital vein, running obliquely across antecubital area to join basilic vein

B. BASILIC VEIN
   1. Origin: medial side of dorsal venous rete (network, arch)
   2. Course: posterior side of medial forearm; below elbow crosses to volar side, ascends in groove between biceps and pronator teres muscles, continues along medial border of biceps, pierces deep fascia just below midarm, and runs along medial side of brachial artery to lower border of teres major, where it joins brachial vein
   3. Termination: joins brachial to form axillary vein
   4. Tributaries from:
      a. Medial side of palmar network of hand
      b. Posterior and medial side of forearm and lower arm
      c. Median cubital vein, from cephalic vein

C. ACCESSORY CEPHALIC VEIN: often small, arising either from dorsal network on medial hand or a network on dorsal forearm. It ends in cephalic vein, below elbow

D. MEDIAN ANTEBRACHIAL VEIN arises in palmar venous network, ascends on ulnar side of forearm, and ends in either basilic or median cubital veins

III. Clinical considerations

A. VEINS OF UPPER EXTREMITY are most often used for either drawing of blood samples or administration of drugs, fluids, or nutriments. The median cubital or the basilic veins, because of the strong support given by the bicipital aponeurosis, are most frequent sites

B. DUE TO THE PROXIMITY of the lateral antebrachial cutaneous nerve to the cephalic vein and the medial antebrachial cutaneous nerve to the basilic vein, care must be used to avoid these if venipuncture of these veins is attempted

C. THE HAND, like the rest of the upper extremity, is drained by 2 sets of veins, a *superficial group* (found in the superficial fascia) and a *deep set* (associated with the arteries)
   1. The superficial venous system is the more important of the 2 sets because it is larger, and most of the finger tributaries and lymphatics accompany its tributaries
   2. A few small veins are found in the tight subcutaneous fascia of the palm and palmar aspect of the fingers, but do not compare in size or number with those in the loose areolar subcutaneous tissue of the dorsum of the hand and fingers
   3. In the hand, forearm, and arm, the superficial and deep sets of veins anastomose with each other via a variable number of communicating or perforating veins

SUPERFICIAL VEINS

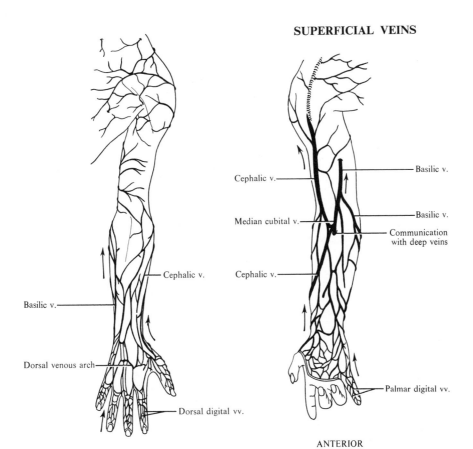

Cephalic v.

Basilic v.

Dorsal venous arch

Dorsal digital vv.

POSTERIOR

Cephalic v.

Median cubital v.

Cephalic v.

Basilic v.

Basilic v.

Communication with deep veins

Palmar digital vv.

ANTERIOR

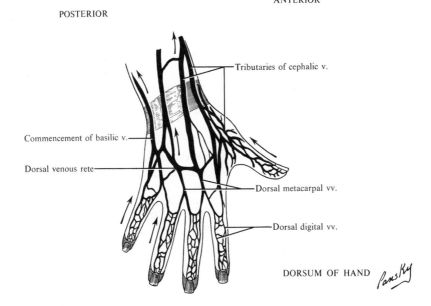

Tributaries of cephalic v.

Commencement of basilic v.

Dorsal venous rete

Dorsal metacarpal vv.

Dorsal digital vv.

DORSUM OF HAND

106. LYMPHATICS OF THE UPPER EXTREMITY

I. Superficial vessels

A. BEGIN IN NETWORK ON HAND. The plexuses of the palm and palmar surface of the fingers are more numerous than on dorsum

B. FROM FINGERS, vessels on either side of digit go toward dorsum of hand

C. FROM PALM may go toward fingers, wrist, or either side of hand, but most of the lymph from the thenar and hypothenar areas flows toward the vessels in the subcutaneous loose areolar tissue, the so-called dorsal subcutaneous space. Most of the lymph from the dorsum leaves via lymph vessels accompanying the cephalic and basilic veins

D. RUN BOTH IN FRONT OF AND BEHIND WRIST, forming radial, median, and ulnar groups of vessels, which travel with the cephalic, median antebrachial, and basilic veins
1. Some of the ulnar vessels end in epitrochlear nodes
2. Some of the radial vessels go to deltopectoral nodes
3. Most go to lateral axillary nodes

II. Deep vessels run with deep arteries and veins.
Most of the central palmar deep lymph vessels proceed deeply to join the lymph vessels associated with the superficial and deep venous arches. From here, lymph channels follow the venae comitantes of the radial and ulnar arteries

A. IN FOREARM, 4 sets: along the radial, ulnar, dorsal, and volar interosseous vessels
1. Some terminate in nodes along brachial artery
2. Most end in lateral axillary nodes

III. Lymph nodes of upper extremity

A. SUPERFICIAL
1. Supratrochlear above medial epicondyle, medial to basilic vein: receives afferents from medial fingers, palm, and forearm. Efferents go to deep nodes
2. Deltopectoral in deltopectoral groove, along cephalic vein: receive afferents from vessels of radial side that follow cephalic vein. Efferents go to either subclavian or inferior cervical nodes

B. DEEP
1. A few are sometimes found in forearm along radial, ulnar, and interosseous vessels and in arm along medial side of brachial artery
2. Axillary
   a. Lateral (brachial) located on medial and posterior aspects of axillary artery. Receive afferents from all upper extremity except those nodes around cephalic vein. Efferents go to central and subclavicular nodes
   b. Pectoral (anterior) located at lower border of pectoralis minor muscle. Receive afferents from anterior and lateral thorax, central and lateral mammary gland. Efferents go to central and subclavicular nodes
   c. Subscapular (posterior) located along course of subscapular artery. Receive afferents from lower back of neck and posterior wall of thorax. Efferents go to central nodes
   d. Central located toward apex of axilla. Receive afferents from all the above nodes. Efferents go to subclavicular
   e. Apical (subclavicular) located behind and above pectoralis minor muscle. Receive afferents from deltopectoral nodes; a direct connection with upper mammary gland and all other axillary nodes. Efferents form the subclavian trunk, which enters jugular trunk, subclavian vein, or thoracic duct (a few go to inferior cervical nodes)

IV. Clinical considerations

A. AXILLARY LYMPH NODES: a patient with an infection or malignancy in the upper extremity may complain of tenderness and swelling in the axilla. This is due to involvement of the axillary lymph nodes, especially the lateral group along the axillary artery, since these filter lymph from much of the upper limb

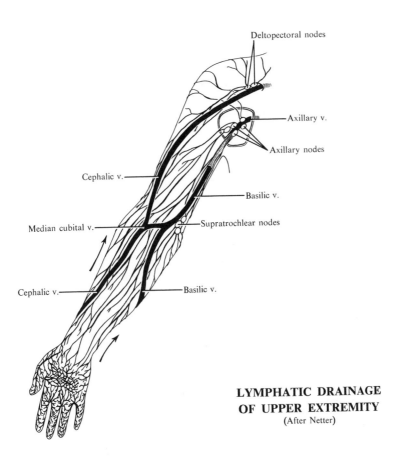

Deltopectoral nodes

Axillary v.

Axillary nodes

Cephalic v.

Basilic v.

Median cubital v.

Supratrochlear nodes

Cephalic v.

Basilic v.

**LYMPHATIC DRAINAGE
OF UPPER EXTREMITY**
(After Netter)

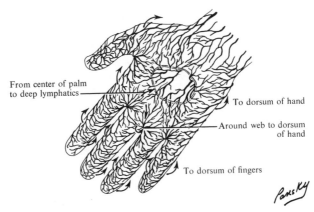

From center of palm
to deep lymphatics

To dorsum of hand

Around web to dorsum
of hand

To dorsum of fingers

Pansky

107. BONES OF THE SHOULDER GIRDLE

I. Clavicle (collar bone)

A. PARTS: sternal extremity, acromial (lateral) extremity, and body
 1. Sternal extremity articulates with sternum and has an articular facet for the first rib
 2. Acromial extremity: flat and rough, has area for articulation with acromion on undersurface. Does not reach "point of shoulder"—formed by acromion
 3. Body
  a. Medial part is rounded; lateral part is flattened
  b. Double curvature: convex anteriorly at medial end and concave anteriorly at lateral end
  c. Lateral part has an upper surface and undersurface, the latter showing the *conoid tubercle* and *trapezoid line*. It also has an anterior and posterior border
  d. Medial part has anterior, superior, and posterior borders; anterior, posterior, and inferior surfaces. The impression for the costoclavicular ligament and groove for the subclavius muscle appear on the inferior surface

II. Scapula

A. FLAT AND TRIANGULAR with 2 surfaces, 3 borders, and 3 angles
 1. Costal (anterior) surface concave—subscapular fossa
 2. Posterior surface divided by *spine* into *supra-* and *infraspinous fossae*
 3. Spine of scapula ends in the flattened *acromion*
  a. *Great notch of scapula* joins the infra- and supraspinous fossae
 4. The *scapular notch* is in the superior border
 5. At the lateral angle is located the *glenoid cavity*, with its *supra-* and *infraglenoid tuberosities*. Medial to the cavity is the *coracoid process*
  a. The *neck* is a constriction just medial to the glenoid cavity

III. Ossification

Location	When Appears	When Closes
A. CLAVICLE, from 3 centers (ossified before any other bone in body)		
Medial (in body)	5th–6th fetal week	
Lateral (in body)	5th–6th fetal week	Close by 25th year
In sternal end	18th–20th year	
B. SCAPULA, from seven or more centers		
Body	8th fetal week	15th year
Middle coracoid	15th–18th month	15th year
Upper glenoid	10th year	16th–18th year
Root of coracoid	14th–20th year	By 25th year
Base of acromion	14th–20th year	By 25th year
Inferior angle	14th–20th year	By 25th year
End of acromion	14th–20th year	By 25th year
Vertebral border	14th–20th year	By 25th year

IV. Clinical considerations

A. FRACTURES OF THE CLAVICLE usually occur in the medial part, with medial and downward displacement of the distal fragment due to the weight of the shoulder and the pull of the pectoralis major muscle

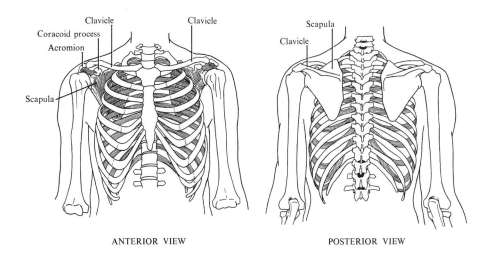

ANTERIOR VIEW   POSTERIOR VIEW

RT. CLAVICLE

SUPERIOR SURFACE

INFERIOR SURFACE

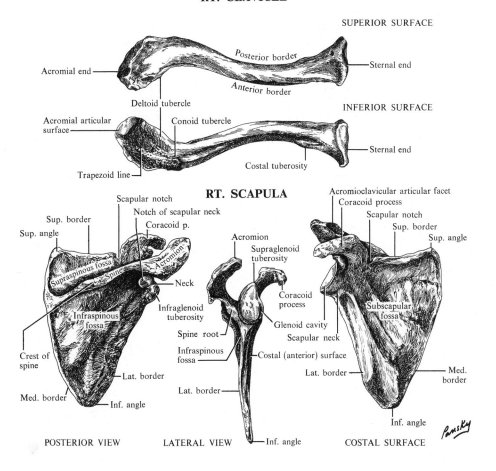

RT. SCAPULA

POSTERIOR VIEW   LATERAL VIEW   COSTAL SURFACE

108. ARTICULATIONS OF THE SHOULDER GIRDLE

I. Definition: bones through which upper extremity is attached to trunk

II. Bones (see p. 232)

III. Articulations

A. ACROMIOCLAVICULAR
1. Type: arthrodial, plane type of synovial
2. Bones: acromion of scapula with clavicle
3. Movement: gliding, rotation of scapula on clavicle
4. Ligaments
   a. Articular capsule completely surrounds articular areas
   b. Superior acromioclavicular covers joint cephalically
   c. Inferior acromioclavicular covers joint caudally
   d. Articular disk frequently absent or incomplete
   e. Coracoclavicular not part of joint but keeps clavicle against acromion
      i. Trapezoid from upper coracoid process to oblique line on caudal surface of clavicle. Limits ventral rotation of scapula
      ii. Conoid from base of coracoid process to coracoid tuberosity on caudal surface of clavicle. Limits dorsal rotation of scapula

B. STERNOCLAVICULAR (see p. 325)
1. Type: arthrodial, saddle type of synovial joint
2. Bones: clavicle with manubrium of sternum and cartilage of 1st rib
3. Movements: gliding, with some motion in almost any direction
4. Ligaments
   a. Articular capsule surrounds articulation
   b. Anterior sternoclavicular from clavicle to ventral surface of manubrium
   c. Posterior sternoclavicular between dorsal surfaces of clavicle and manubrium
   d. Interclavicular joins cephalic surface of the 2 clavicles and limits depression of shoulder
   e. Costoclavicular from first costal cartilage to costal tuberosity on caudal surface of clavicle. Limits elevation of shoulder
   f. Articular disk between sternum and clavicle and attached to clavicle and sternum. Helps limit depression of shoulder

IV. Muscles acting on shoulder girdle

A.	*Elevation*	*Depression*	*Protraction**	*Retraction†*
	Upper trapezius	Subclavius	Serratus anterior	Rhomboids
	Levator scapulae	Pect. minor	Pect. minor	Middle and lower
	Sternocleidomastoid	Lower trapezius		trapezius
	Rhomboids			

B. *Rotation of Scapula*	*Up*	*Down*
	Upper and lower trapezius	Levator scapulae
	Serratus anterior	Rhomboids
		Pectoralis major and minor
		Latissimus dorsi

*Protraction: drawing forward or ventral.
†Retraction: drawing backward or dorsal.

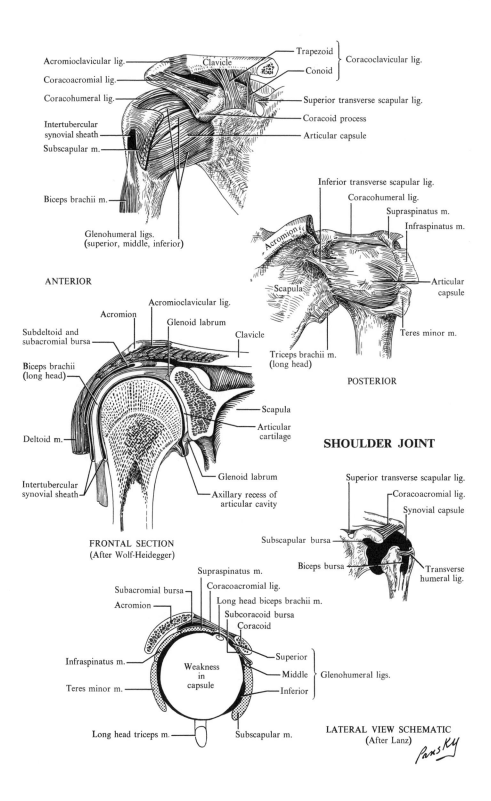

Acromioclavicular lig.

Clavicle

Trapezoid
Conoid
} Coracoclavicular lig.

Coracoacromial lig.

Coracohumeral lig.

Superior transverse scapular lig.

Coracoid process

Intertubercular
synovial sheath

Articular capsule

Subscapular m.

Biceps brachii m.

Glenohumeral ligs.
(superior, middle, inferior)

ANTERIOR

Inferior transverse scapular lig.

Coracohumeral lig.

Supraspinatus m.

Infraspinatus m.

Acromion

Scapula

Articular
capsule

Teres minor m.

Triceps brachii m.
(long head)

POSTERIOR

Acromioclavicular lig.

Acromion

Glenoid labrum

Subdeltoid and
subacromial bursa

Clavicle

Biceps brachii
(long head)

Scapula

Articular
cartilage

Deltoid m.

Glenoid labrum

Intertubercular
synovial sheath

Axillary recess of
articular cavity

FRONTAL SECTION
(After Wolf-Heidegger)

SHOULDER JOINT

Superior transverse scapular lig.

Coracoacromial lig.

Synovial capsule

Subscapular bursa

Biceps bursa

Transverse
humeral lig.

Supraspinatus m.

Coracoacromial lig.

Subacromial bursa

Long head biceps brachii m.

Acromion

Subcoracoid bursa

Coracoid

Infraspinatus m.

Weakness
in
capsule

Superior

Middle

Inferior

} Glenohumeral ligs.

Teres minor m.

Long head triceps m.

Subscapular m.

LATERAL VIEW SCHEMATIC
(After Lanz)

Pansky

109. HUMERUS

I. General: largest and longest bone of upper extremity

II. Parts: proximal extremity, shaft, and distal extremity
A. PROXIMAL: head, anatomical neck, greater tubercle, lesser tubercle, intertubercular groove or sulcus (bicipital groove), crests of the tubercles (are lips of groove), and surgical neck
B. SHAFT
    1. Borders: lateral and medial
    2. Surfaces: anterolateral, anteromedial, and posterior
    3. Special features: radial sulcus, deltoid tuberosity
    4. Lateral supracondylar ridge
C. DISTAL: medial and lateral epicondyles, capitulum, radial fossa, ulnar sulcus, trochlea, coronoid fossa, olecranon fossa, lateral and medial supracondylar ridges

III. Ossification of humerus: from 8 centers

Part	When Appears	When Closes
Body	8th week, fetal	As joined by the following
Head	1st year	
Greater tubercle	3rd year	Join body in 16th–17th year
Lesser tubercle	5th year	All form 1 center in 6th year
Capitulum	2nd year	
Trochlea	12th year	16th–17th year
Lateral epicondyle	13th year	All form 1 center and join body at same time
Medial epicondyle	5th year	18th year

IV. Clinical considerations
A. SURGICAL NECK—constriction just below the tubercles. Frequent site of fractures
B. ALTHOUGH TRAUMA often determines how bones will be displaced when broken, the displacement due to muscle insertion is also important, especially on the shorter of the two broken pieces. This is an important consideration in reduction of fractures
C. FRACTURES OF THE MIDDLE PART of shaft may injure radial nerve
D. TRAUMA AT MEDIAL EPICONDYLE may injure ulnar nerve
E. COMPLETE FRACTURES of the humerus usually show some overriding and other displacement of the ends as a result of muscular pull; rotatory and angular displacements of the upper fragment are characteristic of fractures of the surgical neck, and angular displacement is characteristic of lower fractures
F. EXPOSURE OF THE HUMERAL SHAFT, in surgery, is difficult because the bone is not subcutaneous anywhere, and important vessels and nerves are closely related to it
G. TRAUMATIC SEPARATION of the proximal epiphysis of the humerus can occur in people under 20 years of age because this epiphysis does not fuse with the body until about the 20th year in males and approximately 2 years earlier in females. The head displacement is similar to that seen in fractures through the surgical neck of the humerus in elderly people

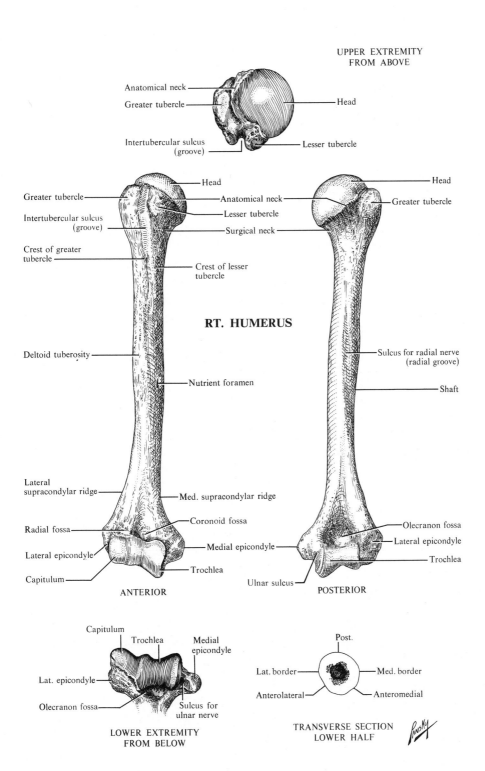

UPPER EXTREMITY
FROM ABOVE

Anatomical neck

Greater tubercle

Head

Intertubercular sulcus
(groove)

Lesser tubercle

Head

Greater tubercle

Anatomical neck

Lesser tubercle

Intertubercular sulcus
(groove)

Surgical neck

Crest of greater
tubercle

Crest of lesser
tubercle

RT. HUMERUS

Deltoid tuberosity

Sulcus for radial nerve
(radial groove)

Nutrient foramen

Shaft

Lateral
supracondylar ridge

Med. supracondylar ridge

Radial fossa

Coronoid fossa

Olecranon fossa

Lateral epicondyle

Medial epicondyle

Lateral epicondyle

Capitulum

Trochlea

Trochlea

Ulnar sulcus

ANTERIOR

POSTERIOR

Capitulum

Trochlea

Medial
epicondyle

Post.

Lat. epicondyle

Lat. border

Med. border

Olecranon fossa

Sulcus for
ulnar nerve

Anterolateral

Anteromedial

LOWER EXTREMITY
FROM BELOW

TRANSVERSE SECTION
LOWER HALF

110. MUSCLES OF THE CHEST WALL

I.

Name	Origin	Insertion	Action	Nerve
Platysma (see p. 52)	Fascia over pect. maj. & deltoid	Mandible Skin of lower face	Draws lower lip down and back Opens jaws Pulls up skin from clavicle	Facial (VII)
Pectoralis major (see p. 250)	Medial clavicle Lat. sternum to 7th cost. cart. Costal carts. 2–6 Apon. of ext. abd. oblique	Crest of greater tubercle of humerus	Flexes, adducts, and rotates arm medially In climbing draws body upward	Med. and lat. ant. thoracic
Pectoralis minor	Upper outer surface, ribs 3–5	Coracoid process of scapula	Draws scapula down and forward	Med. ant. thoracic
Subclavius	First rib cartilage	Subclavian groove of clavicle	Draws shoulder down and forward	N. to sub-clavius
Serratus anterior	Outer surface and upper border of ribs 1–8	Med. angle, vertebral border, and inf. angle of scapula	Rotates scapula, raising point of shoulder Draws scapula forward	Long thoracic
Subscapularis (see p. 250)	Medial part, subscapular fossa	Lesser tubercle of humerus Capsule of shoulder joint	Rotates arm medially, helps in adduction, abduction, flexion, and extension	Upper and lower sub-scapular
Teres major (see p. 250)	Dorsal surface inf. angle of scapula	Crest of lesser tubercle of humerus	Adducts, extends, and rotates arm medially	Lower sub-scapular

II. Special features

A. THE DIRECTION OF THE FIBERS of the pectoralis major should be noted: from clavicular origin, down and laterally; from sternum, horizontally; from lower ribs, upward and laterally

B. THE ACTION OF THE SERRATUS ANTERIOR in the rotation of the scapula makes possible abduction of the arm of more than 90°

C. WHEN THE HANDS ARE FIXED, as in gripping an object over the head, the pectoralis major helps to draw the body toward the hands

D. THE APPARENT INCONGRUITY in the action of the subscapularis in performing opposing action depends on position of the arm; if extended, the muscle flexes; if flexed, it helps to extend, etc.

MUSCLES OF CHEST WALL

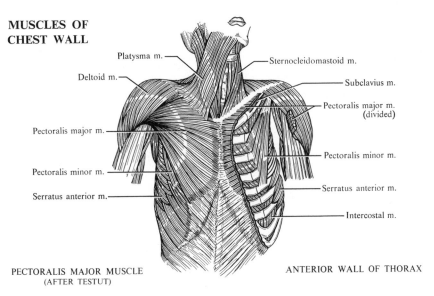

Platysma m.
Deltoid m.
Pectoralis major m.
Pectoralis minor m.
Serratus anterior m.

Sternocleidomastoid m.
Subclavius m.
Pectoralis major m. (divided)
Pectoralis minor m.
Serratus anterior m.
Intercostal m.

ANTERIOR WALL OF THORAX

PECTORALIS MAJOR MUSCLE
(AFTER TESTUT)

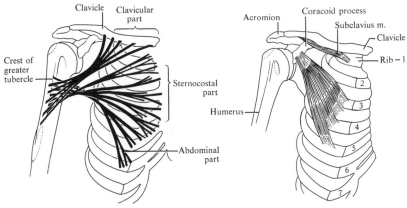

Clavicle
Clavicular part
Crest of greater tubercle
Sternocostal part
Abdominal part

Acromion
Coracoid process
Subclavius m.
Clavicle
Rib – 1
2
3
4
5
6
7
Humerus

PECTORALIS MINOR MUSCLE

SERRATUS ANTERIOR MUSCLE

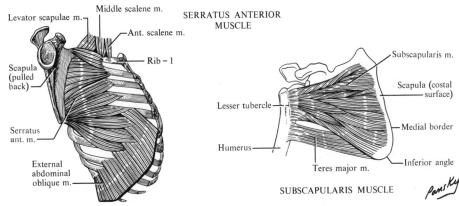

Levator scapulae m.
Middle scalene m.
Ant. scalene m.
Rib – 1
Scapula (pulled back)
Serratus ant. m.
External abdominal oblique m.

Subscapularis m.
Scapula (costal surface)
Lesser tubercle
Medial border
Humerus
Inferior angle
Teres major m.

SUBSCAPULARIS MUSCLE

Pansky

– 239 –

111. PARTS OF THE AXILLARY ARTERY AND THE BRACHIAL PLEXUS

I. Axillary artery

A. ORIGIN: subclavian artery becomes axillary artery at lateral border of first rib
B. TERMINATION: becomes brachial artery at lower border of tendon of teres major muscle
C. DIVISIONS
   1. First part from first rib to upper border of pectoralis minor muscle
   2. Second part lies behind pectoralis minor muscle
   3. Third part from lower border of pectoralis minor muscle to lower border of tendon of teres major muscle

II. Brachial plexus

A. ORIGIN: chiefly, anterior primary divisions of cervical nerves 5–8 plus first thoracic nerve
B. COMPONENTS: roots, trunks, and cords
C. MANNER OF FORMATION
   1. Roots (the anterior primary divisions) of C5 and C6 join to form *upper trunk;* C7 continues by itself as *middle trunk;* C8 and T1 join to form *lower trunk.* This occurs near lateral border of the scalenes in the posterior triangle of neck
   2. Trunks are short and split into anterior and posterior divisions. The anterior divisions of the upper and middle trunks form the *lateral cord;* the anterior part of the lower trunk becomes the *medial cord;* the posterior divisions of all trunks form the *posterior cord*
   3. The cords give rise to terminal branches
D. BRANCHES FROM

	Part	Spinal Segment	Part	Spinal Segment
	1. Roots		2. Trunks	
	a. Dorsal scapular	C5	a. N. to subclavius	C5, 6
	b. Long thoracic	C5, 6, 7	b. Suprascapular	C5, 6
	3. Cords		4. Terminal	
	a. Pectorals (lat. and med.)	C5, 6, 7, 8, T1	a. Musculocutaneous (lateral cord)	C5, 6, 7
Post.	b. Subscapular (lower and upper)	C5, 6	b. Median (lateral and medial cord)	C6, 7, 8, T1
	c. Thoracodorsal	C5, 6, 7	c. Ulnar (medial cord)	C8, T1
	d. Axillary	C5, 6		
Med.	e. Med. brach. cutan.	C8, T1	d. Radial (posterior cord)	C5, 6, 7, 8, T1
	f. Med. antebrach. cutan.	C8, T1		

III. Special features

A. TWO ROOTS of the brachial plexus lie above and two below the level of the cricoid cartilage
B. THE TRUNKS of the plexus lie between the scalene muscles. The muscles do not lie lateral to the midpoint of the clavicle, and the same may be said of the trunks
C. THE DIVISIONS of the plexus lie adjacent to the 3rd part of the subclavian artery
D. THE CORDS of the plexus extend from the midpoint of the clavicle to the inferomedial aspect of the coracoid process of the scapula

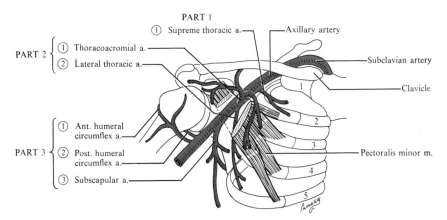

PART 1
① Supreme thoracic a.

PART 2 { ① Thoracoacromial a.
         ② Lateral thoracic a.

PART 3 { ① Ant. humeral circumflex a.
         ② Post. humeral circumflex a.
         ③ Subscapular a.

Axillary artery
Subclavian artery
Clavicle
Pectoralis minor m.

BRANCHES OF THE AXILLARY ARTERY

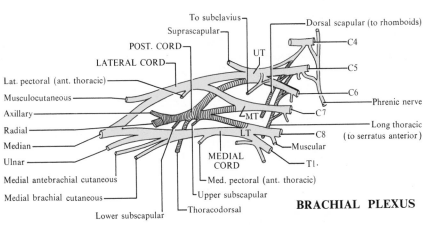

To subclavius
Suprascapular
POST. CORD
LATERAL CORD
Lat. pectoral (ant. thoracic)
Musculocutaneous
Axillary
Radial
Median
Ulnar
Medial antebrachial cutaneous
Medial brachial cutaneous
Lower subscapular

Dorsal scapular (to rhomboids)
C4
UT
C5
C6
C7
Phrenic nerve
MT
Long thoracic (to serratus anterior)
LT
C8
Muscular
MEDIAL CORD
T1.
Med. pectoral (ant. thoracic)
Upper subscapular
Thoracodorsal

BRACHIAL PLEXUS

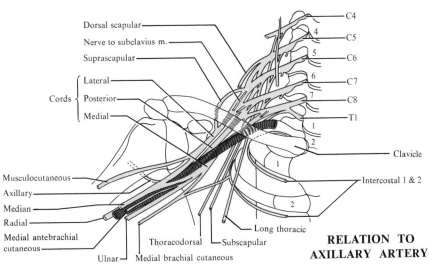

Dorsal scapular
Nerve to subclavius m.
Suprascapular
Lateral
Cords { Posterior
        Medial
Musculocutaneous
Axillary
Median
Radial
Medial antebrachial cutaneous
Ulnar
Thoracodorsal
Medial brachial cutaneous
Subscapular
Long thoracic

C4
C5
C6
C7
C8
T1
Clavicle
Intercostal 1 & 2

RELATION TO AXILLARY ARTERY

112. THE AXILLARY REGION

I. The axilla
A. A SPACE, pyramidal in shape, between the arm and thoracic wall
B. BOUNDARIES
   1. Apex: directed upward and medialward, ending in the *cervicoaxillary canal,* which leads into the posterior triangle of neck
   2. Base: formed by axillary fascia and skin
   3. Anterior wall: pectoralis major and minor muscles, clavipectoral fascia
   4. Posterior wall: subscapularis, teres major, latissimus dorsi muscles
   5. Medial wall: first 4 ribs and intercostal muscles, upper part of serratus anterior muscle
   6. Lateral wall: humerus, coracobrachialis and biceps muscles
C. CONTENTS
   1. Axillary vessels and their branches
   2. Brachial plexus and its branches
   3. Lymph nodes, embedded in fat

II. The fascia

A. AXILLARY: investing layer extending from the pectoralis major to latissimus dorsi muscles, arching inward to form the hollow of the armpit. It is continuous with the fasciae covering the muscles bounding the axilla
B. CLAVIPECTORAL (CORACOCLAVICULAR): attached above in front and behind the subclavius muscle. At lower border of this muscle, these 2 sheets form a single layer, which extends downward and laterally to the border of the pectoralis minor muscle. Here, the fascia splits to surround that muscle and again forms a single sheet at its lower border, which is continuous with the axillary fascia as the *suspensory ligament*
   1. Costocoracoid ligament: that part of the above fascia, greatly strengthened, which lies in front of the subclavius muscle, extending from the first rib to the coracoid process
   2. Coracoclavicular fascia: that portion of the clavipectoral fascia lying between the pectoralis minor and subclavius muscles. Is pierced by thoracoacromial vessels, cephalic vein, and lateral anterior thoracic nerve
C. AXILLARY SHEATH: at apex of axilla, fascia of first two ribs in first interspace becomes continuous with scalene fascia. Here this fascia forms a tubular sheath for vessels and nerves entering the axilla. Under subclavius and pectoralis minor muscles, it is adherent to the clavipectoral fascia

III. Special features
A. THE AXILLA houses the great vessels and nerves of the limb, which are closely grouped together and enclosed in a layer of fascia (*axillary fascia*)
B. MOST OF THE NERVES appearing in the axilla are the lower part of the brachial plexus and its major branches, and closely surround the axillary artery
C. THE AXILLARY ARTERY is conveniently thought of as the central structure of the axilla, with axillary vein closely related to it
D. LYMPH NODES of the axilla lie in the looser connective tissue that serves as padding about the vital structures and are particularly related to the blood vessels
E. THE AREA THROUGH WHICH the apex of the axilla reaches the neck is called the *cervicoaxillary canal,* bounded anteriorly by the clavicle, posteriorly by the scapula, and medially by the 1st rib. A majority of structures coursing between neck and axilla or the reverse do so via this canal. The clavicle holds the shoulder laterally to permit this. Infections also may follow this pathway

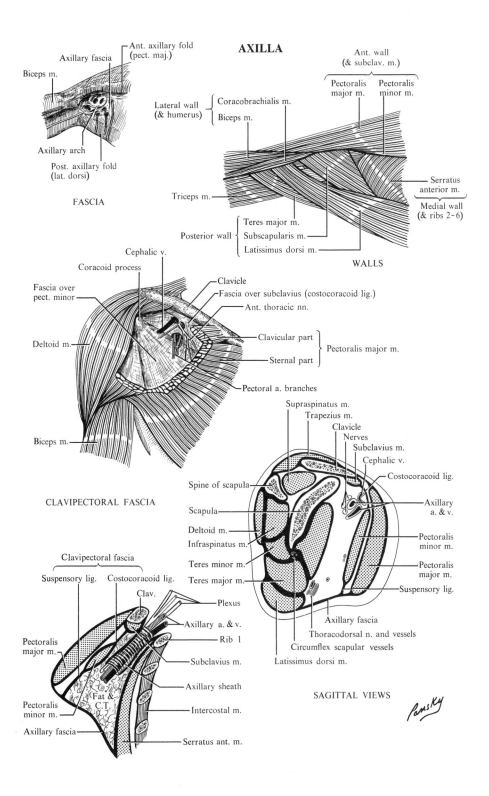

AXILLA

FASCIA

Ant. axillary fold (pect. maj.)
Axillary fascia
Biceps m.
Axillary arch
Post. axillary fold (lat. dorsi)

WALLS

Ant. wall (& subclav. m.)
Pectoralis major m.
Pectoralis minor m.
Lateral wall (& humerus)
Coracobrachialis m.
Biceps m.
Serratus anterior m.
Medial wall (& ribs 2-6)
Triceps m.
Teres major m.
Posterior wall
Subscapularis m.
Latissimus dorsi m.

CLAVIPECTORAL FASCIA

Cephalic v.
Coracoid process
Fascia over pect. minor
Clavicle
Fascia over subclavius (costocoracoid lig.)
Ant. thoracic nn.
Deltoid m.
Clavicular part
Sternal part
Pectoralis major m.
Pectoral a. branches
Biceps m.

Clavipectoral fascia
Suspensory lig.
Costocoracoid lig.
Clav.
Plexus
Axillary a. & v.
Rib 1
Subclavius m.
Axillary sheath
Intercostal m.
Serratus ant. m.
Pectoralis major m.
Pectoralis minor m.
Axillary fascia
Fat & C.T.

SAGITTAL VIEWS

Supraspinatus m.
Trapezius m.
Clavicle
Nerves
Subclavius m.
Cephalic v.
Costocoracoid lig.
Spine of scapula
Axillary a. & v.
Scapula
Pectoralis minor m.
Deltoid m.
Infraspinatus m.
Teres minor m.
Pectoralis major m.
Teres major m.
Suspensory lig.
Axillary fascia
Thoracodorsal n. and vessels
Circumflex scapular vessels
Latissimus dorsi m.

Pansky

113. VESSELS AND NERVES OF AXILLA

I. Axillary artery

A. RELATIONS

1. Part I

> *Anterior*
> Subclav. and pect. minor mm., clavipect. fascia,
> lat. ant. thoracic n., thoracoacromial vessels, cephalic v.
> *Lateral* **Axillary I** *Medial*
> Brachial plexus Axillary v.
> *Posterior*
> First intercost. and serratus ant. mm.,
> Long thoracic n.

2. Part II

> *Anterior*
> Pect. major and minor mm.
> *Lateral* *Medial*
> Lat. cord of plexus **Axillary II** Med. cord brachial plexus,
> axillary v.
> *Posterior*
> Post. cord of plexus, subscapular m.

3. Part III

> *Anterior*
> Skin, fascia, pect. major m.,
> medial head of median n.
> *Lateral* *Medial*
> Coracobrach. m., median **Axillary III** Med. antebrach. cutaneous and
> and musculocutaneous nn. ulnar nn., axillary v.
> *Posterior*
> Subscap. m., tendons of teres maj. and latissimus
> dorsi mm., radial and axillary nn.

B. BRANCHES

1. Part I: supreme (superior) thoracic
2. Part II: thoracoacromial and lateral thoracic
3. Part III: subscapular; anterior and posterior circumflex humeral

II. Origin and distribution of some nerves from brachial plexus

Name	Distribution
Long thoracic	Serratus ant. m.
N. to subclavius	Subclavius m.
Lat. pectoral	Pect. major m.
Med. pectoral	Pect. major and minor mm.
Subscapular	Subscap. and teres major mm.
Thoracodorsal	Latissimus dorsi m.
Axillary	Deltoid, teres minor mm., and skin of arm
Musculocutaneous	Muscles of arm and skin of forearm

III. Axillary vein begins at union of basilic and brachial veins and terminates at first rib as subclavian vein. Lies medial to and partly overlaps axillary artery with the medial cord, median, ulnar, and medial anterior pectoral nerves between them. Receives tributaries corresponding to branches of artery plus the cephalic vein

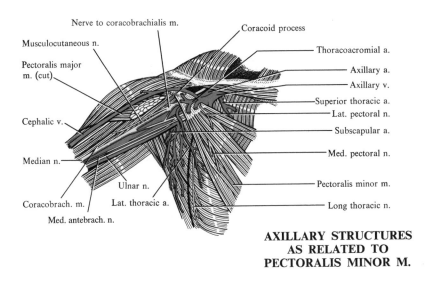

AXILLARY STRUCTURES
AS RELATED TO
PECTORALIS MINOR M.

AXILLA
(After Lanz)

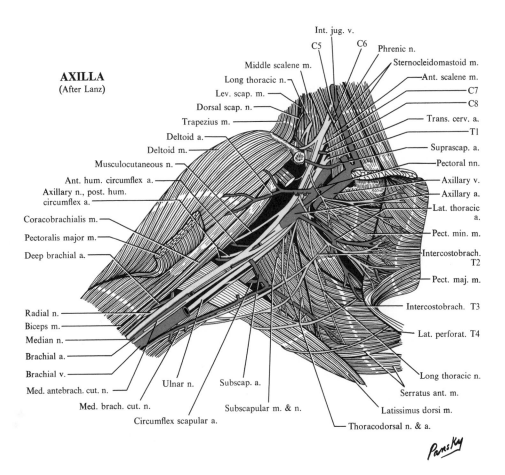

114. MUSCLES OF THE SHOULDER

I.

Name	Origin	Insertion	Action	Nerve
Deltoid (see p. 250)	Ant. border and upper surface of clavicle Lat. margin, upper surface of acromion Lower border of spine of scapula	Deltoid tubercle of humerus	Abducts arm Ant. fibers flex and rotate arm medially Post. fibers extend and rotate arm laterally	Axillary
Supra-spinatus	Medial part of supraspinous fossa	Highest impression, great tubercle of humerus	Abducts arm	Supra-scapular
Infra-spinatus	Medial part of infraspinous fossa	Middle impression, great tubercle of humerus	Rotates arm laterally Upper fibers abduct Lower fibers adduct	Supra-scapular
Teres minor	Upper dorsal axillary border of scapula	Lowest impression, great tubercle of humerus United to joint capsule	Rotates arm laterally and adducts it	Axillary
Subscapularis (see p. 238) Teres major (see p. 238)				

II. Special features

A. THE DELTOID MUSCLE is responsible for the roundness of the shoulder
   1. Although the deltoid is a strong abductor of the humerus, it cannot initiate this movement since, when the arm is by the side, the line of pull of the intermediate part of the muscle is parallel to the humeral shaft. Thus, the supraspinatus is important in early phases of abduction, but both muscles act together throughout arm abduction

B. CERTAIN SPACES BETWEEN ADJOINING MUSCLES or parts of muscle and bone should be noted, for they permit passage of vessels and nerves
   1. Triangular space, a triangular hiatus whose base is lateral; bounded by the long head of the triceps, the teres minor, and teres major
   2. Quadrilateral space, formed by the teres minor, the teres major, the humerus, and the long head of the triceps
   3. Triangular interval, formed by the teres major, lateral head of the triceps, and long head of the triceps

C. TENDONS of the supraspinatus, infraspinatus, and teres major all strengthen the shoulder joint

D. THE SUPRASPINATUS, INFRASPINATUS, AND TERES MINOR MUSCLES are sometimes referred to as the great "sit" muscles of the shoulder: s for supra, i for infra, and t for teres minor, which all "sit" on the great tubercle of the humerus

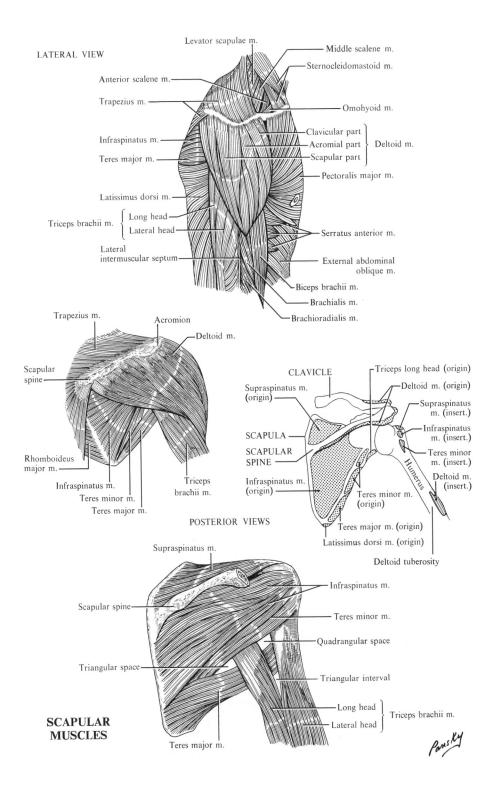

LATERAL VIEW

Levator scapulae m.

Middle scalene m.

Sternocleidomastoid m.

Anterior scalene m.

Trapezius m.

Omohyoid m.

Clavicular part
Acromial part ⎱ Deltoid m.
Scapular part

Infraspinatus m.

Teres major m.

Pectoralis major m.

Latissimus dorsi m.

Triceps brachii m. ⎰ Long head
Lateral head

Lateral
intermuscular septum

Serratus anterior m.

External abdominal
oblique m.

Biceps brachii m.

Brachialis m.

Brachioradialis m.

Trapezius m.

Acromion

Deltoid m.

Scapular
spine

CLAVICLE

Triceps long head (origin)

Supraspinatus m.
(origin)

Deltoid m. (origin)

Supraspinatus
m. (insert.)

SCAPULA

Infraspinatus
m. (insert.)

SCAPULAR
SPINE

Teres minor
m. (insert.)

Rhomboideus
major m.

Infraspinatus m.
(origin)

Deltoid m.
(insert.)

Infraspinatus m.

Teres minor m.

Teres major m.

Triceps
brachii m.

Teres minor m.
(origin)

Teres major m. (origin)

Latissimus dorsi m. (origin)

Deltoid tuberosity

POSTERIOR VIEWS

Supraspinatus m.

Infraspinatus m.

Scapular spine

Teres minor m.

Quadrangular space

Triangular space

Triangular interval

Long head
Lateral head ⎱ Triceps brachii m.

**SCAPULAR
MUSCLES**

Teres major m.

Pansky

– 247 –

115. COLLATERAL CIRCULATION AROUND THE SHOULDER

I. Collateral circulation for ligation of or obstruction in any artery is important and is accomplished through the anastomosing of arteries that arise from the main channel above the stoppage with those that arise below that point

II. Collateral circulation for the ligation of the axillary artery

Arteries Above Ligature	*Arteries Below Ligature*
A. IN AXILLARY 1, above the origin of the thoracoacromial artery	
1. Suprascapular (from thyrocervical trunk)	Circumflex scapular (from subscapular)
2. Descending (deep) ramus (from trans. cervical of thyrocervical trunk)	Circumflex scapular
3. Descending scapular (dorsal scapular) (from subclavian or thyrocervical trunk)	Circumflex scapular
4. Intercostals (from aorta)	Thoracodorsal (from axillary 2)
B. IN AXILLARY 2, between origins of thoracoacromial and lateral thoracic	
1. Suprascapular—as above	Circumflex scapular—as above
2. Descending ramus—as above	Circumflex scapular—as above
3. Descending scapular—as above	Circumflex scapular—as above
4. Intercostals (from aorta)	Lateral thoracic (from axillary 2) and thoracodorsal (from subscapular)
5. Pectoral branch (from thoraco-acromial)	Lateral thoracic (from axillary 2)
C. IN AXILLARY 2, between lateral thoracic and subscapular	
1. Suprascapular—as above	As above
2. Descending ramus—as above	As above
3. Descending scapular—as above	As above
4. Lateral thoracic (from axillary 2)	Thoracodorsal (from subscapular)
5. Acromial branch (thoracoacromial)	Post. circumflex humeral (axillary 3)
D. IN AXILLARY 3, below the subscapular and the 2 circumflex humeral	
1. Ant. and post. circumflex humeral (from axillary 3)	Deep brachial (from brachial)

III. Sites where anastomoses occur
A. BETWEEN THE SUPRASCAPULAR AND CIRCUMFLEX SCAPULAR in infraspinous fossa
B. BETWEEN DESCENDING SCAPULAR AND CIRCUMFLEX SCAPULAR along medial border of scapula
C. BETWEEN PECTORAL BRANCHES, INTERCOSTALS, LATERAL THORACIC, OR THORACODORSAL on thoracic walls
D. BETWEEN ACROMIAL AND POSTERIOR CIRCUMFLEX HUMERAL around acromion
E. BETWEEN CIRCUMFLEX HUMERAL AND DEEP BRACHIAL: in or under triceps muscle

IV. Special features
A. THE CIRCUMFLEX SCAPULAR ARTERY leaves axilla through the triangular space
B. THE POSTERIOR CIRCUMFLEX HUMERAL ARTERY leaves the axilla via the quadrilateral space with the axillary nerve
C. SUPRASCAPULAR NERVE AND SUPRASCAPULAR ARTERY run together, but the nerve goes under and the artery passes over the transverse scapular ligament
D. THE DORSAL SCAPULAR NERVE descends with the descending branch of the transverse cervical artery

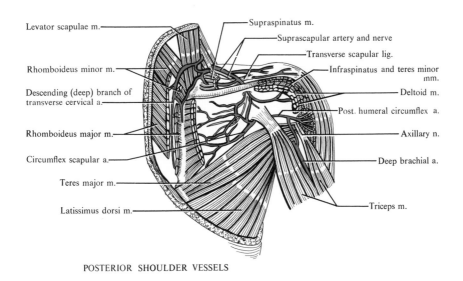

Levator scapulae m.

Suprascapular m.

Suprascapular artery and nerve

Transverse scapular lig.

Rhomboideus minor m.

Infraspinatus and teres minor mm.

Descending (deep) branch of transverse cervical a.

Deltoid m.

Post. humeral circumflex a.

Rhomboideus major m.

Axillary n.

Circumflex scapular a.

Deep brachial a.

Teres major m.

Latissimus dorsi m.

Triceps m.

POSTERIOR SHOULDER VESSELS

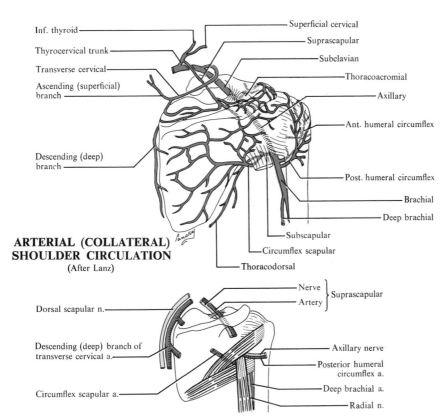

Inf. thyroid

Superficial cervical

Thyrocervical trunk

Suprascapular

Transverse cervical

Subclavian

Ascending (superficial) branch

Thoracoacromial

Axillary

Ant. humeral circumflex

Descending (deep) branch

Post. humeral circumflex

Brachial

Deep brachial

ARTERIAL (COLLATERAL) SHOULDER CIRCULATION
(After Lanz)

Subscapular

Circumflex scapular

Thoracodorsal

Nerve

Artery

} Suprascapular

Dorsal scapular n.

Descending (deep) branch of transverse cervical a.

Axillary nerve

Posterior humeral circumflex a.

Circumflex scapular a.

Deep brachial a.

Radial n.

116. SHOULDER JOINT

I. Type: enarthrodial (ball and socket)

II. Bones: spherical head of humerus and glenoid cavity of scapula

III. Movements: flexion, extension, abduction, adduction, medial rotation, lateral rotation, and circumduction

IV. Ligaments
A. ARTICULAR CAPSULE from edge of glenoid fossa and glenoidal labrum to anatomical neck of humerus. Strengthened: anteriorly, subscapular muscle; posteriorly, infraspinatus and teres minor muscle; above, supraspinatus muscle; below, long head of triceps
B. CORACOHUMERAL from coracoid to greater tubercle
C. GLENOHUMERAL: usually 3 bands
   1. From medial side of glenoid cavity to lower part of lesser tubercle
   2. From lower edge of glenoid cavity to lower anatomical neck
   3. From above glenoid rim near root of coracoid process along medial side of biceps tendon to a depression above lesser tubercle of humerus
D. TRANSVERSE HUMERAL bridges intertubercular sulcus (groove)
E. GLENOIDAL LABRUM: fibrocartilage attached to edges of glenoid fossa to deepen it and protect bone. Above, it is continuous with biceps tendon (long head)

V. Synovial membrane: from glenoid cavity over labrum; lines inside of capsule and is reflected on anatomical neck to articular cartilage. Encloses tendon of long head of biceps muscle in a tubular sheath, intertubercular synovial sheath

VI. Muscles acting on joint

Flexion	Extension	Abduction	Adduction	Med. Rotation	Lat. Rotation
Pect. major (clav. head)	Latissimus dorsi	Deltoid (as whole)	Pect. major (as whole)	Pect. major (as whole)	Infraspinatus
Deltoid (ant. fibers)	Teres major	Supra-spinatus	Latissimus dorsi	Latissimus dorsi	Teres minor
Coraco-brach.	Deltoid (post. fibers)		Teres major	Teres major	Deltoid (post. fibers)
Biceps (long head)	Triceps (long head)		Subscap-ularis	Subscap-ularis	
			Triceps (long head)	Deltoid (ant. fibers)	

VII. Clinical considerations
A. MOST OF THE STRENGTH OF THE JOINT is provided by muscle tendons that run behind, above, and in front of the capsule (see p. 243)
B. WEAKEST PART IS BELOW, where dislocation is easiest, especially when arm is abducted. In this position the humeral head breaks through the capsule below the glenoid cavity
C. THE UPPER SURFACE of the musculotendinous cuff is separated from the overlying acromion, coracoacromial ligament, and the upper part of the deltoid muscle by the *subacromial bursa* (there may be a lateral *subdeltoid bursa,* but the two are usually fused). Calcium deposits in the tendinous floor of the bursa may cause disability of the shoulder, provoking a *bursitis,* which makes movements painful

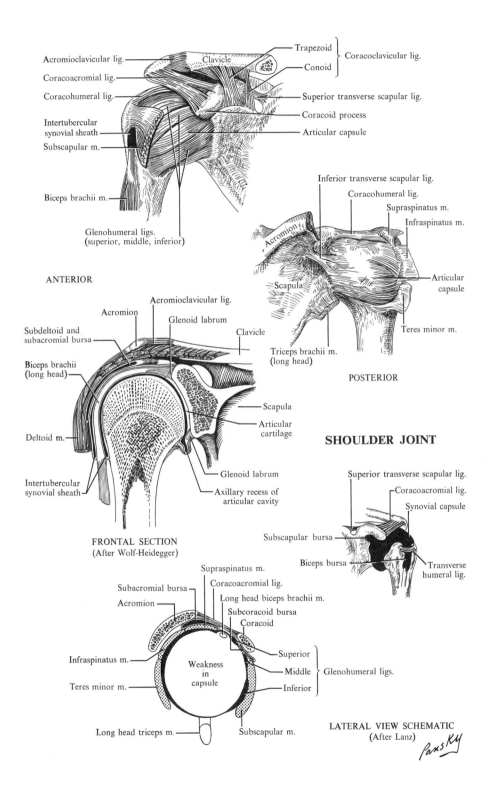

Acromioclavicular lig.
Coracoacromial lig.
Coracohumeral lig.
Intertubercular synovial sheath
Subscapular m.
Biceps brachii m.
Glenohumeral ligs. (superior, middle, inferior)

Clavicle
Trapezoid
Conoid
Coracoclavicular lig.
Superior transverse scapular lig.
Coracoid process
Articular capsule

ANTERIOR

Inferior transverse scapular lig.
Coracohumeral lig.
Supraspinatus m.
Infraspinatus m.
Acromion
Scapula
Articular capsule
Teres minor m.
Triceps brachii m. (long head)

POSTERIOR

Acromion
Acromioclavicular lig.
Glenoid labrum
Clavicle
Subdeltoid and subacromial bursa
Biceps brachii (long head)
Deltoid m.
Intertubercular synovial sheath
Scapula
Articular cartilage
Glenoid labrum
Axillary recess of articular cavity

FRONTAL SECTION
(After Wolf-Heidegger)

SHOULDER JOINT

Superior transverse scapular lig.
Coracoacromial lig.
Synovial capsule
Subscapular bursa
Biceps bursa
Transverse humeral lig.

Supraspinatus m.
Coracoacromial lig.
Long head biceps brachii m.
Subcoracoid bursa
Coracoid
Subacromial bursa
Acromion
Infraspinatus m.
Teres minor m.
Weakness in capsule
Superior
Middle
Inferior
Glenohumeral ligs.
Long head triceps m.
Subscapular m.

LATERAL VIEW SCHEMATIC
(After Lanz)

Pansky

117. MUSCLES OF THE ARM

I.

Name	Origin	Insertion	Action	Nerves
Coraco-brachialis	Apex, coracoid proc. of scapula	Middle shaft of humerus	Flexes and adducts arm	Musculo-cutaneous
Biceps				
Long head	Supraglenoid tuberosity	Tuberosity of radius; bicipital aponeurosis to fascia of forearm	Flexes arm and forearm	Musculo-cutaneous
Short head	Apex, coracoid proc. of scapula		Supinates hand	Musculo-cutaneous
Brachialis	Lower half of front of humerus	Tuberosity and coronoid process of ulna	Flexes forearm	Musculo-cutaneous, radial, median
Triceps				
Long head	Infraglenoid tuberosity of scapula	Post. and upper olecranon Fascia of forearm	Extends and adducts arm (long head only)	Radial
Lat. head	Post. humerus above radial groove Lat. side of humerus	Post. and upper olecranon Fascia of forearm	Extends forearm	Radial
Med. head	Post. humerus below radial groove	Post. and upper olecranon Fascia of forearm	Extends forearm	Radial

II. Special features
A. TENDON OF ORIGIN OF LONG HEAD OF BICEPS runs over head of humerus, ensheathed by a layer of synovial membrane that follows the tendon as far as the surgical neck of humerus. The tendon is held in the intertubercular groove by the *transverse humeral ligament* and a prolongation of the tendon of the pectoralis major muscle
B. THE APONEUROSIS OF THE BICEPS overlies the brachial artery, thus protecting it. Further, it supports the median cubital vein, making it easier to introduce a needle

III. Clinical considerations
A. FRACTURES OF HUMERUS: as noted previously (p. 236), displacement of bone fragments after fracture may be determined by muscle attachments. For example, after fracture through the surgical neck, both medial and lateral rotators are attached to the proximal fragment, therefore, this part is not rotated since they are balanced; the supraspinatus muscle is attached to the proximal piece and is unopposed, therefore, the proximal fragment is abducted; the distal fragment has no lateral rotators attached to it but does have medial rotators, resulting in this part's being medially rotated; the distal part also has attached to it the large and powerful adductors—latissimus dorsi and pectoralis major—causing the distal piece to be adducted; and there will be overriding of the ends of the two pieces because of the pull of such muscles as the biceps and triceps

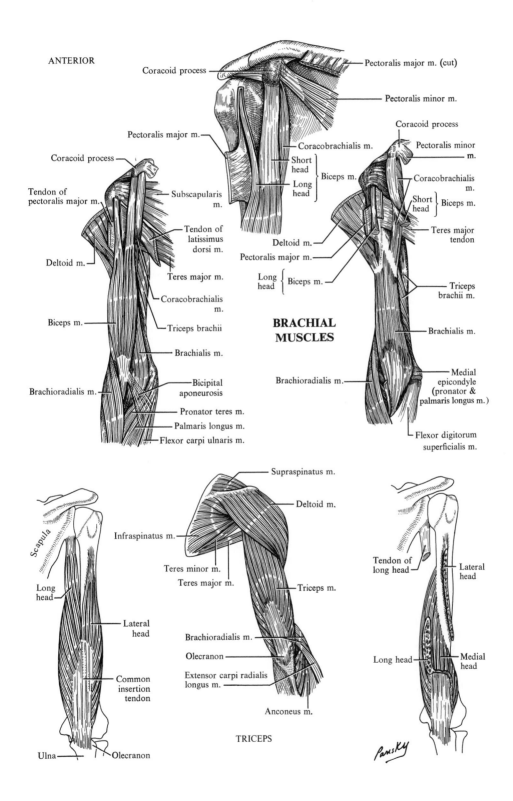

ANTERIOR

Coracoid process

Pectoralis major m. (cut)

Pectoralis minor m.

Coracoid process

Pectoralis major m.

Pectoralis minor m.

Coracobrachialis m.

Short head

Biceps m.

Long head

Coracobrachialis m.

Short head } Biceps m.

Coracoid process

Tendon of pectoralis major m.

Subscapularis m.

Tendon of latissimus dorsi m.

Deltoid m.

Teres major m.

Coracobrachialis m.

Biceps m.

Triceps brachii

Brachialis m.

Brachioradialis m.

Bicipital aponeurosis

Pronator teres m.

Palmaris longus m.

Flexor carpi ulnaris m.

Deltoid m.

Pectoralis major m.

Long head } Biceps m.

Teres major tendon

Triceps brachii m.

Brachialis m.

Medial epicondyle (pronator & palmaris longus m.)

Brachioradialis m.

Flexor digitorum superficialis m.

BRACHIAL MUSCLES

Scapula

Long head

Lateral head

Common insertion tendon

Ulna

Olecranon

Supraspinatus m.

Deltoid m.

Infraspinatus m.

Teres minor m.

Teres major m.

Triceps m.

Brachioradialis m.

Olecranon

Extensor carpi radialis longus m.

Anconeus m.

TRICEPS

Tendon of long head

Lateral head

Long head

Medial head

Pansky

118. VESSELS AND NERVES OF THE ARM

I. Brachial fascia: continuous with axillary, pectoral, and deltoid fasciae above, is attached to the humerus and ulna, and is continuous with the antebrachial fascia below

II. Intermuscular septa: separate posterior and anterior muscle groups
A. LATERAL from brachial fascia to lateral side of humerus
B. MEDIAL from brachial fascia to medial side of humerus

III. Brachial artery is a direct continuation of the axillary artery and terminates as the radial and ulnar arteries

A. RELATIONS

	Anterior	
	Skin, fascia, bicipital aponeurosis,	
	median n., med. cubital v.	
Lateral		*Medial*
Coracobrachialis and	**Brachial artery**	Median, med. antebrach.
biceps mm., median n.		cutaneous, and ulnar nn.;
		basilic v.
	Posterior	
	Triceps and brachialis mm.,	
	radial n., deep brachial a.	

B. BRANCHES (see also p. 269)
1. Deep brachial enters radial sulcus (groove) behind humerus and terminates as radial and middle collateral arteries
2. Superior ulnar collateral runs with ulnar nerve behind medial intermuscular septum
3. Inferior ulnar collateral descends to back of elbow

IV. Nerves (see also p. 241)
A. MEDIAN first lateral to, then crosses, and finally is medial to brachial artery. No branches in arm
B. MUSCULOCUTANEOUS pierces coracobrachialis muscle, sending branches to it, the biceps, and brachialis muscles. Continues as lateral antebrachial cutaneous nerve
C. RADIAL passes in groove on back of humerus with deep brachial artery, pierces lateral intermuscular septum, and divides into superficial and deep branches in front of lateral epicondyle. In arm, gives branches to triceps and to skin (posterior brachial, inferior lateral brachial, and posterior antebrachial cutaneous nerves)
D. ULNAR medial to axillary and brachial arteries to the middle of arm, pierces medial intermuscular septum and runs with superior ulnar collateral artery to groove behind medial epicondyle of humerus. No branches in arm

V. Special features
A. THE BRACHIAL ARTERY lies successively on 3 muscles, gives 3 main branches, is in contact with 3 important nerves, and is associated with 3 veins
1. The 3 muscles that constitute the floor on which the artery runs are (from above downward) the long head of the triceps, the coracobrachialis, and the brachialis
2. The 3 main branches of the artery are the profunda (deep) brachii, the superior ulnar collateral, and the inferior ulnar collateral
3. The 3 important nerves associated with the artery are radial, ulnar, and median
4. The 3 veins associated with the brachial artery are its two venae comitantes (brachial veins) and the basilic
B. THE BIFURCATION of the brachial artery is not constant and may take place at a high level in the arm

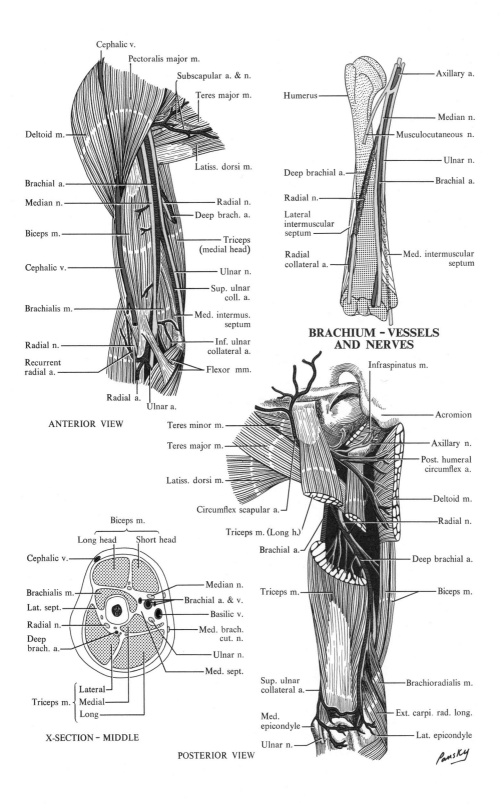

Cephalic v.

Pectoralis major m.

Subscapular a. & n.

Teres major m.

Deltoid m.

Latiss. dorsi m.

Brachial a.

Median n.

Radial n.

Deep brach. a.

Biceps m.

Triceps
(medial head)

Cephalic v.

Ulnar n.

Sup. ulnar
coll. a.

Brachialis m.

Med. intermus.
septum

Radial n.

Inf. ulnar
collateral a.

Recurrent
radial a.

Flexor mm.

Radial a.

Ulnar a.

ANTERIOR VIEW

Humerus

Axillary a.

Median n.

Musculocutaneous n.

Ulnar n.

Deep brachial a.

Brachial a.

Radial n.

Lateral
intermuscular
septum

Radial
collateral a.

Med. intermuscular
septum

BRACHIUM – VESSELS
AND NERVES

Infraspinatus m.

Acromion

Teres minor m.

Teres major m.

Axillary n.

Post. humeral
circumflex a.

Latiss. dorsi m.

Circumflex scapular a.

Deltoid m.

Radial n.

Triceps m. (Long h.)

Brachial a.

Deep brachial a.

Biceps m.

Biceps m.

Long head   Short head

Triceps m.

Cephalic v.

Median n.

Brachialis m.

Brachial a. & v.

Lat. sept.

Basilic v.

Radial n.

Med. brach.
cut. n.

Deep
brach. a.

Ulnar n.

Med. sept.

Lateral
Triceps m. Medial
Long

Sup. ulnar
collateral a.

Brachioradialis m.

Med.
epicondyle

Ext. carpi. rad. long.

Lat. epicondyle

Ulnar n.

Pansky

X-SECTION – MIDDLE

POSTERIOR VIEW

119. BONES OF THE FOREARM

I. Ulna: medial bone of forearm

A. PARTS: proximal extremity, shaft, and distal extremity

1. Proximal: olecranon, trochlear notch with smooth articular surface, and the coronoid process with a tuberosity medially and a radial notch laterally
2. Shaft. Three borders: anterior, posterior, and lateral (interosseous). Three surfaces: anterior, medial, and posterior
3. Distal: head, with styloid process and articular surface

II. Radius: lateral bone of forearm, parallel to above

A. PARTS: proximal extremity, shaft, and distal extremity

1. Proximal: head, neck, and tuberosity; and shallow depression (fovea) on upper head
2. Shaft. Three borders: anterior, posterior, and medial (interosseous). Three surfaces: anterior, posterior, and lateral
3. Distal: large carpal articular surface with ulnar notch on medial side, 3 distinct grooves on dorsal side for tendons, and styloid process

III. Ossification

Location	When Appears	When Closes
A. RADIUS from 3 centers		
Shaft	8th fetal week	——
Distal	2nd year	20th year
Proximal	5th year	17th–18th years
B. ULNA from 3 centers		
Shaft	8th fetal week	——
Head	4th year	20th year
Olecranon	10th year	16th year

IV. Clinical considerations

A. ULNA IS MORE SUBJECT TO FRACTURE DUE TO TRAUMA AT ELBOW; radius is more subject to fracture from falls on hands

B. COLLES' FRACTURE is a fracture of the lower end of the radius with displacement of the hand backward and outward

C. MALUNITED FRACTURE OF THE HEAD OF THE RADIUS often causes severe disability, since rotation of the forearm is restricted, and extension of the elbow may be impaired

D. MALUNION OR EVEN NONUNION OF FRACTURES OF THE OLECRANON often results in little or no limitation of elbow function

E. THE CORRECT POSITION FOR A RADIUS FRACTURED ABOVE the insertion of the pronator teres is with the elbow flexed and the hand supinated. In fractures below the pronator teres, the thumb-up (midprone) position is used with flexion at the elbow. The important rule in all fractures of the radius above the position of a Colles' fracture is to keep the elbow flexed; otherwise, it will be impossible to maintain the forearm in any given position of rotation

F. A REVERSE COLLES' FRACTURE is called *Smith's fracture* and is usually produced by a fall on the back of the hand with the wrist flexed

G. EPIPHYSEAL SEPARATION may be confused with a Colles' fracture, is common in children, and may occur at any time up to the 18th or the 20th year

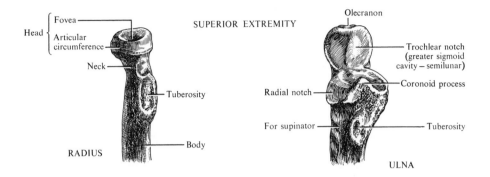

Head { Fovea
Articular circumference
Neck

SUPERIOR EXTREMITY

Tuberosity

Body

RADIUS

Olecranon

Trochlear notch (greater sigmoid cavity – semilunar)

Coronoid process

Radial notch

For supinator

Tuberosity

ULNA

RT. RADIUS AND ULNA

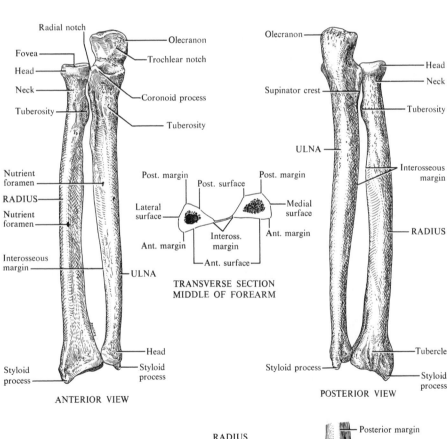

Radial notch

Fovea

Head

Neck

Tuberosity

Nutrient foramen

RADIUS

Nutrient foramen

Interosseous margin

Olecranon

Trochlear notch

Coronoid process

Tuberosity

ULNA

Head

Styloid process

Styloid process

ANTERIOR VIEW

Post. margin

Post. surface

Post. margin

Lateral surface

Medial surface

Ant. margin

Inteross. margin

Ant. margin

Ant. surface

TRANSVERSE SECTION
MIDDLE OF FOREARM

Olecranon

Head

Neck

Supinator crest

Tuberosity

ULNA

Interosseous margin

RADIUS

Tubercle

Styloid process

Styloid process

POSTERIOR VIEW

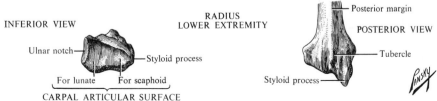

INFERIOR VIEW

Ulnar notch

Styloid process

For lunate   For scaphoid

CARPAL ARTICULAR SURFACE

RADIUS
LOWER EXTREMITY

Posterior margin

POSTERIOR VIEW

Tubercle

Styloid process

120. ANTECUBITAL REGION

I. Cubital fossa: a triangular region at front of elbow

A. BOUNDARIES: base, a line drawn between the 2 humeral epicondyles; lateral side, the brachioradialis muscle; medial side, the pronator teres muscle; roof, the deep fascia strengthened by the aponeurosis of the biceps muscle (see p. 261); and floor, the brachialis and supinator muscles

II. Superficial structures: overlying roof of fossa

A. VEINS
   1. Cephalic runs at lateral edge of fossa along brachialis muscle
   2. Median cubital, from cephalic vein below elbow, upward and medialward to join basilic vein. Crosses superficial to aponeurosis of biceps
   3. Basilic, runs along medial edge of fossa, overlying aponeurosis of biceps muscle as it merges with the antebrachial fascia
   4. Median basilic and median cephalic veins are often present (see opposite page)

B. NERVES
   1. Medial antebrachial cutaneous sends branches both deep and superficial to the basilic vein at the aponeurosis of biceps
   2. Lateral antebrachial cutaneous lies deep to the cephalic vein at elbow and here splits into posterior and anterior branches

III. Deep structures in fossa

A. ARTERIES
   1. Brachial runs through middle of fossa, dividing into its terminal branches, *radial* and *ulnar,* opposite neck of radius
      a. Relations: behind, brachialis muscle; in front, skin and fascia, separated from median cubital vein by the aponeurosis of the biceps muscle; medially, above is median nerve, below is ulnar head of pronator teres muscle; and laterally, the tendon of biceps muscle
   2. Radial lies on tendon of biceps, supinator, and pronator teres muscles
   3. Ulnar passes almost at once beneath pronator teres muscle and leaves fossa

B. NERVES
   1. Median in upper fossa lies close to medial side of brachial artery. Lower down, it passes between heads of pronator teres muscle and is separated from the ulnar artery by the medial head of the muscle
   2. Radial usually lies beneath the brachioradialis muscle and is not in the fossa. It splits beneath this muscle in front of the lateral epicondyle into superficial and deep branches
      a. Superficial branch continues down forearm under brachioradialis muscle
      b. Deep branch curves around lateral side of radius between layers of the supinator muscle to the dorsum of forearm
   3. Ulnar stays behind medial intermuscular septum of arm and passes behind medial epicondyle. Is never related to fossa

IV. Clinical considerations

A. VENIPUNCTURE: this is an extremely common procedure performed when a large blood sample is needed for transfusions, for intravenous feeding, and for intravenous anesthetics. Anatomically, the median cubital vein should be used because it overlies the bicipital aponeurosis which offers some support and also some protection for underlying parts, and it is not accompanied by sizable cutaneous nerves

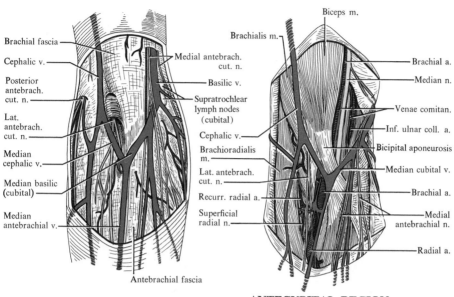

Brachial fascia

Cephalic v.

Posterior antebrach. cut. n.

Lat. antebrach. cut. n.

Median cephalic v.

Median basilic (cubital)

Median antebrachial v.

Medial antebrach. cut. n.

Basilic v.

Supratrochlear lymph nodes (cubital)

Cephalic v.

Brachioradialis m.

Lat. antebrach. cut. n.

Recurr. radial a.

Superficial radial n.

Antebrachial fascia

Biceps m.

Brachialis m.

Brachial a.

Median n.

Venae comitan.

Inf. ulnar coll. a.

Bicipital aponeurosis

Median cubital v.

Brachial a.

Medial antebrachial n.

Radial a.

SUPERFICIAL VEINS

ANTECUBITAL REGION

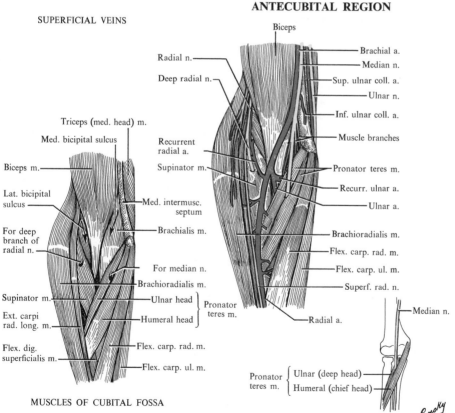

Biceps

Radial n.

Deep radial n.

Triceps (med. head) m.

Med. bicipital sulcus

Biceps m.

Lat. bicipital sulcus

For deep branch of radial n.

Supinator m.

Ext. carpi rad. long. m.

Flex. dig. superficialis m.

Recurrent radial a.

Supinator m.

Med. intermusc. septum

Brachialis m.

For median n.

Brachioradialis m.

Ulnar head
Humeral head
} **Pronator teres m.**

Flex. carp. rad. m.

Flex. carp. ul. m.

Brachial a.

Median n.

Sup. ulnar coll. a.

Ulnar n.

Inf. ulnar coll. a.

Muscle branches

Pronator teres m.

Recurr. ulnar a.

Ulnar a.

Brachioradialis m.

Flex. carp. rad. m.

Flex. carp. ul. m.

Superf. rad. n.

Radial a.

Median n.

Pronator teres m. { **Ulnar (deep head)**
Humeral (chief head) }

MUSCLES OF CUBITAL FOSSA

121. MUSCLES OF THE VOLAR FOREARM

Name	Origin	Insertion	Action	Nerve
I. Superficial group				
Pronator teres	Above med. epicond. Med. side of coronoid process of ulna	Lat. side of radius	Pronates hand	Median
Flex. carpi radialis	Med. epicond.	Bases of second and third metacarpals	Flexes forearm Flexes and abducts hand	Median
Palmaris longus	Med. epicond.	Trans. carpal lig. Palmar aponeurosis	Flexes hand and forearm	Median
Flex. carpi ulnaris	Med. epicond. Med. olecranon Post. border ulna	Pisiform, hamate, and fifth metacarpal	Flexes and adducts hand Flexes forearm	Ulnar
Flex. digit. superficialis	Med. epicond. Med. coronoid proc. of ulna Oblique line, radius	Sides of second phalanx of 4 fingers	Flexes first and second phalanges Flexes hand Flexes forearm	Median
II. Deep group				
Flex. digit. profundus	Upper anterior and med. ulna Med. coronoid proc. and upper posterior ulna Interos. memb.	Bases of terminal phalanges of 4 fingers	Flexes all phalanges Flexes hand	Anterior interosseous of median; ulnar
Flex. pol. longus	Anterior shaft, radius Interos. memb.	Base distal phalanx of thumb	Flexes thumb Flexes and adducts first metacarpal	Anterior interos. of median
Pronator quadratus	Lower anterior shaft, ulna Med. anterior surface, distal ulna	Lower lat. border and lower anterior surface, shaft of radius	Pronates hand	Anterior interos. of median

I. Special features
A. THE FLEXOR MUSCLES of the forearm are concerned with pronation of the forearm, with flexion and abduction at the wrist, and with flexion of the digits
B. THE MOST COMMON VARIATION AMONG THIS GROUP OF MUSCLES is the absence of the palmaris longus muscle (about 12%)

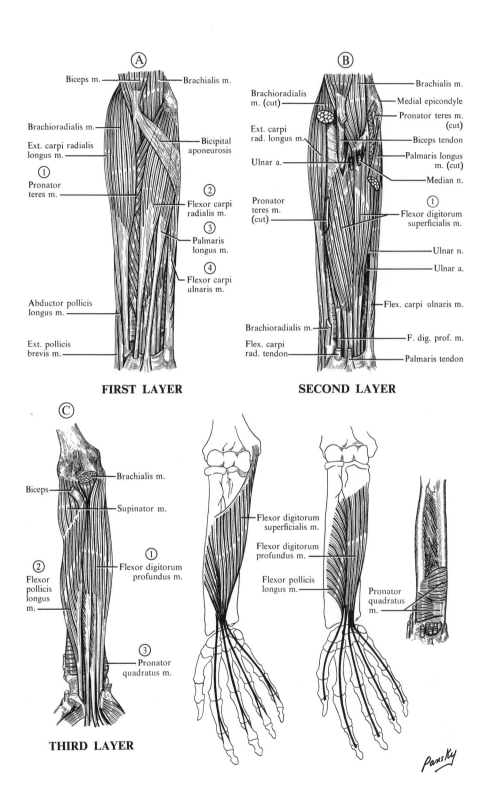

A

Biceps m.

Brachialis m.

Brachioradialis m.

Bicipital aponeurosis

Ext. carpi radialis longus m.

① Pronator teres m.

② Flexor carpi radialis m.

③ Palmaris longus m.

④ Flexor carpi ulnaris m.

Abductor pollicis longus m.

Ext. pollicis brevis m.

FIRST LAYER

B

Brachialis m.

Brachioradialis m. (cut)

Medial epicondyle

Pronator teres m. (cut)

Ext. carpi rad. longus m.

Biceps tendon

Palmaris longus m. (cut)

Ulnar a.

Median n.

Pronator teres m. (cut)

① Flexor digitorum superficialis m.

Ulnar n.

Ulnar a.

Flex. carpi ulnaris m.

Brachioradialis m.

Flex. carpi rad. tendon

F. dig. prof. m.

Palmaris tendon

SECOND LAYER

C

Brachialis m.

Biceps

Supinator m.

① Flexor digitorum profundus m.

② Flexor pollicis longus m.

③ Pronator quadratus m.

Flexor digitorum superficialis m.

Flexor digitorum profundus m.

Flexor pollicis longus m.

Pronator quadratus m.

THIRD LAYER

Pansky

122. DEEP VESSELS AND NERVES OF THE VOLAR FOREARM

I. Radial artery extends from neck of radius to medial side of radial styloid process

A. Relations		
	Anterior	
	Skin and fascia, brachioradialis	
Lateral	**Radial**	*Medial*
Brachioradialis m., superfic. radial n.	**artery**	Pronator teres and flex. carp radialis mm.
	Posterior	
	Tendon biceps, supinator, flex. digit. superfic., pronator teres, flex. pol. long., and pronat. quadrat. mm., radius	

   B. BRANCHES: radial recurrent, muscular, palmar carpal, and superficial palmar

II. Ulnar artery extends from neck of radius to flexor retinaculum at wrist

A. Relations		
	Anterior	
	Skin and fascia, superfic. flexor mm., median n.	
Lateral	**Ulnar**	*Medial*
Flex. digit. superfic. m.	**artery**	Flex. carpi ulnaris m., ulnar n.
	Posterior	
	Brachialis and flex. digit. profundus mm.	

   B. BRANCHES: anterior and posterior ulnar recurrent, common interosseous, and muscular

III. Common interosseous artery: trunk arising from the ulnar artery below radial tuberosity to the upper interosseous membrane, where it divides into:
   A. POSTERIOR INTEROSSEOUS (see p. 266)
   B. ANTERIOR INTEROSSEOUS: down forearm on interosseous membrane to upper border of the pronator quadratus muscle. Here it sends a branch through the interosseous membrane to dorsum to join the posterior interosseous artery, and a small branch continues under the pronator quadratus muscle to the palmar carpal net. Gives muscular and nutrient (radius and ulna) branches and a branch to median nerve

IV. Radial nerve divides into superficial and deep branches
   A. SUPERFICIAL runs beneath brachioradialis muscle just lateral to radial artery. In lower third of forearm, crosses to dorsum. Cutaneous to dorsum of hand and thumb
   B. DEEP (see p. 266)

V. Median nerve passes between heads of pronator teres muscle; lies between flexor digitorum superficialis and profundus muscles; near wrist is more superficial, between tendons of flexor digitorum superficialis and flexor carpi radialis muscles; and then passes deep and medial to the tendon of the palmaris longus muscle
   A. BRANCHES to all superficial muscles except flexor carpi ulnaris. The anterior interosseous branch accompanies the anterior interosseous artery on the interosseous membrane to supply the flexor digitorum profundus, flexor pollicis longus, and pronator quadratus muscles. The palmar branch goes to skin of palm

VI. Ulnar nerve, from behind medial epicondyle enters forearm through flexor carpi ulnaris muscle, continues between this and flexor digitorum profundus, and in the lower half of the forearm, the ulnar artery lies close to its medial side. Both artery and nerve are covered by skin and fascia lateral to the flexor carpi ulnaris muscle
   A. BRANCHES to flexor carpi ulnaris and medial half of the flexor digitorum profundus muscles; palmar cutaneous to medial side of palm and a posterior branch (see p. 298)

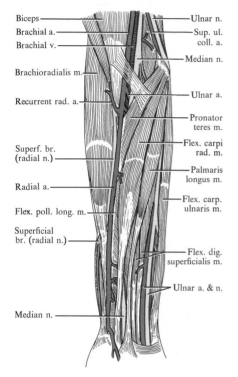

Biceps
Brachial a.
Brachial v.
Brachioradialis m.
Recurrent rad. a.
Superf. br. (radial n.)
Radial a.
Flex. poll. long. m.
Superficial br. (radial n.)
Median n.

Ulnar n.
Sup. ul. coll. a.
Median n.
Ulnar a.
Pronator teres m.
Flex. carpi rad. m.
Palmaris longus m.
Flex. carp. ulnaris m.
Flex. dig. superficialis m.
Ulnar a. & n.

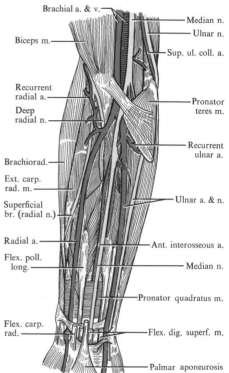

Brachial a. & v.
Biceps m.
Recurrent radial a.
Deep radial n.
Brachiorad.
Ext. carp. rad. m.
Superficial br. (radial n.)
Radial a.
Flex. poll. long.
Flex. carp. rad.

Median n.
Ulnar n.
Sup. ul. coll. a.
Pronator teres m.
Recurrent ulnar a.
Ulnar a. & n.
Ant. interosseous a.
Median n.
Pronator quadratus m.
Flex. dig. superf. m.
Palmar aponeurosis

VOLAR FOREARM
DEEP VESSELS AND NERVES

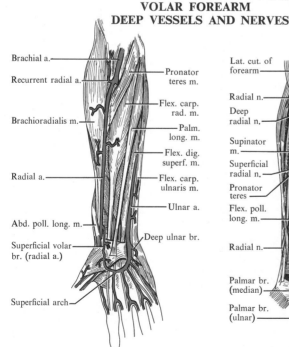

Brachial a.
Recurrent radial a.
Brachioradialis m.
Radial a.
Abd. poll. long. m.
Superficial volar br. (radial a.)
Superficial arch

Pronator teres m.
Flex. carp. rad. m.
Palm. long. m.
Flex. dig. superf. m.
Flex. carp. ulnaris m.
Ulnar a.
Deep ulnar br.

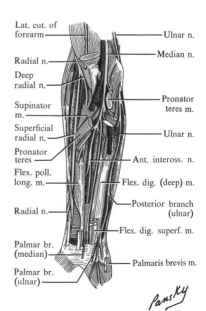

Lat. cut. of forearm
Radial n.
Deep radial n.
Supinator m.
Superficial radial n.
Pronator teres
Flex. poll. long. m.
Radial n.
Palmar br. (median)
Palmar br. (ulnar)

Ulnar n.
Median n.
Pronator teres m.
Ulnar n.
Ant. inteross. n.
Flex. dig. (deep) m.
Posterior branch (ulnar)
Flex. dig. superf. m.
Palmaris brevis m.

Pansky

123. MUSCLES OF THE POSTERIOR FOREARM

Name	Origin	Insertion	Action	Nerve
I. Superficial group				
Brachio-radialis	Lat. supracon-dylar ridge of humerus	Lat. side of base, radial styloid	Flexes forearm	Radial
Ext. carpi radialis long.	Lat. supracon-dylar ridge of humerus	Post. base of second meta-carpal	Extends and abducts hand	Radial
Ext. carpi radialis brev.	Lat. epicondyle of humerus	Post. base of third metacarpal	Extends and abducts hand	Radial
Ext. digit.	Lat. epicondyle of humerus	Mid. base of second and third phalanges of 4 fingers	Extends fingers Extends hand	Deep radial
Ext. digit. minimi	Common exten-sor tendon	Aponeurosis; post., first phalanx of little finger	Extends little finger	Deep radial
Ext. carpi ulnaris	Lat. epicondyle Post. border of ulna	Tubercle at base of fifth metacarpal	Extends and adducts hand	Deep radial
Anconeus	Back of lat. epicondyle	Olecranon, upper posterior ulna	Extends forearm	Radial
II. Deep group				
Supinator	Lat. epicondyle Ulna below radial notch	Radial tubercle Oblique line of radius	Supinates hand	Deep radial
Abductor pollicis longus	Lat. post. ulna Interos. memb. Post. radius	Lat. side of base of first metacarpal	Abducts thumb and hand	Deep radial
Ext. pollicis brev.	Post. radius Interos. memb.	Base of first phalanx of thumb	Extends last phalanx of thumb Abducts hand	Deep radial
Ext. pollicis long.	Lat. side, post. surface of ulna	Base of last phalanx of thumb	Extends first phalanx of thumb Abducts hand	Deep radial
Ext. indicis	Post. shaft of ulna Interos. memb.	Tendon of ext. dig. index finger	Extends and adducts index finger	Deep radial

III. Special features

A. INSERTION OF THE TENDONS of the lumbricales and interossei is in a common extensor aponeurosis on the dorsum of each first phalanx

B. OPPOSITE THE FIRST INTERPHALANGEAL JOINT is a splitting of the extensor aponeurosis into 3 slips, with the middle going to the base of the second phalanx and the 2 lateral going to the base of the third phalanx

C. THE DEEP RADIAL NERVE passes between the 2 layers of the supinator muscle

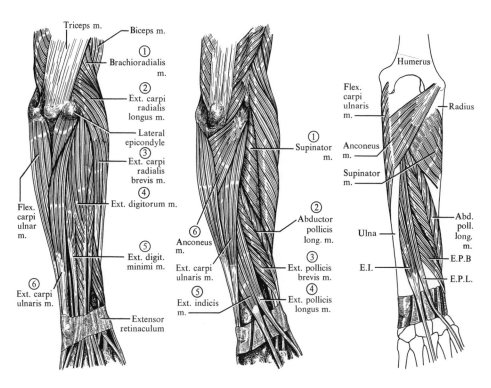

Triceps m. — Biceps m.

① Brachioradialis m.

② Ext. carpi radialis longus m.

Lateral epicondyle

③ Ext. carpi radialis brevis m.

④ Ext. digitorum m.

Flex. carpi ulnar m.

⑤ Ext. digit. minimi m.

⑥ Ext. carpi ulnaris m.

Extensor retinaculum

① Supinator m.

② Abductor pollicis long. m.

⑥ Anconeus m.

Ext. carpi ulnaris m.

③ Ext. pollicis brevis m.

⑤ Ext. indicis m.

④ Ext. pollicis longus m.

Humerus

Flex. carpi ulnaris m.

Radius

Anconeus m.

Supinator m.

Ulna

E.I.

Abd. poll. long. m.

E.P.B

E.P.L.

SUPERFICIAL GROUP OF MUSCLES   DEEP GROUP OF MUSCLES

DORSAL MUSCLES

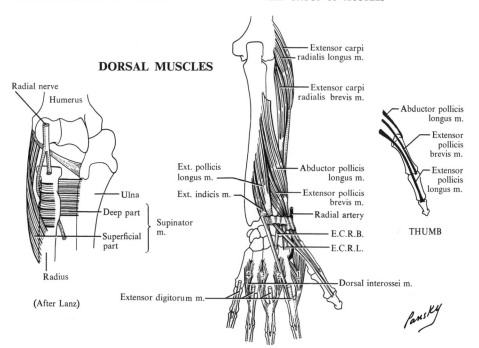

Radial nerve

Humerus

Ulna

Deep part

Superficial part

Supinator m.

Radius

(After Lanz)

Extensor carpi radialis longus m.

Extensor carpi radialis brevis m.

Ext. pollicis longus m.

Ext. indicis m.

Abductor pollicis longus m.

Extensor pollicis brevis m.

Radial artery

E.C.R.B.

E.C.R.L.

Dorsal interossei m.

Extensor digitorum m.

Abductor pollicis longus m.

Extensor pollicis brevis m.

Extensor pollicis longus m.

THUMB

Pansky

124. COMPARTMENTS OF FOREARM, DEEP VESSELS AND NERVES OF POSTERIOR ASPECT

I. **Forearm is divided into posterior and anterior compartments** by medial and lateral intermuscular septa of the antebrachial fascia, the radius, the ulna, and the interosseous membrane

A. ANTEBRACHIAL FASCIA is fused to the ulna throughout its length

B. IN LOWER FOREARM fascia splits into 2 layers: posterior and anterior to the palmaris longus, flexor carpi radialis, and flexor carpi ulnaris muscles

II. **Vessels**

A. POSTERIOR INTEROSSEOUS ARTERY arises from common interosseous artery at upper border of interosseous membrane over which it passes, comes through between the borders of supinator and abductor pollicis longus muscles, and descends through forearm between superficial and deep layers. In lower forearm is joined by one of the terminal branches of the anterior interosseous artery and enters the dorsal carpal network

1. Branches: muscular; interosseous recurrent artery, which ascends between olecranon and lateral epicondyle under anconeus muscle to anastomose with the middle collateral artery, inferior ulnar collateral artery, and posterior ulnar recurrent artery

III. **Nerves**

A. RADIAL NERVE pierces lateral intermuscular septum of arm, runs between brachialis and brachioradialis muscles, and in front of the lateral epicondyle divides into:

1. Superficial branch (see p. 298)

2. Deep: curves around lateral border of radius between layers of supinator muscle, continues between superficial and deep layers of muscles to the middle of forearm, and then, as the *posterior interosseous nerve,* it lies between the interosseous membrane and extensor pollicis longus muscle. It continues to the posterior of the carpus, giving branches to all extensor muscles

IV. **Clinical considerations**

A. DAMAGE TO THE RADIAL NERVE, especially above the elbow, leads to "wrist drop," the inability to extend the hand at the wrist. Extension of the fingers is also impossible. Sensation on the lateral side of the back of the hand is also lost

1. Because of the shortness of the extensor muscles of the digits, however, a person with wrist drop can passively extend his wrist by tightly clenching his fingers

B. LESIONS OF THE DEEP BRANCH OF THE RADIAL may have varying effects, but complete ones abolish extension at the metacarpophalangeal joints of the digits while allowing extension of the wrist through the radial extensors

C. ALL THE EXTENSOR MUSCLES ARE INNERVATED through the radial nerve. The lateral ones (brachioradialis and radial extensors) receive their innervation from the nerve while it lies on the front of the arm or forearm; the posterior ones (extensor digitorum, digiti minimi, and carpi ulnaris) receive their innervation on their deep surfaces from the deep branch of the radial nerve

D. IT IS USUALLY STATED THAT the brachioradialis receives fibers from the 5th and 6th cervical nerves; the radial extensors from about C6 and C7; the extensor digitorum and extensor digiti minimi from C6, C7, and C8; and the ulnar extensor from these or C7 and C8 only

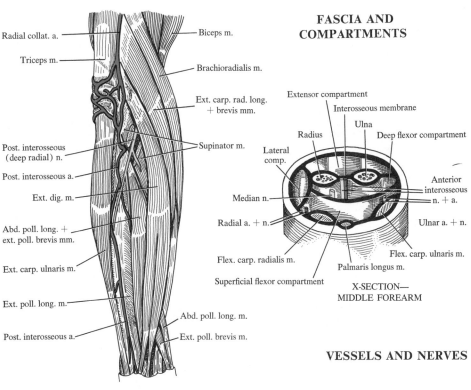

FASCIA AND COMPARTMENTS

Radial collat. a.

Triceps m.

Post. interosseous (deep radial) n.

Post. interosseous a.

Ext. dig. m.

Abd. poll. long. + ext. poll. brevis mm.

Ext. carp. ulnaris m.

Ext. poll. long. m.

Post. interosseous a.

Biceps m.

Brachioradialis m.

Ext. carp. rad. long. + brevis mm.

Supinator m.

Abd. poll. long. m.

Ext. poll. brevis m.

Extensor compartment

Interosseous membrane

Ulna

Deep flexor compartment

Radius

Lateral comp.

Anterior interosseous n. + a.

Median n.

Radial a. + n.

Ulnar a. + n.

Flex. carp. radialis m.

Flex. carp. ulnaris m.

Palmaris longus m.

Superficial flexor compartment

X-SECTION—
MIDDLE FOREARM

VESSELS AND NERVES

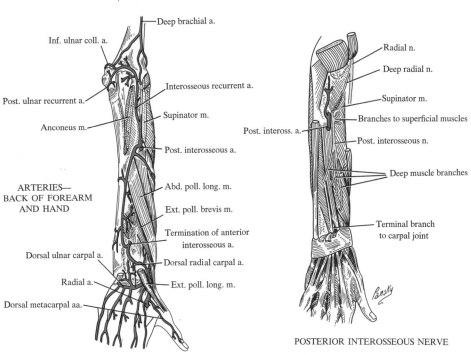

Deep brachial a.

Inf. ulnar coll. a.

Post. ulnar recurrent a.

Anconeus m.

Interosseous recurrent a.

Supinator m.

Post. interosseous a.

ARTERIES—
BACK OF FOREARM
AND HAND

Abd. poll. long. m.

Ext. poll. brevis m.

Termination of anterior interosseous a.

Dorsal ulnar carpal a.

Radial a.

Dorsal metacarpal aa.

Dorsal radial carpal a.

Ext. poll. long. m.

Radial n.

Deep radial n.

Supinator m.

Branches to superficial muscles

Post. inteross. a.

Post. interosseous n.

Deep muscle branches

Terminal branch to carpal joint

POSTERIOR INTEROSSEOUS NERVE

– 267 –

125. COLLATERAL CIRCULATION INVOLVING THE ANASTOMOSIS AROUND THE ELBOW

I. Collateral circulation for ligation of the brachial artery

A. LIGATION BETWEEN ORIGINS OF THE SUPERIOR ULNAR COLLATERAL AND THE PROFUNDA ARTERIES

Arteries Above Ligature	Arteries Below Ligature
1. Radial collateral (from deep brachial)	Radial recurrent (from radial)
2. Medial collateral (from deep brachial	Interosseous recurrent (from posterior interosseous)

B. LIGATION BETWEEN ORIGINS OF SUPERIOR AND INFERIOR ULNAR COLLATERALS

1. Radial collateral, as above	
2. Medial collateral, as above	
3. Sup. ulnar collateral (from brachial)	Inf. ulnar collateral (from brachial)
	Posterior ulnar recurrent (from ulnar)

C. LIGATION BETWEEN ORIGIN OF INFERIOR ULNAR COLLATERAL AND END OF BRACHIAL ARTERY

1. Radial collateral, as above	
2. Medial collateral, as above	Interosseous recurrent (from posterior interosseous)
3. Sup. ulnar collateral, as above	
4. Inf. ulnar collateral (from brachial)	Anterior ulnar recurrent (from ulnar)
	Posterior ulnar recurrent (from ulnar)

II. Sites where anastomoses occur

A. BETWEEN ANTERIOR ULNAR RECURRENT AND INFERIOR ULNAR COLLATERAL in front of the medial epicondyle

B. POSTERIOR ULNAR RECURRENT AND POSTERIOR BRANCHES OF INFERIOR ULNAR COLLATERAL behind medial epicondyle

C. BETWEEN INTEROSSEOUS RECURRENT AND MEDIAL COLLATERAL in interval between lateral epicondyle of humerus and olecranon

D. BETWEEN RADIAL RECURRENT AND RADIAL COLLATERAL in front of the lateral epicondyle

III. Clinical considerations

A. THE ABOVE-MENTIONED ANASTOMOSES, in theory, may preserve the limb distal to the interruption of the main vessel. However, the small anastomotic channels, in reality, may not open in time, especially when the major blood supply is cut off suddenly

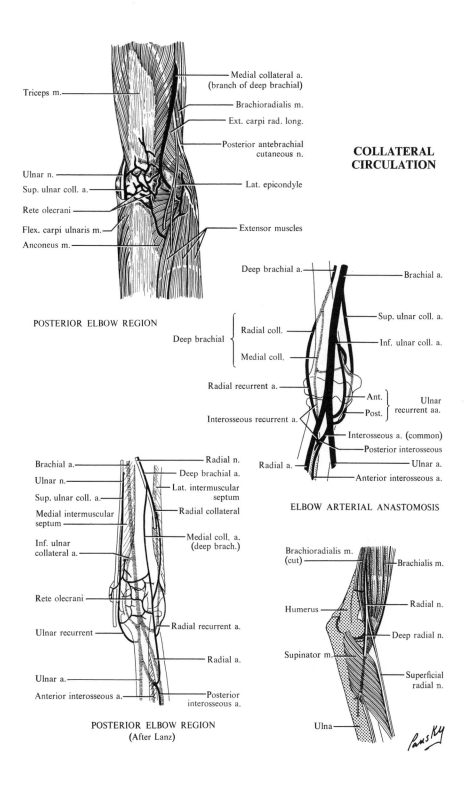

Triceps m.

Medial collateral a. (branch of deep brachial)

Brachioradialis m.

Ext. carpi rad. long.

Posterior antebrachial cutaneous n.

Ulnar n.

Sup. ulnar coll. a.

Lat. epicondyle

Rete olecrani

Flex. carpi ulnaris m.

Extensor muscles

Anconeus m.

COLLATERAL CIRCULATION

POSTERIOR ELBOW REGION

Deep brachial a.

Brachial a.

Sup. ulnar coll. a.

Deep brachial { Radial coll.

Inf. ulnar coll. a.

Medial coll.

Radial recurrent a.

Ant. } Ulnar recurrent aa.

Post.

Interosseous recurrent a.

Interosseous a. (common)

Posterior interosseous

Radial a.

Ulnar a.

Anterior interosseous a.

ELBOW ARTERIAL ANASTOMOSIS

Brachial a.

Radial n.

Ulnar n.

Deep brachial a.

Sup. ulnar coll. a.

Lat. intermuscular septum

Medial intermuscular septum

Radial collateral

Medial coll. a. (deep brach.)

Inf. ulnar collateral a.

Brachioradialis m. (cut)

Brachialis m.

Rete olecrani

Humerus

Radial n.

Deep radial n.

Ulnar recurrent

Radial recurrent a.

Supinator m.

Radial a.

Superficial radial n.

Ulnar a.

Anterior interosseous a.

Posterior interosseous a.

Ulna

POSTERIOR ELBOW REGION (After Lanz)

Pansky

– 269 –

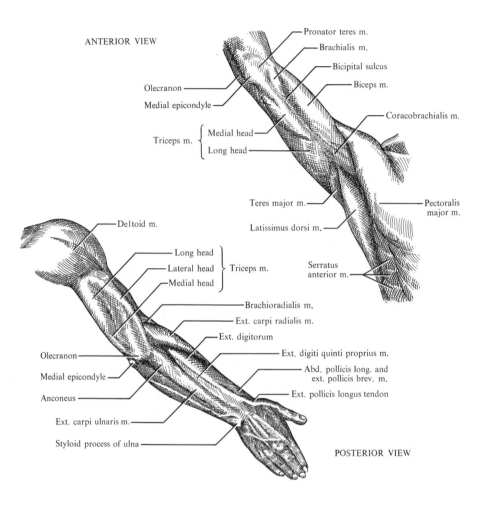

ANTERIOR VIEW

Pronator teres m.

Brachialis m.

Bicipital sulcus

Biceps m.

Olecranon

Medial epicondyle

Coracobrachialis m.

Triceps m. { Medial head

Long head

Teres major m.

Pectoralis major m.

Latissimus dorsi m.

Deltoid m.

Long head

Lateral head } Triceps m.

Medial head

Serratus anterior m.

Brachioradialis m.

Ext. carpi radialis m.

Ext. digitorum

Olecranon

Ext. digiti quinti proprius m.

Medial epicondyle

Abd. pollicis long. and ext. pollicis brev. m.

Anconeus

Ext. pollicis longus tendon

Ext. carpi ulnaris m.

Styloid process of ulna

POSTERIOR VIEW

SURFACE ANATOMY

ANTERIOR VIEW

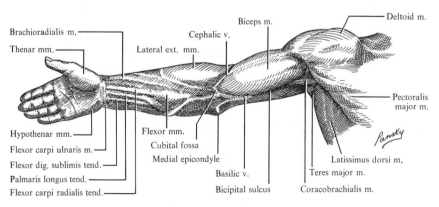

Brachioradialis m.

Thenar mm.

Lateral ext. mm.

Cephalic v.

Biceps m.

Deltoid m.

Pectoralis major m.

Hypothenar mm.

Flexor mm.

Flexor carpi ulnaris m.

Cubital fossa

Flexor dig. sublimis tend.

Medial epicondyle

Palmaris longus tend.

Latissimus dorsi m.

Flexor carpi radialis tend.

Basilic v.

Teres major m.

Bicipital sulcus

Coracobrachialis m.

Pansky

FIGURE 15. **Surface anatomy of upper extremity.**

– 270 –

REGIONAL NERVE BLOCK FOR UPPER EXTREMITY

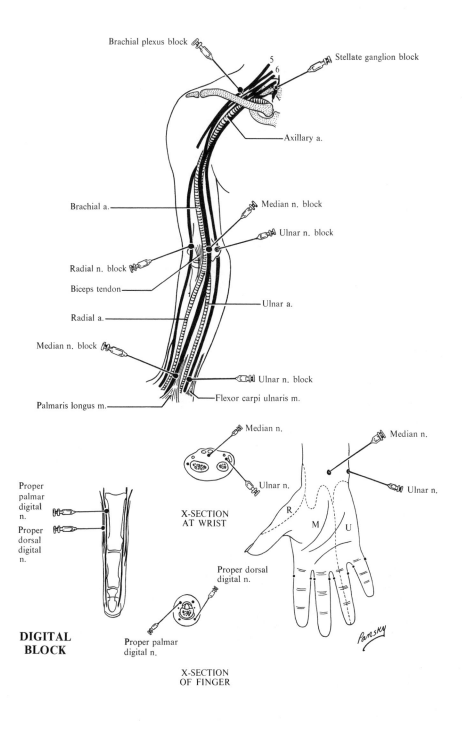

Brachial plexus block

Stellate ganglion block

5
6

Axillary a.

Brachial a.

Median n. block

Ulnar n. block

Radial n. block

Biceps tendon

Ulnar a.

Radial a.

Median n. block

Ulnar n. block

Palmaris longus m.

Flexor carpi ulnaris m.

Median n.

Median n.

Ulnar n.

Ulnar n.

Proper palmar digital n.

Proper dorsal digital n.

X-SECTION
AT WRIST

R

M   U

Proper dorsal digital n.

DIGITAL BLOCK

Proper palmar digital n.

X-SECTION
OF FINGER

FIGURE 16. **Nerve blocks of upper extremity.**

126. THE ELBOW JOINT

I. Humeroulnar
A. TYPE: ginglymus (hinge)
B. BONES: trochlea of humerus with trochlear notch of ulna, capitulum of humerus with fovea of radial head
C. MOVEMENTS: flexion and extension
D. LIGAMENTS
    1. Articular capsule, in front, extends from medial epicondyle and front of humerus above coronoid and radial fossae to the anterior coronoid process, anular ligament, and the collateral ligaments. Behind, it extends from the humerus (behind capitulum) and medial side of trochlea to the margins of the olecranon, posterior part of ulna, and anular ligament
    2. Ulnar collateral extends from front and back of the medial epicondyle to the medial side of the coronoid process and olecranon
    3. Radial collateral extends from below the lateral epicondyle to the anular ligament and lateral margin of the ulna
E. SYNOVIAL MEMBRANE: very extensive. Extends from margins of articular areas of humerus. Lines the 3 fossae of the humerus and the inside of the capsule. It forms a sac between the head of radius, the radial notch, and the anular ligament. Three fat pads exist between the synovia and capsule: over the olecranon fossa, over the coronoid fossa, and over the radial fossa. The former is pressed by the triceps muscle into the fossa during flexion. The other 2 are pressed by the brachialis muscle into their fossae in extension of the forearm
F. MUSCLES ACTING ON THE JOINT

Flexion	Extension
Biceps	Triceps
Brachialis	Anconeus
Brachioradialis	Extensor carpi radialis longus and brevis
Pronator teres	Extensor digitorum
Flexor carpi radialis and ulnaris	Extensor digiti minimi
Palmaris longus	Extensor carpi ulnaris
Flexor digitorum superficialis	Supinator

II. Radioulnar
A. PROXIMAL
    1. Type: trochoid (pivot)
    2. Bones: head of radius and radial notch of ulna
    3. Movements: rotation (pronation, supination)
    4. Ligaments
        a. Anular encircles radial head, attached to ends of radial notch
        b. Quadrate between neck of radius and lower part of radial notch of ulna
    5. Muscles acting on proximal radioulnar joint

Supination	Pronation
Biceps and supinator	Pronator teres
Extensors of thumb	Pronator quadratus

III. Clinical considerations
A. SINCE THE ULNA CANNOT BE ANTERIORLY DISLOCATED without a concomitant fracture, most dislocations at the elbow are posterior; the ulnar nerve is frequently injured
B. DISLOCATION OF THE HEAD OF THE RADIUS ALONE is usually anterior

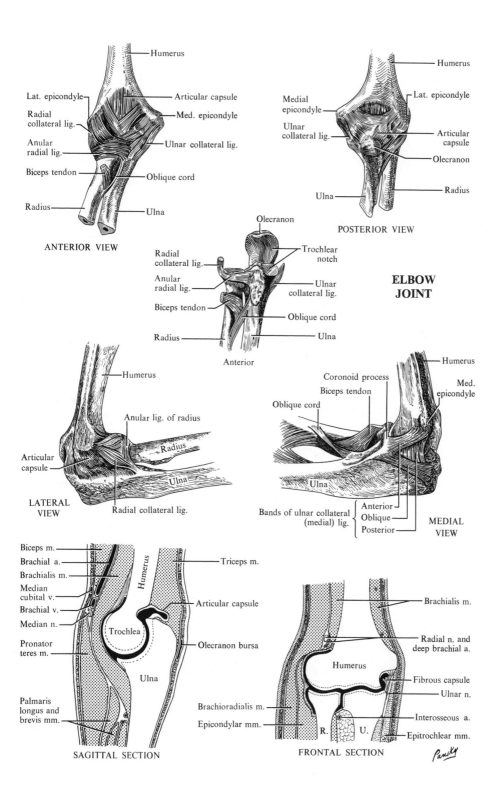

Humerus

Lat. epicondyle

Radial
collateral lig.

Anular
radial lig.

Biceps tendon

Radius

ANTERIOR VIEW

Articular capsule

Med. epicondyle

Ulnar collateral lig.

Oblique cord

Ulna

Medial
epicondyle

Ulnar
collateral lig.

Ulna

POSTERIOR VIEW

Humerus

Lat. epicondyle

Articular
capsule

Olecranon

Radius

Olecranon

Radial
collateral lig.

Anular
radial lig.

Biceps tendon

Radius

Trochlear
notch

Ulnar
collateral lig.

Oblique cord

Ulna

Anterior

**ELBOW
JOINT**

Humerus

Anular lig. of radius

Radius

Articular
capsule

Ulna

LATERAL
VIEW

Radial collateral lig.

Coronoid process

Biceps tendon

Oblique cord

Ulna

Bands of ulnar collateral
(medial) lig.

Humerus

Med.
epicondyle

Anterior
Oblique
Posterior

MEDIAL
VIEW

Biceps m.

Brachial a.

Brachialis m.

Median
cubital v.

Brachial v.

Median n.

Pronator
teres m.

Palmaris
longus and
brevis mm.

SAGITTAL SECTION

Humerus

Triceps m.

Articular capsule

Trochlea

Olecranon bursa

Ulna

Brachialis m.

Radial n. and
deep brachial a.

Fibrous capsule

Ulnar n.

Interosseous a.

Epitrochlear mm.

Brachioradialis m.

Epicondylar mm.

R.   U.

Humerus

FRONTAL SECTION

Pansky

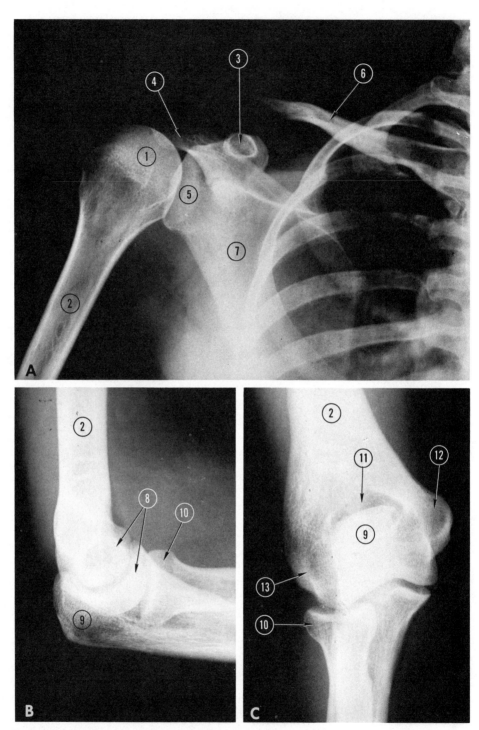

FIGURE 17. **Shoulder and elbow joints. A, Shoulder joint; B, elbow, lateral view; C, elbow AP view.** *1*, Head of humerus; *2*, shaft of humerus; *3*, coracoid process of scapula; *4*, acromion of scapula; *5*, glenoid fossa of scapula; *6*, clavicle; *7*, body of scapula; *8*, capitulum and trochlea of humerus; *9*, olecranon; *10*, head of radius; *11*, olecranon fossa; *12*, medial epicondyle of humerus; *13*, lateral epicondyle of humerus.

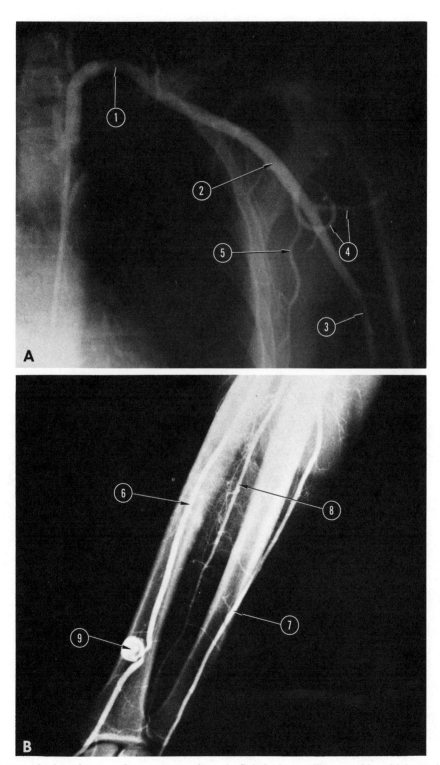

FIGURE 18. **Arteriogram of upper extremity. A, Subclavian, axillary, and brachial arteries; B, forearm.** *1,* Subclavian; *2,* axillary; *3,* brachial; *4,* circumflex humeral; *5,* subscapular; *6,* radial; *7,* ulnar; *8,* anterior interosseous; *9,* aneurysm of radial artery.

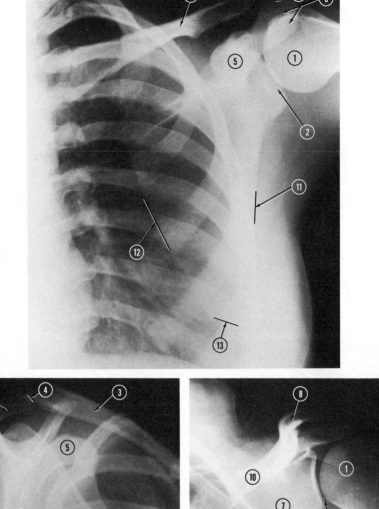

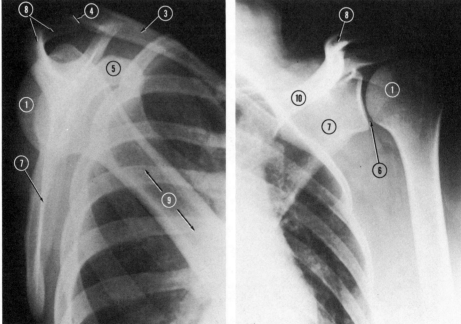

FIGURE 19. **Shoulder joint.** *1*, Head of humerus; *2*, glenoid cavity; *3*, clavicle; *4*, acromioclavicular joint; *5*, coracoid process; *6*, glenohumeral joint; *7*, scapula; *8*, acromion; *9*, shaft of humerus; *10*, scapular spine; *11*, lateral border of scapula; *12*, medial border of scapula; *13*, inferior angle of scapula.

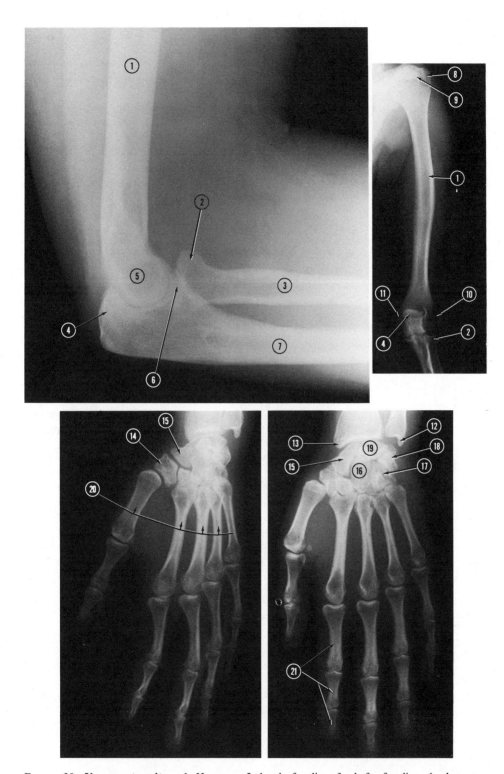

FIGURE 20. **Upper extremity.** *1*, Humerus; *2*, head of radius; *3*, shaft of radius; *4*, olecranon process; *5*, trochlea of humerus; *6*, coronoid process of ulna; *7*, shaft of ulna; *8*, greater tuberosity of humerus; *9*, lesser tuberosity of humerus; *10*, lateral epicondyle; *11*, medial epicondyle; *12*, styloid process of ulna; *13*, styloid process of radius, *14*, trapezium; *15*, scaphoid; *16*, capitate; *17*, hamate; *18*, triquetral; *19*, lunate; *20*, metacarpals; *21*, phalanges (proximal, middle and distal).

127. BONES OF THE WRIST AND HAND

I. Carpals (wrist bones): 8 in number, arranged in proximal and distal rows

A. SCAPHOID (NAVICULAR) (boat-shaped): largest of proximal row, superior and inferior surfaces concave, has a tubercle

B. LUNATE (SEMILUNAR): crescent-shaped, superior surface convex, inferior concave

C. TRIQUETRAL: pyramidal in shape, has an oval articular facet on its palmar surface

D. PISIFORM: small, pear-shaped, oval articular facet in dorsal surface

E. TRAPEZIUM (GREATER MULTANGULAR): palmar surface shows an oblique deep groove

F. TRAPEZOID (LESSER MULTANGULAR): smallest of distal row, wedge-shaped

G. CAPITATE: largest of carpals, at center of wrist and the first to ossify

H. HAMATE: hook-shaped, has a curved, hooklike process (hamulus) from medial side of palmar surface

II. Metacarpals: 5, numbered from lateral to medial

A. COMMON CHARACTERISTICS: each has a base proximally, a shaft, and a head distally

B. OUTSTANDING CHARACTERISTICS OF INDIVIDUAL BONES
   1. First: shorter and stouter, has tubercle on lateral side of base
   2. Second: longest with largest base
   3. Third: slightly smaller than second; styloid process on base
   4. Fourth: base has 2 facets on its lateral side (for 3rd metacarpal) and 1 facet on medial side (for 5th metacarpal)
   5. Fifth: smallest, has tubercle on medial side of base

III. Phalanges: total of 14: 2 for thumb and 3 for each finger. Numbered or named on each finger from proximal to distal: 1 (proximal), 2 (middle), 3 (distal)

A. COMMON CHARACTERISTICS: each has proximal base, tapering body, and distal extremity
   1. First phalanges have oval, concave articular surfaces; second and third show double concavities separated by a ridge at base
   2. First and second phalanges also show 2 condyles separated by a groove

IV. Ossification

A. CARPALS: from 1 center in each, which appear as follows: capitate and hamate, first year; triquetral, third year; lunate and trapezium, fifth year; scaphoid, sixth year; trapezoid, eighth year; pisiform, twelfth year

B. METACARPALS: from 2 centers, body and distal end, which appear and close as follows: body, eighth and ninth fetal week; extremity, third year. They join in twentieth year. First has center in body and base, which for body appears in the eighth to ninth fetal week; base appears in third year, and they join in the twentieth year

C. PHALANGES: from 2 centers, 1 for body, 1 for base. Appear in body in eighth week, in base at 3 to 4 years (in proximal row), and later in distal row. Unite in twentieth year

V. Clinical considerations

A. OSSIFICATION CENTERS: knowledge of the appearance and closure of ossification centers is important both medicolegally and in the radiologic examination of the young

B. FRACTURES: 70–75% of fractures of the carpus include the scaphoid bone. In 10–15%, due to the lack of blood supply to the proximal part, the bone may not heal and necrosis of the scaphoid takes place

C. IN THE PROXIMAL ROW of carpal bones, only the scaphoid and lunate articulate with the radius. Thus, they transmit to the forearm the entire force of a fall on the hand

D. EXCEPT FOR THE PISIFORM, no muscles attach to the proximal row of carpals

E. THE CAPITATE forms the keystone of the carpal arch (carpal tunnel). A fall on the base of the palm forces the head of the capitate upward into the concave waist of the scaphoid, accounting for the frequency with which the latter bone is fractured from a fall on the hand

WRIST AND HAND BONES

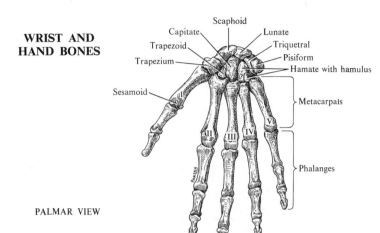

Scaphoid
Capitate
Lunate
Trapezoid
Triquetral
Trapezium
Pisiform
Hamate with hamulus
Sesamoid
Metacarpals
Phalanges

PALMAR VIEW

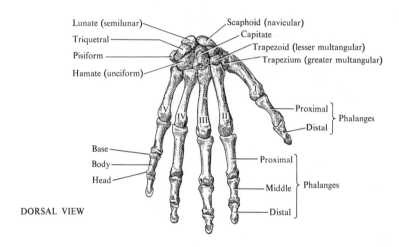

Lunate (semilunar)
Scaphoid (navicular)
Triquetral
Capitate
Pisiform
Trapezoid (lesser multangular)
Hamate (unciform)
Trapezium (greater multangular)
Proximal
Distal
Phalanges
Base
Body
Head
Proximal
Middle
Distal
Phalanges

DORSAL VIEW

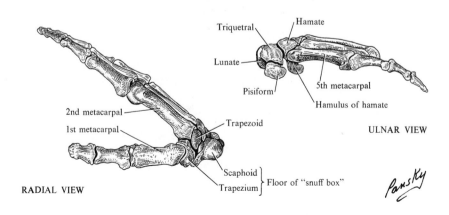

Triquetral
Hamate
Lunate
Pisiform
5th metacarpal
Hamulus of hamate

ULNAR VIEW

2nd metacarpal
1st metacarpal
Trapezoid
Scaphoid
Trapezium
Floor of "snuff box"

RADIAL VIEW

Pansky

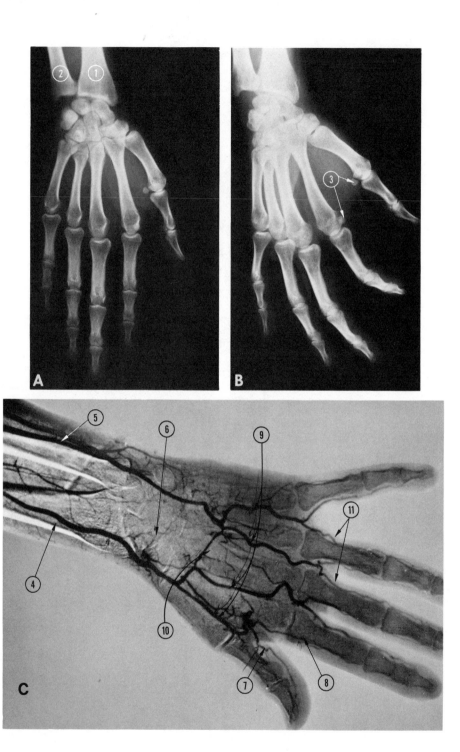

FIGURE 21. **Right hand, adult.   A, AP view; B, oblique view; C, vascular anatomy, subtraction film.**   *1,* Radius; *2,* ulna; *3,* sesamoid bones; *4,* radial artery; *5,* ulnar artery; *6,* carpal branches; *7,* principal artery to thumb; *8,* radial artery to index finger; *9,* common palmar digital arteries; *10,* superficial palmar arch; *11,* proper digital arteries.

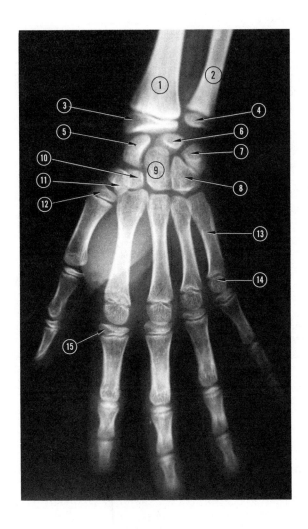

FIGURE 22. **Left hand, child (about 11 years).** *1*, Radius; *2*, ulna; *3*, distal epiphysis of radius; *4*, distal epiphysis of ulna; *5*, scaphoid; *6*, lunate; *7*, triquetral; *8*, hamate; *9*, capitate; *10*, trapezoid; *11*, trapezium; *12*, proximal epiphysis of 1st metacarpal; *13*, 5th metacarpal; *14*, head of 5th metacarpal; *15*, proximal epiphysis of 1st phalanx of index finger.

128. ANTERIOR AND LATERAL ASPECTS OF THE WRIST

I. Antebrachial fascia
A. PALMAR CARPAL LIGAMENT (volar carpal ligament) fascia strengthened by transverse fibers extending between the radial and ulnar styloid processes

II. Vessels and nerves
A. ULNAR ARTERY with ulnar nerve on its medial side is covered by carpal ligament and skin. The flexor carpi ulnaris is just medial to these
B. RADIAL ARTERY is also quite superficial, with the flexor carpi radialis tendon just medial to it. The superficial radial nerve lies close to its lateral side
C. MEDIAN NERVE is overlapped by the tendon of the palmaris longus

III. Tendons
A. TENDON OF PALMARIS LONGUS is superficial and crosses wrist near its middle, with the tendon of the flexor carpi radialis lying close to its lateral side
B. TENDONS OF THE FLEXOR DIGITORUM SUPERFICIALIS are arranged so that those to the middle and ring fingers are superficial to those to the index and little finger
C. TENDONS OF FLEXOR DIGITORUM PROFUNDUS are arranged in order from side to side, deep to the superficial flexor tendons
D. THE TENDON OF THE FLEXOR POLLICIS LONGUS lies in the same plane as the deep flexors, just to radial side of the tendon to the index finger

IV. Special features
A. THE TENDONS of the extensor pollicis brevis and abductor pollicis longus run from the dorsum of the forearm, cross the tendons of the extensor carpi radialis longus and brevis to reach their insertion on the radial side of thumb. The extensor of the thumb also crosses the radial carpal extensors, but is separated from the short extensor of the thumb by a triangular space, the "anatomical snuffbox"
B. THE RADIAL ARTERY winds to the dorsum of the hand by passing under these tendons across the "snuffbox" to reach the base of the first interosseous space. Branches of the superficial radial nerve, especially the medial branch with its first dorsal digital, are related to this area superficially
C. THE PALMAR CARPAL LIGAMENT crosses superficial to all the superficial flexor muscles as well as the ulnar nerve and vessels
D. THE TRANSVERSE CARPAL ARCH (or carpal tunnel) is maintained by the transverse carpal ligament (see p. 290) which blends proximally with the palmar carpal ligament. The transverse carpal ligament bridges the arch between the pisiform and hamate, medially, and the scaphoid and trapezium, laterally. Except for the flexor carpi ulnaris and palmaris longus muscle tendons, all the other flexor tendons, as well as the median nerve, pass through the carpal tunnel created by the transverse carpal ligament. The ulnar artery and nerve pass superficial to the transverse carpal ligament, but deep to the palmar carpal ligament, just on the radial side of the pisiform bone where the pulsations of the artery are felt

V. Clinical considerations
A. THE SUPERFICIAL POSITION OF THE VESSELS, NERVES, AND TENDONS at the wrist makes them exceedingly vulnerable to damage and injury
B. THE SUPERFICIAL POSITION OF THE RADIAL ARTERY to the radius makes it possible to take the pulse at the wrist near the proximal palmar carpal skin crease
C. THE FREE ANASTOMOSES, both from side to side and from posterior to palmar surface, will preserve the hand even if a major vessel is cut. However, both ends of the severed artery usually must be clamped

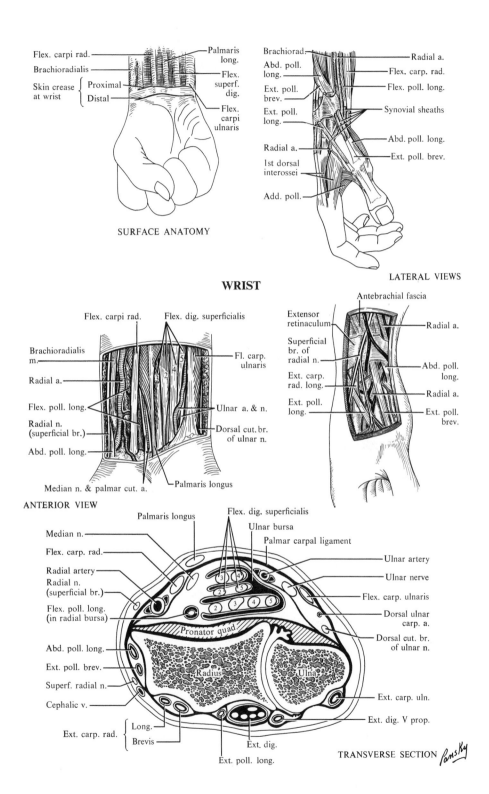

Flex. carpi rad.
Brachioradialis
Skin crease at wrist — Proximal
Distal
Palmaris long.
Flex. superf. dig.
Flex. carpi ulnaris

SURFACE ANATOMY

Brachiorad.
Abd. poll. long.
Ext. poll. brev.
Ext. poll. long.
Radial a.
1st dorsal interossei
Add. poll.
Radial a.
Flex. carp. rad.
Flex. poll. long.
Synovial sheaths
Abd. poll. long.
Ext. poll. brev.

LATERAL VIEWS

WRIST

Flex. carpi rad.
Flex. dig. superficialis
Brachioradialis m.
Radial a.
Flex. poll. long.
Radial n. (superficial br.)
Abd. poll. long.
Fl. carp. ulnaris
Ulnar a. & n.
Dorsal cut. br. of ulnar n.
Median n. & palmar cut. a.
Palmaris longus

ANTERIOR VIEW

Antebrachial fascia
Extensor retinaculum
Superficial br. of radial n.
Ext. carp. rad. long.
Ext. poll. long.
Radial a.
Abd. poll. long.
Radial a.
Ext. poll. brev.

Palmaris longus
Flex. dig. superficialis
Ulnar bursa
Palmar carpal ligament
Median n.
Flex. carp. rad.
Radial artery
Radial n. (superficial br.)
Flex. poll. long. (in radial bursa)
Abd. poll. long.
Ext. poll. brev.
Superf. radial n.
Cephalic v.
Ext. carp. rad. — Long.
Brevis
Ulnar artery
Ulnar nerve
Flex. carp. ulnaris
Dorsal ulnar carp. a.
Dorsal cut. br. of ulnar n.
Pronator quad
Radius
Ulna
Ext. carp. uln.
Ext. dig. V prop.
Ext. dig.
Ext. poll. long.

TRANSVERSE SECTION *Pansky*

129. SUPERFICIAL SURFACE OF THE HAND

I. **Dorsum of hand:** skin is thin and loose in contrast with the tight relationship of the skin of palm and its underlying fascia and palmar aponeurosis. Tissue is so lax that a fold of skin can be grasped and elevated several centimeters off the underlying deep fascia
A. DORSAL SUBCUTANEOUS SPACE lies between skin and deep fascia over the extensor tendons and is due to looseness of the areolar tissue
   1. The sensory nerves, veins, and lymphatics course through this loose areolar layer
   2. Most of the lymph from the palmar aspects of the fingers, web areas, hypothenar and thenar eminences flows into the lymph channels and lacunae found in the loose areolar layer on the dorsum of the hand
      a. Accounts for lymphedematous swelling seen on the back of the hand even when the infection is on palmar surface of fingers, web space, hypothenar, or thenar areas
B. DORSAL SUBAPONEUROTIC SPACE lies between the fused deep dorsal fascia and extensor tendons, which form its roof, and the interossei mm. and metacarpal bones which form its floor. The latter structures form a barrier to the palm, thus, this space is infrequently involved in hand infections
C. EXTENSOR TENDONS stand out clearly (to about knuckles) when wrist is extended and fingers are abducted
D. KNUCKLES, visible when hand is made into a fist, represent the heads of the metacarpal bones
E. HAIR is seen on the dorsum of the hand and proximal parts of the digits
F. DORSAL VENOUS NETWORK is clearly seen on the dorsum of the hand

II. **Skin of palm:** thick, coarse, and more vascular than on dorsum
A. CONTAINS NO HAIR or sebaceous glands, but sweat glands are numerous
B. BOUND TO THE PALMAR APONEUROSIS by fibrous septa or fasciculi between which is found granular fat

III. **Palmar flexion creases**
A. RADIAL LONGITUDINAL (vertical) partially encircles thenar eminence. For movement of opposition of thumb (formed by short muscle of thumb)
B. MIDPALMAR begins at distal transverse crease and ends on hypothenar eminence (formed by the short muscles of the 5th digit)
C. PROXIMAL TRANSVERSE begins on radial border of palm, in common with A and superficial to head of 2nd metacarpal bone. Extends medially and slightly proximally across palm, superficial to shafts of 3rd, 4th, and 5th metacarpals. Useful for movement of index finger. Marks convexity of superficial palmar arterial arch
D. DISTAL TRANSVERSE begins near cleft between index and middle fingers and crosses the palm superficial to heads of 2nd, 3rd, and 4th metacarpals. Useful in movement of medial three digits (marks metacarpal heads)

IV. **Digital flexion creases:** each of the medial four digits usually has 3 transverse flexion creases. All deepen with flexion of the digits
A. PROXIMAL: at root of finger, about 2 cm distal to metacarpophalangeal joint
B. DISTAL lies proximal to the distal interphalangeal joint
C. MIDDLE lies over the proximal interphalangeal joint

V. **The thumb** has only two flexion creases. The proximal crosses the thumb obliquely, proximal to the 1st metacarpophalangeal joint. The distal lies proximal to the interphalangeal joint

VI. **Clinical considerations**
A. SIMIAN CREASE is the one transverse palmar crease seen in the hand of people with Down's syndrome
B. THE SKIN is firmly bound to the underlying connective tissue and structures deep to it at the skin creases. Thus, these are to be avoided in surgical incisions on the palmar surface or fingers. In addition, scar tissue in the area of the creases may limit the range of motion

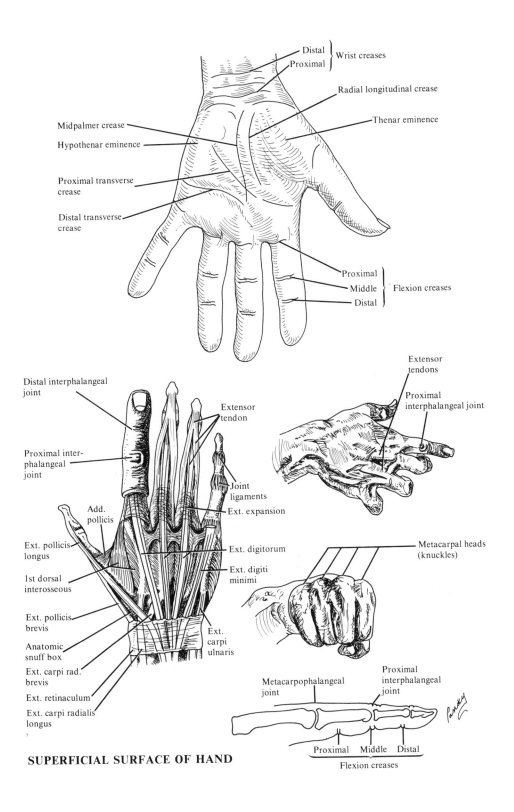

Distal ⎤ Wrist creases
Proximal ⎦

Radial longitudinal crease

Thenar eminence

Midpalmer crease

Hypothenar eminence

Proximal transverse crease

Distal transverse crease

Proximal ⎤
Middle ⎬ Flexion creases
Distal ⎦

Distal interphalangeal joint

Extensor tendon

Proximal inter-phalangeal joint

Joint ligaments

Add. pollicis

Ext. expansion

Ext. pollicis longus

Ext. digitorum

1st dorsal interosseous

Ext. digiti minimi

Ext. pollicis brevis

Anatomic snuff box

Ext. carpi ulnaris

Ext. carpi rad. brevis

Ext. retinaculum

Ext. carpi radialis longus

Extensor tendons

Proximal interphalangeal joint

Metacarpal heads (knuckles)

Metacarpophalangeal joint

Proximal interphalangeal joint

Proximal Middle Distal
Flexion creases

SUPERFICIAL SURFACE OF HAND

130. VESSELS OF THE WRIST AND HAND

I. **Arteries** are the distal ends of the radial and ulnar arteries

A. RADIAL ARTERY crosses distal end of radius at wrist and winds around trapezium and 1st metacarpal bone to reach dorsal surface. Here it pierces 1st dorsal interosseus m. to reach space between 1st and 2nd metacarpal bones. It enters palm between the oblique and transverse heads of adductor pollicis m. The deep arch is formed by joining a smaller branch from the ulnar a. The deep arch lies anterior to the interossei mm. and metacarpal bones and posterior to combined tendons of flex. dig. profundus and superficialis mm. and their attached lumbricals. The deep arch is accompanied by the deep br. of the ulnar n. and is more proximally located than the superficial arch. Branches at wrist and in hand are:

1. Palmar radial carpal: runs deep to flexor tendons and forms anterior carpal arch by joining a similar branch from ulnar a.
2. Superficial palmar: crosses wrist superficial to or through thenar mm. to terminate by helping to form the superficial palmar arch
3. Dorsal carpal branch: crosses wrist and helps form dorsal carpal arch and network with branches from ulnar and ant. interosseous arteries
   a. Three dorsal metacarpal arteries arise from this network, course down on the 2nd, 3rd, and 4th interosseus mm. and then bifurcate into the dorsal digital brs. of the middle, ring, and little fingers. They communicate with palmar digital brs. of superficial palmar arch
4. First dorsal metacarpal: arises before radial enters 1st dorsal interosseus m. Divides into 2 brs. to supply adjacent sides of thumb and index finger
5. Princeps pollicis: arises after radial penetrates 1st dorsal interosseus m. and runs deep to oblique and transverse head of adductor pollicis m. Sends two brs. to thumb on either side of flexor pollicis longus tendon
6. Radialis indicis: arises from deep arch, passes between 1st dorsal interosseus and transverse head of adductor pollicis to supply radial side of index finger
7. Deep palmar arch: continuation of radial a. in hand is considered to be the deep arch. The latter has 3 sets of branches: a *recurrent,* which passes over front of wrist joint to anastomose with anterior carpal arch; *perforating brs.* pass through spaces between metacarpals to join dorsal metacarpal aa.; and *palmar metacarpal brs.,* three in number, which cross interosseus mm. of medial 3 spaces and join the common digital brs. of the superficial arch before they bifurcate

B. ULNAR ARTERY passes anterior to flexor retinaculum and terminates by dividing into deep and superficial branches. The branches in the wrist and hand are:

1. Palmar carpal: joins with that of radial to form *palmar carpal arch*
2. Dorsal ulnar carpal: passes deep to flexor carpi ulnaris m., crosses ulnar side of wrist joint, and continues in a plane deep to the ext. carpi ulnaris to join a similar branch of the radial a. to form the *dorsal carpal arch*
3. Deep branch: passes between abductor digiti minimi and flexor digiti minimi mm. and then deep to the opponens digiti minimi to join the deep palmar arch
4. Superficial palmar arch: is a direct continuation of the ulnar a. It lies superficial to the tendons in hand as well as lumbricales. Branches of the ulnar and median nerves are behind it. It lies distal to the deep palmar arch. Its branches are 4 *common palmar digital arteries,* the 1st of which goes to the ulnar side of the little finger, and the other 3 pass between the long flexor tendons. At the finger webs, they divide into 2 branches, the *proper digital arteries,* one for each of the contiguous sides of the fingers

II. **Veins:** all arteries are accompanied by venae comitantes

III. **Clinical considerations:** in case of wounds of hand involving arteries, the extensive anastomoses make necessary the ligation of both ends of severed vessels

ARTERIES OF HAND

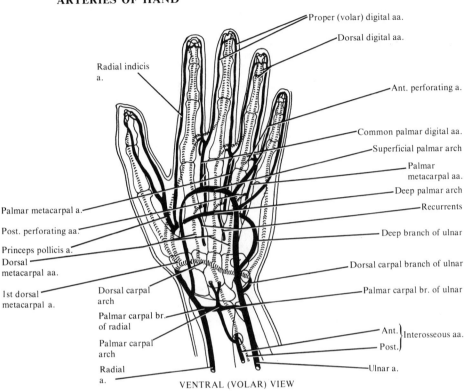

Proper (volar) digital aa.

Dorsal digital aa.

Radial indicis a.

Ant. perforating a.

Common palmar digital aa.

Superficial palmar arch

Palmar metacarpal aa.

Deep palmar arch

Recurrents

Palmar metacarpal a.

Post. perforating aa.

Princeps pollicis a.

Dorsal metacarpal aa.

Deep branch of ulnar

Dorsal carpal branch of ulnar

1st dorsal metacarpal a.

Dorsal carpal arch

Palmar carpal br. of ulnar

Palmar carpal br. of radial

Ant.
Post. Interosseous aa.

Palmar carpal arch

Radial a.

Ulnar a.

VENTRAL (VOLAR) VIEW

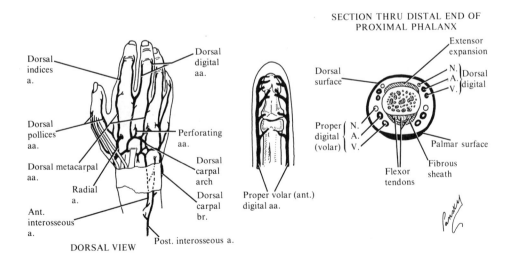

SECTION THRU DISTAL END OF
PROXIMAL PHALANX

Extensor expansion

Dorsal indices a.

Dorsal digital aa.

Dorsal surface

N.
A. Dorsal digital
V.

Dorsal pollices aa.

Perforating aa.

Proper digital (volar) N.
A.
V.

Palmar surface

Dorsal metacarpal aa.

Dorsal carpal arch

Radial a.

Flexor tendons

Fibrous sheath

Ant. interosseous a.

Dorsal carpal br.

Proper volar (ant.) digital aa.

Post. interosseous a.

DORSAL VIEW

131. NERVES OF THE HAND

I. **The nerves of the hand** are branches of the radial, ulnar, and median nerves
A. RADIAL: the cutaneous or superficial branches of the nerve supply the posterior surface of the radial side of the hand, the thumb, and the proximal ends of the 2nd, 3rd, and radial side of the 4th digit. The deep radial does not enter the hand
B. ULNAR descends on the flexor digitorum profundus (in forearm), gives a *dorsal branch* to the back of the hand on the ulnar side, a variable *cutaneous branch* (palmar cutaneous) to the palm, and enters the palm lateral to the pisiform and medial to the hook of the hamate on the anterior surface of the flexor retinaculum. It then divides into superficial and deep branches
　1. Superficial branch: innervates the palmaris brevis muscle and then passes distally where it divides into two *common palmar digital branches*
　　a. One common digital continues distally to innervate the medial (ulnar) side of the little finger as a *proper palmar digital*
　　b. The other common palmar digital follows the 4th lumbrical muscle to the web between the ring and little fingers where it divides into *proper digital branches* for contiguous sides of those fingers
　　c. The proper palmar digital nerves also wind around the fingers to reach the dorsal surface of the middle and distal phalanges of the digits they supply
　2. Deep branch: follows the deep branch of the ulnar artery between the abductor digiti minimi and flexor digiti minimi and deep to the opponens digiti minimi. It supplies these muscles and crosses the palm with the deep palmar arch. It gives branches to all interosseus mm., 3rd and 4th lumbrical mm., joints and bones of hand, and terminates by supplying the oblique and transverse heads of the adductor pollicis muscle
C. MEDIAN descends into hand deep to the flexor retinaculum and immediately divides into lateral and medial divisions which appear in the palm of the hand superficial to the tendons but deep to the superficial palmar arch
　1. Lateral division: gives off a *recurrent branch* to the three thenar eminence muscles and then divides into three *common digital nerves*
　　a. Two of the common digital nerves pass on the anterior surface of the adductor pollicis m. to be distributed to the thumb as proper digital brs. (along with princeps pollicis a.) on either side of the tendon of the flexor pollicis longus m.
　　b. The remaining common digital branch passes on the surface of the 1st lumbrical to become the proper digital nerve on the radial side of the index finger (it supplies the 1st lumbrical)
　2. Medial division: divides into two *common digital nerves*. These pass distally on the superficial surfaces of the 2nd and 3rd lumbrical mm. They supply the 2nd lumbrical m. Each of the common digitals divides into two proper digital brs. which supply contiguous sides of the index and middle fingers and of the middle and ring fingers
　3. All the digital nerves give off branches which wind around the fingers to the dorsal side of the middle and distal phalanges of the digits they supply
　4. The median nerve is the motor nerve of the thenar eminence muscles as well as the 1st and 2nd lumbrical mm. It is the sensory nerve to the anterior surface of the palm, the thumb, index finger, middle finger, and radial side of ring finger. Its cutaneous nerves extend onto the dorsal surface of the middle and terminal phalanges

II. **Clinical considerations**
A. KNOWLEDGE OF POSITION of the proper digital nerves makes possible adequate nerve blocks for surgery of the digits
B. THE MAJOR MUSCULAR BRANCH of the median arises from the lateral cutaneous br. to the thumb just distal to the transverse carpal ligament

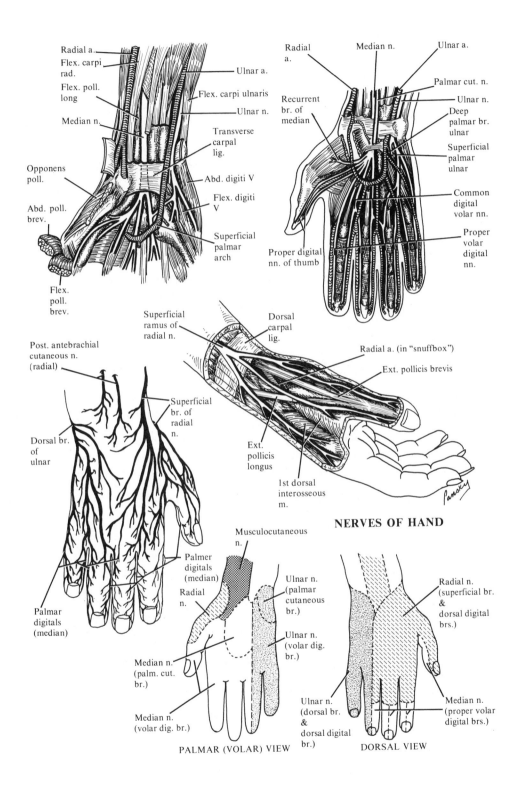

Radial a.
Flex. carpi rad.
Flex. poll. long
Median n.
Opponens poll.
Abd. poll. brev.
Flex. poll. brev.

Ulnar a.
Flex. carpi ulnaris
Ulnar n.
Transverse carpal lig.
Abd. digiti V
Flex. digiti V
Superficial palmar arch

Radial a.
Median n.
Ulnar a.
Recurrent br. of median
Palmar cut. n.
Ulnar n.
Deep palmar br. ulnar
Superficial palmar ulnar
Common digital volar nn.
Proper volar digital nn.
Proper digital nn. of thumb

Post. antebrachial cutaneous n. (radial)
Dorsal br. of ulnar
Palmar digitals (median)

Superficial ramus of radial n.
Dorsal carpal lig.
Radial a. (in "snuffbox")
Ext. pollicis brevis
Superficial br. of radial n.
Ext. pollicis longus
1st dorsal interosseous m.

NERVES OF HAND

Musculocutaneous n.
Palmer digitals (median)
Radial n.
Ulnar n. (palmar cutaneous br.)
Ulnar n. (volar dig. br.)
Median n. (palm. cut. br.)
Median n. (volar dig. br.)

PALMAR (VOLAR) VIEW

Radial n. (superficial br. & dorsal digital brs.)
Ulnar n. (dorsal br. & dorsal digital br.)
Median n. (proper volar digital brs.)

DORSAL VIEW

Pansky

– 289 –

132. DEEP PALMAR FASCIA

I. The deep palmar fascia is continuous proximally with the antebrachial fascia and at the borders of the palm with the fascia on the dorsum of the hand. It is thin over the thenar and hypothenar eminences, but is thick in the palm, where it forms the *palmar aponeurosis*. It continues distally onto the fingers by helping to form the *fibrous digital sheaths*

A. THE PALMAR FASCIA is in intimate contact with the thenar and hypothenar muscles which helps prevent the formation of a space over the muscles within which pus could accumulate

B. THE PALMAR APONEUROSIS: strong, heavy, dense, well-defined triangular layer of the deep fascia in the middle of the palm. Its apex (proximally) is anchored to the flexor retinaculum (transverse carpal ligament), and its base (distally) ends opposite the metacarpophalangeal joints where it divides at the roots of the fingers as 4 heavy, longitudinal bands or fiber slips which fuse with the fibrous digital sheaths. The aponeurosis consists of 2 layers: longitudinal and transverse

    1. Longitudinal: is more superficial and represents an extension of the tendon of the palmaris longus to the fingers. It forms bands distally over the long flexor tendons and fuses to the fibrous digital sheaths

    2. Transverse: lies deeper and is attached to the flexor retinaculum. These fibers are especially prominent between the diverging longitudinal bands and extend as far distally as the heads of the metacarpals to form the *superficial transverse metacarpal ligament*

C. THE PALMAR APONEUROSIS COVERS the long tendons of the digital muscles as they pass through the hand and also covers the arteries and nerves destined for the fingers

    1. The vessels and nerves escape from the aponeurosis to proceed to the web between each two digits before dividing to reach contiguous sides of the two digits

D. IN THE DISTAL PART of the palm, septa extend from the deep aspect of the palmar aponeurosis. They form the sides of anular fibrous canals for passage of the ensheathed flexor tendons and lumbrical muscles, as well as the blood vessels and nerves. These septa appear unattached on their deep aspect, but just distal to the ends of the midpalmar and thenar spaces, they actually attach to the deep transverse metacarpal ligament

    1. A particular septum, from the aponeurosis to the 3rd metacarpal bone, separates the thenar space (bursa) from the midpalmar space (bursa)

E. THE LATERAL EDGES of the palmar aponeurosis meet the deep fascia surrounding the thenar and hypothenar muscles

II. Clinical considerations

A. THE GREAT MAJORITY OF HAND INFECTIONS, excluding those of the fingers, are found in the hollow of the palm between or distal to the hypothenar and thenar eminences

B. DUPUYTREN'S CONTRACTURE is a disease, often familial, of the palmar aponeurosis, of unknown origin but with hereditary predisposition apparently involving many genetic factors. It manifests itself as a progressive fibrosis which typically produces abnormal bands of fibrous tissue that extend from the aponeurosis to the bases of the phalanges and pull one or more digits into marked flexion at the metacarpophalangeal joints so that they cannot be straightened. This may cause surprisingly little disability, but if the middle and index fingers are affected, the hand is virtually useless

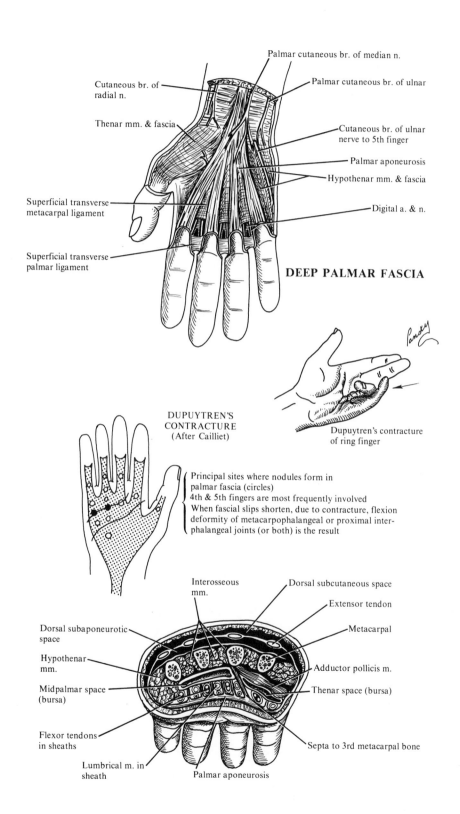

Palmar cutaneous br. of median n.

Cutaneous br. of radial n.

Palmar cutaneous br. of ulnar

Thenar mm. & fascia

Cutaneous br. of ulnar nerve to 5th finger

Palmar aponeurosis

Hypothenar mm. & fascia

Superficial transverse metacarpal ligament

Digital a. & n.

Superficial transverse palmar ligament

DEEP PALMAR FASCIA

DUPUYTREN'S CONTRACTURE
(After Cailliet)

Dupuytren's contracture of ring finger

Principal sites where nodules form in palmar fascia (circles)
4th & 5th fingers are most frequently involved
When fascial slips shorten, due to contracture, flexion deformity of metacarpophalangeal or proximal inter-phalangeal joints (or both) is the result

Interosseous mm.

Dorsal subcutaneous space

Extensor tendon

Dorsal subaponeurotic space

Metacarpal

Hypothenar mm.

Adductor pollicis m.

Midpalmar space (bursa)

Thenar space (bursa)

Flexor tendons in sheaths

Septa to 3rd metacarpal bone

Lumbrical m. in sheath

Palmar aponeurosis

I. Muscles of hand, arranged in 3 groups: thenar, hypothenar, and intermediate

Name	Origin	Insertion	Action	Nerve
A. THENAR (muscles of the thumb)				
Abductor pol. brev.	Flex. retinac. Tuberos. of scaphoid Trapezium	Lat. side, base of first phalanx of thumb	Abducts thumb	Median
Opponens pollicis	Flex. retinac. Trapezium	Radial side, first metacarpal	Abducts, flexes, rotates first metacarpal	Median
Flex. pol. brev.	Flex. retinac. Trapezium	Lat. side of base of first phalanx of thumb Med. side of base of first phalanx	Flexes and adducts thumb	Median and deep ulnar
Adductor pollicis				
Obliq.	Capitate Bases, second and third metacarpals	Med. side, base, first phalanx of thumb	Adducts thumb	Deep ulnar
Trans.	Shaft, third metacarpal	Med. side, base, first phalanx of thumb	Adducts thumb	Deep ulnar
B. HYPOTHENAR				
Palmaris brev.	Flex. retinac. Palmar aponeurosis	Skin, med. border of palm	Builds up hypothenar eminence	Ulnar
Abduct. dig. minimi	Pisiform Tendon of flex. carp. ulnaris	Med. side, base of first phalanx of small finger	Abducts small finger Flexes first phalanx	Ulnar
Flex. digit. minimi brev.	Hamulus of hamate Flex. retinac.	Med. side, base of first phalanx of little finger	Flexes first phalanx	Ulnar
Opponens digit. minimi	Hamulus of hamate Flex. retinac.	Med. side, shaft of fifth metacarpal	Opposes the fifth finger	Ulnar

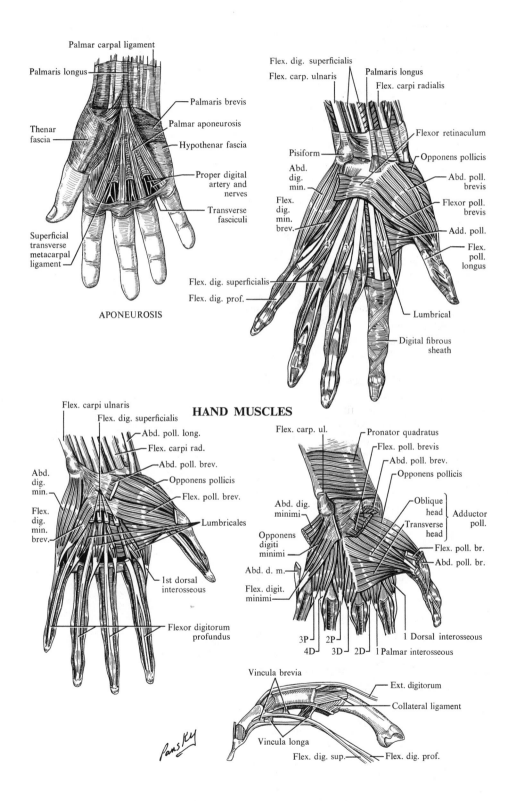

Palmar carpal ligament

Palmaris longus

Thenar fascia

Superficial transverse metacarpal ligament

Palmaris brevis

Palmar aponeurosis

Hypothenar fascia

Proper digital artery and nerves

Transverse fasciculi

APONEUROSIS

Flex. dig. superficialis

Flex. carp. ulnaris

Palmaris longus

Flex. carpi radialis

Pisiform

Abd. dig. min.

Flex. dig. min. brev.

Flexor retinaculum

Opponens pollicis

Abd. poll. brevis

Flexor poll. brevis

Add. poll.

Flex. poll. longus

Flex. dig. superficialis

Flex. dig. prof.

Lumbrical

Digital fibrous sheath

HAND MUSCLES

Flex. carpi ulnaris

Flex. dig. superficialis

Abd. poll. long.

Flex. carpi rad.

Abd. poll. brev.

Opponens pollicis

Flex. poll. brev.

Abd. dig. min.

Flex. dig. min. brev.

Lumbricales

1st dorsal interosseous

Flexor digitorum profundus

Flex. carp. ul.

Pronator quadratus

Flex. poll. brevis

Abd. poll. brev.

Opponens pollicis

Abd. dig. minimi

Opponens digiti minimi

Abd. d. m.

Flex. digit. minimi

Oblique head

Transverse head

Adductor poll.

Flex. poll. br.

Abd. poll. br.

3P   2P

4D   3D   2D

1 Dorsal interosseous

1 Palmar interosseous

Vincula brevia

Ext. digitorum

Collateral ligament

Vincula longa

Flex. dig. sup.

Flex. dig. prof.

-293-

Name	Origin	Insertion	Action	Nerve
C. INTERMEDIATE				
Lumbricales 1 and 2	Radial side deep flexor tendons to index and middle finger	Tendon, ext. digit. communis, on dorsum of first phalanx, index, and middle fingers	Flex first phalanx, extend second and third phalanges	Median
Lumbricales 3 and 4	Adjoining sides, deep flexor tendons to middle and ring fingers	Tendon, ext. digitorum, on dorsum of first phalanx, ring, and little fingers	Flex first phalanx, and extend distal phalanges, ring and little fingers	Deep ulnar
Palmar interossei	1. Med. side, second metacarpal	Med. side, base first phalanx, index finger Tendon ext. digitorum	All 3 adduct fingers, flex first phalanx, extend second and third	Deep ulnar
	2. Lat. side, fourth metacarpal	Same as above for ring finger		
	3. Lat. side, fifth metacarpal	Same as above for little finger		
Dorsal interossei	1. Adjacent sides, second and first metacarpals	Lat. side, base first phalanx, index finger Ext. digit.	All 4 abduct fingers, flex first phalanx, extend second and third	Deep ulnar
	2. Adjacent sides, second and third metacarpals	Lat. side, base first phalanx, middle finger Ext. digit.		
	3. Adjacent sides, third and fourth metacarpals	Med. side, base first phalanx, middle finger Ext. digit		
	4. Adjacent sides, fourth and fifth metacarpals	Med. side, base first phalanx, ring finger Tendon ext. digit. com.		

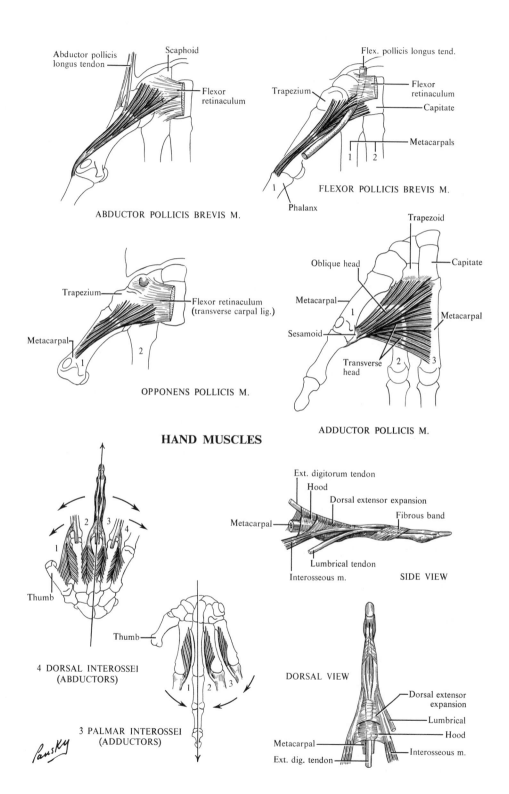

Abductor pollicis
longus tendon

Scaphoid

Flexor
retinaculum

ABDUCTOR POLLICIS BREVIS M.

Flex. pollicis longus tend.

Trapezium

Flexor
retinaculum

Capitate

Metacarpals

1   2

1

Phalanx

FLEXOR POLLICIS BREVIS M.

Trapezium

Flexor retinaculum
(transverse carpal lig.)

Metacarpal

1

2

OPPONENS POLLICIS M.

Trapezoid

Oblique head

Capitate

Metacarpal

1

Sesamoid

Metacarpal

Transverse
head

2   3

ADDUCTOR POLLICIS M.

HAND MUSCLES

2   3   4

1

Thumb

4 DORSAL INTEROSSEI
(ABDUCTORS)

Thumb

1   2   3

3 PALMAR INTEROSSEI
(ADDUCTORS)

Ext. digitorum tendon

Hood

Dorsal extensor expansion

Fibrous band

Metacarpal

Lumbrical tendon

Interosseous m.

SIDE VIEW

DORSAL VIEW

Dorsal extensor
expansion

Lumbrical

Hood

Metacarpal

Interosseous m.

Ext. dig. tendon

135. THE SYNOVIAL BURSAE, COMPARTMENTS AND SPACES IN THE HAND

I. **Bursae:** synovial compartments for flexor tendons consisting of a sheath which is a double-layered, elongated sac composed of a synovial membrane and containing synovial fluid (in contrast to fibrous sheath outside of synovial sheath which serves to hold tendon onto phalanges and their joints during flexion)

A. RADIAL: smaller, begins proximal to flexor retinaculum, encloses tendon of flexor pollicis longus, and extends to base of terminal phalanx of thumb

B. ULNAR: large, begins proximal to flexor retinaculum, encloses tendons of flexor digitorum superficialis and profundus, and continues to terminal phalanx of little finger. For other fingers, ends near middle of metacarpals

   1. Arrangement of tendons
      a. Tendons of superficial flexors are paired. Those for the third and fourth fingers lie superficial to those of the second and fifth fingers. Each pair is covered medially, dorsally, and ventrally by a synovial membrane, the pairs separated by a space. Distally, in hand, paired arrangement disappears—tendons lie side by side
      b. The tendons of the deep flexors are enclosed in the same sheet, separated by a space from the deeper pair of superficial tendons

II. **Fascial clefts (spaces):** lines of separation between fascial membranes

A. MIDDLE PALMAR: between deep surface of flexor tendons and fascia covering the interossei. Closed laterally by septum attached to third metacarpal, medially by hypothenar septum to fifth metacarpal

   1. Lumbrical canals: fascial tubes leading out of middle palmar cleft. The canals prolong the midpalmar cleft to the lateral side of fingers. They contain the lumbrical tendons and the proper digital vessels and nerves

B. THENAR: overlies fascia of adductor pollicis muscle. Medial boundary is same as lateral septum for middle cleft; lateral is thenar septum and first metacarpal. Proximally, it ends at flexor retinaculum, distally, at transverse head of the adductor

III. **Fascial compartments:** area enclosed by fascial membranes

A. THENAR (thenar eminence): closed proximally by attachments of thenar muscles to flexor retinaculum and carpals, distally by insertion of muscles to first phalanx of thumb, dorsally by subcutaneous border of first metacarpal, medially by attachment of thenar septum to adductors. Contains short thumb muscles, except adductor; tendon of flexor pollicis longus; palmar branch of radial artery

B. HYPOTHENAR (hypothenar eminence): proximally and distally closed as A above; laterally is the hypothenar septum, which ends between flexor digiti minimi muscles and third palmar interosseus
   Contents: all short muscles of small finger

C. CENTRAL (intermediate): between the thenar and hypothenar septa. Ventrally is the palmar aponeurosis, dorsally is the deep fascia on the interosseus mm. and adductor m. of the thumb. This compartment is divided into 2 areas: a ventral containing nerves and vessels, and a dorsal occupied by the tendons of the long muscles of the forearm plus their tendon sheaths
   Contents: lumbrical muscles, superficial volar arch, palmar branch of median nerve, superficial branch of ulnar nerve, and all the flexor tendons

D. INTEROSSEUS-ADDUCTOR: enclosed by the dorsal and palmar interosseous fascia and closed proximally and distally by the origins and insertions of the interosseus muscles
   Contents: palmar and dorsal interossei, adductor pollicis muscles, metacarpals 2–4, deep volar arch, and deep branch of the ulnar nerve

IV. **Clinical considerations**

A. TENOSYNOVITIS is an inflammation of the tendon sheaths contained within the bursae

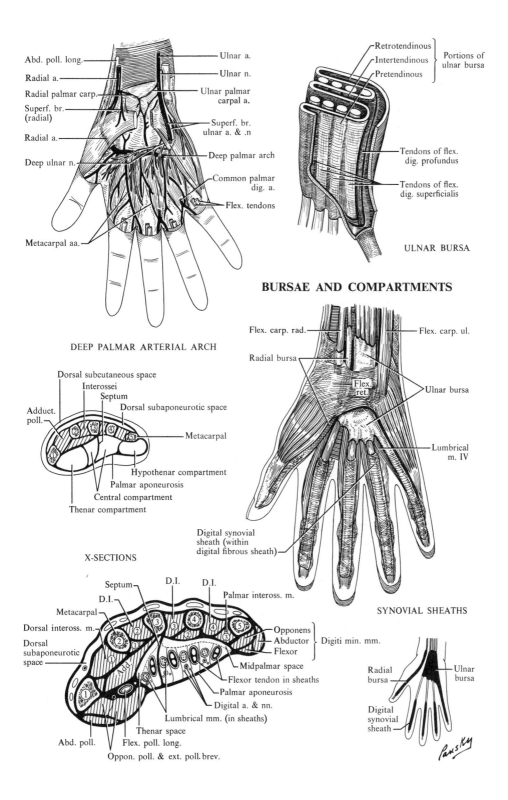

Abd. poll. long.

Radial a.

Radial palmar carp.

Superf. br. (radial)

Radial a.

Deep ulnar n.

Metacarpal aa.

Ulnar a.

Ulnar n.

Ulnar palmar carpal a.

Superf. br. ulnar a. & .n

Deep palmar arch

Common palmar dig. a.

Flex. tendons

DEEP PALMAR ARTERIAL ARCH

Retrotendinous
Intertendinous
Pretendinous } Portions of ulnar bursa

Tendons of flex. dig. profundus

Tendons of flex. dig. superficialis

ULNAR BURSA

BURSAE AND COMPARTMENTS

Flex. carp. rad.

Radial bursa

Flex. ret.

Flex. carp. ul.

Ulnar bursa

Lumbrical m. IV

Digital synovial sheath (within digital fibrous sheath)

SYNOVIAL SHEATHS

Dorsal subcutaneous space
Interossei
Septum
Dorsal subaponeurotic space

Adduct. poll.

Metacarpal

Hypothenar compartment

Palmar aponeurosis

Central compartment

Thenar compartment

X-SECTIONS

Septum
D.I.
D.I.
D.I.
Palmar inteross. m.

Metacarpal

Dorsal inteross. m.

Dorsal subaponeurotic space

Add.

Opponens
Abductor } Digiti min. mm.
Flexor

Midpalmar space

Flexor tendon in sheaths

Palmar aponeurosis

Digital a. & nn.

Lumbrical mm. (in sheaths)

Thenar space

Abd. poll.

Flex. poll. long.

Oppon. poll. & ext. poll. brev.

Radial bursa

Ulnar bursa

Digital synovial sheath

Pansky

– 297 –

136. RELATIONS AT DORSUM OF WRIST WITH VESSELS AND NERVES TO DORSUM OF HAND

I. The ligaments

A. EXTENSOR RETINACULUM: antebrachial fascia strengthened by circular fibers. Attached to styloid of ulna, the triquetral and the pisiform bones, and the radius. Between the medial and lateral attachments of the retinaculum, osseofibrous canals are formed for the tendons of the dorsal antebrachial muscles

B. CANALS FOR TENDONS: 6 in number, from lateral to medial
1. On lateral side of radial styloid for abductor pollicis longus and extensor pollicis brevis
2. Dorsal to styloid for extensor carpi radialis longus and brevis
3. On middorsum of radius for extensor pollicis longus
4. Medial to 3, above, for extensor digitorum and extensor indicis
5. Between radius and ulna for extensor digiti minimi
6. Between ulnar head and styloid for extensor carpi ulnaris

C. SYNOVIAL SHEATHS for all extensor tendons begin just above the retinaculum. They terminate either proximal to the heads of the metacarpals or at the metacarpophalangeal joints

II. Nerves

A. DORSAL OF ULNAR gives 2 dorsal digital branches and a metacarpal communicating branch
1. Most medial digital branch to medial side of little finger; the other supplies adjacent sides of ring and little fingers
2. Metacarpal branch to skin of dorsal hand on ulnar side

B. SUPERFICIAL BRANCH of radial gives a lateral and medial branch
1. Lateral to radial side and ball of thumb
2. Medial has 4 dorsal digital branches: first to medial thumb, second to lateral side of index finger, third to adjacent sides of index and middle fingers, and fourth to adjacent side of middle and ring fingers

C. POSTERIOR ANTEBRACHIAL CUTANEOUS (OF RADIAL NERVE) may extend to the middle part of the proximal hand

D. DORSAL BRANCHES OF ULNAR AND LATERAL ANTEBRACHIAL CUTANEOUS (OF MUSCULOCUTANEOUS) reach the wrist but do not extend onto hand

III. Arteries

A. DORSAL CARPAL NET formed by dorsal carpal branch of radial, dorsal carpal branch of ulnar, and posterior interosseous. This gives:
1. Three dorsal metacarpals, which run on second, third, and fourth dorsal interossei. Each divides into dorsal digital branches to adjacent sides of index, middle, ring, and small fingers
   a. Superior perforating from dorsal metacarpals to deep palmar arch
   b. Inferior perforating from dorsal metacarpals to common palmar digitals

B. FIRST DORSAL METACARPAL, from radial, divides to supply the adjacent sides of thumb and index finger

C. FIFTH DORSAL METACARPAL, from ulnar, to medial side of little finger

DORSUM OF HAND AND WRIST

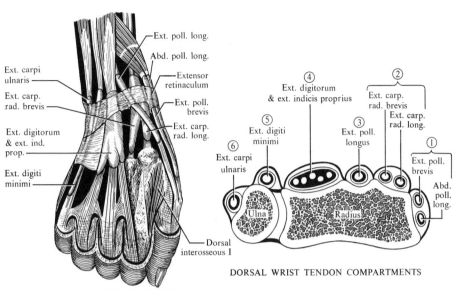

SYNOVIAL TENDON SHEATHS

Ext. poll. long.

Abd. poll. long.

Ext. carpi ulnaris

Ext. carp. rad. brevis

Ext. digitorum & ext. ind. prop.

Ext. digiti minimi

Extensor retinaculum

Ext. poll. brevis

Ext. carp. rad. long.

Dorsal interosseous I

DORSAL WRIST TENDON COMPARTMENTS

④ Ext. digitorum & ext. indicis proprius

② Ext. carp. rad. brevis / Ext. carp. rad. long.

⑤ Ext. digiti minimi

③ Ext. poll. longus

⑥ Ext. carpi ulnaris

① Ext. poll. brevis / Abd. poll. long.

Ulna

Radius

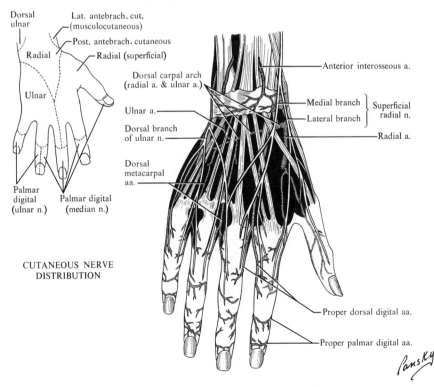

CUTANEOUS NERVE DISTRIBUTION

Dorsal ulnar

Lat. antebrach. cut. (musculocutaneous)

Radial

Post. antebrach. cutaneous

Radial (superficial)

Ulnar

Palmar digital (ulnar n.)

Palmar digital (median n.)

Anterior interosseous a.

Dorsal carpal arch (radial a. & ulnar a.)

Medial branch

Lateral branch

Superficial radial n.

Ulnar a.

Dorsal branch of ulnar n.

Radial a.

Dorsal metacarpal aa.

Proper dorsal digital aa.

Proper palmar digital aa.

Pansky

137. FASCIA, TENDON SHEATHS, NERVES, AND VESSELS OF THE FINGERS

I. Digital tendon sheaths: fibrous bands that form osseofibrous canals with palmar surface of phalanges to hold tendons of flexor digitorum superficialis and profundus (see p. 293)

A. AT MIDDLE OF FIRST AND SECOND PHALANX, greatly strengthened by circular fibers—*digital vaginal ligaments*

B. OVER JOINTS, sheaths thinner, fibers circular or obliquely crossed (*cruciate*)

II. The vincula tendina

A. SHORT: 2 for each finger: (1) from superficial tendon to first interphalangeal joint and head of first phalanx, (2) from deep tendon to second interphalangeal joint and head of second phalanx

B. LONG: 2 for each finger: (1) under side of profunda to the superficialis, (2) superficial tendon to proximal end of first phalanx

III. Extensor aponeurosis formed by tendons of extensor digitorum. At metacarpophalangeal joints, tendon is bound laterally with the collateral ligaments of this joint. Thus, the aponeurosis acts as dorsal ligaments, which are lacking, per se, in this joint. After crossing joint, tendon broadens to cover dorsal aspect of first phalanx, here reinforced by tendons of lumbricales and interossei. At first interphalangeal joint, splits into 3 slips, 2 lateral and 1 intermediate. The intermediate inserts in base of second phalanx. The 2 lateral reunite before they reach the second interphalangeal joint, to insert into base of third phalanx. This aponeurosis also serves as dorsal ligaments to interphalangeal joints

IV. Synovial sheaths

A. EXTENSORS terminate in proximal third of hand and are not found in the fingers

B. FLEXORS
1. On thumb and little finger these sheaths continuous with the so-called *radial* and *ulnar bursae,* respectively. For fingers 2, 3, and 4, start at heads of their respective metacarpals. They continue distally to insertion of deep flexors

V. Arteries

A. PROPER PALMAR DIGITAL ARTERIES extend on either side of finger, dorsal to the corresponding nerves. Cross anastomoses in fingertips and at interphalangeal joints occur, and anastomotic branches to dorsal digitals are present

B. DORSAL DIGITAL ARTERIES run just ventral to their corresponding nerves but reach only to the first interphalangeal joint. The palmar digitals supply the second and third phalanges and nail bed

VI. Nerves

A. PROPER PALMAR DIGITAL NERVES lie just ventral to arteries; they extend to end of finger supplying the skin of the ball, the nail bed, and the skin over the dorsum of the third phalanx

B. DORSAL DIGITAL NERVES lie just above corresponding artery and extend to near the second interphalangeal joint

VII. Special features

A. AS THE EXTENSOR TENDONS REACH THE DORSAL ASPECT of the metacarpophalangeal joints, they expand laterally over the sides of the joints in addition to their central thicker portion. This expansion of the extensor tendon is called the *extensor hood*

B. THE FASCIA forms a pad (called ''pulp'') of considerable thickness on the distal phalanx; here the retinacula cutis are so arranged that they guide superficial infections to deep positions. Infection of the pulp of the terminal phalanx is called a *felon* or *whitlow*

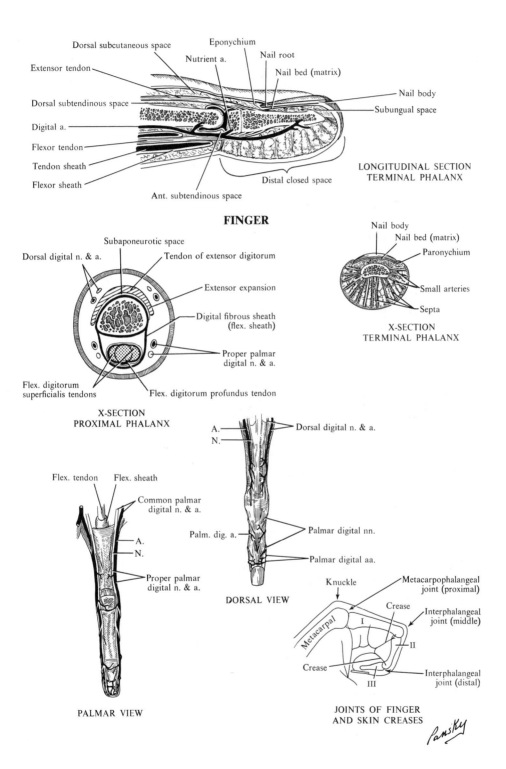

Dorsal subcutaneous space

Eponychium

Nutrient a.

Nail root

Nail bed (matrix)

Extensor tendon

Nail body

Dorsal subtendinous space

Subungual space

Digital a.

Flexor tendon

Tendon sheath

Flexor sheath

Distal closed space

LONGITUDINAL SECTION
TERMINAL PHALANX

Ant. subtendinous space

FINGER

Nail body

Nail bed (matrix)

Paronychium

Subaponeurotic space

Dorsal digital n. & a.

Tendon of extensor digitorum

Extensor expansion

Small arteries

Digital fibrous sheath
(flex. sheath)

Septa

Proper palmar
digital n. & a.

X-SECTION
TERMINAL PHALANX

Flex. digitorum
superficialis tendons

Flex. digitorum profundus tendon

X-SECTION
PROXIMAL PHALANX

A.

N.

Dorsal digital n. & a.

Flex. tendon   Flex. sheath

Common palmar
digital n. & a.

Palm. dig. a.

Palmar digital nn.

A.

N.

Palmar digital aa.

Proper palmar
digital n. & a.

Knuckle

Metacarpophalangeal
joint (proximal)

DORSAL VIEW

Crease

Interphalangeal
joint (middle)

Metacarpal

I

II

Crease

Interphalangeal
joint (distal)

III

PALMAR VIEW

JOINTS OF FINGER
AND SKIN CREASES

Pansky

138. JOINTS OF FOREARM, WRIST, AND HAND

I. Radioulnar
A. PROXIMAL (see p. 272)
B. MIDDLE considered a syndesmosis with very slight movement. Shafts of radius and ulna joined by oblique cord and interosseous membrane
C. DISTAL
   1. Type: trochoid (pivot)
   2. Bones: head of ulna and ulnar notch of radius
   3. Movements: rotation (pronation, supination)
   4. Ligaments
      a. Articular capsule with palmar and dorsal radioulnar ligaments
      b. Articular disk

II. Muscles acting on radioulnar joints (see p. 272)

III. Wrist joint (radiocarpal articulation)
A. TYPE: condyloid
B. BONES: distal radius and articular disk with proximal row of carpals
C. MOVEMENTS: flexion, extension, abduction, adduction
D. LIGAMENTS
   1. Palmar radiocarpal
   2. Dorsal radiocarpal
   3. Ulnar carpal collateral
   4. Radial carpal collateral

IV. Muscles acting on wrist joint

Flexion	Extension	Abduction	Adduction
Flex. carp. uln.	Ext. carp. rad. long.	Flex. carp. rad.	Flex. carp. uln.
Flex. carp. rad.	Ext. carp. rad. brev.	Ext. carp. rad. long.	Ext. carp. uln.
Palm. long.	Ext. carp. uln.	Ext. carp. rad. br.	Ext. indic. prop.
Flex. dig. superfic.	Ext. dig. communis	Ext. pol. long.	
Flex. dig. prof.		Ext. pol. brev.	

V. Intercarpal articulation
A. ARTICULATIONS OF CARPALS with each other: arthrodial joints with dorsal, palmar, and interosseous ligaments
B. ARTICULATIONS BETWEEN THE 2 ROWS OF CARPALS (the *midcarpal joint*): a hinge joint, acting with wrist. Has dorsal, palmar, and collateral ligaments

VI. Carpometacarpal
A. BETWEEN FIRST METACARPAL AND TRAPEZIUM: a saddle joint, permitting flexion, extension, abduction, adduction, and opposition
B. ALL OTHERS between distal row of carpals and bases of metacarpals: arthrodial, with dorsal, palmar, and interosseous ligaments

VII. Intermetacarpal among bases of metacarpals: dorsal, palmar, and interosseous ligaments
A. DEEP TRANSVERSE METACARPAL LIGAMENTS join heads of metacarpals

VIII. Metacarpophalangeal joints: condyloid, joined by 1 palmar and 2 collateral ligaments

IX. Interphalangeal joints: hinge joints, with 1 palmar and 2 collateral ligaments

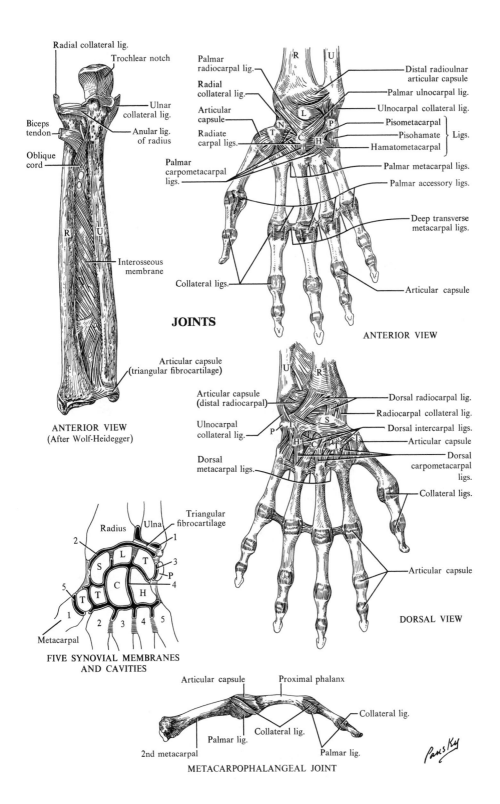

Radial collateral lig.

Trochlear notch

Ulnar collateral lig.

Biceps tendon

Anular lig. of radius

Oblique cord

R U

Interosseous membrane

ANTERIOR VIEW
(After Wolf-Heidegger)

JOINTS

Palmar radiocarpal lig.

Radial collateral lig.

Articular capsule

Radiate carpal ligs.

Palmar carpometacarpal ligs.

R U

L

N P

T C H

Distal radioulnar articular capsule

Palmar ulnocarpal lig.

Ulnocarpal collateral lig.

Pisometacarpal

Pisohamate } Ligs.

Hamatometacarpal

Palmar metacarpal ligs.

Palmar accessory ligs.

Deep transverse metacarpal ligs.

Collateral ligs.

Articular capsule

ANTERIOR VIEW

Articular capsule (triangular fibrocartilage)

Articular capsule (distal radiocarpal)

Ulnocarpal collateral lig.

Dorsal metacarpal ligs.

U R

P

S

H C T

Dorsal radiocarpal lig.

Radiocarpal collateral lig.

Dorsal intercarpal ligs.

Articular capsule

Dorsal carpometacarpal ligs.

Collateral ligs.

Articular capsule

DORSAL VIEW

Radius   Ulna   Triangular fibrocartilage

2   1

L   T

S   3

C   P

T   T   H   4

5

1   2   3   4   5

Metacarpal

FIVE SYNOVIAL MEMBRANES AND CAVITIES

Articular capsule   Proximal phalanx

Collateral lig.

2nd metacarpal

Palmar lig.   Collateral lig.

Palmar lig.

METACARPOPHALANGEAL JOINT

Pansky

– 303 –

139. SURFACE ANATOMY OF UPPER EXTREMITY

I. Arteries

A. AXILLARY: its medial end can be compressed against the 2nd rib by pressure applied below the middle of the clavicle just medial to the coracoid process of the scapula. The line extends laterally from here to the medial border of the biceps at level of the posterior axillary fold

B. BRACHIAL: its line extends downward along the medial side of the biceps (along bicipital groove) to a point in the cubital fossa midway between the humeral epicondyles and terminates about 2.5 cm below this point opposite the head of the radius

C. RADIAL: this can be indicated by a line beginning 1 in. below the center of the line joining the humeral epicondyles to a point just medial to the styloid process of the radius. From this point, the vessel curves around the lateral side of the wrist to reach the proximal end of the first intermetacarpal space on the back of the hand

D. ULNAR: beginning at the terminal point of the brachial artery, the line runs obliquely downward to reach the ulnar side of the forearm about halfway between the elbow and wrist. Thence, the line continues downward to a point just lateral to the pisiform bone

1. Pulsations of the ulnar artery can be felt lateral to the pisiform bone. Just distal to this point, it divides into its larger branch to form most of the superficial arch and the smaller branch which forms the lesser part of the deep arch

E. SUPERFICIAL PALMAR ARCH: beginning just to the lateral side of the pisiform bone, the line arches distally and laterally across the palm. The convexity of the arch reaches a level of the medial border of the fully extended and abducted thumb

F. DEEP PALMAR ARCH: this is indicated by an almost directly transverse line about 1.0 cm proximal to the superficial arch

II. Nerves

A. MEDIAN: in the arm, its line coincides with that for the brachial artery; in the lower cubital fossa, it lies at the medial side of ulnar artery; just below the cubital fossa, it crosses anterior to the ulnar artery and continues in the forearm on a line drawn down the center of the anterior surface of the forearm; it continues down the wrist along the middle of its anterior surface between the tendons of the flexor carpi radialis and palmaris longus

B. ULNAR: in the upper arm, it follows the line of the brachial artery, thence, it passes deep and runs in a groove just behind the medial epicondyle of the humerus; in the forearm, it shifts anteriorly, and at the junction of the middle and upper thirds of the forearm, it joins the ulnar artery and follows its line into the wrist

C. RADIAL: in upper part of arm, it lies on the medial side of the humerus; at the junction of the upper and middle thirds of the humerus, it passes behind the humerus in the radial sulcus (groove) on the posterior aspect of that bone; at the junction of the middle and lower thirds of arm, it reaches the lateral side of the arm and continues downward across the lateral side of the elbow joint, lateral to the cubital fossa

ARTERIES

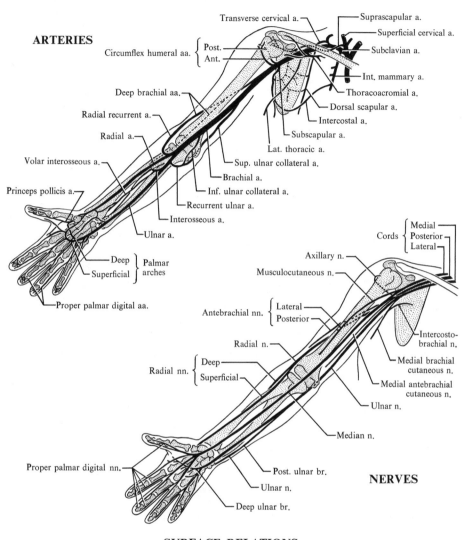

Transverse cervical a.
Suprascapular a.
Superficial cervical a.
Circumflex humeral aa. { Post. Ant.
Subclavian a.
Int. mammary a.
Thoracoacromial a.
Deep brachial aa.
Dorsal scapular a.
Intercostal a.
Radial recurrent a.
Subscapular a.
Radial a.
Lat. thoracic a.
Volar interosseous a.
Sup. ulnar collateral a.
Brachial a.
Princeps pollicis a.
Inf. ulnar collateral a.
Recurrent ulnar a.
Interosseous a.
Ulnar a.
Deep } Palmar
Superficial } arches
Proper palmar digital aa.

Cords { Medial Posterior Lateral

Axillary n.
Musculocutaneous n.
Antebrachial nn. { Lateral Posterior
Intercosto-brachial n.
Radial n.
Medial brachial cutaneous n.
Radial nn. { Deep Superficial
Medial antebrachial cutaneous n.
Ulnar n.
Median n.
Proper palmar digital nn.
Post. ulnar br.
Ulnar n.
Deep ulnar br.

NERVES

SURFACE RELATIONS

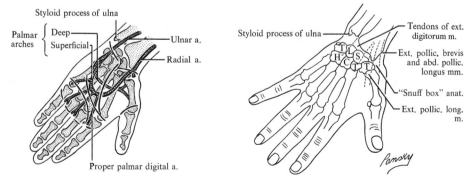

Palmar arches { Deep Superficial
Styloid process of ulna
Ulnar a.
Radial a.
Proper palmar digital a.

Styloid process of ulna
Tendons of ext. digitorum m.
Ext. pollic. brevis and abd. pollic. longus mm.
"Snuff box" anat.
Ext. pollic. long. m.

140. CLINICAL CONSIDERATIONS: BRACHIAL PLEXUS

I. **Introduction:** the brachial plexus can be injured by stretching, wounding, or by disease in the neck and axilla. Injury may involve roots, trunks, divisions, cords, or cord branches and result in loss of muscle movement (paralysis), as well as loss of cutaneous sensation or feeling (anesthesia). Paralysis can be complete (no movement) or incomplete (weak movement)

II. **Upper brachial plexus injuries** (Erb's or Duchenne-Erb palsy) generally result from marked separation of the neck and shoulder as a result of a blow to the region or a birth injury due to stretching of the nerves at delivery

A. DORSAL AND VENTRAL ROOTS of spinal nerves C5–C6 may be pulled out of the spinal cord, resulting in a loss of function of the dorsal axial muscles over area of the back supplied by dorsal primary rami and paralysis of muscles and loss of sensation in the upper limb related to the ventral rami

   1. This type of brachial birth palsy results in paralysis of the muscles of the upper arm, particularly, the deltoid, biceps brachii, brachioradialis, supraspinatus, infraspinatus, teres minor, and supinator
      a. Deltoid and supraspinatus paralysis: arm cannot be abducted from side
      b. Flexors of elbow are paralyzed
      c. Loss of biceps and supinator: supination of forearm is greatly weakened
      d. Weakness of pectoralis major (lateral pectoral nn.) and latissimus dorsi (C6–C8) and loss of infraspinatus (C5–C6) lead to medial shoulder rotation

B. IF THE UPPER TRUNK is torn or severed as it emerges between the anterior and middle scalene muscles, one sees loss of flexion, abduction, and lateral rotation of shoulder joint as well as loss of elbow joint flexion. The limb hangs by the side in medial rotation at the shoulder with the forearm pronated, in the so-called "waiter's tip" position. There is also anesthesia over an area on the lateral side of the arm

C. IF THE LESION is confined to C5, one sees no sensory change. If both C5 and C6 are involved, the loss of sensation is seen on the lateral upper limb

D. INCORRECT USE OF CRUTCHES (crutch palsy) may compress axillary structures, particularly the posterior cord. The radial nerve is predominantly affected, leading to paralysis of the triceps, anconeus, and wrist extensors, with the inability to extend elbow, wrist joint, and fingers, resulting in a *wrist drop*

III. **Lower brachial plexus injuries** are not as common as upper plexus injuries and result from a sudden upward pull of the shoulder (at birth and in adult injuries), which injures the lower trunk (C8–T1) with the paralysis and anesthesia generally affecting the area supplied by the ulnar nerve

A. KLUMPKE'S PALSY (paralysis, Klumpke-Déjérine syndrome): atrophic paralysis of the forearm and small muscles of the hand with paralysis of the cervical sympathetic system (predominantly occurs with a birth injury due to arm being forcibly abducted over the head)
   1. Paralysis of small muscles of hand and weakness of the flexors of the fingers and thumb. There is also reduced sensation along the ulnar side of the arm, forearm, and hand
   2. Similar to ulnar nerve paralysis plus addition of paralysis of thenar muscles and flexors of the digits

IV. **Cervical rib syndrome:** a cervical rib would exert pressure on the lower trunk of the plexus and compress nerves as well as blood vessels supplying them (anoxia of the axons). Tingling and numbness are seen in the upper extremity which are relieved when the pressure is removed

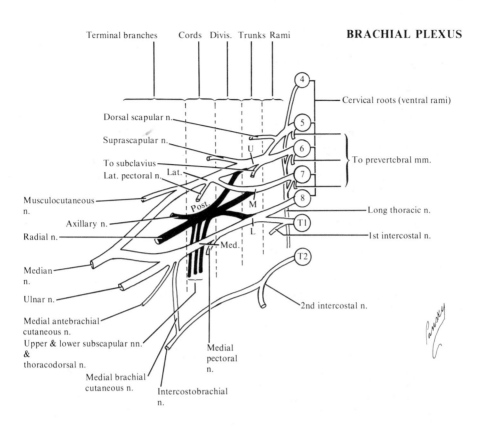

Terminal branches   Cords   Divis.   Trunks   Rami

BRACHIAL PLEXUS

Cervical roots (ventral rami)

Dorsal scapular n.

Suprascapular n.

To subclavius

Lat. pectoral n.   Lat.

To prevertebral mm.

Musculocutaneous n.

Axillary n.

Radial n.

Long thoracic n.

1st intercostal n.

Median n.

Ulnar n.

2nd intercostal n.

Medial antebrachial cutaneous n.

Upper & lower subscapular nn. & thoracodorsal n.

Medial pectoral n.

Medial brachial cutaneous n.

Intercostobrachial n.

Pulled prolapsed fetal arm; lower trunk or entire plexus may be avulsed from cord

Head forceably pulled to side in delivery; roots of upper trunk (C5, 6) of plexus may be pulled out

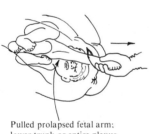

Neck and shoulder violently separated in fall may produce upper brachial plexus injury (C5, 6)

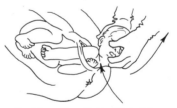

"Tip-taking" position of upper limb with an upper brachial plexus injury

141. SPECIFIC NERVE INJURY—BRACHIAL PLEXUS (C5–T1) BRANCHES

I. **Dorsal scapular nerve** (C5) innervates rhomboids and levator scapulae muscles. Injury leads to impairment of scapular medial rotation

II. **Long thoracic nerve** (C5–C7) innervates serratus anterior muscle. Injury leads to paralysis of serratus anterior muscle, resulting in a "winged scapula" where the vertebral border and inferior angle of the scapula become very prominent because the scapula is no longer held tightly against the chest wall. This is accentuated when the patient pushes against a wall with both hands
   A. IN ADDITION, flexing or abducting of the arm above 45° from the body is difficult because the muscle also helps to protract and rotate the scapula, enabling the glenoid fossa to face upward in these motions

III. **Suprascapular nerve** (C5–C6) innervates supraspinatus and infraspinatus muscles and shoulder joint. Injury hinders initiation of abduction at shoulder (somewhat questionable) as well as weakens lateral rotation at shoulder

IV. **Pectoral nerves** (lateral from C5–C7 to pectoralis major m., medial from C8–T1 to pectoralis minor and part of major muscles). Injury results in severe loss of flexion of the shoulder joint

V. **Nerve to subclavius muscle** (C5): injury not easily detected. May see loss of cutaneous sensation (anesthesia) over area just below clavicle

VI. **Upper subscapular** (C5–C6 to subscapularis m.) and lower subscapular (C5–C6 to the subscapularis and teres major mm.): injury may impair shoulder joint function due to loss of muscles which help hold the head of the humerus to the glenoid fossa. Loss of the teres major weakens shoulder extension as well

VII. **Thoracodorsal** (C6–C8): injury would paralyze the latissimus dorsi and thus weaken the actions of extension, adduction, and medial rotation of the shoulder. Along with the teres major would particularly affect shoulder extension

VIII. **Axillary nerve** (C5–C6) innervates the deltoid and teres minor muscles
   A. PARALYSIS OF THE NERVE leads to an impaired ability to abduct the arm from the side more than ±30 degrees (some abduction is possible via the supraspinatus muscle, but it is unable to initiate abduction), and there is weakness of flexion and extension of the shoulder
   B. ONE SEES A LOSS of shoulder roundness due to atrophy of the deltoid muscle
   C. LOSS OF SENSATION is over lateral side of the proximal arm (over deltoid area)

IX. **Musculocutaneous nerve** (C5–C6) innervates the biceps brachii, coracobrachialis, and brachialis muscles
   A. INJURY CAUSES A CONSIDERABLE interference (great loss) with the power of flexion of the elbow, which is not complete because the brachioradialis and flexor muscles of the forearm (arising from the two epicondyles) are still intact
   B. THE BRANCHES TO THE BRACHIALIS by the radial nerve are thought to be sensory, thus, if the musculocutaneous nerve is cut, the muscle is usually completely paralyzed
   C. SUPINATION OF THE FOREARM is weakened by the loss of the biceps brachii which becomes wasted and atrophic
   D. FLEXION AT THE SHOULDER is slightly involved due to loss of the coracobrachialis and long head of the biceps brachii
   E. SENSORY LOSS is of little import, and one sees sensation on the lateral aspect of the forearm lost (lateral antebrachial cutaneous nerve), but it depends on the extent of overlapping of adjacent nerves

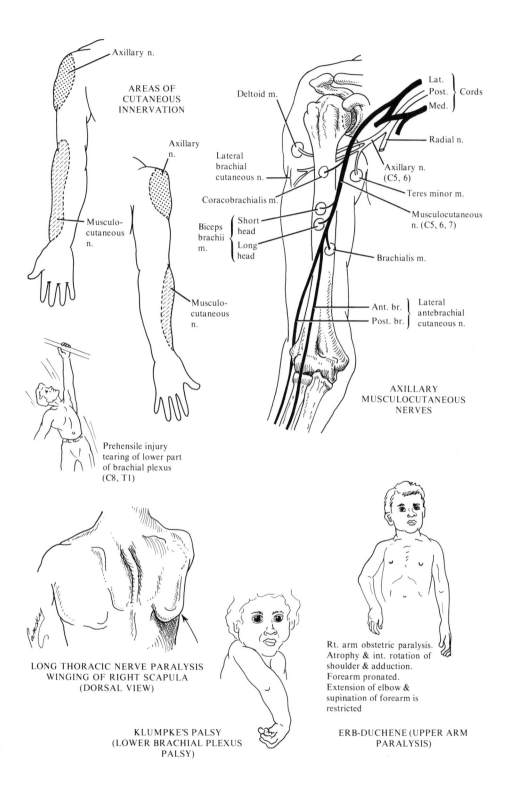

Axillary n.

AREAS OF
CUTANEOUS
INNERVATION

Deltoid m.

Lat.
Post. } Cords
Med.

Radial n.

Axillary
n.

Lateral
brachial
cutaneous n.

Coracobrachialis m.

Axillary n.
(C5, 6)

Teres minor m.

Musculocutaneous
n. (C5, 6, 7)

Biceps
brachii
m.
{ Short head
Long head

Musculo-
cutaneous
n.

Brachialis m.

Ant. br. } Lateral
Post. br. } antebrachial
cutaneous n.

Musculo-
cutaneous
n.

AXILLARY
MUSCULOCUTANEOUS
NERVES

Prehensile injury
tearing of lower part
of brachial plexus
(C8, T1)

LONG THORACIC NERVE PARALYSIS
WINGING OF RIGHT SCAPULA
(DORSAL VIEW)

KLUMPKE'S PALSY
(LOWER BRACHIAL PLEXUS
PALSY)

Rt. arm obstetric paralysis.
Atrophy & int. rotation of
shoulder & adduction.
Forearm pronated.
Extension of elbow &
supination of forearm is
restricted

ERB-DUCHENE (UPPER ARM
PARALYSIS)

142. SPECIFIC NERVE INJURY—RADIAL NERVE (C5–T1)

I. **Injury to nerve proximal** to origin of triceps muscle results in paralysis of the triceps m. (elbow is flexed and cannot be extended against resistance, but extremity hangs down straight at side due to gravity overcoming the tone of the flexors); absence of triceps reflex; paralysis of brachioradialis, supinator (supination, however, is only weakened, not abolished, because the biceps is still functional), and extensors of wrist, thumb, and fingers (wrist is flexed, forearm is pronated, thumb is flexed and adducted, and there is a tendency for the fingers to be flexed, but not severely, due to loss of extensor digitorum)

A. CHARACTERISTIC CLINICAL SIGN is *wrist drop,* where the hand is flexed at the wrist and lies flaccid with an inability to extend hand at wrist and digits at the metacarpophalangeal joints, and lateral movements of the hand are difficult because the ulnar and radial extensors are paralyzed

B. THE ADDUCTED POSITION of the thumb and position of the already flexed hand make flexion of the fingers difficult. The thumb cannot be extended or abducted because of paralysis of the abductor pollicis longus and extensor pollicis longus and brevis

C. THE INTERPHALANGEAL JOINTS can be weakly extended via action of the intact lumbrical and interossei mm. (median and ulnar nn.)

D. WHILE THE HAND is in its "dropped" position, the metacarpophalangeal joints are not, as one might expect, flexed, but extended, due to passive stretching of the extensor tendons by the flexion of the wrist. As a result, the grip is poor

   1. When the wrist is passively dorsiflexed, the metacarpophalangeal joints flex under the pull of the flexor muscles, and the hand curls up with the thumb coming across the palm. In this position, the flexor muscles can take firm hold of anything put in their grasp, and the hand becomes useful. Thus, by wearing a "cork-up" splint, dorsiflexion can be obtained. The hand, however, is not fully functional since the fingers cannot open properly to get hold of their objective, which must be placed into the hand between the half-closed fingers

E. THE BULGE OF THE DORSAL FOREARM group of muscles (extensor-supinator group) is flattened or hollowed

F. THERE IS AN ABSENCE of the periosteal reflex or tapping of the radius

G. THERE IS ANESTHESIA of the dorsolateral aspects of the distal end of the arm, posterior surface of the forearm, dorsum of the hand, and proximal phalanges on the radial side

   1. The radial nerve has only a small region of exclusive cutaneous supply on the hand, and the extent of anesthesia is minimal even in serious nerve injury (confined to small area on lateral aspect of dorsum of hand)

H. NOTE: the radial nerve supplies no muscles in the hand itself

II. **Injury to nerve in radial groove:** this is the most likely site of injury, where the nerve is affected by a fracture of the humerus. This injury does not normally affect branches to the triceps since these come off very early in the course of the nerve

A. THERE IS TOTAL PARALYSIS of extensor muscles of forearm and hand. Inability to extend hand at wrist indicates damage of nerve proximal to the elbow. Extension of wrist is impossible (see I.A), and hand assumes a flexed position (see discussion above). Sensory loss may be minimal because area of exclusive radial supply is small due to overlapping innervation. In addition, sensory loss on back of arm and forearm is not vital

III. **Injury of nerve in forearm** as it passes through the supinator muscle occurs with deep wounds and affects the deep radial nerve. Results in inability to extend thumb as well as the metacarpophalangeal joints of the fingers. There is no loss of sensation because deep radial n. is entirely a muscular and articular nerve. The nerve supply to the ext. carpi radialis longus is intact (arises from radial before its superficial and deep branches), as is that of the supinator and extensors

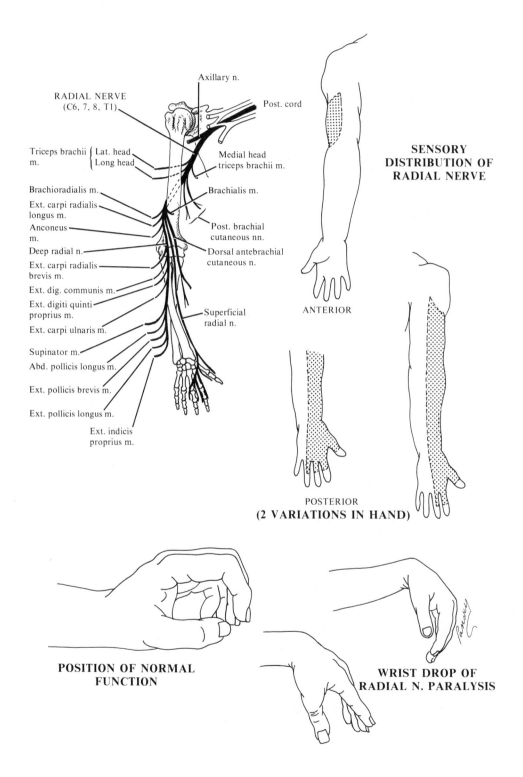

Axillary n.

RADIAL NERVE
(C6, 7, 8, T1)

Post. cord

Triceps brachii
m.

Lat. head
Long head

Medial head
triceps brachii m.

Brachioradialis m.

Brachialis m.

Ext. carpi radialis
longus m.

Anconeus
m.

Post. brachial
cutaneous nn.

Deep radial n.

Dorsal antebrachial
cutaneous n.

Ext. carpi radialis
brevis m.

Ext. dig. communis m.

Ext. digiti quinti
proprius m.

Superficial
radial n.

Ext. carpi ulnaris m.

Supinator m.

Abd. pollicis longus m.

Ext. pollicis brevis m.

Ext. pollicis longus m.

Ext. indicis
proprius m.

**SENSORY
DISTRIBUTION OF
RADIAL NERVE**

ANTERIOR

POSTERIOR
(2 VARIATIONS IN HAND)

**POSITION OF NORMAL
FUNCTION**

**WRIST DROP OF
RADIAL N. PARALYSIS**

143. SPECIFIC NERVE INJURY—ULNAR NERVE (C8, T1)

I. Injury at wrist: hand takes up a striking and characteristic posture called a "claw hand"

A. THE FINGERS are hyperextended at the metacarpophalangeal joints and flexed at the interphalangeal joints because of paralysis of the interossei and medial two lumbrical muscles which upsets the balance of strength (power) on these joints
　　1. The two radial fingers are less flexed at the interphalangeal joints than the two medial ones because the two lateral lumbricals are usually intact, being innervated by the median nerve

B. THE TENDONS of the flexor digitorum profundus to the two medial fingers are paralyzed, so no flexion is possible at the distal interphalangeal joints of the ring and little fingers. The superficialis tendons are operable (median n.), so the proximal interphalangeal joints can be flexed

C. THE SMALL MUSCLES of the little finger are all paralyzed and wasted, but the small muscles of the thumb are intact except for the adductor pollicis (in 10% of the population, they are all paralyzed because the ulnar nerve also supplies the thenar muscles)

D. THERE IS IMPAIRED ADDUCTION AND ABDUCTION of the fingers due to paralysis of the interossei muscles. Adduction of the thumb is lost, but other movements are normal
　　1. The interosseous spaces are empty of their muscles due to atrophy and the metacarpal bones protrude under the skin

E. PRODUCES A SENSORY CHANGE in the medial part of the hand and in the ring and little fingers

F. TESTS FOR ULNAR WEAKNESS
　　1. Make patient abduct little finger against resistance (abductor digiti minimi)
　　2. Grip a card between finger and thumb without bending tip of finger (tests adductor pollicis and 1st dorsal interosseus mm.)

II. Injury of nerve at or near elbow: since the ulnar innervates no muscles in the arm, the shoulder and elbow are unaffected

A. INJURIES HERE paralyze the flexor carpi ulnaris and the medial portion (ulnar 1/2) of the flexor digitorum profundus. One sees ulnar deviation of the wrist weakened, and hand is slightly extended and abducted in position. In addition, the motor loss is predominantly one of fine movements of the fingers, as described above, with injury at the wrist
　　1. There are impaired flexion and adduction of wrist with impaired movement of the thumb, ring, and little fingers
　　2. There is the inability to abduct and adduct the medial four digits due to loss of the strength of the interossei muscles
　　3. The fingers can still be flexed, and the grip is strong except for the two medial fingers
　　4. Heavy tools can be carried and used, but sewing or knitting, piano playing, writing, etc., are difficult, even though it is possible to oppose the thumb to the fingers

B. SENSORY LOSS IN ULNAR NERVE PARALYSIS is not of great importance and involves the two least important fingers and the medial side of the palm

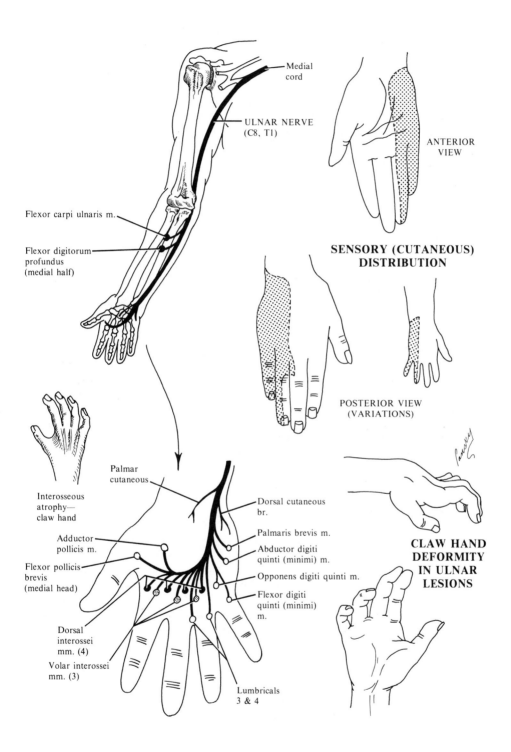

Medial cord

ULNAR NERVE (C8, T1)

Flexor carpi ulnaris m.

Flexor digitorum profundus (medial half)

ANTERIOR VIEW

SENSORY (CUTANEOUS) DISTRIBUTION

POSTERIOR VIEW (VARIATIONS)

Interosseous atrophy— claw hand

Palmar cutaneous

Dorsal cutaneous br.

Palmaris brevis m.

Adductor pollicis m.

Abductor digiti quinti (minimi) m.

Opponens digiti quinti m.

Flexor pollicis brevis (medial head)

Flexor digiti quinti (minimi) m.

Dorsal interossei mm. (4)

Volar interossei mm. (3)

Lumbricals 3 & 4

CLAW HAND DEFORMITY IN ULNAR LESIONS

144. SPECIFIC NERVE INJURY—MEDIAN NERVE (C5–T1)

I. Nerve injury at wrist: the nerve is most vulnerable as it crosses the wrist just proximal to the flexor retinaculum (suicide attempts, putting the hand through a window, etc.). Loss of function of the palmaris longus is not missed

A. THE SHORT MUSCLES of the thumb will usually be put out of action with the exception of the adductor pollicis muscle (ulnar n.)

B. THE THUMB, deprived of tone of short abductor, cannot abduct at rt. angles to palm, and due to loss of opponens pollicis, it cannot be opposed to fingers. It tends to be pulled into plane of palm by intact adductor and 1st dorsal interosseus

C. THENAR EMINENCE is grossly wasted (atrophies with a flattened eminence—"ape hand")

D. THE LONG FLEXOR of thumb is working (supplied by median nerve in forearm)

E. THE LATERAL (radial) 2 lumbrical mm. are usually paralyzed, affecting the ability of the index and middle fingers to make independent movements, but their power is not much affected since the long flexors and interossei are still intact

F. SENSORY LOSS: areas of sensation lost (including deep) generally include lateral side of palm, palmar surface of thumb, index, middle, and 1/2 of ring finger, including the nail beds over the middle and terminal phalanges of these fingers

    1. The sense of joint movement is abolished in the interphalangeal joints of the thumb, index, and middle fingers, but is partly compensated for by the activity of the stretch receptors in the flexor and extensor muscles

II. Carpal tunnel syndrome: compression of the median nerve (due to inflammation of flexor retinaculum, arthritis, or tenosynovitis) as it passes through the osseofibrous carpal tunnel along with the tendons of the long digital muscles. Results in paresthesia (tingling), anesthesia (loss of tactile sensation), or hypesthesia (diminished sensation) in skin areas related to the thumb, index, middle, and lateral 1/2 of ring fingers. Palm may be saved due to palmar cutaneous branch arising superficial to flexor retinaculum. Also see a progressive loss of strength and coordination in thumb with diminished use of thumb, index, and middle fingers as nerve is compressed. Relieved by partial or complete division of the flexor retinaculum

III. Nerve injury above elbow: muscles in arm are not involved; muscles in forearm and hand are involved since nerve supplies them only after it has entered forearm

A. ALL FLEXORS OF WRIST are out of action except the flexor carpi ulnaris and medial two tendons of the flexor digitorum profundus

    1. Flexion and abduction of wrist are considerably weakened but not abolished

    2. Thumb condition similar to nerve injury at wrist, but long flexor now is also paralyzed, and thumb cannot flex at all. It is practically useless. It can be used like a lobster claw to nip articles between its medial surface and index finger (intact adductor and 1st dorsal interosseus muscles)

    3. Long flexor tendons to index and middle fingers are paralyzed. These fingers cannot be flexed at the interphalangeal joints, though flexion at the metacarpophalangeal joints is still possible due to intact interossei (ulnar n.). However, movement is greatly weakened due to loss of lumbricals 1 and 2

    4. The superficialis tendons to ring and little fingers are inoperable, but profundus tendons are intact, permitting flexion at all joints of these fingers

    5. Pronation in the forearm is impossible (pronators inoperable)

    6. Sensory loss is the same as that seen when the nerve is interrupted at wrist

    7. The grip of such a hand is greatly deficient, though small objects can be held between medial 2 fingers and palm. Nothing heavy can be carried. Without thumb, no precision grip is possible. Hand is unable to do manual labor or be used in dressing. Writing is possible with pen between thumb and index finger or between 2 fingers and using the shoulder muscles. Patient holds cigarette between ring and little finger and may burn himself without being aware of it as a result of the sensory loss

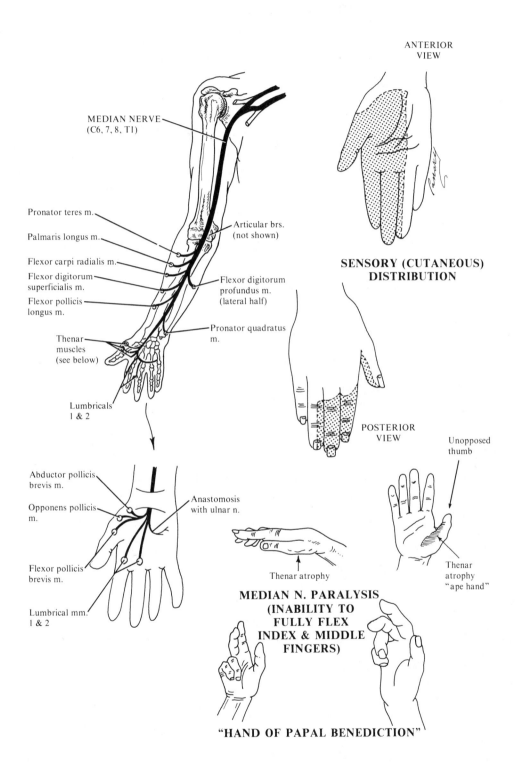

ANTERIOR
VIEW

MEDIAN NERVE
(C6, 7, 8, T1)

Pronator teres m.

Palmaris longus m.

Flexor carpi radialis m.

Flexor digitorum
superficialis m.

Flexor pollicis
longus m.

Articular brs.
(not shown)

Flexor digitorum
profundus m.
(lateral half)

Pronator quadratus
m.

Thenar
muscles
(see below)

Lumbricals
1 & 2

SENSORY (CUTANEOUS)
DISTRIBUTION

POSTERIOR
VIEW

Abductor pollicis
brevis m.

Opponens pollicis
m.

Anastomosis
with ulnar n.

Flexor pollicis
brevis m.

Lumbrical mm.
1 & 2

Thenar atrophy

Unopposed
thumb

Thenar
atrophy
"ape hand"

MEDIAN N. PARALYSIS
(INABILITY TO
FULLY FLEX
INDEX & MIDDLE
FINGERS)

"HAND OF PAPAL BENEDICTION"

145. SUMMARY OF MUSCLE-NERVE RELATIONSHIPS IN UPPER EXTREMITY

Segment	Muscle	Named Nerve
C5–C6	Biceps brachii	Musculocutaneous
C5–C6	Brachialis	Musculocutaneous and radial
C5–C6	Brachioradialis	Radial
C5–C6	Coracobrachialis	Musculocutaneous
C5–C6	Deltoid	Axillary
C5–C6	Infraspinatus	Suprascapular
C5–C6	Subscapularis	Upper and lower subscapular
C5–C6	Supraspinatus	Suprascapular
C5–C6	Teres major	Lower subscapular
C5–C6	Teres minor	Axillary
C6–C7	Abd. pollicis longus	Deep radial
C6–C7	Ext. carpi radialis brevis	Deep radial
C6–C7	Ext. carpi radialis longus	Radial
C6–C7	Ext. pollicis brevis	Deep radial
C6–C7	Flex. carpi radialis	Median
C6–C7	Palmaris longus	Median
C6–C7	Pronator teres	Median
C6–C7	Supinator	Deep radial
C6–C7	Ext. carpi ulnaris	Deep radial
C6–C7	Ext. digitorum	Deep radial
C6–C7	Ext. indicis	Deep radial
C6–C7	Ext. pollicis longus	Deep radial
C6–C7	Triceps brachii	Radial
C7–C8	Anconeus	Radial
C7–C8	Ext. digiti minimi	Deep radial
C7–C8	Flex. digitorum profundus	Median and ulnar
C7–C8	Flex. digitorum superficialis	Median
C8–T1	Abd. digiti minimi	Deep ulnar
C8–T1	Abd. pollicis brevis	Median
C8–T1	Add. pollicis	Deep ulnar
C8–T1	Flex. carpi ulnaris	Ulnar
C8–T1	Flex. digiti minimi	Deep ulnar
C8–T1	Flex. pollicis brevis	Median
C8–T1	Flex. pollicis longus	Median
C8–T1	Interossei	Deep ulnar
C8–T1	Lumbricales	Deep ulnar and median
C8–T1	Opponens digiti minimi	Deep ulnar
C8–T1	Opponens pollicis	Median
C8–T1	Palmaris brevis	Superficial ulnar
C8–T1	Pronator quadratus	Median

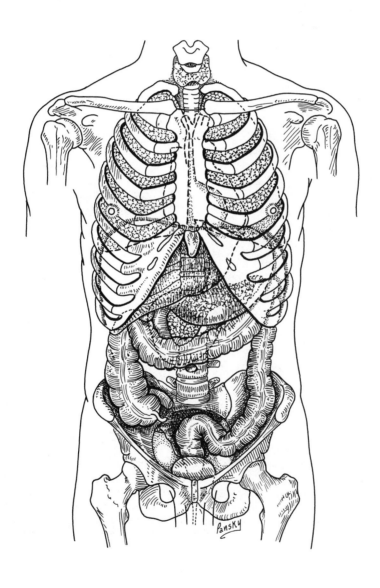

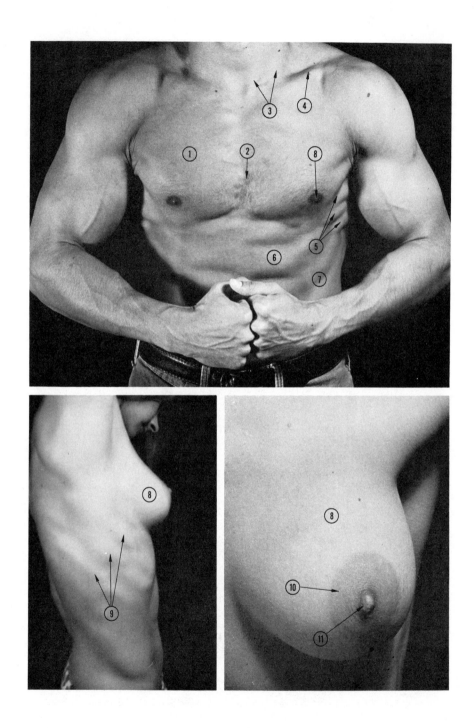

FIGURE 23. **Surface anatomy of thorax.** *1,* Pectoralis major m.; *2,* sternum; *3,* sternocleidomastoid m.; *4,* clavicle; *5,* serratus anterior m.; *6,* rectus abdominis m.; *7,* external oblique m.; *8,* mammary gland; *9,* ribs; *10,* areola; *11,* nipple.

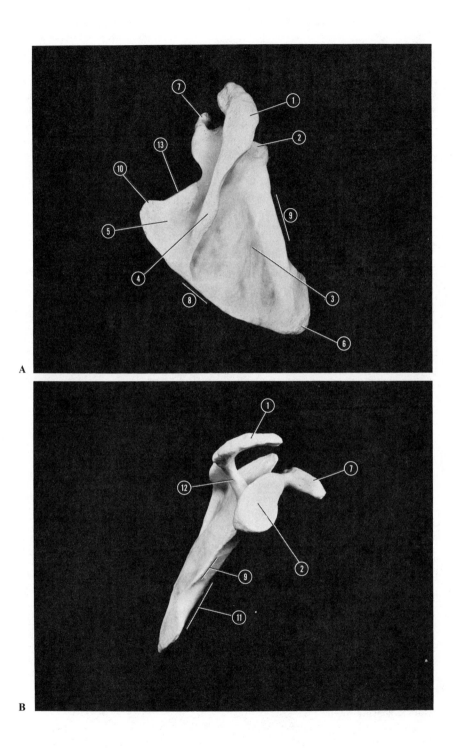

FIGURE 24. **Scapula. A, Posterior view; B, medial view.** *1*, Acromion; *2*, glenoid cavity; *3*, infraspinous fossa; *4*, spine; *5*, supraspinous fossa; *6*, inferior angle; *7*, coracoid process; *8*, medial border; *9*, lateral border; *10*, superior angle; *11*, subscapular fossa; *12*, spine root; *13*, superior border.

146. THE MAMMARY GLAND

I. **Location and extent:** vertically from 2nd to 6th or 7th rib and horizontally from lateral border of sternum to midaxillary line. Mammary papilla (nipple) is just below center of gland at the 4th interspace. Superolateral part frequently projects toward axilla as an axillary tail (of Spence) in relation to pectoralis major m. and pectoral axillary lymph nodes

II. **Structure:** each breast is made up of 15 to 20 separate lobes, further divided into lobules and acini, arranged like segments of a citrus fruit. Each gland forms a lobe and is divided from its neighbor by connective tissue septa
   A. TYPE OF GLAND: compound tubuloalveolar glands, each emptying onto the nipple
   B. ENTIRE GLAND lies in superficial fascia with glands embedded in fat (round contour)
      1. Deep surface is separated from muscle fascia by loose connective tissue
   C. SUSPENSORY LIGAMENTS (of Cooper) run from corium of skin, vertically through gland to deepest part of superficial fascia. Well developed over upper breast

III. **Ducts:** each of the tubuloalveolar glands opens into a *lactiferous duct* that leads into a dilated area, the *lactiferous sinus*
   A. AMPULLAE: another name for lactiferous sinuses (dilations of duct beneath areola)
   B. AT THE BASE OF NIPPLE, ducts narrow down, change direction from horizontal to vertical, and run to summit of nipple

IV. **Nipple:** skin wrinkled; contains smooth muscle fibers that contract on tactile stimulation inducing firmness and prominence

V. **Areola:** pigmented area surrounding nipple for a distance of 1.0 to 2.0 cm; pink before pregnancy, but turns brown after pregnancy and remains this color
   A. AREOLAR GLANDS OF MONTGOMERY: sebaceous glands that enlarge in pregnancy and secrete an oily substance to provide a protective lubricant for areola and nipple. Produce little irregularities or small projections in areolae

VI. **Blood supply**
   A. ARTERIES
      1. From axillary artery: supreme thoracic, pectoral branch of thoracoacromial, and lateral thoracic arteries, which give rise to the external mammary artery
      2. From intercostal arteries: cutaneous branches in 3rd, 4th, and 5th spaces
      3. From internal thoracic artery: perforating branches in 2nd to 4th interspaces
   B. VEINS: from anastomotic circle around papilla. Branches pass peripherally from this to axillary, internal thoracic, lateral thoracic, and upper intercostal veins

VII. **Nerves:** from anterior and lateral cutaneous branches of 2nd to 6th intercostal nerves

VIII. **Lymphatic drainage:** originates from an extensive perilobular plexus
   A. PRINCIPAL: 2 trunks, which follow the lactiferous ducts to the areola to form a subareolar plexus. Lymph drains into the pectoral group of axillary nodes
   B. SECONDARY
      1. From medial side of gland to parasternal nodes. Some cross to opposite gland
      2. From upper gland to apical axillary nodes or to supraclavicular nodes
      3. From caudal gland, vessels pass to abdominal lymph nodes

IX. **Clinical considerations**
   A. CANCER OF GLANDULAR TISSUE causes a dimpling of the skin due to stress on the connective tissue septa. If ducts are invaded, nipple may invaginate or invert
   B. BREAST CANCER: the most common form of cancer in women; spreads via vascular and lymphatic channels as well as along fibrous tissue to deeper structures. Radical mastectomy is often done to remove the primary tumor, underlying fascia, pectoral muscles, and axillary lymph nodes (most common site of metastases)

BREAST

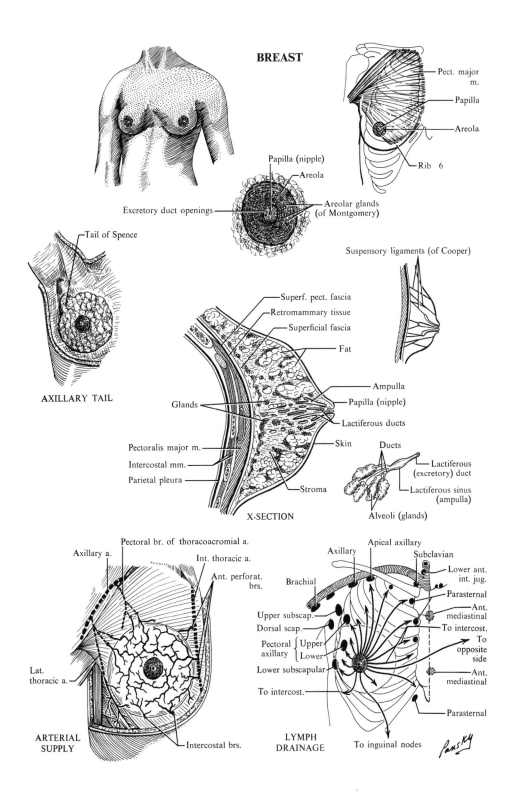

Tail of Spence

Papilla (nipple)

Areola

Excretory duct openings

Areolar glands (of Montgomery)

Pect. major m.

Papilla

Areola

Rib 6

AXILLARY TAIL

Suspensory ligaments (of Cooper)

Superf. pect. fascia
Retromammary tissue
Superficial fascia
Fat
Ampulla
Papilla (nipple)
Lactiferous ducts
Skin
Stroma

Glands

Pectoralis major m.
Intercostal mm.
Parietal pleura

Ducts
Lactiferous (excretory) duct
Lactiferous sinus (ampulla)
Alveoli (glands)

X-SECTION

Axillary a.
Pectoral br. of thoracoacromial a.
Int. thoracic a.
Ant. perforat. brs.

Lat. thoracic a.

ARTERIAL SUPPLY

Intercostal brs.

Apical axillary
Axillary
Subclavian
Brachial
Lower ant. int. jug.
Parasternal
Ant. mediastinal
To intercost.
To opposite side
Ant. mediastinal
Parasternal

Upper subscap.
Dorsal scap.
Pectoral axillary { Upper / Lower }
Lower subscapular
To intercost.

LYMPH DRAINAGE

To inguinal nodes

Pansky

147. STERNUM AND THORACIC CAGE AS A WHOLE

I. **Bony thorax:** shaped like a truncated cone; flattened from before backward. Circular at birth and 2 years after, and breathing is abdominal (diaphragmatic) since circumference remains constant. Oval-shaped in adult form with breathing becoming intercostal (thoracic)

A. ANTERIOR WALL: sternum and 1st ten pairs of ribs with their costal cartilages

B. LATERAL WALLS: formed by ribs that slope obliquely downward and forward

C. POSTERIOR WALL: made up of 12 thoracic vertebrae and ribs as far as their angles

D. SUPERIOR APERTURE (inlet): kidney-shaped; slopes down and forward; formed by body of 1st thoracic vertebra, 1st ribs and cartilages, and manubrium sterni

E. INFERIOR APERTURE (outlet): large; irregular; formed by 12th thoracic vertebra, lowest ribs, 7–12th costal cartilages and xiphisternal joint. Occupied by diaphragm

II. **Sternum (breast bone):** elongated flat bone, covered by skin, superficial fascia, and periosteum, except where pectoralis and sternocleidomastoid mm. arise from it

A. PARTS (3)

  1. Manubrium: at level with 3rd and 4th thoracic vertebrae. Roughly quadrangular; has 2 surfaces (anterior and posterior) and 4 borders (superior, inferior, and 2 lateral). Both surfaces tend to be smooth and concave

    a. Jugular (suprasternal) notch: in middle of superior border, flanked by oval articular surfaces for clavicles (clavicular notches)

    b. Inferior border: rough; normally covered with cartilage for articulation with body (usually a symphysis with fibrocartilage and ligaments joining bones)

    c. Lateral borders: marked above by depression for 1st costal cartilage and below by small articular facet, which, with similar one on body forms a joint with 2nd costal cartilage (synchondrosis) which is strong but flexible

  2. Body: long and narrow; lies opposite 5–9th thoracic vertebrae; has 2 surfaces (anterior and posterior) and 4 borders (superior, inferior, and 2 lateral)

    a. Anterior surface has 3 transverse ridges at the level of the articular depressions for the third, fourth, and fifth costal cartilages

    b. Superior border: oval; articulates with manubrium at sternal angle (of Louis); angle opposite disk between vertebrae T4–5 and about 5 cm below jugular notch

    c. Each lateral border: at superior angle, small notch for 2nd costal cartilage. Below this, 4 costal notches for cartilages of ribs 3–6. At inferior angle is a small facet which, with adjoining one in xiphoid, holds 7th rib cartilage

    d. Inferior border: narrow, for articulation with xiphoid

  3. Xiphoid: smallest, most variable and most caudal. May be bifid. Has 2 surfaces (anterior and posterior) and 3 borders (superior and 2 lateral)

    a. At cephalic end, has small depression for part of 7th costal cartilage

    b. Xiphoid is cartilaginous at birth, during infancy, and early childhood

B. OSSIFICATION, from 6 centers: 1 manubrium, 4 body, and 1 xiphoid

Name	When Appear	When Closed
Manubrium	6th fetal month	25th year
1st body	6th fetal month	25th year
2nd and 3rd body	7th fetal month	25th year
4th body	1st year postnatal	Puberty
Xiphoid	5th to 18th year	30 to after 40 years

III. **Clinical considerations**

A. PIGEON BREAST (pectus carinatum): sternum projects forward and downward

B. FUNNEL CHEST (pectus excavatum): sternum pushed back by rib overgrowth

C. SERIOUS DEFORMITIES of the chest are almost predominantly congenital in origin and are always associated with overgrowth of the ribs

D. SLIGHT DEFORMITIES and asymmetries of the thorax are common, cause no disability

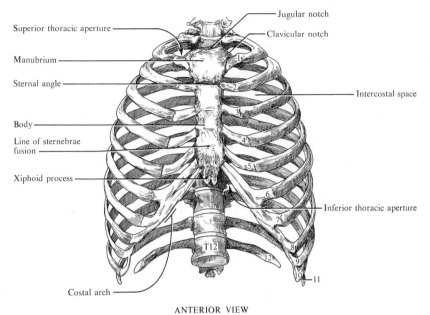

Jugular notch
Clavicular notch

Superior thoracic aperture

Manubrium

Sternal angle

Intercostal space

Body

Line of sternebrae fusion

Xiphoid process

Inferior thoracic aperture

T12

Costal arch

ANTERIOR VIEW

THORACIC CAGE

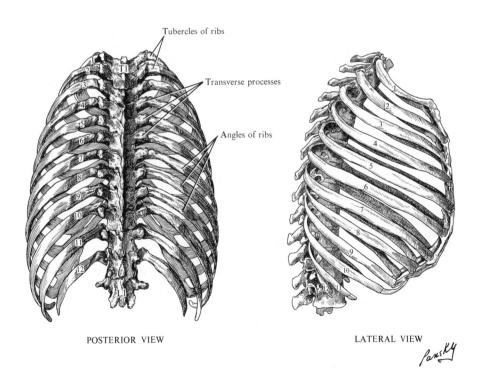

Tubercles of ribs

Transverse processes

Angles of ribs

T1

L1

POSTERIOR VIEW

LATERAL VIEW

Pansky

148. STERNOCLAVICULAR, STERNOCOSTAL, AND INTRASTERNAL ARTICULATIONS

I. Sternoclavicular joint (see p. 234)

II. Sternocostal articulations
A. BETWEEN FIRST RIB AND STERNUM
   1. Type: synarthrosis (synchondrosis)
B. BETWEEN STERNUM AND RIBS 2–7
   1. Type: arthrodial
   2. Movements: gliding
   3. Ligaments
      a. Articular capsule
      b. Radiate sternocostal: from dorsal and ventral side of sternal ends of cartilages to dorsal and ventral side of sternum
      c. Intra-articular sternocostal: usually found with second rib only, where ligament runs from end of cartilage to the fibrocartilage between manubrium and body of sternum
      d. Costoxiphoid: join dorsal and ventral surfaces of seventh costal cartilage to dorsal and ventral sides of xiphoid process
C. INTERCHONDRAL between costal cartilages of ribs 6–8 (sometimes through 10) join together at adjacent articular facets
   1. Ligaments
      a. Articular capsules
      b. Interchondral, running from 1 cartilage to another
D. COSTOCHONDRAL between depression in medial end of rib and lateral end of costal cartilage. These are synarthroses, joined by periosteum

III. Intrasternal
A. BETWEEN MANUBRIUM AND BODY OF STERNUM
   1. Type: synarthrosis with fibrocartilage between the edges of bone, strengthened by fibrous tissue (probably periosteum). In some cases, a synovial cavity is found. In old age, ossification sometimes is found (a true synostosis)
B. BETWEEN XIPHOID AND BODY OF STERNUM
   1. Type: synarthrosis (synchondrosis)
   2. In general, by fifteenth year cartilage is replaced by bone in a synostosis

IV. Clinical considerations
A. DISLOCATION OF THE ACROMIOCLAVICULAR JOINT is a common injury (shoulder separation)
B. DISLOCATION OF THE STERNAL END OF THE CLAVICLE occurs much less frequently than acromioclavicular dislocation
C. STERNOCLAVICULAR DISLOCATION results from a violent fall or blow upon the shoulder: sternoclavicular ligaments are ruptured, and the intra-articular fibrocartilage remains attached to the clavicle

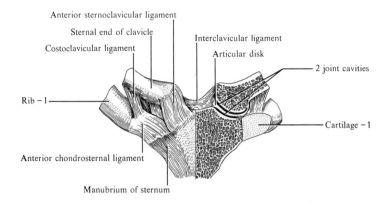

Anterior sternoclavicular ligament

Sternal end of clavicle

Costoclavicular ligament

Interclavicular ligament

Articular disk

2 joint cavities

Rib – 1

Cartilage – 1

Anterior chondrosternal ligament

Manubrium of sternum

STERNOCLAVICULAR

ARTICULATIONS – ANTERIOR VIEWS

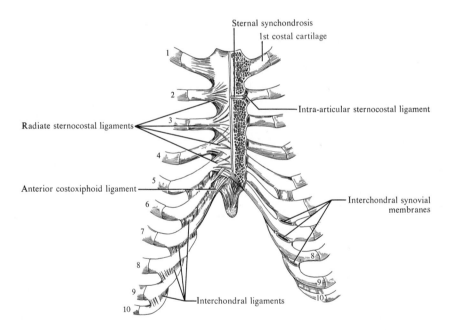

Sternal synchondrosis

1st costal cartilage

Intra-articular sternocostal ligament

Radiate sternocostal ligaments

Anterior costoxiphoid ligament

Interchondral synovial membranes

Interchondral ligaments

STERNOCOSTAL AND INTERCHONDRAL

149. THE RIBS

I. Classification
A. TRUE RIBS (VERTEBROCOSTAL) (1–7) articulate behind with the vertebrae and in front, through cartilages, with the sternum
B. FALSE RIBS
    1. Vertebrochondral (8–10) join the vertebrae behind, but ventrally join the costal cartilages of the ribs above
    2. Floating (11–12): the ventral ends of the ribs are free

II. General characteristics of a "typical" rib: each rib has a dorsal (*vertebral*) and a ventral (*sternal*) extremity with an intervening body (*shaft*)
A. DORSAL EXTREMITY
    1. Head: has an articular surface divided into 2 parts by a ridge
    2. Neck: about 1 in. in length, just lateral to the head
    3. Tubercle: at junction of neck and body. Has an articular facet
B. BODY (SHAFT): thin, flat, and arched with 2 surfaces (external and internal) and 2 borders (superior and inferior)
    1. Angle of rib: an oblique ridge just beyond the tubercle on the external surface
    2. Costal groove: along and just above the inferior border, on the internal surface
C. VENTRAL EXTREMITY ends in an oval concavity for costal cartilage

III. Atypical ribs
A. FIRST RIB has greatest curvature and is the shortest
    1. Flattened in a plane different from the others so that there are superior and inferior surfaces with lateral and medial borders
    2. Head has no division of its articular facet
    3. Tubercle is very thick and prominent
    4. There is no angle
    5. Upper surface shows 2 shallow grooves (for subclavian a. and v.), separated by the *scalene tubercle*
    6. No costal groove
    7. Anterior extremity is larger and thicker than other ribs
B. SECOND RIB: longer than first and intermediate in form between the first rib and the typical rib. Has the same curvature as the first rib, but is not quite as flattened. The angle is slight and close to the tubercle
C. TENTH RIB: like a typical rib except that it has a single articular facet on its head
D. ELEVENTH AND TWELFTH RIBS
    1. Have a single articular facet on head
    2. Have neither neck nor tubercle
    3. Ventral ends are pointed
    4. Eleventh has a very shallow costal groove; twelfth has none

IV. Ossification
A. RIBS 1–10 OSSIFY FROM 4 CENTERS: for body, for head, for articular part of tubercle, and for nonarticular part of tubercle. They appear in body near angle in the eighth fetal week (seen first in ribs 6 and 7). Centers for the head and tubercles appear between the sixteenth and twentieth years and unite with the body in the twenty-fifth year

V. Special features
A. THE ARTICULATIONS of the ribs merit little description. Most have 2 *costovertebral joints,* one between the head of the rib and adjacent vertebral bodies, the other between the tubercle and the transverse process (*costotransverse joints*). See page 324 for description of the sternocostal joints

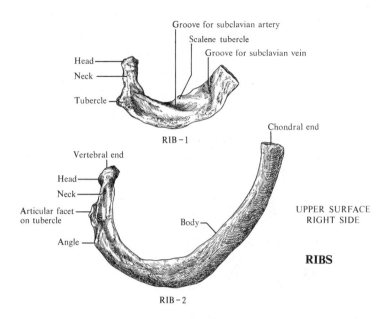

Groove for subclavian artery

Scalene tubercle

Groove for subclavian vein

Head

Neck

Tubercle

RIB-1

Chondral end

Vertebral end

Head

Neck

Articular facet
on tubercle

Angle

Body

UPPER SURFACE
RIGHT SIDE

RIBS

RIB-2

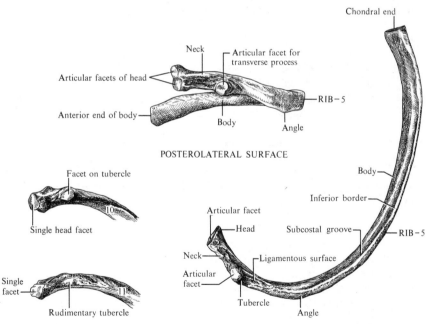

Chondral end

Neck

Articular facet for
transverse process

Articular facets of head

Anterior end of body

RIB-5

Body

Angle

POSTEROLATERAL SURFACE

Facet on tubercle

10

Single head facet

Body

Inferior border

Articular facet

Head

Subcostal groove

RIB-5

Single
facet

11

Neck

Articular
facet

Ligamentous surface

Rudimentary tubercle

Tubercle

Angle

Single
facet

12

UNDERSURFACE
RIGHT SIDE

SEEN FROM BELOW, RIGHT SIDE

Pansky

150. MUSCLES, DEEP VESSELS, AND NERVES OF THE THORACIC WALL

I. Muscles (see p. 239)

A. INTERCOSTALS: all innervated by intercostal nerves. All draw ribs together, raise ribs
   1. External (11 pairs). Origin: lower border of rib; insertion: in upper border of rib below. Extend from tubercle of rib to costal cartilage. External intercostal membrane replaces muscle from costal cartilages to sternum
   2. Internal (11 pairs). Origin: from inner surface of rib; insertion: in upper border of rib below. Extend from sternum to angles of rib. Internal intercostal membrane replaces muscle between angle of rib and vertebrae

B. SUBCOSTAL: in lower thorax only. Origin: from inner surface near angle of rib; insertion: 2 or 3 ribs below. Action: when last rib is fixed by quadratus lumborum muscle, it lowers ribs

C. TRANSVERSE THORACIC: on inner chest wall. Origin: from posterior body and xiphoid of sternum and sternal ends of costal cartilages of ribs 4–6; insertion: on lower border and costal cartilages of ribs 2–6. Action: draws ribs down

D. LEVATORES COSTARUM (12 pairs): in posterior thorax. Origin: from ends and transverse processes of vertebrae C7–T11; insertion: into outer surface of rib immediately below origin. Action: raise ribs; extend, laterally flex, and rotate the vertebral column to opposite side

E. SERRATUS POSTERIOR SUPERIOR: on posterior of upper thorax (see p. 195). Origin: from ligamentum nuchae, supraspinal ligament, and spines of vertebrae C7–T3; insertion: on upper borders of ribs 2–5, lateral to angles. Action: raises ribs

F. SERRATUS POSTERIOR INFERIOR: on posterior of lower thorax (see p. 195). Origin: from supraspinous ligament and spines of vertebrae T11–L3; insertion: on lower border of ribs 9–12, lateral to angles. Action: lowers ribs

II. Intercostal arteries

A. AORTIC (POSTERIOR) INTERCOSTALS: 9 pairs, which run in lower 9 interspaces. Arise from dorsal side of aorta and divide into 2 rami
   1. Anterior ramus. In general, has vein above and nerve below. Branches: to muscles, lateral cutaneous, and collateral intercostal, which run along border of rib below to join branches of the internal thoracic artery. The mammary branches arise in spaces 3–5
   2. Dorsal ramus. Runs backward between necks of ribs. Branches: to spinal cord, muscles, and skin of dorsum

B. INTERCOSTAL BRANCHES OF INTERNAL THORACIC ARTERY: 2 in each of upper 6 interspaces. Anastomose with anterior rami of aortic intercostals. Have perforating branches to muscle, skin, and mammary gland

C. SUPREME (HIGHEST) INTERCOSTAL ARTERY arises from costocervical trunk and gives intercostal branches to first 2 interspaces

III. Intercostal nerves

A. REPRESENTED BY INTERCOSTAL PART of first thoracic and anterior primary divisions of other thoracic nerves. Course similar to artery up to middle of rib; then it lies with the internal intercostal muscle as far as the costal cartilage; then it runs between internal intercostal muscle and pleura. Branches: anterior and lateral cutaneous and muscular

IV. Clinical considerations

A. BECAUSE THE CHIEF VESSELS AND NERVES are at the lower border of each rib, it is customary in passing a needle through the thoracic wall (*thoracentesis* or *paracentesis*) to place the needle close to the lower rather than the upper rib, and thus avoid injury to these nerves and vessels

DEEP THORACIC MUSCLES

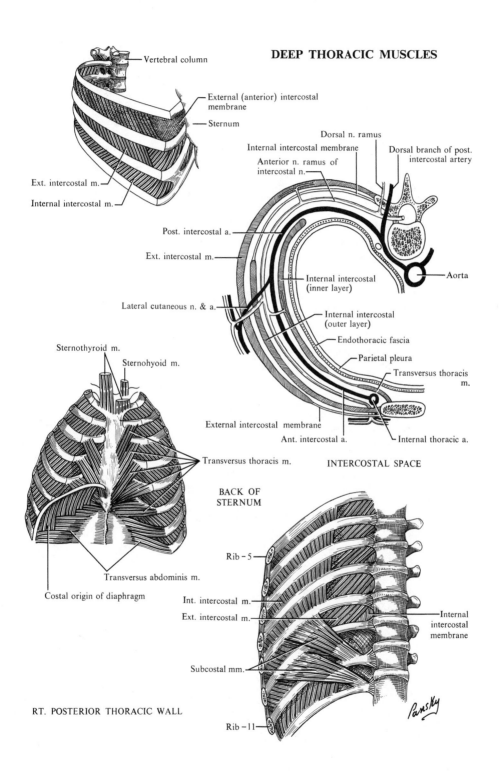

Vertebral column

External (anterior) intercostal membrane

Sternum

Dorsal n. ramus

Internal intercostal membrane

Anterior n. ramus of intercostal n.

Dorsal branch of post. intercostal artery

Ext. intercostal m.

Internal intercostal m.

Post. intercostal a.

Ext. intercostal m.

Internal intercostal (inner layer)

Aorta

Lateral cutaneous n. & a.

Internal intercostal (outer layer)

Endothoracic fascia

Parietal pleura

Transversus thoracis m.

Sternothyroid m.

Sternohyoid m.

External intercostal membrane

Ant. intercostal a.

Internal thoracic a.

INTERCOSTAL SPACE

Transversus thoracis m.

BACK OF STERNUM

Transversus abdominis m.

Costal origin of diaphragm

Rib – 5

Int. intercostal m.

Ext. intercostal m.

Internal intercostal membrane

Subcostal mm.

RT. POSTERIOR THORACIC WALL

Rib – 11

Pansky

151. SURFACE PROJECTIONS OF LUNGS AND PLEURA

I. Lungs
A. APEX: same for both lungs, about 2.5 cm above medial end of clavicle
B. MEDIAL MARGINS
   1. Right descends in a slight curve from the sternoclavicular joint and crosses the midsternal line at the sternal angle, descends just to left of midline to sixth costosternal junction
   2. Left descends in slight curve from sternoclavicular joint, stays to left of midline, parallel and close to the right as far as level of fourth costal cartilage, passes to the left along this cartilage to the parasternal line, descends across the fifth costal cartilage, and curves medially to upper sixth costal cartilage to the left of its costosternal union
C. INFERIOR MARGINS: same for both lungs. Extend along length of sixth cartilage, cross sixth costochondral union, curve downward and lateralward to upper eighth rib in midaxillary line, continue to ninth or tenth rib in scapular line, then medially to eleventh costovertebral joint
D. POSTERIOR MARGINS: same for both lungs. Extend cephalically from eleventh costovertebral joint, on either side of vertebral spines to T2, then curve laterally to apex
E. FISSURES
   1. Oblique (interlobar) starts 6 cm below apex, at level of third rib. With arm at side, this fissure crosses the back from third thoracic spine to scapular spine, then caudally around thorax to end of sixth rib
   2. Horizontal (accessory) in right lung. Starting at the point where the oblique fissure crosses the midaxillary line, this fissure runs nearly horizontally to the ventral margin at level of fourth rib

II. The pleurae: except for the inferior limits, the pleurae and lungs are similar enough to make complete repetition unnecessary
A. RIGHT SIDE: from the xiphisternal junction downward across the sternocostal union, crosses the eighth costochondral articulation in the mammary line, the tenth rib in the midaxillary line, then runs posteriorly and medially to the spinous process of the twelfth thoracic vertebrae
B. LEFT SIDE: same as right to level of fourth costal cartilage. Curves downward, just to the left of the sternal margin to sixth costal cartilage, then crosses the seventh costal cartilage, and from there is same as right. Although the divergence of the pleura to the left of the midline on the left side is more pronounced than on the right, it does not exhibit a cardiac notch as does the left lung

III. Clinical consideration
A. THE FACT THAT THE LUNGS do not go as low as the lowest limits of the pleurae creates a potential space, which makes possible the draining of blood or fluid from the pleural cavity without endangering the lungs (see p. 332). To drain the space (thoracentesis), one must not go below the 7th, 9th, or 11th ribs in the midclavicular, midaxillary, or posterior scapular lines, respectively, since these points are below the diaphragm. Insert needle at upper border of rib to avoid intercostal vessels, and direct upward to avoid puncturing diaphragm
B. THE BASE OF THE LUNG refers clinically not to the anatomic base, which is related to the diaphragm and, therefore, is inferior, but to the lower limits of the posterior surface of the lower lobe. To listen to the base of the lung, apply the stethoscope to the posterior chest wall about at a level with the 10th thoracic vertebra

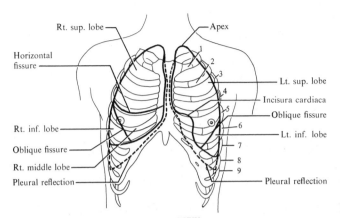

Rt. sup. lobe — — Apex

Horizontal fissure — 1

2

3

— Lt. sup. lobe

4 — Incisura cardiaca

5 — Oblique fissure

Rt. inf. lobe — 6 — Lt. inf. lobe

Oblique fissure — 7

Rt. middle lobe — 8

9

Pleural reflection — — Pleural reflection

ANTERIOR VIEW

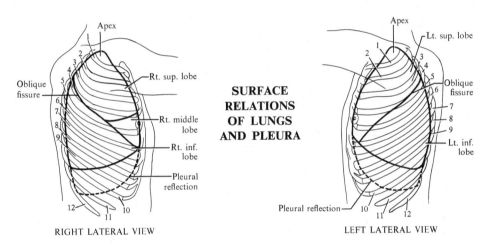

Apex

1
2
3
4

Oblique fissure — 5 — Rt. sup. lobe

6
7
8 — Rt. middle lobe
9

— Rt. inf. lobe

— Pleural reflection

12   11   10

RIGHT LATERAL VIEW

**SURFACE
RELATIONS
OF LUNGS
AND PLEURA**

Apex — Lt. sup. lobe

1
2
3
4
5
6

Oblique fissure

7
8
9

— Lt. inf. lobe

Pleural reflection — 10
11   12

LEFT LATERAL VIEW

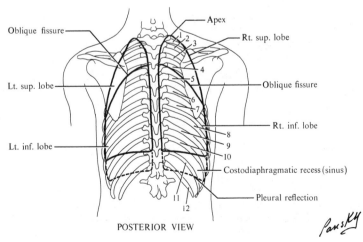

Oblique fissure — — Apex

1 2
3 — Rt. sup. lobe

4

Lt. sup. lobe — 5 — Oblique fissure

6
7

— Rt. inf. lobe

8
9
Lt. inf. lobe — 10

— Costodiaphragmatic recess (sinus)

11 — Pleural reflection
12

POSTERIOR VIEW

Pansky

152. LUNGS

I. Apex: rises 2.5 to 5 cm above first sternocostal articulation

II. Diaphragmatic surface (base): concave, where it rests on convexity of diaphragm

III. Surfaces and their impressions
A. COSTAL: convex to conform to thoracic wall
B. MEDIASTINAL
1. Cardiac impression for pericardium, deeper on left lung to form so-called *cardiac notch*
2. Hilum above cardiac impression
3. Above hilum, on right lung, is arched groove for azygos vein
4. Groove for superior vena cava and right brachiocephalic vein, some distance from apex toward ventral side of right lung
5. Groove for right subclavian artery lies behind this and nearer apex
6. Posterior to hilum is the groove for esophagus on right lung
7. Groove for arch of aorta is above hilum and descending groove for thoracic aorta behind hilum on left lung
8. Groove for subclavian artery runs upward and lateralward from groove for aortic arch on left lung
9. Anterior and below groove for subclavian artery is groove for the left brachiocephalic vein on left lung

IV. Borders
A. INFERIOR: sharp, where it separates base and costal surface; blunt at mediastinal border
B. POSTERIOR (VERTEBRAL): broad and round to fit in the groove lateral to the vertebrae
C. ANTERIOR: thin and sharp; overlaps pericardium on right almost straight; on left, cardiac notch

V. Fissures and lobes
A. LEFT divided into 2 lobes, superior and inferior, by oblique fissure
B. RIGHT divided into 3 lobes by 2 fissures. One fissure, similar to left lung, separates inferior from middle and superior lobes. The other, more horizontal, separates the superior and middle lobes

VI. Root structures at the hilum
A. RIGHT LUNG: pulmonary artery lies ventral to bronchus; 1 pulmonary vein lies just caudal to artery; the other vein is caudal to the bronchus
B. LEFT LUNG: pulmonary artery lies most cephalic; 1 vein lies ventral to bronchus, both being caudal to artery; the other pulmonary vein is caudal to bronchus

VII. Pleura
A. VISCERAL covers lungs and dips into fissures
B. PARIETAL lines thoracic walls. Named by location: costal, diaphragmatic, mediastinal, and cervical (cupula)
C. REFLECTIONS
1. Pulmonary ligament: a fold where the mediastinal pleura, reflected on dorsal and ventral sides of lung root, is continued downward toward diaphragm
2. Phrenicocostal recess (sinus): where costal pleura extends into groove between diaphragm and chest wall
3. Costomediastinal recess (sinus): where costal and mediastinal pleurae meet

LUNGS AND PLEURAE

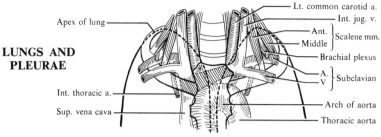

Apex of lung

Lt. common carotid a.
Int. jug. v.
Ant.
Middle } Scalene mm.
Brachial plexus
A.
V } Subclavian
Arch of aorta
Thoracic aorta

Int. thoracic a.
Sup. vena cava

TOPOGRAPHIC RELATIONS OF LUNG APEX

MEDIASTINAL SURFACES

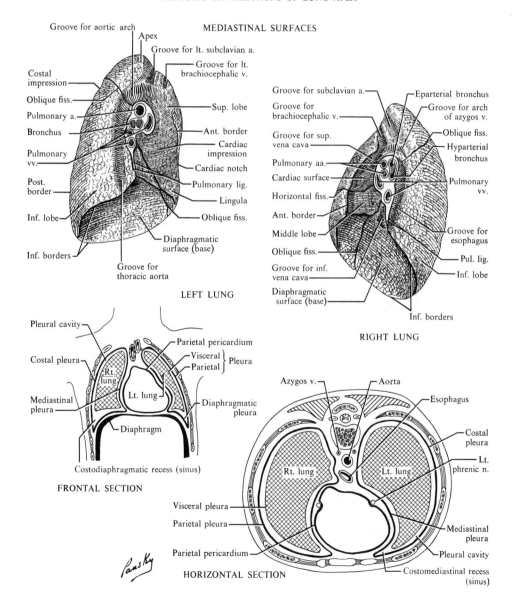

Groove for aortic arch
Apex
Groove for lt. subclavian a.
Groove for lt. brachiocephalic v.

Costal impression
Oblique fiss.
Pulmonary a.
Bronchus
Pulmonary vv.
Post. border
Inf. lobe
Inf. borders

Sup. lobe
Ant. border
Cardiac impression
Cardiac notch
Pulmonary lig.
Lingula
Oblique fiss.
Diaphragmatic surface (base)

Groove for thoracic aorta

LEFT LUNG

Groove for subclavian a.
Groove for brachiocephalic v.
Groove for sup. vena cava
Pulmonary aa.
Cardiac surface
Horizontal fiss.
Ant. border
Middle lobe
Oblique fiss.
Groove for inf. vena cava
Diaphragmatic surface (base)

Eparterial bronchus
Groove for arch of azygos v.
Oblique fiss.
Hyparterial bronchus
Pulmonary vv.
Groove for esophagus
Pul. lig.
Inf. lobe

Inf. borders

RIGHT LUNG

Pleural cavity
Costal pleura
Mediastinal pleura

Parietal pericardium
Visceral } Pleura
Parietal
Rt. lung
Lt. lung
Diaphragmatic pleura
Diaphragm

Costodiaphragmatic recess (sinus)

FRONTAL SECTION

Azygos v.
Aorta
Esophagus
Costal pleura
Lt. phrenic n.
Rt. lung
Lt. lung
Mediastinal pleura
Visceral pleura
Parietal pleura
Pleural cavity
Parietal pericardium
Costomediastinal recess (sinus)

Pansky

HORIZONTAL SECTION

153. BRONCHOPULMONARY SEGMENTS

I. Primary bronchi (see p. 339)

A. RIGHT: 3 secondary bronchi come off here, 1 for each of the 3 lobes. The branch to the superior lobe lies above the pulmonary artery (*eparterial*). The branches to the middle and inferior lobes are below the artery (*hyparterial*). Each bronchus then branches according to its bronchopulmonary segments
   1. Superior lobe bronchus: 3 branches
   2. Middle lobe bronchus: 2 branches
   3. Inferior lobe bronchus: 5 branches

B. LEFT: 2 secondary bronchi come off here, 1 for each lobe. Both lie below the artery, thus are *both hyparterial*
   1. Superior lobe bronchus: 2 major branches, a superior and an inferior
      a. Superior division bronchus: 2 branches
      b. Inferior division bronchus: 2 branches
   2. Inferior lobe bronchus: 4 or 5 branches

C. EACH PRINCIPAL BRONCHUS divides into *secondary lobar bronchi* (two on left, three on right), each of which supplies a lobe of the lung. Each lobar bronchus then divides into *tertiary segmental bronchi* which supply specific sections of the lung called *bronchopulmonary segments*

D. A PROJECTING, KEEL-LIKE RIDGE is seen at the inferior end of the trachea, marking the origins of the right and left bronchi, and is called the *carina*. The mucous membrane at the carina is a very sensitive area of the tracheobronchial tree and is associated with the *cough reflex*. The carina is considered to be a last line of defense before lung is entered

II. Segments: each segment of the lung is pyramidal in shape with an apex that points toward the root of the lung and its base at the pleural surface. Each segment has its own segmental bronchus, artery, and vein. The segments are separated by connective tissue septa which are continuous with the visceral pleura

A. RIGHT LUNG
   1. Superior lobe: apical, posterior, and anterior segments
   2. Middle lobe: lateral and medial segments
   3. Inferior lobe: superior; medial, anterior, lateral, and posterior basal segments

B. LEFT LUNG
   1. Superior lobe
      a. Superior division: apical-posterior and anterior segments
      b. Inferior division: superior and inferior (lingual) segments
   2. Inferior lobe: superior; anterior, medial, lateral, and posterior basal segments

III. Clinical considerations

A. SINCE THE BRONCHOPULMONARY SEGMENTAL TERRITORIES are relatively distinct, it is theoretically possible to isolate the individual segments without having to remove an entire lobe of a lung. This has been accomplished in a number of surgical procedures

B. THE UPPER LOBE of the lung is related to more of the chest wall than meets the eye. Much of the upper lobe is related not only to the anterior chest between the clavicle and nipple but also to the lateral chest wall and can be examined here

C. THE MIDDLE LOBE is an anterior part of the right lung. It is not related to the posterior chest wall. It relates to the anterior chest wall between the 4th and 6th costal cartilages

D. POSTURAL DRAINAGE: there are many times in the practice of medicine when the assistance of gravity is sought in the care of the chest; for example, when it is necessary to promote drainage from all or part of the bronchial tree, the posture of the patient is adjusted to bring maximal influence of gravity to bear on the bronchus to be drained

E. ALTHOUGH EACH BRONCHOPULMONARY SEGMENT is supplied by its own nerve, artery, and vein, one should note that during surgical resection of the segments, the planes between them are crossed by branches of pulmonary veins and sometimes by pulmonary arteries. The bronchial arteries also run through the interlobular septa to supply the visceral pleura

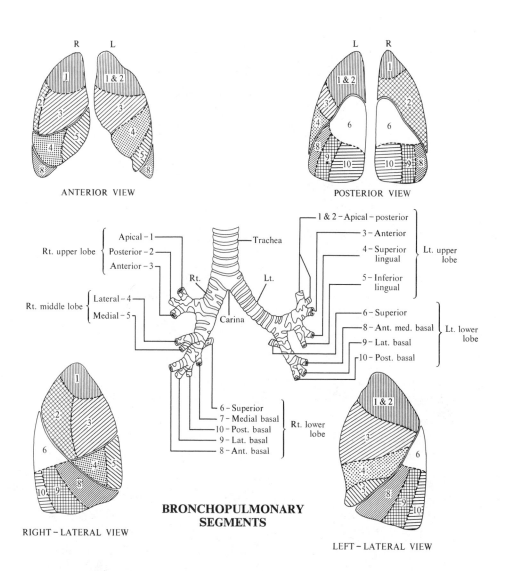

ANTERIOR VIEW

POSTERIOR VIEW

Rt. upper lobe
Apical – 1
Posterior – 2
Anterior – 3

Trachea

Rt.

Lt.

Carina

Rt. middle lobe
Lateral – 4
Medial – 5

1 & 2 – Apical – posterior

3 – Anterior

4 – Superior lingual

5 – Inferior lingual

Lt. upper lobe

6 – Superior

8 – Ant. med. basal

9 – Lat. basal

10 – Post. basal

Lt. lower lobe

6 – Superior
7 – Medial basal
10 – Post. basal
9 – Lat. basal
8 – Ant. basal

Rt. lower lobe

BRONCHOPULMONARY SEGMENTS

RIGHT – LATERAL VIEW

LEFT – LATERAL VIEW

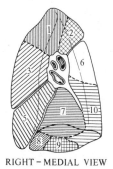

RIGHT – MEDIAL VIEW

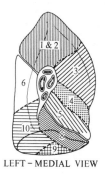

LEFT – MEDIAL VIEW

– 335 –

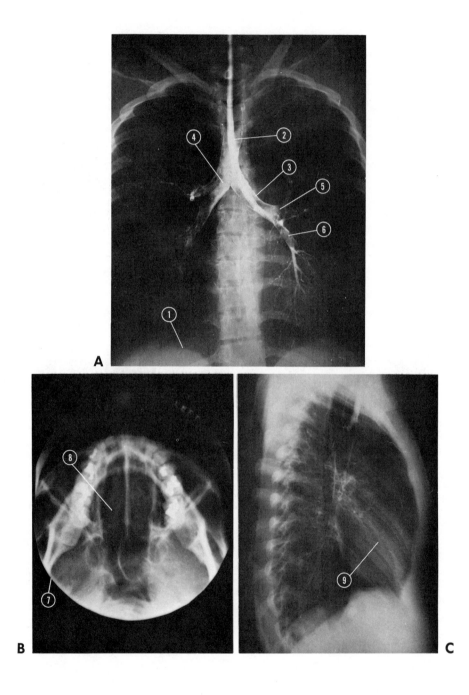

FIGURE 25. **Respiratory system and bronchography. A, Normal bronchogram; B, mouth; C, chest, lateral view.** *1,* Diaphragm; *2,* trachea; *3,* left primary bronchus; *4,* right primary bronchus; *5,* left upper secondary bronchus; *6,* left lower secondary bronchus; *7,* maxilla; *8,* hard palate; *9,* heart.

RAP   RL

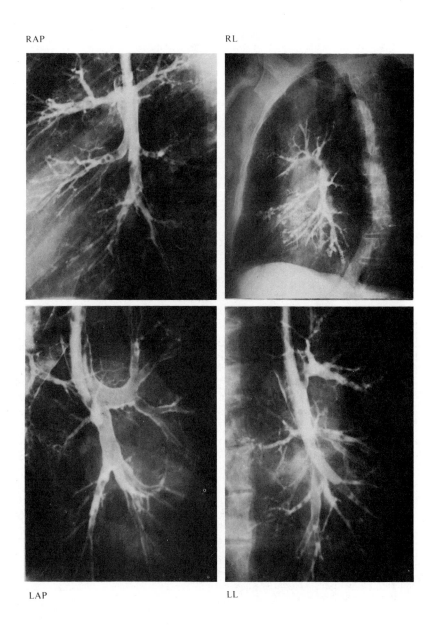

LAP   LL

FIGURE 26. **Series of bronchograms from both the left and right sides.** RAP and LAP-right and left anterior-posterior views, respectively; RL and LL-right and left lateral views, respectively.

154. CIRCULATION AND INNERVATION OF THE LUNGS

I. Intrapulmonary bronchi
A. SECONDARY BRONCHI to lobes enter lung substance to become intrapulmonary. Immediately begin to subdivide, the smaller divisions being known as bronchioles
B. BRONCHIOLES
   1. Terminal (lobular) bronchioles: the last division that has semblance of typical bronchial structure, are not respiratory in function. Each divides into 2 or more respiratory bronchioles
   2. Respiratory bronchioles, which are somewhat similar to the terminal but with a variable number of respiratory units (single alveoli for respiration): these give rise to alveolar ducts
   3. Alveolar ducts: these are similar to 2, above, but with a greater number of respiratory units. These, in turn, lead into alveolar sacs
   4. Alveolar sacs: these are structures whose walls are composed of *pulmonary alveoli*. It is around these that most of respiration occurs. (NOTE: Small circular spaces, called *atria*, are sometimes spoken of as existing between the alveolar ducts and sacs)

II. Lobulation
A. PRIMARY LOBULE begins with the alveolar duct and includes all atria, alveolar sacs, and alveoli coming from it
B. SECONDARY LOBULE made up of several primaries

III. Blood supply
A. SYSTEMIC, through bronchial arteries, which arise from the aorta or upper intercostal arteries and accompany the bronchial tree as far as the respiratory bronchiole. Supplies walls of bronchi and bronchioles
B. PULMONARY, through the right and left pulmonary arteries from the pulmonary trunk. Follows the bronchial tree, behind the bronchi. Forms a capillary net around the alveoli

IV. Venous drainage
A. SYSTEMIC arises from the territory supplied by the bronchial arteries. Vessels converge to form 1 vein at the root of each lung. These end on the right in the azygos, on the left in the supreme intercostal vein
B. PULMONARY arises chiefly in the capillary net around the alveoli with a very small amount from the bronchial system. These veins pass through the lung substance independent of bronchi and pulmonary arteries. They converge to form 2 veins at the root of each lung. These terminate in the left atrium

V. Lymphatics: 2 plexuses. *Superficial,* under the pleura, from tissue along the alveolar ducts and pleura. These curve around the borders of the lungs and terminate in nodes (bronchopulmonary) at the hilum. The *deep* vessels follow the bronchial tree, pass through the hilum to the *bronchopulmonary nodes*. From the latter, vessels pass to the *tracheopulmonary nodes* which lie around the trachea and principal bronchi. The afferents of these nodes arise from the lungs, bronchi, trachea, heart, and posterior mediastinum. The efferents join those of the internal thoracic and anterior mediastinum to form the *bronchomediastinal trunks,* which usually terminate on each side at the junction of subclavian and internal jugular veins (left may join thoracic duct)

VI. Nerves: the innervation of the lungs is by both vagal and sympathetic fibers via the anterior and posterior pulmonary plexuses (see p. 359). The pulmonary arteries are apparently innervated by sympathetic fibers only, the smooth muscles of the bronchi by parasympathetic fibers, and the bronchial glands are innervated by sympathetic fibers. All afferent fibers from the lung are vagal fibers

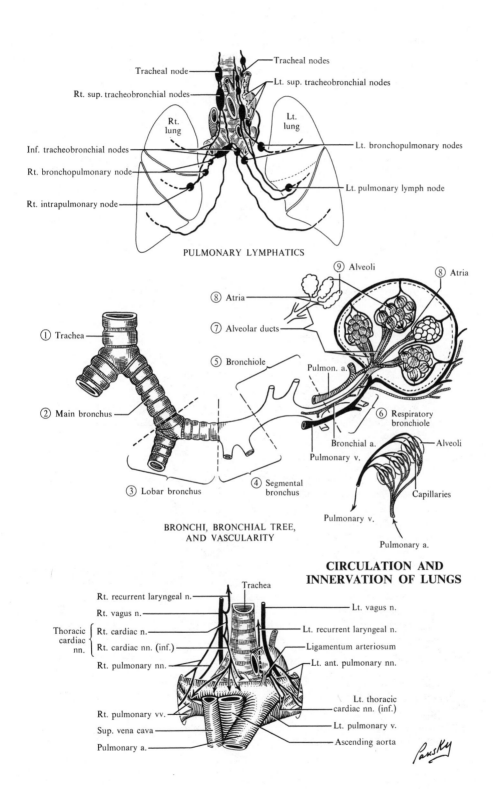

Tracheal node

Rt. sup. tracheobronchial nodes

Rt.
lung

Inf. tracheobronchial nodes

Rt. bronchopulmonary node

Rt. intrapulmonary node

Tracheal nodes

Lt. sup. tracheobronchial nodes

Lt.
lung

Lt. bronchopulmonary nodes

Lt. pulmonary lymph node

PULMONARY LYMPHATICS

① Trachea

② Main bronchus

③ Lobar bronchus

⑤ Bronchiole

④ Segmental
bronchus

⑨ Alveoli

⑧ Atria

⑧ Atria

⑦ Alveolar ducts

Pulmon. a.

⑥ Respiratory
bronchiole

Bronchial a.

Pulmonary v.

Alveoli

Capillaries

Pulmonary v.

Pulmonary a.

BRONCHI, BRONCHIAL TREE,
AND VASCULARITY

**CIRCULATION AND
INNERVATION OF LUNGS**

Trachea

Rt. recurrent laryngeal n.

Rt. vagus n.

Thoracic
cardiac
nn.

Rt. cardiac n.

Rt. cardiac nn. (inf.)

Rt. pulmonary nn.

Lt. vagus n.

Lt. recurrent laryngeal n.

Ligamentum arteriosum

Lt. ant. pulmonary nn.

Rt. pulmonary vv.

Sup. vena cava

Pulmonary a.

Lt. thoracic
cardiac nn. (inf.)

Lt. pulmonary v.

Ascending aorta

Pansky

155. THE HEART, SURFACE PROJECTIONS, GREAT VESSELS, AND CORONARIES

I. Surface projections of the heart
A. APEX: in fifth interspace, approximately 8 cm from midsternal line
B. BASE: slightly oblique line at level of third costal cartilage projecting 2 cm to left and 1 cm to right of lateral border of sternum
C. INFERIOR (DIAPHRAGMATIC) BORDER: right end under sixth costosternal junction; line slopes down across xiphisternal junction to apex
D. RIGHT BORDER begins at right end of base, curves slightly to right, reaching 2.5 cm from sternal margin in fourth interspace, ends at right end of inferior border
E. LEFT BORDER curves upward and medialward from apex to left end of base line

II. Superficial features of heart
A. CORONARY SULCUS: groove separating atria from ventricles
B. ANTERIOR AND POSTERIOR INTERVENTRICULAR (LONGITUDINAL) SULCI divide ventricles into right and left
   1. Incisura apices cordis: notch near apex where the longitudinal sulci meet
C. APEX points down and to the left
D. BASE faces toward right, upward and backward
E. SURFACES: sternocostal beneath sternum and ribs, diaphragmatic against diaphragm
F. MARGINS: right (acute) runs from diaphragmatic to base; left (obtuse) runs from base to apex

III. Coronary vessels (see p. 360)
A. ARTERIES
   1. Right originates in right aortic sinus, runs to right, beneath right auricle to coronary sulcus
      a. Posterior descending branch down posterior interventricular sulcus to apex
      b. Marginal branch follows right margin to apex
   2. Left originates in left aortic sinus, runs under left auricle and divides
      a. Anterior descending branch runs in anterior interventricular sulcus to apex
      b. Circumflex runs in left part of coronary sulcus, curving around to posterior interventricular sulcus
B. VEINS
   1. Coronary sinus: large vessel in posterior part of coronary sulcus, receives most veins of heart and ends in right atrium
      a. Great cardiac vein starts at apex, ascends in anterior interventricular sulcus, and ends in coronary sinus
      b. Small cardiac vein on right side of coronary sulcus opens into end of sinus
         i. Right marginal: along right margin of heart
      c. Middle cardiac vein from apex ascends in posterior longitudinal sulcus to sinus
      d. Posterior vein of left ventricle on diaphragmatic surface, to sinus
      e. Oblique vein of left atrium descends on dorsum of atrium to sinus
   2. Veins opening directly into atrium (thebesian veins)
      a. Anterior (3 or 4) from right ventricular wall
      b. Smallest (very small) from cardiac muscle, usually ending in right atrium, but some terminate in ventricles

IV. Clinical considerations
A. OBSTRUCTION OF A CORONARY ARTERY can lead to anoxia of the heart area supplied, resulting in spasmodic contractions (heart attack), and eventually in death
B. 55–60% of cases, the rt. coronary artery supplies SA node
C. 85–90% of cases, the rt. coronary artery supplies AV node

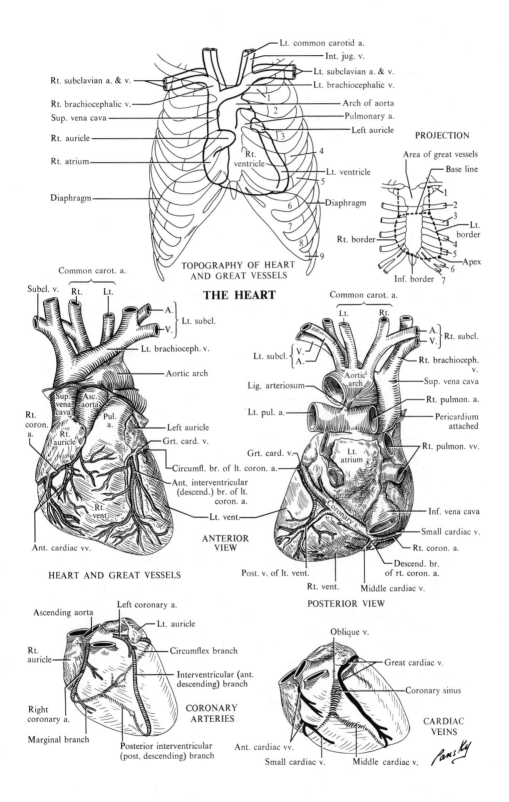

Lt. common carotid a.

Int. jug. v.

Rt. subclavian a. & v.

Lt. subclavian a. & v.

Lt. brachiocephalic v.

Rt. brachiocephalic v.

Arch of aorta

Sup. vena cava

Pulmonary a.

Rt. auricle

Left auricle

Rt. atrium

Rt. ventricle

Lt. ventricle

Diaphragm

Diaphragm

TOPOGRAPHY OF HEART AND GREAT VESSELS

PROJECTION

Area of great vessels

Base line

Rt. border

Lt. border

Apex

Inf. border

THE HEART

Common carot. a.

Subcl. v.

Rt.   Lt.

A.

V.

} Lt. subcl.

Lt. brachioceph. v.

Aortic arch

Sup. vena cava

Asc. aorta

Pul. a.

Rt. coron. a.

Rt. auricle

Left auricle

Grt. card. v.

Circumfl. br. of lt. coron. a.

Ant. interventricular (descend.) br. of lt. coron. a.

Lt. vent.

Ant. cardiac vv.

Rt. vent.

ANTERIOR VIEW

HEART AND GREAT VESSELS

Common carot. a.

Lt.   Rt.

A.

V.

} Rt. subcl.

Lt. subcl. {
V.
A.

Rt. brachioceph. v.

Lig. arteriosum

Aortic arch

Sup. vena cava

Lt. pul. a.

Rt. pulmon. a.

Pericardium attached

Lt. atrium

Rt. pulmon. vv.

Grt. card. v.

Inf. vena cava

Small cardiac v.

Coronary s.

Rt. coron. a.

Descend. br. of rt. coron. a.

Post. v. of lt. vent.

Rt. vent.

Middle cardiac v.

POSTERIOR VIEW

Ascending aorta

Left coronary a.

Lt. auricle

Rt. auricle

Circumflex branch

Interventricular (ant. descending) branch

Right coronary a.

CORONARY ARTERIES

Marginal branch

Posterior interventricular (post. descending) branch

Oblique v.

Great cardiac v.

Coronary sinus

CARDIAC VEINS

Ant. cardiac vv.

Small cardiac v.

Middle cardiac v.

Pansky

156. THE PERICARDIUM

I. Components
A. FIBROUS: tough, fibrous sac, closed above by attachments to the great vessels. Outer layer is attached to
   1. Manubrium and xiphoid process of sternum by the superior and inferior *sternopericardial ligaments*
   2. Vertebral column by the *vertebropericardial ligament*
   3. Diaphragm by a wide area of fibers attached to the central tendon of the diaphragm. The *phrenicopericardial ligament* is a thickened band near the inferior vena cava
   4. Pleura on either side where the fibrous pericardium and the mediastinal pleura meet. This union is loose and permits the phrenic nerve and vessels to run between the pleura and the pericardium
B. SEROUS: smooth membrane with a mesothelial layer that lines the fibrous sac and also covers the surface of the heart
   1. Visceral pericardium (epicardium) covers the entire surface of the heart and extends along its great vessels for 3 cm, where it is reflected onto the inner surface of the fibrous pericardium
   2. Parietal pericardium lines the inner surface of the fibrous pericardium

II. Pericardial cavity: the potential space between the parietal and visceral pericardium. The pericardial surfaces are in contact, covered with a watery fluid to allow for freedom of heart movement during its contractions

III. The mesocardia: reflections of the epicardium along the great vessels and onto the fibrous pericardial sac
A. ARTERIAL: tubular prolongation on aorta and pulmonary trunk
B. VENOUS: extension along the venae cavae and pulmonary veins
   1. Due to the arrangement of the veins, an inverted U-shaped pocket called the *oblique pericardial sinus* is formed
   2. The narrow, pericardial-lined groove between IIIA and B, above, is the *transverse pericardial sinus*

IV. Clinical considerations
A. IN INJURY OR DISEASE, fluid may accumulate in the pericardial cavity so that the parietal and visceral layers become greatly separated. This may be withdrawn by anterior thoracic approach, just above the 6th costal cartilage in the 5th interspace, one finger's breadth from the left side of the sternum; by the abdominal approach, passing needle upward and to the left between the xiphoid process and the left costal margin, directed toward the nipple; posterior thoracic approach, the needle being inserted over the 8th rib, midway between the inferior angle of scapula and midline, directing it toward the middle of the chest
B. IN SOME CASES OF CORONARY OCCLUSION, attempts have been made to create adhesions between the layers of the pericardium in the hope that the vessels in the pericardium could help supply the heart
C. THE TERM "PERICARDIAL SAC" refers to the fibrous plus parietal layer of serous pericardium. It does not have a very rich blood supply. It surrounds the heart loosely, and if the heart gradually enlarges, so does the sac
D. AS A RESULT OF SEVERE PERICARDITIS, the pericardium becomes greatly thickened and adherent to the heart (constrictive pericarditis)

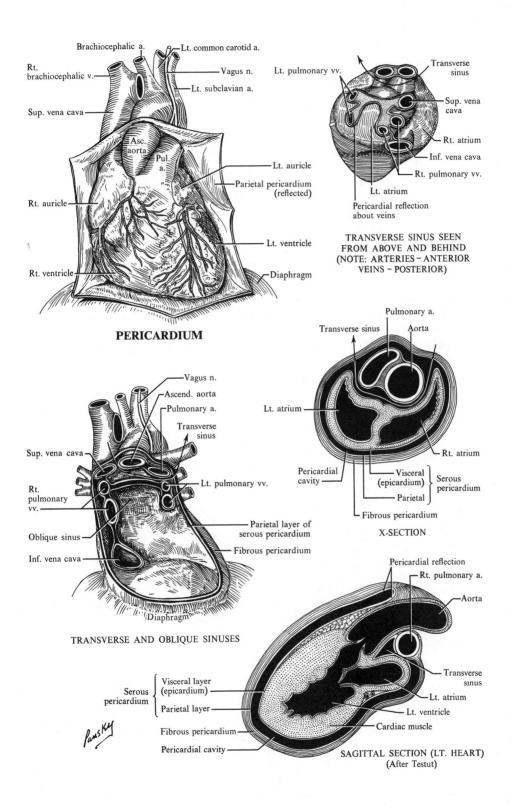

Brachiocephalic a.

Lt. common carotid a.

Rt. brachiocephalic v.

Vagus n.

Lt. subclavian a.

Sup. vena cava

Asc. aorta

Pul. a.

Lt. auricle

Parietal pericardium (reflected)

Rt. auricle

Lt. ventricle

Rt. ventricle

Diaphragm

PERICARDIUM

Lt. pulmonary vv.

Transverse sinus

Sup. vena cava

Rt. atrium

Inf. vena cava

Rt. pulmonary vv.

Lt. atrium

Pericardial reflection about veins

TRANSVERSE SINUS SEEN
FROM ABOVE AND BEHIND
(NOTE: ARTERIES – ANTERIOR
VEINS – POSTERIOR)

Vagus n.

Ascend. aorta

Pulmonary a.

Transverse sinus

Sup. vena cava

Lt. pulmonary vv.

Rt. pulmonary vv.

Oblique sinus

Parietal layer of serous pericardium

Fibrous pericardium

Inf. vena cava

Diaphragm

TRANSVERSE AND OBLIQUE SINUSES

Pulmonary a.

Aorta

Transverse sinus

Lt. atrium

Rt. atrium

Pericardial cavity

Visceral (epicardium)

Parietal

Serous pericardium

Fibrous pericardium

X-SECTION

Pericardial reflection

Rt. pulmonary a.

Aorta

Transverse sinus

Lt. atrium

Lt. ventricle

Cardiac muscle

Serous pericardium

Visceral layer (epicardium)

Parietal layer

Fibrous pericardium

Pericardial cavity

SAGITTAL SECTION (LT. HEART)
(After Testut)

– 343 –

157. INTERIOR OF THE HEART

I. Right atrium: divided into 2 parts by ridge, the *crista terminalis* (creates an external impression between venae cavae, the *sulcus terminalis*)
A. AURICLE: blind pocket
    1. Lined by parallel ridges, the *pectinate muscles*
B. SINUS OF VENAE CAVAE (VENARUM): principal cavity
    1. Lining is smooth
    2. Openings
        a. Superior vena cava from above
        b. Inferior vena cava from below
            i. Valve of inferior vena cava, a crescent-shaped fold along anterior and left margin of cava
        c. Coronary sinus opens between orifice of inferior vena cava and atrioventricular orifice
            i. Valve, a fold attached along right and inferior borders of opening
        d. Small coronary veins (see p. 340)
        e. Atrioventricular orifice, oval-shaped opening into right ventricle
C. INTERATRIAL SEPTUM: the posterior wall of atrium
    1. Fossa ovalis, oval depression, where septum is thin
    2. Limbus of the fossa ovalis: the thick margin of fossa, distinct except below

II. Left atrium: like the right, has a smooth-walled portion and a smaller muscular auricle. The 4 pulmonary veins open into the smooth-walled portion

III. Right ventricle occupies most of sternocostal surface from coronary to anterior longitudinal sulci
A. ATRIOVENTRICULAR (TRICUSPID) VALVE (see p. 346): leads to inflow limb of blood flow
B. PARTS, separated by ridge, supraventricular crest
    1. Conus arteriosus (infundibulum): smooth cephalic part leading into pulmonary trunk (outflow limb)
    2. The supraventricular crest is a thick muscular ridge lying between the venous or inflowing part of the ventricle and the outflowing part of the ventricle (conus arteriosus or infundibulum)
    3. Ventricle proper
        a. Trabeculae carneae: irregular bundles of muscle projecting on inner surface. Of 3 types:
            i. Ridges along wall
            ii. Extension across ventricular cavity: septomarginal trabecula (moderator band), an enlarged bundle near apex
            iii. Papillary muscles, conical projections into cavity. *Anterior* (largest) partly from anterior and septal walls; *posterior* from numerous parts on posterior wall; and *septal* near septal end of supraventricular crest
        b. Chordae tendineae: small tendinous bands from ventricular wall, at apices of papillary muscles, attached to cusps of atrioventricular valves
        c. Interventricular septum: set obliquely; the right convex side encroaches on the right ventricle
            i. Muscular part: thickest, major part of septum
            ii. Membranous part: thin upper part near atrium

IV. Left ventricle: small part of sternocostal and much of diaphragmatic surface
A. LEFT ATRIOVENTRICULAR (MITRAL OR BICUSPID) VALVE (see p. 346)
B. STRUCTURE SIMILAR TO RIGHT except trabeculae carneae more dense. Two large sets of papillary muscles, *anterior* and *posterior*, are attached to their respective walls
C. AORTA with its semilunar valve
D. THE LOWER, or inflowing, part of the cavity communicates with the left atrium. The upper and anterior part is the *aortic vestibule*, the walls of which are fibrous and lead into the aorta

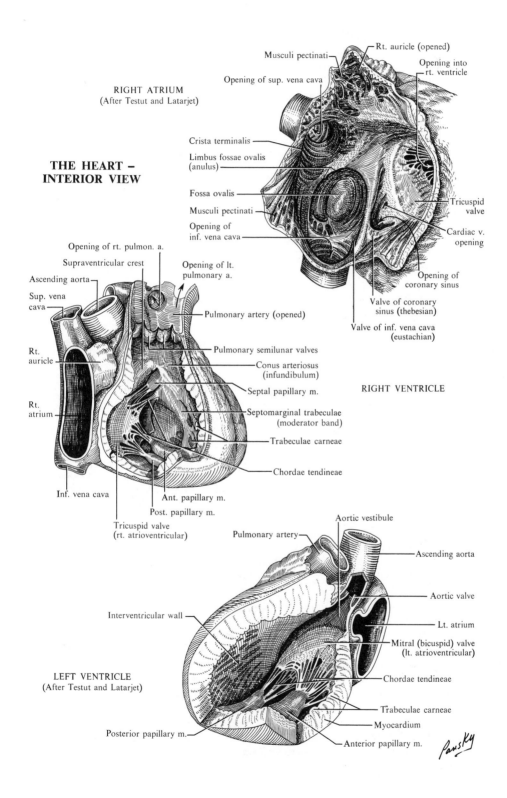

THE HEART –
INTERIOR VIEW

RIGHT ATRIUM
(After Testut and Latarjet)

Musculi pectinati

Rt. auricle (opened)

Opening of sup. vena cava

Opening into
rt. ventricle

Crista terminalis

Limbus fossae ovalis
(anulus)

Fossa ovalis

Musculi pectinati

Opening of
inf. vena cava

Tricuspid
valve

Cardiac v.
opening

Opening of
coronary sinus

Valve of coronary
sinus (thebesian)

Valve of inf. vena cava
(eustachian)

RIGHT VENTRICLE

Opening of rt. pulmon. a.

Supraventricular crest

Ascending aorta

Sup. vena
cava

Opening of lt.
pulmonary a.

Pulmonary artery (opened)

Pulmonary semilunar valves

Conus arteriosus
(infundibulum)

Septal papillary m.

Septomarginal trabeculae
(moderator band)

Trabeculae carneae

Chordae tendineae

Rt.
auricle

Rt.
atrium

Inf. vena cava

Ant. papillary m.

Post. papillary m.

Tricuspid valve
(rt. atrioventricular)

Aortic vestibule

Pulmonary artery

Ascending aorta

Aortic valve

Lt. atrium

Mitral (bicuspid) valve
(lt. atrioventricular)

Chordae tendineae

Trabeculae carneae

Myocardium

Anterior papillary m.

Interventricular wall

LEFT VENTRICLE
(After Testut and Latarjet)

Posterior papillary m.

Pansky

158. THE VALVES OF THE HEART

I. **Surface projections of valves** (for valve sounds, see opposite page)
A. PULMONARY: upper border of angle of third costosternal union, on left
B. AORTIC: slightly below and medial to above, opposite articulation of third costosternal joint
C. LEFT ATRIOVENTRICULAR: just to left of midline, opposite fourth costosternal articulation
D. RIGHT ATRIOVENTRICULAR: near sternal margin on right, opposite fourth interspace

II. **Structure**
A. ATRIOVENTRICULAR
   1. Right (tricuspid): set in oval aperture 4 cm in longest diameter. Composed of 3 fibrous cusps, thick near attached border, thin near free edges. All cusps are attached to ring (anulus fibrosus) of dense tissue around orifice
      a. Anterior cusp: largest, attached to ventral wall near infundibulum. Receives chordae tendineae from anterior and septal papillary muscles
      b. Posterior cusp: from curved ventricular wall where sternocostal and diaphragmatic surfaces become continuous. Receives chordae from both anterior and posterior papillary muscles
      c. Septal cusp (medial): from septal wall. Receives chordae from posterior and septal papillary muscles
   2. Left (bicuspid, mitral): set in ring smaller than right. Composed of 2 cusps
      a. Anterior (aortic): placed anteriorly, attached near aortic orifice. Receives chordae from both papillary muscles of ventricle
      b. Posterior: smaller, attached to posterior wall. Receives chordae from both papillary muscles
B. ARTERIAL
   1. Pulmonary semilunar: composed of 3 semilunar cusps (anterior, right, and left) set in a circular orifice at top of the infundibulum
      a. Each cusp attached along a curve that is convex caudally
      b. Sinus: pocket between (behind) cusp and vessel wall at origin of pulmonary artery
      c. Commissure: point where attachments of adjacent cusps come together
      d. Nodule: a thickened area in center of free margin of cusp
      e. Lunula: crescent-shaped, thin part of cusp lying between free margin and a thickened band of fibers in the cusp that arches downward from nodule to commissure
   2. Aortic semilunar (right, left, and posterior cusps): same structure as above except nodules thicker, lunulae more distinct, coronary aa. rise from 2 sinuses

III. **Cardiac skeleton** consists of fibrous or fibrocartilaginous tissue that surrounds the atrioventricular and semilunar openings, gives attachments to the valves and the muscle layers, and is continuous with the roots of the aorta and pulmonary trunk, as well as membranous part of the interventricular septum

IV. **Clinical considerations**
A. IT HAS BEEN NOTED that, in general, the free margins of all valves (those that approximate when the valves are closed) are thin. If these become thickened and inflexible or damaged through disease, a close fit will be impossible. Thus, blood can flow back into the chamber from which it has come
B. STENOSIS: a narrowing or constriction of an orifice. Most frequently occurs in pulmonary or aortic opening
C. HEART SOUNDS: 2 audible sounds occur during each heartbeat: *lubb-dupp*. The first is low-pitched and of long duration, caused by closure of both atrioventricular valves and the contraction of the ventricular muscle. The second is short, sharp, and high-pitched and is caused by closure of the semilunar valves

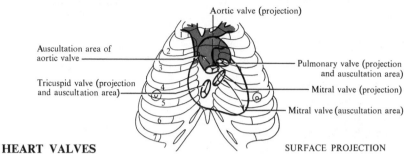

Aortic valve (projection)

Auscultation area of aortic valve

Pulmonary valve (projection and auscultation area)

Tricuspid valve (projection and auscultation area)

Mitral valve (projection)

Mitral valve (auscultation area)

HEART VALVES   SURFACE PROJECTION

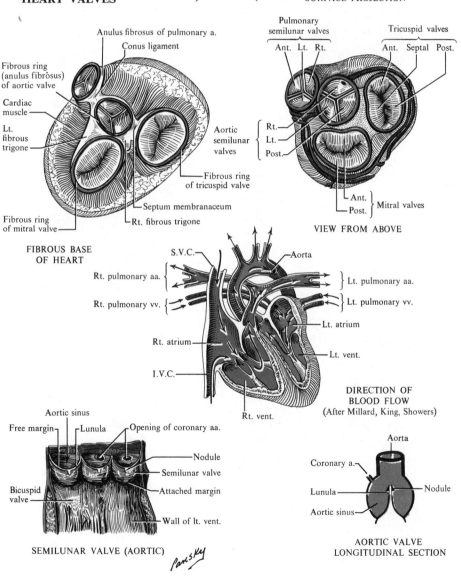

Anulus fibrosus of pulmonary a.

Conus ligament

Fibrous ring (anulus fibrosus) of aortic valve

Cardiac muscle

Lt. fibrous trigone

Aortic semilunar valves

Fibrous ring of tricuspid valve

Septum membranaceum

Fibrous ring of mitral valve

Rt. fibrous trigone

FIBROUS BASE OF HEART

Pulmonary semilunar valves

Ant. Lt. Rt.

Tricuspid valves

Ant. Septal Post.

Rt.
Lt.
Post.

Ant.
Post. } Mitral valves

VIEW FROM ABOVE

S.V.C.

Aorta

Rt. pulmonary aa.

Lt. pulmonary aa.

Rt. pulmonary vv.

Lt. pulmonary vv.

Lt. atrium

Rt. atrium

Lt. vent.

I.V.C.

Rt. vent.

DIRECTION OF BLOOD FLOW
(After Millard, King, Showers)

Aortic sinus

Free margin

Lunula

Opening of coronary aa.

Nodule

Semilunar valve

Bicuspid valve

Attached margin

Wall of lt. vent.

SEMILUNAR VALVE (AORTIC)

Aorta

Coronary a.

Nodule

Lunula

Aortic sinus

AORTIC VALVE LONGITUDINAL SECTION

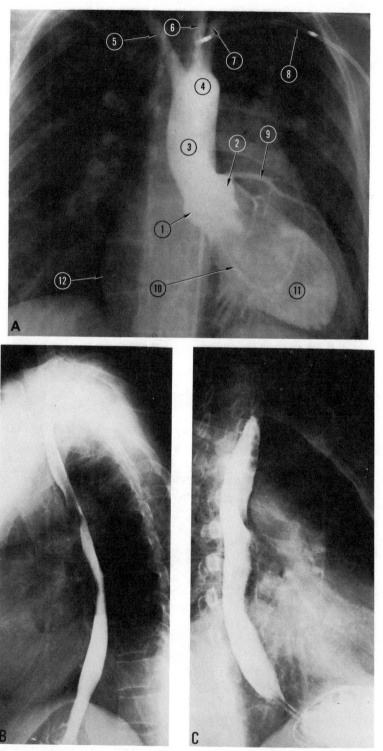

FIGURE 27. **Ascending aorta, coronary arteries, and esophagus. A, Ascending aorta and coronaries; B, Thoracic esophagus, lateral view; C, Thoracic esophagus, oblique view.** *1,* Right aortic sinus; *2,* left aortic sinus; *3,* ascending aorta; *4,* arch of aorta; *5,* brachiocephalic artery; *6,* common carotid artery; *7,* subclavian artery; *8,* catheter; *9,* left coronary artery; *10,* right coronary artery; *11,* cavity of left ventricle; *12,* right margin of heart.

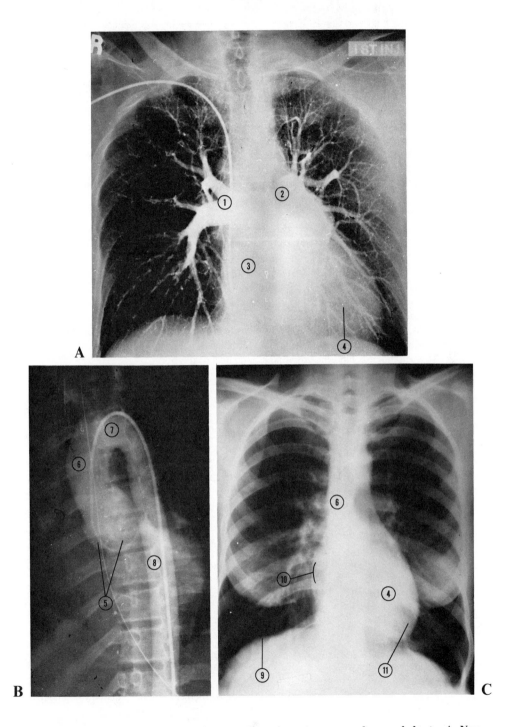

FIGURE 28. **Normal pulmonary arteriogram, thoracic aortogram, and normal chest.** **A, Normal pulmonary arteriogram, AP view; B, thoracic aortogram; C, normal chest.** *1,* Right pulmonary artery; *2,* left pulmonary artery; *3,* right ventricle; *4,* left ventricle; *5,* aortic valves; *6,* ascending aorta; *7,* aortic arch; *8,* thoracic aorta; *9,* diaphragm; *10,* right border of heart; *11,* apex of heart.

159. THE MEDIASTINUM

I. **Definition:** a thick partition in the thorax, bounded laterally by the pleurae, anteriorly by the sternum, and posteriorly by the vertebral column

II. **Divisions**
   A. SUPERIOR: above by plane of first rib, below by horizontal line at level of sternal angle. This passes through the disk between fourth and fifth thoracic vertebrae
   B. INFERIOR: above by superior mediastinum, below by diaphragm
      1. Anterior: above by lower border of superior, below by diaphragm, anteriorly by body of sternum and transverse thoracic muscle, posteriorly by pericardium
      2. Middle: above and below as the anterior. Its posterior and anterior limits are the fibrous pericardium
      3. Posterior: same upper and lower limits as anterior and middle. Lies between the pericardium and the vertebral column. Caudally extends to level of twelfth thoracic vertebra behind diaphragm

III. **Superior mediastinum:** contents and relations
   A. MOST ANTERIOR, the origins of the sternohyoid and sternothyroid muscles
   B. THYMUS GLAND, usually only fibrous and fatty remnants
   C. VESSELS
      1. Arteries: in lower part, the arch of the aorta; above the arch are the brachiocephalic, left common carotid, and left subclavian arteries
      2. Veins: to the right side and below is the superior vena cava; above and anterior to the arteries are the brachiocephalic veins. Termination of azygos v.
      3. Thoracic duct: behind the aortic arch and subclavian artery
   D. VISCERA
      1. Trachea in midline, posterior to major vessels
      2. Esophagus posterior and slightly to left of trachea
   E. NERVES
      1. Both vagus nerves related to the brachiocephalic artery on the right and to the subclavian artery on the left
      2. Cardiac nerves
      3. Left recurrent nerve, near groove between trachea and esophagus
      4. Both phrenic nerves
   F. MOST POSTERIORLY, lower part of longus colli muscle

IV. **Anterior mediastinum**
   A. NO MAJOR STRUCTURES: areolar tissue, transverse thoracic muscle, a few lymph nodes; small blood vessels and lymphatics

V. **Middle mediastinum**
   A. PERICARDIUM AND HEART (see p. 342)
   B. GREAT VESSELS
      1. Superior vena cava
      2. Ascending aorta
      3. Pulmonary trunk with the origins of the right and left pulmonary arteries
   C. NERVES
      1. Phrenic: most lateral, actually along the side of the pericardium, thus on the border of (rather than in) the middle mediastinum

VI. **Posterior mediastinum** (see p. 366)

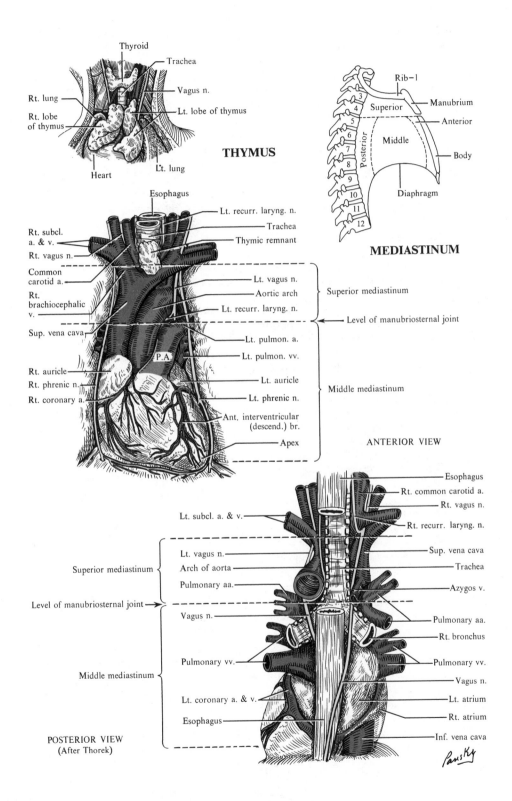

Thyroid

Trachea

Rt. lung

Vagus n.

Rt. lobe
of thymus

Lt. lobe of thymus

Heart

Lt. lung

THYMUS

Rib-1

3
4   Superior   Manubrium
5   Anterior
6   Middle
Posterior
7   Body
8
9
10
11
12   Diaphragm

MEDIASTINUM

Esophagus

Rt. subcl.
a. & v.

Lt. recurr. laryng. n.

Trachea

Thymic remnant

Rt. vagus n.

Common
carotid a.

Lt. vagus n.

Rt.
brachiocephalic
v.

Aortic arch

Lt. recurr. laryng. n.

Superior mediastinum

Sup. vena cava

Level of manubriosternal joint

P.A

Lt. pulmon. a.

Lt. pulmon. vv.

Rt. auricle

Rt. phrenic n.

Lt. auricle

Rt. coronary a.

Lt. phrenic n.

Middle mediastinum

Ant. interventricular
(descend.) br.

Apex

ANTERIOR VIEW

Esophagus

Rt. common carotid a.

Rt. vagus n.

Lt. subcl. a. & v.

Rt. recurr. laryng. n.

Lt. vagus n.

Sup. vena cava

Superior mediastinum

Arch of aorta

Trachea

Pulmonary aa.

Azygos v.

Level of manubriosternal joint

Vagus n.

Pulmonary aa.

Rt. bronchus

Pulmonary vv.

Pulmonary vv.

Vagus n.

Middle mediastinum

Lt. coronary a. & v.

Lt. atrium

Rt. atrium

Esophagus

Inf. vena cava

POSTERIOR VIEW
(After Thorek)

Pansky

– 351 –

160. LYMPHATICS OF THE MEDIASTINUM

I. Introduction: mediastinal lymph node involvement is common in diseases of the chest wall, breast, diaphragm, and intra-abdominal or retroperitoneal structures. Patterns of nodal enlargement offer a clue to the origin or nature of a disease process

A. INTRATHORACIC LYMPH NODES are usually divided into 2 major anatomic groups
  1. Parietal mediastinal nodes: drain thoracic wall and extrathoracic tissues
  2. Visceral mediastinal nodes: are concerned with intrathoracic structures

B. EXTENSIVE CONNECTIONS exist between parietal and visceral nodal groups, as well as within the visceral group itself. Three mediastinal zones (compartments) are described

II. Anterior mediastinal zone (compartment): two lymph node chains are seen

A. STERNAL (ANTERIOR PARIETAL OR INTERNAL THORACIC) GROUP is found extrapleurally
  1. Nodes are scattered along int. thoracic aa. and behind ant. intercostal spaces and costal cartilages. A few are directly retrosternal. Nodes vary in number
  2. Afferents drain ant. abdominal and thoracic walls, ant. diaphragm, and medial parts of breast. Afferents communicate with the ant. mediastinal visceral and cervical nodes
  3. Major efferents channels are: the rt. lymphatic or bronchomediastinal duct, on the right, and the thoracic duct, on the left

B. ANTERIOR MEDIASTINAL (PREVASCULAR) GROUP belongs to the visceral collection
  1. Majority of nodes are grouped along sup. vena cava and brachiocephalic vein, on the right, and in front of the aorta and carotid artery, on the left
  2. Afferents drain mediastinal structures including: pericardium, part of heart, thymus, thyroid, diaphragmatic and mediastinal pleura, and the hila of the lungs, bilaterally
  3. Efferents drain into the bronchomediastinal trunk (or thoracic duct)

III. Posterior mediastinal zone (compartment) contains parietal and visceral groups

A. INTERCOSTAL (POST. PARIETAL) GROUP lies in the intercostal spaces and paravertebral areas
  1. Both groups drain intercostal spaces, parietal pleura, and vertebral column
  2. They communicate with the post. mediastinal visceral group and have efferent channels that drain to thoracic duct from the upper thorax and to cisterna chyli from lower thorax

B. POSTERIOR MEDIASTINAL (VISCERAL) GROUP lies along lower esophagus and thoracic aorta
  1. Afferents drain post. diaphragm, pericardium, esophagus, and lower lobes of lungs. They communicate with bifurcation group of tracheobronchial nodes
  2. Efferents drain predominantly via the thoracic duct

IV. Middle mediastinal zone (compartment) includes parietal and visceral portions

A. PARIETAL GROUP found mainly around pericardial attachment to the diaphragm
  1. Afferents drain diaphragm and parts of liver
  2. Efferents pass ant. to sternal or ant. mediastinal nodes and post. to post. med. nodes

B. TRACHEOBRONCHIAL, bifurcation (carinal), and bronchopulmonary nodes are the most important in thorax and freely communicate with each other. Only 1 group may be involved
  1. Tracheobronchial nodes: scattered along lateral aspects of trachea and in the tracheobronchial angle, being more numerous on the rt. than on the lt.
    a. The azygos node: constant node in rt. tracheobronchial angle near azygos v.
    b. Afferents: from bifurcation and bronchopulmonary nodes, trachea, esophagus, & rt. and lt. lungs. Communicate with ant. and post. visceral mediastinal nodes
    c. Efferents: to bronchomediastinal trunk, on rt., to thoracic duct, on lt.
  2. Bifurcation (carinal) nodes: along ant. and inferior aspects of carina. Extend downward bilaterally along main bronchi to lower lobe bronchi
    a. Afferents: from bronchopulmonary nodes, ant. and post. mediastinal nodes, heart, pericardium, esophagus, and directly from both lungs
    b. Efferents drain to the right tracheobronchial group of nodes
  3. Bronchopulmonary nodes: lie in angles between hilar bronchial bifurcations
    a. Afferents: from pulmonary lobes
    b. Efferents: drain to bifurcation (carina) and tracheobronchial nodes

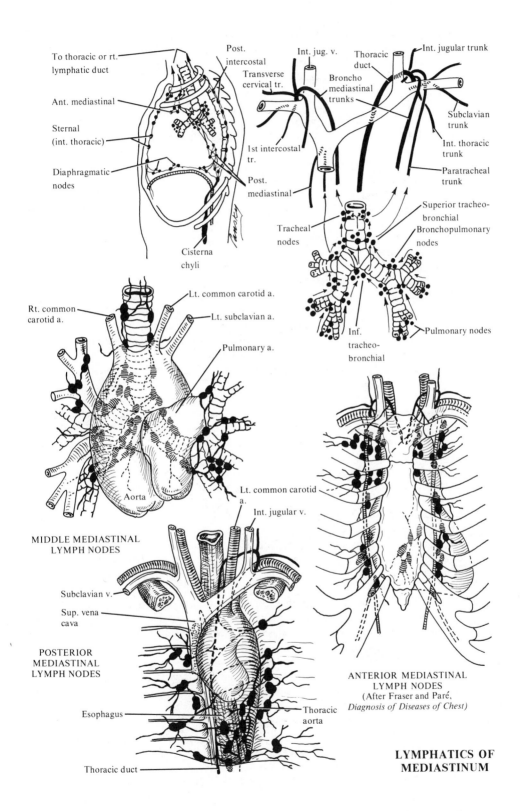

To thoracic or rt. lymphatic duct

Post. intercostal

Int. jug. v.

Thoracic duct

Int. jugular trunk

Ant. mediastinal

Transverse cervical tr.

Broncho mediastinal trunks

Sternal (int. thoracic)

Subclavian trunk

1st intercostal tr.

Int. thoracic trunk

Diaphragmatic nodes

Post. mediastinal

Paratracheal trunk

Cisterna chyli

Tracheal nodes

Superior tracheo-bronchial

Bronchopulmonary nodes

Rt. common carotid a.

Lt. common carotid a.

Lt. subclavian a.

Pulmonary a.

Inf. tracheo-bronchial

Pulmonary nodes

Aorta

MIDDLE MEDIASTINAL LYMPH NODES

Lt. common carotid a.

Int. jugular v.

Subclavian v.

Sup. vena cava

POSTERIOR MEDIASTINAL LYMPH NODES

Esophagus

Thoracic aorta

ANTERIOR MEDIASTINAL LYMPH NODES
(After Fraser and Paré, *Diagnosis of Diseases of Chest*)

Thoracic duct

LYMPHATICS OF MEDIASTINUM

161. CONDUCTING SYSTEM OF HEART

I. Introduction: consists of specialized cardiac muscle fibers (nodal and Purkinje fibers) and is responsible for initiating and maintaining the normal cardiac rhythm, resulting in coordination of atrial and ventricular contractions. Consists of sinuatrial (sinoatrial) (SA) node, atrioventricular (AV) node and bundle (of His), and its two limbs (fasciculi, branches or crura), and the subendocardial plexuses of Purkinje fibers in the ventricles where they terminate

II. Sinuatrial (sinoatrial) (SA) node: narrow, horseshoe-shaped, and found in upper end of sulcus terminalis of right atrium and extending medially in front of superior cava opening

A. BULK OF NODE lies on sinus venarum side of crista terminalis and extends through the atrial wall thickness from epicardium to endocardium

B. ITS FUSIFORM nodal fibers have a higher rate of intrinsic rhythmic contraction than any cardiac m., thus, the cardiac cycle contraction of heart is here initiated

C. CALLED "PACEMAKER" of heart because normally other parts of the heart contract at the rate imposed on them by the node

D. SINCE NODAL FIBERS are in contact with cells of neighboring myocardial fibers, the impulse for cardiac contraction, beginning in node, is conducted throughout the atria by ordinary atrial myocardial fibers, arriving eventually at the AV node

  1. There are no Purkinje fibers in the atrial walls

  2. It was generally thought that no specialized nodal pathways connected the 2 nodes. However, it has been shown that internodal cell-to-cell conduction takes place more rapidly along 3 functional interatrial pathways or muscle fiber bundles

    a. Anterior pathway: via ant. interatrial band (Bachmann's bundle) which passes directly between nodes (thru atrial septum) and synchronizes the 2 atria

    b. Middle and posterior internodal pathways: mixture of ordinary myocardial cells as well as some more specialized in conduction

III. Atrioventricular (AV) node*: smaller than SA node with broader, more cylindrical fibers. Lies above opening of coronary sinus, embedded in myocardial fibers of atrial septum. Cells composing nodal fibers are contiguous with cells of atrial septal muscles as well as with those composing fibers of the AV bundle of His

A. IS A REGION of slow conductance which ensures the needed delay between atrial and ventricular contractions, allowing time for the ventricles to fill

IV. AV bundle (of His) passes upward from AV node in the rt. fibrous trigone beneath attachment of septal cusp of rt. AV valve to reach posterior margin of membranous ventricular septum, and then turns forward below it. Here it divides into the rt. and lt. limbs or bundles with the division straddling the upper end of muscular part of interventricular septum

A. THE LT. LIMB descends as a flattened band beneath the endocardium on the lt. side of the septum and divides into 2 or more strands (ant. and post.) that descend to ventricular apex

  1. The strands divide into a variable no. of branches which pass into the trabeculae carneae to the papillary mm. where they break up into plexus formations of Purkinje fibers from which branches pass beneath the endocardium to all parts of the left ventricle to become continuous with myocardial fibers at varying depths

B. THE RT. LIMB is rounded and a continuation of the bundle; passes toward the apex embedded in the muscle near the rt. surface of the septum, but later lies beneath endocardium. It enters the *septomarginal trabecula* (moderator band) which conveys it to the base of the ant. papillary m. Here it breaks into a Purkinje fiber plexus from which branches pass beneath endocardium to all parts of the rt. ventricle and become continuous with myocardium fibers

C. THE AV NODE, bundle and limbs, and ventricular Purkinje plexuses are surrounded by a delicate connective tissue sheath that insulates them from the myocardium

V. Vascular supply of conducting system: see illustrations on p. 361

---

*3 parts of AV node are identified, based on characteristic electrophysiologic responses.

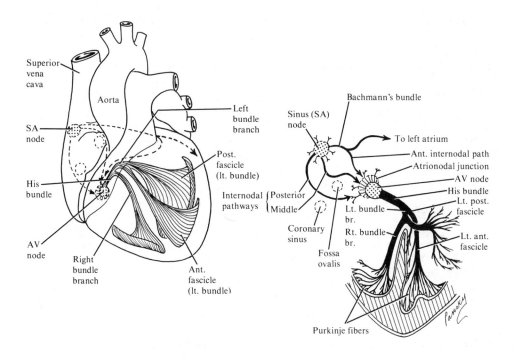

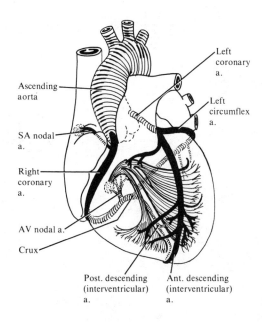

VASCULAR SUPPLY OF CONDUCTING SYSTEM

1. SA NODE: nodal a. from right coronary a.
2. AV NODE: post. interventricular or
   right coronary a.
3. RIGHT BUNDLE OF HIS
a. Proximal part: by branches of
   ant. interventricular, AV nodal
   artery or both
b. Distal part: by branches of
   ant. interventricular a.
4. LEFT BUNDLE OF HIS AND ANT. FASCICLE
see 3a above
5. LEFT BUNDLE POST. FASCICLE
branches of AV nodal a., post.
interventricular a. and
circumflex a.

162. THE ELECTROCARDIOGRAM AND SOME CLINICAL CONSIDERATIONS

I. Introduction: as excitation spreads over the heart and dissipates, an electrical field is produced whose changes in magnitude and direction, in time, can be sensed on the body surface and reflected in changes in potential differences measured between various body surface sites. The *electrocardiogram* (*ECG* or *EKG*) represents such potential differences as a function of time. It is an indicator of *heart excitation, not contraction*

  A. ELECTRODES ARE ATTACHED to the rt. arm and lt. leg to give a bipolar recording from the body surface in the direction of the long axis of the heart

  B. ONE SEES BOTH positive and negative deflections or waves

  C. THE DISTANCE BETWEEN 2 waves is called a *segment* (e.g., PQ segment extends from the end of P wave to beginning of the QRS complex)

  D. AN INTERVAL COMPRISES both waves and segments (e.g., the PQ interval, from beginning of the P wave to beginning of the QRS wave)

    1. PQ interval: time from atrial excitation to ventricular excitation (less than 0.2 sec)

    2. QT interval depends on heart rate (as rate increases, the interval decreases)

    3. RR interval, between the peaks of two successive R waves, corresponds to period of the beat cycle, and is the reciprocal of the beat rate (60/RR interval(s) = beats/min)

II. P wave: first wave, represents spread of excitation over 2 atria (atrial depolarization)

  A. DURING SUBSEQUENT PQ SEGMENT, atria as a whole are excited

  B. AS EXCITATION IN ATRIA DIES OUT, the 1st deflection in ventricular part of curve begins (from beginning of Q to end of T wave)

III. QRS complex: second wave, occurs about 0.1–0.2 sec after 1st wave and is the expression of the spread of excitation over both ventricles (ventricular depolarization)

IV. ST segment: analogous to PQ segment, indicates excitation of the ventricular myocardium

V. T wave: final wave; recovery in ventricles from excitation (ventricular repolarization)

  A. NO MANIFESTATION of atrial repolarization is seen because it occurs during ventricular depolarization and is masked by the QRS complex

VI. From the T to the next P wave, both atria and ventricles are inactive

VII. U wave may follow T wave, corresponds to "dying out" of excitation in terminal branches of the conducting system

VIII. Clinical considerations

  A. ANGINA PECTORIS: when coronary aa. narrow (atherosclerosis), blood supply to myocardium, sufficient at rest, may be inadequate when work load on heart increases. The accumulation of metabolites in the hypoxic muscle activates afferent nerves that travel via sympathetic nerves to the CNS and cause precordial pain (angina pectoris) that is relieved by rest

    1. Recurrent episodes of angina at rest, without changes in heart rate, blood pressure, or ventricular contractility, are due to constriction of coronary artery (vasospasm)

  B. MYOCARDIAL INFARCTION: sudden insufficiency of arterial or venous blood supply due to emboli, thrombi, vessel torsion, or pressure that produces a macroscopic area of necrosis

    1. Complications: most deaths in early hours after MI are due to ventricular fibrillation or cardiac arrest; ventricular aneurysm can be due to continued bulging of a fibrotic myocardium; heart block may be due to impairment of blood supply of conducting tissue

  C. TACHYCARDIA: increased heart rate resulting from increased adrenergic activation of the SA node (emotional stress or reflex response to the fall in blood pressure)

  D. BRADYCARDIA: decreased heart rate, usually under 60 beats/min

  E. BUNDLE BRANCH BLOCK: complete branch block seen when the impulse migrating down from the AV node via bundle of His is obstructed by a lesion in the rt. or lt. bundle

ELECTROCARDIOGRAPHIC
WAVES, INTERVALS,
AND
SEGMENTS

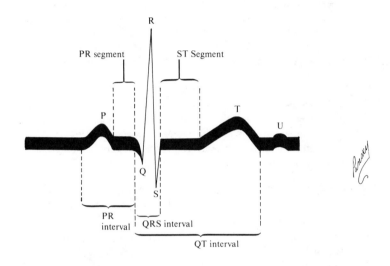

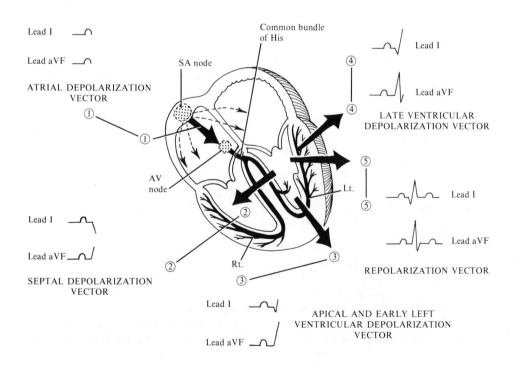

163. EXTRINSIC INNERVATION OF THE HEART AND THE CARDIAC PLEXUSES

I. Introduction: the heart is influenced by autonomic control via the sympathetic and parasympathetic (vagal) systems and the cardiac plexuses through which both supply the heart. The efferent preganglionic sympathetics arise in the upper 4 or 5 thoracic segments of the spinal cord and synapse in the 3 cervical and upper 4 or 5 thoracic sympathetic ganglia. From these, postganglionic fibers form cardiac nerves. The efferent parasympathetic fibers are derived from the dorsal nucleus of the vagus and run in cardiac branches of the vagus to synapse in cells of the cardiac plexuses and in the atrial walls. Postganglionics supply the heart

A. IMPULSES IN THE ADRENERGIC sympathetic nerves increase heart rate (chronotropic effect) and the force of cardiac contraction (inotropic effect). Those in the cholinergic vagal cardiac fibers decrease heart rate

B. A SIGNIFICANT COMPONENT of sympathetic vasoconstrictor tone exists in coronary arterioles, and the release of this tone contributes to vasodilation under stress conditions. The effects of antiadrenergic drugs and of chronic cardiac denervation suggest that sympathetic nerves augment coronary flow mainly by increasing metabolic demand. The heart does not possess sympathetic vasodilator nerves as seen in skeletal muscle. Cardiac parasympathetic vasodilator nerves exist, but their influence on resistance is small, and they have no known function

II. The cardiac plexus: found at base of heart, divided into a *superficial* (*ventral*) and a *deep* (*dorsal*) portion, which are closely connected. Small ganglia are found in the plexus, the cardiac ganglion being the largest and most consistent

A. SUPERFICIAL (VENTRAL) PART lies below the aortic arch, anterior to the right pulmonary artery and to the right of the ligamentum arteriosum
   1. Formation: left superior cervical sympathetic cardiac branch (from superior cervical ganglion), and the lower of the two left vagal cervical cardiac branches
   2. Distribution: to deep part of plexus, to right coronary plexus, and to the left anterior pulmonary plexus

B. DEEP (DORSAL) PART lies in front of the tracheal bifurcation above the division of the pulmonary trunk and posterior arch of the aorta
   1. Formation: receives rt. superior, middle, and inferior cervical sympathetic cardiac nerves, the left middle and inferior cervical sympathetic cardiac nerves, the visceral rami from the upper 4 or 5 thoracic sympathetic ganglia, the superior and inferior cervical and thoracic branches of the right vagus, and the superior and thoracic branches of the left vagus
   2. Distribution
      a. Branches from the right half of the deep part
         i. Pass in front of rt. pulmonary a., go to the rt. anterior pulmonary plexus, and continue on to form the rt. coronary plexus to supply the right atrium and ventricle
         ii. Pass behind the rt. pulmonary a., to the rt. atrium, and continue on to help form part of the left coronary plexus
      b. Branches from the left half of the deep plexus connect with the superficial part, give filaments to the left atrium, to the left ant. pulmonary plexus, and continue on to form the greater part of the left coronary plexus. The latter supplies the left atrium and ventricle

III. All cardiac branches of the vagus and sympathetics contain both afferent and efferent fibers, except for the cardiac branch of the superior cervical sympathetic ganglion which contains efferent postganglionic fibers only

A. MOST PAIN FROM THE HEART is carried in visceral afferent fibers contained in sympathetic nerves. Pain is referred to the medial side of the left forearm and hand since this area is covered by nerves from the 8th cervical and 1st thoracic spinal cord segments, the same that receive afferents from the heart

B. THE VAGUS CARRIES visceral afferent fibers from the aortic bodies for reflex control of respiration and from associated arteries for reflex control of heart output

INNERVATION OF HEART

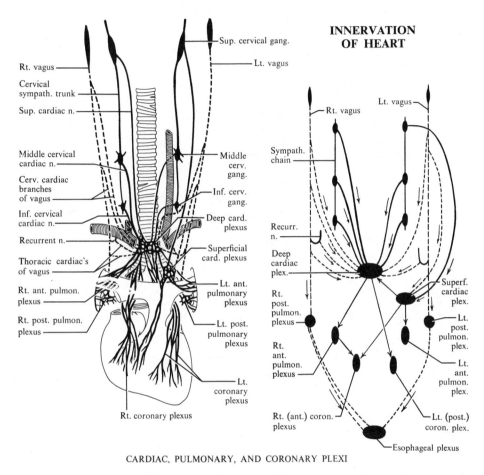

Rt. vagus — **Lt. vagus**

Sup. cervical gang.

Cervical sympath. trunk

Sup. cardiac n.

Middle cervical cardiac n.

Cerv. cardiac branches of vagus

Inf. cervical cardiac n.

Recurrent n.

Thoracic cardiac's of vagus

Rt. ant. pulmon. plexus

Rt. post. pulmon. plexus

Rt. coronary plexus

Middle cerv. gang.

Inf. cerv. gang.

Deep card. plexus

Superficial card. plexus

Lt. ant. pulmonary plexus

Lt. post. pulmonary plexus

Lt. coronary plexus

Sympath. chain

Recurr. n.

Deep cardiac plex.

Rt. post. pulmon. plexus

Rt. ant. pulmon. plexus

Rt. (ant.) coron. plexus

Superf. cardiac plex.

Lt. post. pulmon. plex.

Lt. ant. pulmon. plex.

Lt. (post.) coron. plex.

Esophageal plexus

CARDIAC, PULMONARY, AND CORONARY PLEXI

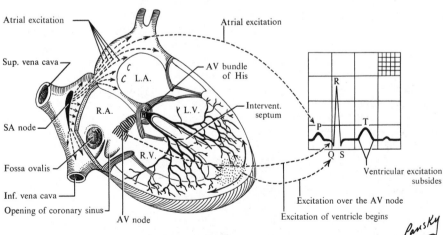

Atrial excitation

Atrial excitation

Sup. vena cava

AV bundle of His

L.A.

R.A.

L.V.

Intervent. septum

SA node

R.V.

Fossa ovalis

Inf. vena cava

Opening of coronary sinus

AV node

Ventricular excitation subsides

Excitation over the AV node

Excitation of ventricle begins

P R T Q S

Pansky

CORRELATION OF ECG WITH CONDUCTION MECHANISM
(After Millard, King, and Showers)

164. SPECIFIC CORONARY ARTERIAL BLOOD SUPPLY TO HEART

I. Introduction: right and left coronary arteries supply the heart and arise from the ventral and left aortic sinuses, respectively. The first few orders of coronary branches supply the epicardium, and subsequent branches penetrate the myocardium. The arteries are supplied by autonomic and sensory fibers of the coronary plexuses. There is no sharp demarcation line between ventricular distribution of the arteries, and preponderance of either right or left coronary supply is variable. Most of the blood in the arteries returns to the heart chambers via veins: some directly by sinusoids in the myocardium or by small branches of arterioles in the endocardium that open directly into the heart chambers. There are also extracardiac anastomoses by which blood may leave

II. Right coronary arises from the anterior (right) aortic sinus

A. EMERGES BETWEEN the pulmonary trunk and right auricle, running in the coronary groove to the back of the heart where it anastomoses with the left coronary

    1. Its first part gives branches to the right ventricle, the first branch supplying the conus arteriosus (this may arise directly from the aorta and is then called the *conus artery*)

    2. A constant marginal branch descends along the right ventricle to the apex

    3. Commonly, the first part of the right coronary gives off a *sinus node artery* which runs upward and medially to supply the right atrium and then circles the level of the opening of the superior vena cava to enter the SA node (the nodal artery may arise from the right coronary, left coronary, or one of its branches)

    4. In its course, the right coronary gives additional branches to the right atrium and right ventricle, and when it reaches the posterior interventricular groove, it gives off several branches, one of which is the *posterior interventricular artery* (it may arise from the left coronary artery)

       a. The posterior interventricular artery courses in the posterior interventricular groove to the apical region where it supplies the adjacent parts of both ventricles and a part of the interventricular septum

       b. The AV nodal artery usually arises from the 1st part of the posterior interventricular artery and sometimes from the right coronary itself

    5. Finally, the right coronary continues across the posterior interventricular groove and anastomoses with the circumflex branch of the left coronary artery

III. Left coronary artery arises from the left aortic sinus behind the pulmonary artery

A. IT COURSES BETWEEN the pulmonary trunk and left auricle, gives off an *anterior interventricular (descending) artery* which supplies the left atrium and continues into the left part of the coronary groove as the *circumflex branch* which anastomoses with the right coronary behind the heart

    1. The anterior interventricular descends in the interventricular groove to the apex of the heart, turns around the apex, and ascends for a variable distance in the posterior interventricular groove where it meets the branches of the posterior interventricular artery. It supplies both ventricles and is the major blood supply to the interventricular septum

    2. The circumflex branch supplies the adjacent parts of the left ventricle (a marginal branch is relatively constant), the left atrium, and often the septum

       a. Posteriorly, in the groove, the circumflex may cross the interventricular groove, supply the AV node, and even give rise to the *posterior interventricular (descending) artery*

IV. There are numerous small arterial and precapillary anastomoses in most heart areas. They are inadequate to provide good collateral circulation if the coronaries or major branches are occluded, but can enlarge during an occlusion that develops slowly

A. MANY VARIATIONS and anomalies of the coronary circulation have been recorded

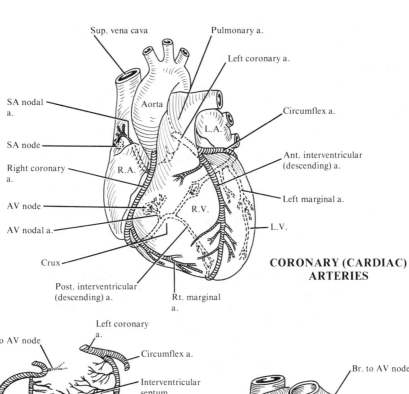

Sup. vena cava

Pulmonary a.

Left coronary a.

SA nodal a.

Aorta

SA node

Circumflex a.

Right coronary a.

L.A.

Ant. interventricular (descending) a.

R.A.

AV node

Left marginal a.

AV nodal a.

R.V.

Crux

L.V.

Post. interventricular (descending) a.

Rt. marginal a.

CORONARY (CARDIAC) ARTERIES

Left coronary a.

Br. to AV node

Circumflex a.

Br. to AV node

Interventricular septum

Rt. coronary a.

Rt. coronary a.

Ant. interventricular a.

Crux

Post. interventricular a.

BLOOD SUPPLY OF INTERVENTRICULAR SEPTUM

Circumflex a.

POSTERIOR VIEW

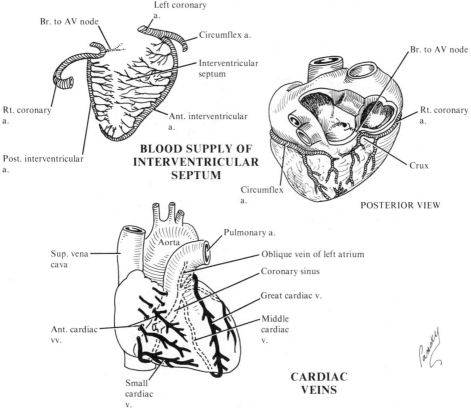

Aorta

Pulmonary a.

Sup. vena cava

Oblique vein of left atrium

Coronary sinus

Great cardiac v.

Ant. cardiac vv.

Middle cardiac v.

Small cardiac v.

CARDIAC VEINS

165. TRACHEA AND BRONCHI

I. Trachea (windpipe)

A. EXTENT: from C6 to upper T5 or T6 vertebrae, where it divides into primary bronchi

B. DIMENSIONS: 9–15 cm long (slightly 1/2 length of esophagus), 2.0–2.5 cm in diameter, larger in the male than in the female, quite mobile

C. SHAPE: nearly cylindrical, slightly flattened posteriorly

D. STRUCTURE: cartilage, muscle, connective tissue, mucous membrane, and glands
   1. Hyaline cartilage: 16–20 "rings" or C-shaped bars, incomplete posteriorly but filled in by muscle and connective tissue where trachea adjoins esophagus.
      a. Sometimes 2 or more are completely or partly fused, sometimes ends are bifid
   2. Trachea has longitudinal elastic fibers that permit it to stretch and descend with lungs during inspiration and recoil with lungs during expiration
   3. Carina: a ridge in the midline, on the inside of trachea, at its bifurcation. Formed by a backward and downward projection of the last tracheal cartilage

E. RELATIONS
   1. In neck: anteriorly, from cephalic to caudal: isthmus of thyroid, inferior thyroid vv., sternohyoid and sternothyroid mm., cervical fascia, and anastomotic veins between the 2 ant. jugular veins. Laterally: common carotid (in sheath), lobes of the thyroid, inf. thyroid aa., and recurrent nn. Posteriorly: the esophagus
   2. In thorax: anteriorly, from superficial to deep: manubrium of sternum, thymus, lt. brachiocephalic v., arch of aorta, brachiocephalic and lt. common carotid aa., and deep cardiac plexus. Laterally: on rt., pleura, rt. vagus, and brachiocephalic a.; on lt., lt. recurrent n., arch of aorta, lt. common carotid, and subclavian aa.

F. SURFACE PROJECTIONS
   1. Bifurcation of trachea behind the sternal angle
   2. Upper part of primary bronchus in second interspace at edge of sternum
   3. Arch of aorta crosses trachea behind manubrium, opposite first interspace

II. Blood supply, lymphatic drainage, and nerves of trachea

A. SUPPLIED MAINLY BY the inferior thyroid arteries; in addition, receives branches from the superior thyroid, bronchial, and internal thoracic arteries

B. DRAINED BY the inferior thyroid veins

C. LYMPHATIC VESSELS drain into adjacent nodes (cervical, tracheal, and tracheobronchial)

D. INNERVATED BY vagus and recurrent laryngeal nerves and sympathetics

III. Primary bronchi

A. RIGHT: 2.5 cm long, wider and shorter and more vertical than left, makes a smaller angle with axis of trachea than does the left bronchus
   1. Azygos v. arches over it; pulmonary a. is at first below it, than anterior to it

B. LEFT: 5.0 cm long, longer and narrower than rt., diverges from tracheal axis at a greater angle than does the right bronchus
   1. Passes under arch of aorta and anterior to esophagus and thoracic aorta

C. BOTH ARE MOBILE, elastic, and have cartilaginous rings which become plates when the bronchi become intrapulmonary at the roots of the lungs

D. BLOOD SUPPLY, lymphatic drainage, and nerve supply
   1. Supplied by bronchial arteries and drained by bronchial veins
   2. Lymphatic vessels drain into bronchopulmonary and tracheobronchial nodes
   3. Nerves similar to trachea and arrive via cardiac and pulmonary plexi

IV. Clinical considerations

A. FOREIGN OBJECTS more likely enter rt. bronchus (larger and presents a less acute angle)

B. EMERGENCY TRACHEOTOMY needed when rima glottidis is closed completely due to spasmodic contraction of laryngeal mm. after severe irritation of mucosa (foreign body, alcohol, etc.). An opening is made in midline, between lower parts of muscular triangles. Below 4th tracheal ring, isthmus of thyroid is avoided.

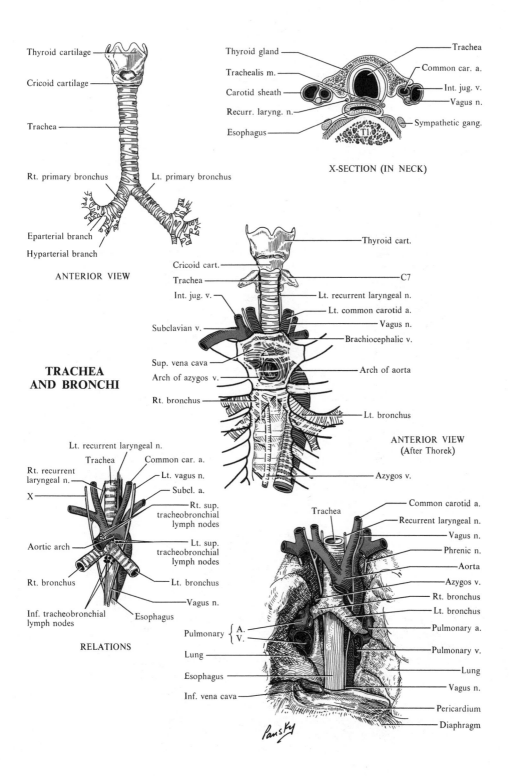

Thyroid cartilage

Cricoid cartilage

Trachea

Rt. primary bronchus   Lt. primary bronchus

Eparterial branch

Hyparterial branch

ANTERIOR VIEW

**TRACHEA
AND BRONCHI**

X-SECTION (IN NECK)

Thyroid gland
Trachealis m.
Carotid sheath
Recurr. laryng. n.
Esophagus

Trachea
Common car. a.
Int. jug. v.
Vagus n.
Sympathetic gang.

T1

Thyroid cart.

Cricoid cart.
Trachea
Int. jug. v.

Subclavian v.

Sup. vena cava
Arch of azygos v.

Rt. bronchus

C7
Lt. recurrent laryngeal n.
Lt. common carotid a.
Vagus n.
Brachiocephalic v.

Arch of aorta

Lt. bronchus

ANTERIOR VIEW
(After Thorek)

Azygos v.

Lt. recurrent laryngeal n.
Trachea   Common car. a.
Rt. recurrent
laryngeal n.
X   Lt. vagus n.
                 Subcl. a.
                 Rt. sup.
                 tracheobronchial
                 lymph nodes
Aortic arch   Lt. sup.
                 tracheobronchial
                 lymph nodes
Rt. bronchus
                 Lt. bronchus
Inf. tracheobronchial   Vagus n.
lymph nodes   Esophagus

RELATIONS

Trachea

Common carotid a.
Recurrent laryngeal n.
Vagus n.
Phrenic n.
Aorta
Azygos v.
Rt. bronchus
Lt. bronchus
Pulmonary a.

Pulmonary { A.
            V.
Lung
Esophagus
Inf. vena cava

Pulmonary v.
Lung
Vagus n.
Pericardium
Diaphragm

Pansky

166. THE ESOPHAGUS

I. **Extent:** from pharynx at level of cricoid cartilage (sixth cervical vertebra) to stomach at level of tenth thoracic vertebra

II. **Course:** generally vertically downward but with 2 curvatures. Deviates to left, remains to left through root of neck, gradually reaches midline at fifth thoracic vertebrae, again shifts toward left as it moves anteriorly toward esophageal hiatus in diaphragm. Has posteroanterior flexures corresponding to curves of vertebral column

III. **Relations**
  A. CERVICAL PORTION
    1. Anterior: trachea. The recurrent nerves ascend in the groove between the trachea and the esophagus
    2. Posterior: vertebral column and prevertebral musculature
    3. Lateral: thyroid gland and carotid sheath
  B. THORACIC PORTION
    1. Anterior: trachea, left bronchus, pericardium, left vagus, and diaphragm
    2. Posterior: vertebral column, longus colli muscle, right aortic intercostal arteries, thoracic duct, right vagus, and aorta (at caudal end)
    3. Left side: aortic arch, left subclavian artery, thoracic duct, pleura, and descending aorta
    4. Right side: azygos vein, pleura, right vagus, and thoracic duct

IV. **Blood supply**
  A. ARTERIES
    1. Cephalically: inferior thyroid branch of thyrocervical trunk
    2. Small branches of thoracic aorta (4 or 5)
    3. Bronchial arteries
    4. Ascending branch from left gastric artery
    5. Ascending branch from inferior phrenic artery
  B. VEINS drain into inferior thyroid, azygos, hemiazygos, and gastric veins
    1. Drainage into gastric veins is one link between portal and systemic systems (see p. 418)

V. **Nerve supply**
  A. RECURRENT NERVES supply the striated muscle in upper third of organ
  B. PARASYMPATHETIC
    1. Esophageal branches: from vagi, above root of lung
    2. Esophageal plexus: below root of lung, vagi split into several bundles, which form a network around esophagus. Postganglionic sympathetic fibers also enter plexus
  C. SYMPATHETIC
    1. From upper 4 or 5 thoracic ganglia directly or through branches from cardiac or aortic plexuses
    2. From lower thoracic ganglia, possibly through splanchnic nerves

VI. **Clinical considerations**
  A. THE VEINS of the esophagus include a plexus on its surface and another in the submucosa. Both sets of vessels anastomose below the diaphragm with veins of the stomach (portal system veins). In consequence, they may become greatly dilated when there is portal hypertension
    1. One cause of death from portal hypertension is esophageal hemorrhage from rupture of varicose submucosal veins

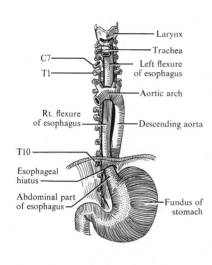

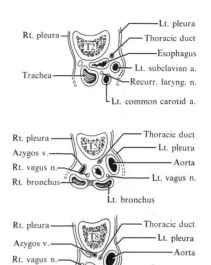

ESOPHAGUS

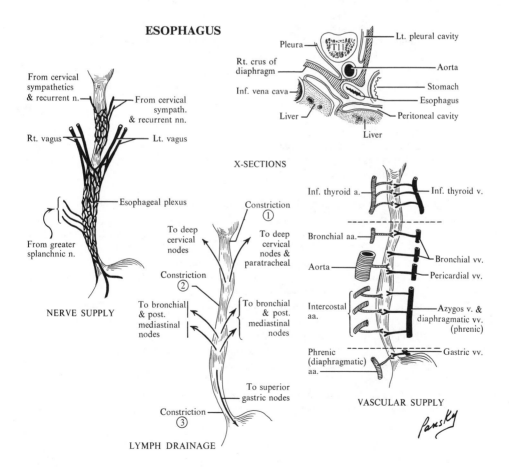

X-SECTIONS

NERVE SUPPLY

LYMPH DRAINAGE

VASCULAR SUPPLY

167. THE POSTERIOR MEDIASTINUM AND THORACIC SYMPATHETICS

I. Vessels

A. THORACIC AORTA (as well as the origins of the intercostal arteries): most posterior structure and situated on the left side

B. AZYGOS VEIN: most posterior structure to the right along with the terminations of the intercostal veins
   1. Hemiazygos vein ascends posteriorly on left and then crosses to join the azygos at the level of T9, passing behind the esophagus, thoracic duct, and aorta

C. THORACIC DUCT lies in front of the bodies of the vertebrae, to the left of midline

II. Viscera

A. ESOPHAGUS descends along the right side of the thoracic aorta and then crosses in front and to the left of the aorta

B. TRACHEA runs in the midline ventral to the esophagus to the level of the upper fifth thoracic vertebra, where it branches into primary bronchi
   1. Right principal bronchus (short and wide): azygos vein arches over it, pulmonary artery at first lies inferior to it and then ventral to it, and the pulmonary vein lies below the artery and bronchus
   2. Left principal bronchus (long and narrow) crosses anterior to esophagus and thoracic aorta, runs at first cephalic, then posterior, and finally caudal to the pulmonary artery. The pulmonary veins are anterior and caudal to the bronchus

C. NERVES
   1. Both vagi
      a. Right runs along trachea behind root of lung through the posterior pulmonary plexus and then to posterior surface of esophagus
      b. Left passes between aorta and left pulmonary artery, then posterior to the root of lung, and finally passes through the posterior pulmonary plexus and down on the anterior surface of esophagus
   2. Splanchnic nerves
      a. Greater arises in cord segments 5–9 and passes through the thoracic sympathetic ganglia and along vertebral border
      b. Lesser arises in cord segments 10–11 and passes through thoracic sympathetic ganglia
      c. Lowest (least, imus) (inconstant) passes through the last thoracic chain ganglion

III. Thoracic portion of sympathetic is not considered to lie in mediastinum

A. COMPONENTS
   1. Series of fusiform ganglia, up to 12 in number, corresponding to each thoracic nerve. Usually, there are fewer due to fusion of 2 or more. In upper thorax, they lie against neck of ribs; more caudally, they lie at sides of vertebrae
   2. Interganglionic cords interconnect the ganglia longitudinally. These cross the intercostal vessels anteriorly. The cords and ganglia are covered by costal pleura

B. PREGANGLIONIC FIBERS arise in the intermediolateral cell column of entire thoracic cord and leave cord through ventral roots of spinal nerves. They leave the spinal nerves through *white rami communicantes* to join thoracic chain. These may synapse in a ganglion at level of origin; they may pass cephalad or caudad, terminating in other ganglia of chain; or they may ascend to cervical trunk or descend into abdomen. Some merely pass through the thoracic chain and go directly to viscera

C. POSTGANGLIONICS: from the thoracic chain they may go as *gray rami* back to the spinal nerves or reach the cardiac, pulmonary, or esophageal plexuses directly

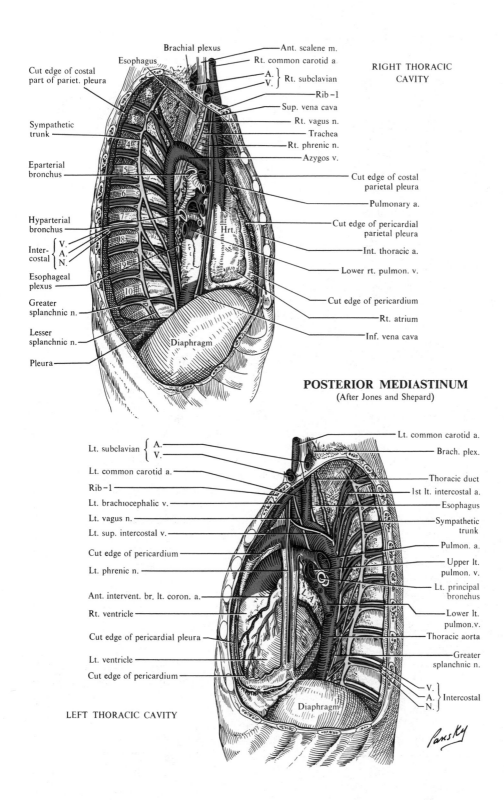

Brachial plexus — — Ant. scalene m.
Esophagus — — Rt. common carotid a.
Cut edge of costal — — A. V. } Rt. subclavian
part of pariet. pleura — — Rib –1
— — Sup. vena cava
— — Rt. vagus n.
Sympathetic — — Trachea
trunk — — Rt. phrenic n.
— — Azygos v.
Eparterial
bronchus — — Cut edge of costal
parietal pleura
— — Pulmonary a.
Hyparterial — — Cut edge of pericardial
bronchus — parietal pleura
Inter- { V. — — Int. thoracic a.
costal { A.
N. — — Lower rt. pulmon. v.
Esophageal
plexus — — Cut edge of pericardium
Greater — — Rt. atrium
splanchnic n. — — Inf. vena cava
Lesser
splanchnic n.
Diaphragm
Pleura

RIGHT THORACIC CAVITY

POSTERIOR MEDIASTINUM
(After Jones and Shepard)

Lt. subclavian { A. — — Lt. common carotid a.
V. — — Brach. plex.
Lt. common carotid a. — — Thoracic duct
Rib –1 — — 1st lt. intercostal a.
Lt. brachiocephalic v. — — Esophagus
Lt. vagus n. — — Sympathetic
trunk
Lt. sup. intercostal v. — — Pulmon. a.
Cut edge of pericardium — — Upper lt.
pulmon. v.
Lt. phrenic n. — — Lt. principal
bronchus
Ant. intervent. br. lt. coron. a. — — Lower lt.
Rt. ventricle — pulmon. v.
— — Thoracic aorta
Cut edge of pericardial pleura — — Greater
splanchnic n.
Lt. ventricle
Cut edge of pericardium — V.
A. } Intercostal
N.
Diaphragm

LEFT THORACIC CAVITY

Pansky

– 367 –

168. THE VENOUS SYSTEM OF THE THORACIC WALL

I. Sympathetic trunk (see p. 366)

II. Veins

A. AZYGOS
1. Origin: in abdomen from the right ascending lumbar vein
2. Course: passes into thorax through aortic hiatus, ascends along right side of vertebral column; at level of fourth thoracic vertebra, it arches over the root of the lung to end in the superior vena cava
3. Tributaries: subcostal; all right intercostal veins, the cephalic 3 or 4 of which form a common stem—the *right superior (highest) intercostal vein;* esophageal, pericardial, rt. bronchial

B. HEMIAZYGOS
1. Origin: from the left ascending lumbar vein of abdomen
2. Course: through crus of diaphragm, ascends along left side of vertebral column to ninth thoracic vertebra, crosses this behind the aorta, esophagus, and thoracic duct to end in azygos vein
3. Tributaries: lower 4 or 5 left intercostal veins, esophageal and mediastinal veins

C. ACCESSORY HEMIAZYGOS
1. Origin: from 3 or 4 left intercostals above those drained by hemiazygos vein
2. Course: descends along left side of vertebral column to end in hemiazygos vein
3. Tributaries: 2 to 5 intercostal. Sometimes left bronchial vein

D. LEFT SUPERIOR (HIGHEST) INTERCOSTAL
1. Origin: from the most cephalic intercostal veins on the left side
2. Course: crosses arch of aorta and ends in the left brachiocephalic vein
3. Tributaries: 2 to 5 intercostal veins, left bronchial vein

III. Thoracic duct (see p. 433): common trunk for entire body except right side of head, neck, thorax, and right upper extremity (these areas drain via the *right lymphatic duct,* a 1- to 2-cm vessel that empties into the venous system at or near the junction of the rt. internal jugular and rt. subclavian veins)

A. ORIGIN: from the cephalic end of the cisterna chyli at the aortic hiatus
B. COURSE: ascends in posterior mediastinum on vertebral bodies, crossing the intercostal arteries and hemiazygos vein. Esophagus and pericardium lie in front. At the level of the fifth thoracic vertebra, it enters the superior mediastinum to pass behind aortic arch and subclavian artery between esophagus and left pleura. In the neck, it arches above clavicle, crosses anterior to subclavian artery, vertebral vessels, and thyrocervical trunk. The left common carotid artery, vagus nerve, and internal jugular vein run anterior to it
C. TERMINATION: junction of left subclavian and internal jugular veins
D. TRIBUTARIES: from upper lumbar nodes, from lymph nodes in lower posterior intercostal spaces, nodes in posterior intercostal spaces of left side, posterior mediastinal nodes, left jugular and subclavian trunks

IV. Clinical considerations

A. SINCE THE THORACIC DUCT IS FILLED with colorless or white lymph, depending on the fat content, it is not as easily seen as is a blood vessel, and it may be injured inadvertently during surgery in the posterior mediastinum. It also may be ruptured by violence
1. Rupture or surgical section of the duct and the pleura overlying it allows the duct to empty its lymph into the pleural cavity (*chylothorax*) at the rate of about 60 to 190 ml per hour, resulting in collapse of the lung and pressure on the heart

B. THE HEMIAZYGOS SYSTEM varies markedly in its development from person to person

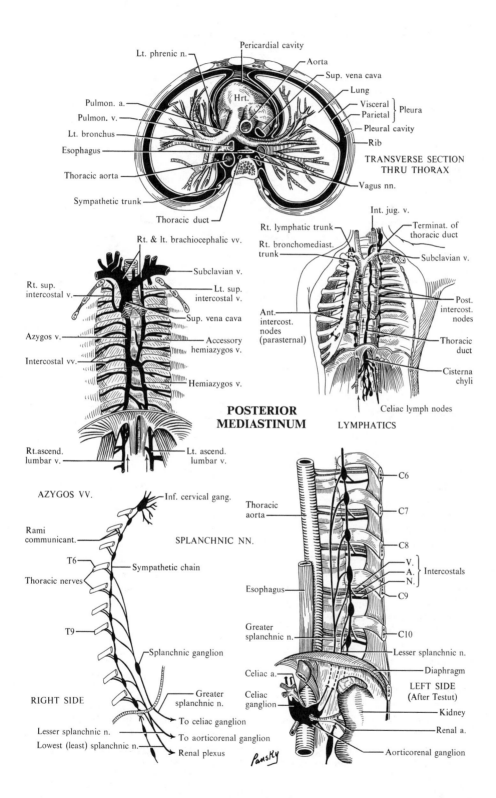

Lt. phrenic n.
Pericardial cavity
Aorta
Sup. vena cava
Lung
Visceral } Pleura
Parietal }
Pleural cavity
Rib
Pulmon. a.
Pulmon. v.
Lt. bronchus
Esophagus
Hrt.
Thoracic aorta
Sympathetic trunk
Thoracic duct
Vagus nn.

**TRANSVERSE SECTION
THRU THORAX**

Rt. & lt. brachiocephalic vv.
Subclavian v.
Rt. sup. intercostal v.
Lt. sup. intercostal v.
Sup. vena cava
Azygos v.
Accessory hemiazygos v.
Intercostal vv.
Hemiazygos v.
Rt. ascend. lumbar v.
Lt. ascend. lumbar v.

AZYGOS VV.

Int. jug. v.
Rt. lymphatic trunk
Terminat. of thoracic duct
Rt. bronchomediast. trunk
Subclavian v.
Ant. intercost. nodes (parasternal)
Post. intercost. nodes
Thoracic duct
Cisterna chyli
Celiac lymph nodes

**POSTERIOR
MEDIASTINUM**

LYMPHATICS

Inf. cervical gang.
Rami communicant.
Sympathetic chain
T6
Thoracic nerves
T9
Splanchnic ganglion
Greater splanchnic n.
To celiac ganglion
To aorticorenal ganglion
Renal plexus
Lesser splanchnic n.
Lowest (least) splanchnic n.

SPLANCHNIC NN.

RIGHT SIDE

Thoracic aorta
C6
C7
C8
V.
A. } Intercostals
N.
Esophagus
C9
Greater splanchnic n.
C10
Lesser splanchnic n.
Celiac a.
Diaphragm
Celiac ganglion
Kidney
Renal a.
Aorticorenal ganglion

LEFT SIDE
(After Testut)

Pansky

169. SUMMARY OF RESPIRATORY MOVEMENTS

I. Movements of the thorax in respiration are based on several anatomic peculiarities

A. THE ANTERIOR ENDS AND MIDDLE BODIES of ribs lie at a more caudal level than their posterior ends

B. THE CURVE OF EACH SUCCESSIVE RIB is greater than that of the one above it. Thus, when the ribs are pulled upward, the diameters of the thorax increase, resulting in increased volume and decreased pressure in the thorax

C. RIBS 1 AND 2 are less mobile than the others and act as a unit with the manubrium. When tension is placed on these, the entire unit is raised, thus increasing the diameter in the superior portion. Raising and fixing of ribs 1 and 2 make possible greater elevation of the ribs below this level, an important feature in forced inspiration

D. THE FREQUENCY OF MOVEMENT of the joints of the thorax is more often than any other joint combinations

E. THE RANGE OF MOVEMENT of any one thoracic joint is small, but any disorder that reduces mobility of the joints interferes with respiration

F. THE 2ND TO 6TH RIBS each moves around 2 axes: at the costovertebral joint in a side-to-side axis, resulting in raising and lowering the ribs' sternal ends, increasing the anteroposterior diameter of thorax (elevates and moves sternum forward in a pump-handle movement); and at the same joint in a front-to-back axis, resulting in a depression or elevation of the middle rib which increases the transverse diameter of the thorax (a bucket-handle movement)

   1. Movements of the 7th to 10th ribs occur about similar axes, but since their cartilages turn up, an elevation of the anterior ends of these ribs tends to be associated with backward movement of the sternum (a pump-handle movement)

II. The action of the muscles in respiration*

A. QUIET RESPIRATION

Raise Ribs	*Lower Ribs*
External intercostal muscles (ribs 3–10)	No muscle, passive

B. DEEP RESPIRATION

Raise Ribs	*Lower Ribs*
External intercostal muscles	No muscle, passive
Scalene muscles	
Sternocleidomastoid muscles	
Levator costarum muscles	
Serratus posterior superior muscles	

C. FORCED RESPIRATION

Raise Ribs	*Lower Ribs*
All the muscles listed above for deep respiration. The levator scapulae, trapezius, and rhomboids raise and fix the scapula so that the pectoral muscles and serratus anterior muscles can raise ribs	Quadratus lumborum, internal intercostals, subcostals, transverse thoracic, and serratus posterior inferior muscles

III. Thoracic volume

A. A MOVEMENT OF ONLY A FEW MM of the bony cage forward, upward, or laterally can increase the volume of the cage by almost one-half liter (volume of air entering or leaving the lungs in quiet respiration)

B. THE DESCENT OF THE DIAPHRAGM, which increases the height of the thoracic cavity, is the other major factor in increasing the thoracic volume

*For the action of the diaphragm in respiration, see p. 436.

BREATHING MECHANISM

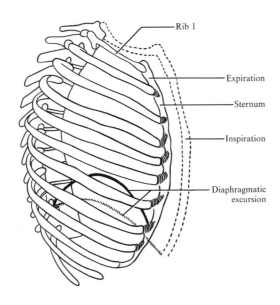

Rib 1

Expiration

Sternum

Inspiration

Diaphragmatic
excursion

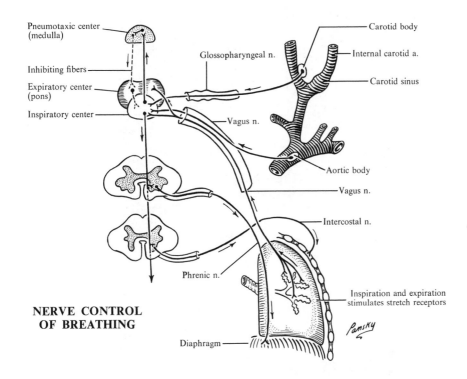

Pneumotaxic center
(medulla)

Carotid body

Glossopharyngeal n.

Internal carotid a.

Inhibiting fibers

Carotid sinus

Expiratory center
(pons)

Inspiratory center

Vagus n.

Aortic body

Vagus n.

Intercostal n.

Phrenic n.

Inspiration and expiration
stimulates stretch receptors

NERVE CONTROL
OF BREATHING

Diaphragm

Pansky

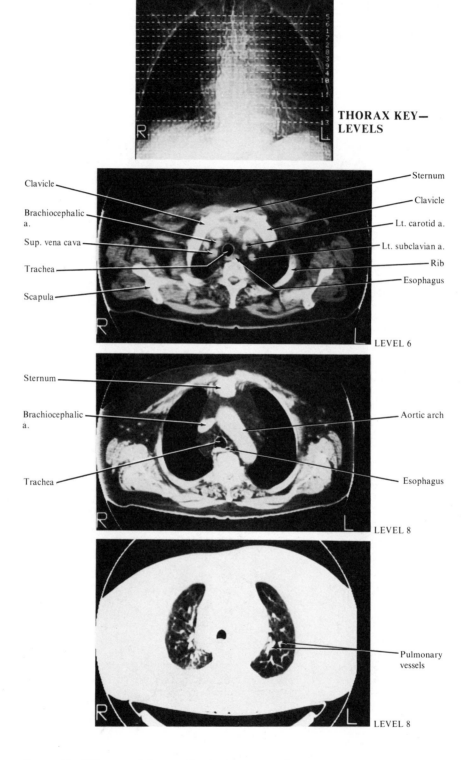

THORAX KEY—LEVELS

Clavicle

Brachiocephalic a.

Sup. vena cava

Trachea

Scapula

Sternum

Clavicle

Lt. carotid a.

Lt. subclavian a.

Rib

Esophagus

LEVEL 6

Sternum

Brachiocephalic a.

Trachea

Aortic arch

Esophagus

LEVEL 8

Pulmonary vessels

LEVEL 8

FIGURE 29. **CT scans of thorax. Key levels and levels 6 and 8.**

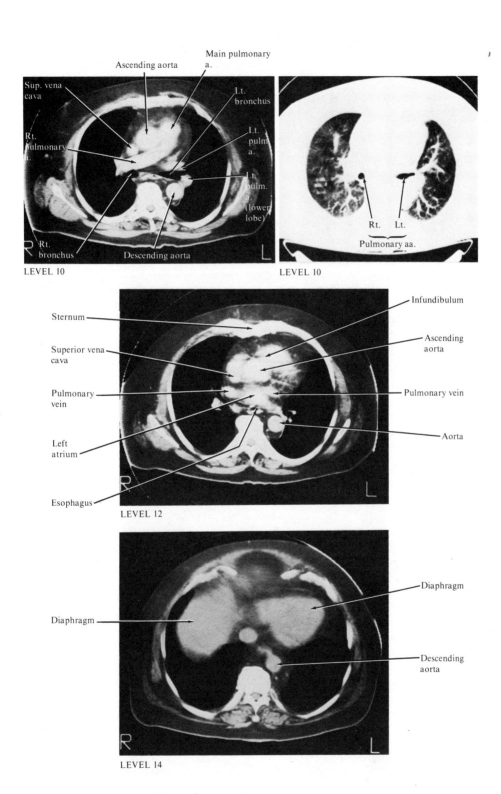

FIGURE 30. **CT scans of thorax. Levels 10, 12, and 14.**

Abdomen and Pelvis

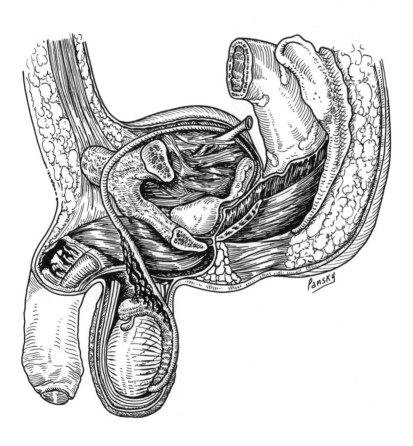

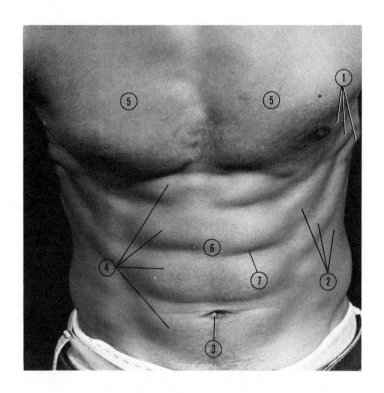

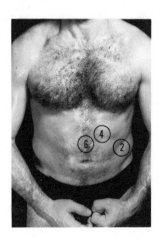

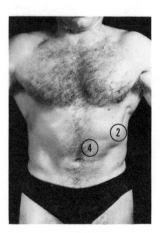

FIGURE 31. **Surface anatomy of abdomen.** *1*, Serratus anterior m.; *2*, external oblique m.; *3*, umbilicus; *4*, rectus abdominis m.; *5*, pectoralis major m.; *6*, linea alba; *7*, tendinous intersections.

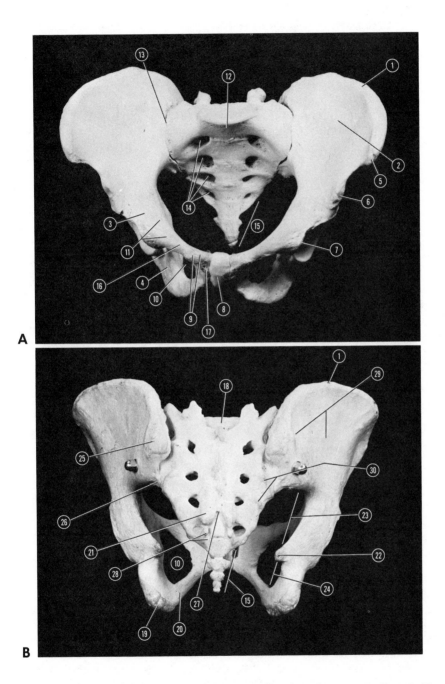

FIGURE 32. **A, The pelvis from in front; B, posterior view of pelvis.** *1*, Iliac crest; *2*, ilium; *3*, pubis; *4*, ischium; *5*, anterior superior iliac spine; *6*, anterior inferior iliac spine; *7*, acetabulum; *8*, symphysis pubis; *9*, pubic tubercle and crest; *10*, obturator foramen; *11*, iliopubic eminence and arcuate line; *12*, promontory; *13*, sacroiliac joint; *14*, pelvic or anterior sacral foramina; *15*, coccygeal bones; *16*, pecten pubis; *17*, inferior ramus of pubis; *18*, vertebral (sacral) canal; *19*, ischial tuberosity; *20*, ramus of ischium; *21*, articular tubercle; *22*, ischial spine; *23*, greater sciatic notch; *24*, less sciatic notch; *25*, posterior superior iliac spine; *26*, posterior inferior iliac spine; *27*, sacral hiatus; *28*, sacral and coccygeal cornua; *29*, gluteal lines; *30*, transverse tubercles of sacrum.

170. TOPOGRAPHY OF THE ABDOMEN

I. Surface lines are all vertical on the thorax
A. MIDSTERNAL: in midline
B. LATERAL STERNAL: line along the lateral margin of the sternum
C. MIDCLAVICULAR: line drawn caudally from point halfway between the middle of the jugular notch and the tip of the acromion
D. PARASTERNAL: line about halfway between A and C, above

II. Surface lines or planes: on abdomen
A. HORIZONTAL
  1. Transpyloric: halfway between jugular notch and upper border of symphysis; crosses tips of 9th costal cartilages anteriorly and lower 1st lumbar vertebra posteriorly
  2. Transtubercular (intertubercular): at level of iliac tubercles; crosses the body of the fifth lumbar vertebra posteriorly
  3. Subcostal: joins lowest point of costal margin on each side; indicates inferior margin of 10th costal cartilages; lies at level of intervertebral disk between L2 and L3 vertebrae
  4. Transumbilical: passes through the umbilicus
  5. Interspinous: passes through rt. and lt. ant. sup. iliac spines and sacral promontory
  6. Supracrestal: passes between highest points of the iliac crests at level of spinous process of L4 vertebra
B. VERTICAL
  1. Midsagittal: directly in the midline of the body
  2. Midclavicular or lateral (2): drawn upward from middle of inguinal ligament to middle of clavicles

III. Zones created by lines or planes: on abdomen
A. EPIGASTRIC: above transpyloric, between the 2 midclavicular lines
B. HYPOCHONDRIAC (RT. AND LT.): above transpyloric, but lateral to rt. and lt. vertical
C. UMBILICAL: between transpyloric, transtubercular, and the 2 vertical lines
D. LATERAL (RT. AND LT.): between the 2 horizontal lines, but lateral to the vertical lines
E. HYPOGASTRIC: below transtubercular line and between vertical lines
F. ILIAC (RT. AND LT.): below transtubercular, but lateral to the vertical lines

IV. Surface projections and locations of certain major viscera
A. STOMACH: *cardiac orifice* is behind 7th costal cartilage, 2.5 cm lateral to lt. border of sternum; *pylorus* is on transpyloric line, 1 cm to the right of the midline
B. DUODENUM: *superior part* lies on the transpyloric line to the right of the midline; *duodenojejunal flexure* lies on the transpyloric line, 2.5 cm to the left of the midline
C. ILEOCECAL JUNCTION: just below and medial to the junction of the right vertical and transtubercular lines
D. BASE OF APPENDIX: on the right vertical line at the level of the anterior iliac spine
E. LIVER: upper limit of right lobe is at the xiphisternal junction, in midline; this line is continued to the right to the 5th costal cartilage in the midclavicular line and then curves to the right and down to the 7th rib at the side of the thorax. The right border continues downward to a point 1 cm below the costal arch. The upper left limit extends from the xiphisternal point to the left 6th costal cartilage, 5 cm from the midline. The lower limiting line runs upward, parallel to, and 1 cm below the thoracic border to the 9th costal cartilage. It then extends obliquely upward toward the left, crossing the midline just above the transpyloric plane to the 8th costal cartilage. From here, a line, slightly convex to the left, is drawn to the upper left limit
F. FUNDUS OF GALLBLADDER: just behind the 9th right costal cartilage
G. UMBILICUS: most obvious marking on abdominal wall; a puckered scar representing former site of attachment of umbilical cord in fetus. Position varies, but is usually at level of intervertebral disk between L3 and L4

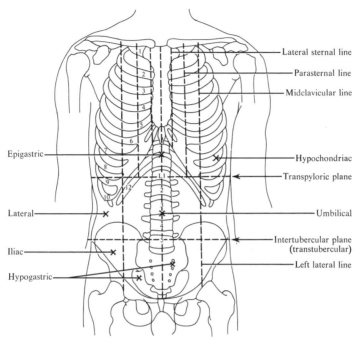

Lateral sternal line
Parasternal line
Midclavicular line

Epigastric
Hypochondriac
Transpyloric plane

Lateral
Umbilical

Intertubercular plane
(transtubercular)

Iliac
Left lateral line

Hypogastric

**SURFACE LINES AND REGIONS
OF ABDOMEN AND THORAX**

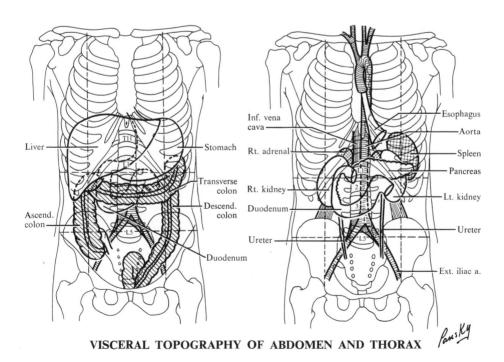

Liver
Stomach

Transverse
colon

Descend.
colon

Ascend.
colon

Duodenum

Inf. vena
cava
Esophagus

Aorta

Rt. adrenal
Spleen

Pancreas

Rt. kidney
Lt. kidney

Duodenum

Ureter
Ureter

Ext. iliac a.

VISCERAL TOPOGRAPHY OF ABDOMEN AND THORAX

Pansky

171. SUPERFICIAL VEINS OF THORAX AND ABDOMEN

I. Thorax has superficial veins, which communicate freely with those on abdominal wall

A. VENTRAL AND LATERAL CHEST WALL are drained by
1. Lateral thoracic vein receives tributaries from the ventral wall, including the venous plexus of the mammary gland. Terminates in the axillary vein
2. Costoaxillary vein receives tributaries mainly from lateral aspect of chest. Terminates in the lateral thoracic vein
3. Thoracoepigastric vein drains lower lateral wall. Terminates in both the lateral thoracic and superficial epigastric veins

B. MEDIAL CHEST WALL is drained by vessels that penetrate the intercostal spaces to terminate in
1. The internal thoracic (mammary) veins and their anterior intercostal tributaries
2. Superior epigastric vein

II. Abdomen has extensive venous plexuses, which communicate with other systems, especially with those of thorax, thigh, and genital region

A. INFERIOR LATERAL PARTS are drained by
1. Superficial circumflex iliac vein, which ends in the great saphenous vein

B. MOST OF THE ABDOMINAL WALL is drained by
1. Superficial epigastric vein, which also directly communicates with the lateral thoracic vein through the thoracoepigastric vein. It terminates in the great saphenous vein

C. LOWER AND MEDIAL PART OF THE ABDOMEN AND THE EXTERNAL GENITALIA are drained by
1. Superficial external pudendal vein, which terminates in the great saphenous vein

III. Clinical considerations

A. THERE IS AN EXTENSIVE anastomosing network of superficial veins in the tela subcutanea of the abdominal wall
1. The veins above the umbilicus anastomose with those directly below it and through the thoracoepigastric v. which extends along the lateral aspect of the abdominal wall from the inguinal region to the axilla
   a. They also communicate with the paraumbilical venous plexus which is connected to the portal vein through the round ligament of the liver
   b. The veins above the umbilicus drain to the superior vena cava by connections with the intercostal and internal thoracic veins or those going to the axilla
2. The veins of the abdominal wall are direct tributaries of the caval system of veins
3. The veins below the umbilicus drain into the femoral and then into the inferior vena cava

B. OCCLUSION OF THE INFERIOR VENA CAVA by an abdominal tumor or disease causing liver enlargement: venous blood may bypass the block by flowing through the above-mentioned superficial venous plexuses, especially the link that exists between the superficial epigastric through the thoracoepigastric veins leading into the lateral thoracic vein and eventually terminating in the axillary system of veins

C. AT THE UMBILICUS, the superficial epigastric vein connects with veins that are, in turn, connected with the portal system. In consequence, when either the portal system or the inferior vena cava is obstructed at certain levels, the veins around the umbilicus that connect the two systems may become enlarged and tortuous, forming the *caput medusae*
1. If the reversal of blood flow progresses, it will produce varicose dilation of other veins of the abdominal wall, which anastomose with the paraumbilical veins
2. A similar situation ensues when there is obstruction of the superior vena cava
3. The anastomosis between the veins above and below the umbilicus permits the backing up of blood into the abdominal venous network in an effort to establish a collateral venous circulation that can convey blood to the vena cava which is not obstructed. The abdominal and superficial thoracic veins become markedly dilated, tortuous, and varicosed
   a. The examiner can determine which vena cava (sup. or inf.) is obstructed and the direction of blood flow by stripping a segment of dilated vein between his fingers. When the fingers are lifted alternately, the direction from which the vein refills most rapidly indicates the direction of blood flow

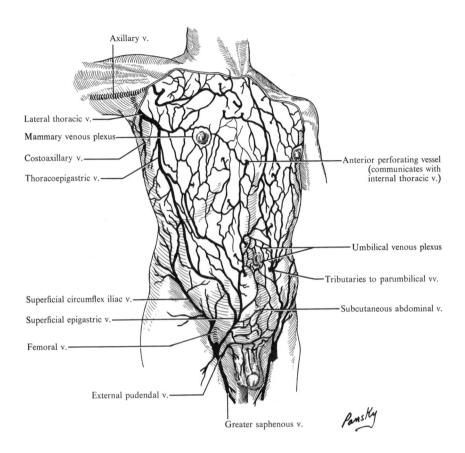

Axillary v.

Lateral thoracic v.

Mammary venous plexus

Costoaxillary v.

Thoracoepigastric v.

Anterior perforating vessel
(communicates with
internal thoracic v.)

Umbilical venous plexus

Tributaries to parumbilical vv.

Superficial circumflex iliac v.

Superficial epigastric v.

Subcutaneous abdominal v.

Femoral v.

External pudendal v.

Greater saphenous v.

Pansky

**SUPERFICIAL VEINS OF
ANTERIOR THORAX AND ABDOMEN**
(After Jones and Shepard)

172. THE MUSCLES OF THE ABDOMEN

I.

Name	Origin	Insertion	Action	Nerve
External abdominal oblique	Lower 8 ribs	Aponeurosis to linea alba from xiphoid to symphysis	Compresses abdomen Flexes and laterally rotates spine Depresses ribs	Intercost. 8–12 Iliohypogastric Ilioinguinal
Internal abdominal oblique	Lat. inguinal lig. Ant. iliac crest Lumbar aponeurosis (of thoracolumbar fascia)	Lower border of ribs 9–12 With aponeurosis to linea alba Pubis and pectineal line	Same as above	Same as above
Transversus abdominis	Lat. inguinal lig. Iliac crest Thoracolumbar fascia Cartilages of lower ribs	Through aponeur. to linea alba Pubis and pectineal line	Compresses abdomen Depresses ribs	Intercost. 7–12 and same as above
Rectus abdominis	Crest of pubis Interpubic lig.	Cartilages of ribs 5–7 Xiphoid process	Compresses abdomen Flexes spine	Intercost. 7–12
Quadratus lumborum	Iliolumbar lig. Post. iliac crest	12th rib Tips of transverse processes L1–4	Fixes last 2 ribs Flexes spine laterally	T12, L1
Psoas major	Trans. processes of L1–5 Sides of bodies and fibrocartilage T12–L5	Lesser trochanter of femur	Flexes thigh and rotates it laterally Flexes spine and bends spine laterally	L2–3
Iliacus	Upper iliac fossa Iliac crest Ant. sacroiliac lig. Base of sacrum	Tendon of psoas maj. and directly to lesser trochanter	Flexes and laterally rotates thigh	Femoral (L2–3)
Pyramidalis	Pubis and ant. pubic lig.	Linea alba	Tenses linea alba	T12
Cremaster	Middle inguinal lig. Int. oblique m.	Tubercle and crest of pubis	Draws up testis	Genitofemoral

II. Special features

A. ACTION OF THE PSOAS MAJOR and iliacus muscles as rotators is controversial. When the limb is free, the iliopsoas probably acts as a lateral rotator of thigh; when the limb is fixed, it may help to swing the pelvis as the other limb goes forward

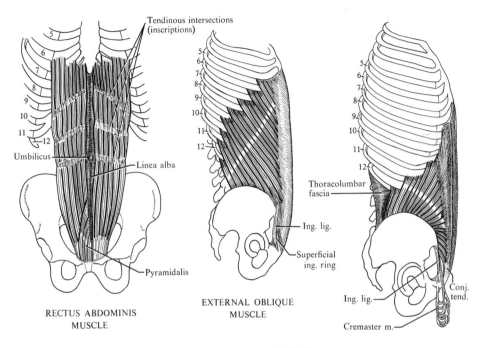

Tendinous intersections (inscriptions)

5
6
7
8
9
10
11
12
Umbilicus
Linea alba
Pyramidalis

RECTUS ABDOMINIS MUSCLE

5
6
7
8
9
10
11
12
Ing. lig.
Superficial ing. ring

EXTERNAL OBLIQUE MUSCLE

5
6
7
8
9
10
11
12
Thoracolumbar fascia
Ing. lig.
Cremaster m.
Conj. tend.

INTERNAL OBLIQUE MUSCLE

ABDOMINAL MUSCLES

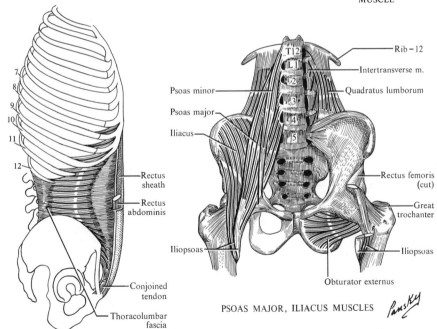

7
8
9
10
11
12
Rectus sheath
Rectus abdominis
Conjoined tendon
Thoracolumbar fascia

TRANSVERSUS ABDOMINIS MUSCLE

T12
L1
2
3
4
5
Psoas minor
Psoas major
Iliacus
Iliopsoas
Rib – 12
Intertransverse m.
Quadratus lumborum
Rectus femoris (cut)
Great trochanter
Iliopsoas
Obturator externus

PSOAS MAJOR, ILIACUS MUSCLES

173. ABDOMINAL APONEUROSIS, RECTUS SHEATH, AND DEEP EPIGASTRIC VESSELS

I. The aponeurosis of the external abdominal oblique muscle is a strong membrane of collagenous fibers, which are the major part of the tendon of insertion of that muscle. It covers the entire abdomen and forms part of the anterior wall of the rectus sheath

A. WHERE THE FIBERS FROM BOTH SIDES interlace in the midline, the *linea alba* is formed

B. INGUINAL LIGAMENT is caudal to the border of the aponeurosis and extends from the anterior superior iliac spine to the pubic tubercle. It is the lowest part of the external oblique aponeurosis

C. LACUNAR LIGAMENT: where the medial end of the inguinal ligament is curved under the spermatic cord to attach to the pectineal line beyond the tubercle

D. REFLECTED INGUINAL LIGAMENT: where the fibers pass upward and medially from the bony attachment of the lacunar ligament to reach the linea alba

E. SUPERFICIAL (SUBCUTANEOUS) INGUINAL RING: triangular gap in the aponeurosis, above and lateral to the pubis for the passage of the spermatic cord. Its sides are the *lateral* and *medial crura*

F. LINEA SEMILUNARIS: slight surface depressions that indicate the lateral borders of the rectus abdominis muscles

G. LINEA TRANSVERSAE: transverse grooves overlying the *tendinous insertions* of the rectus abdominis muscles, extend laterally from the linea alba

II. Rectus sheath

A. BASIC FORMATION: composed of the aponeurosis of the external and internal oblique and transversus abdominis muscles. At the lateral border of the rectus abdominis, *posterior* and *anterior layers* are formed, which pass on either surface of the rectus muscle

B. BECAUSE OF THE PECULIARITIES OF ARRANGEMENT, the sheath is best described regionally
   1. Cephalic part (above arcuate line)
      a. Aponeurosis of the external oblique passes anterior
      b. Aponeurosis of internal oblique splits, 1 layer passes anterior, the other posterior
      c. Aponeurosis of transversus runs entirely posterior to rectus
      d. Transversalis fascia runs posterior to rectus
   2. Caudal part (below arcuate line)
      a. Aponeurosis of external oblique passes anterior to rectus
      b. Aponeurosis of internal oblique passes anterior to rectus
      c. Aponeurosis of transversus passes anterior to rectus
      d. Transversalis fascia remains posterior to rectus

C. ARCUATE LINE (SEMICIRCULAR LINE OF DOUGLAS). This is a line on the posterior wall of the rectus sheath below which only the transversalis fascia makes up the posterior layer of the rectus sheath; usually midway between umbilicus and pubic crest

D. STRUCTURES IN SHEATH: sup. and inf. epigastric vessels, terminal parts of lower 5 intercostal and subcostal vessels and nerves

III. Epigastric arteries

A. SUPERIOR descends as 1 of the terminal branches of the internal thoracic artery, passes anterior to the diaphragm, and enters rectus sheath. In the sheath, it is at first behind and then enters the muscle. It anastomoses with the inferior epigastric artery

B. INFERIOR arises from the external iliac artery, above the inguinal ligament, courses upward and medially, pierces transversalis fascia, and enters the posterior rectus sheath. As it ascends, it lies medial to the deep inguinal (internal abdominal) ring

IV. Special features

A. THERE HAVE BEEN MANY DIVERSE OPINIONS on the exact anatomy and on the preferred names of structures in the region of insertion of the internal oblique and transversus muscles, particularly on whether a *conjoined tendon or falx inguinalis* (in the sense of a combination of the two mentioned muscles) is ever formed lateral to the lateral border of the rectus muscle

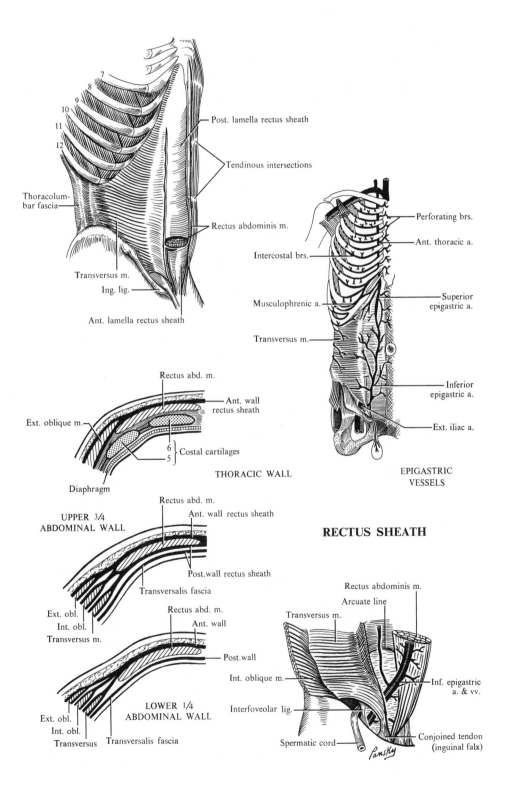

7
8
9
10
11
12

Post. lamella rectus sheath

Tendinous intersections

Thoracolumbar fascia

Rectus abdominis m.

Transversus m.

Ing. lig.

Ant. lamella rectus sheath

Perforating brs.

Ant. thoracic a.

Intercostal brs.

Musculophrenic a.

Transversus m.

Superior epigastric a.

Inferior epigastric a.

Ext. iliac a.

Rectus abd. m.

Ant. wall rectus sheath

Ext. oblique m.

6
5 } Costal cartilages

Diaphragm

THORACIC WALL

EPIGASTRIC VESSELS

UPPER 3/4 ABDOMINAL WALL

Rectus abd. m.

Ant. wall rectus sheath

Post.wall rectus sheath

Transversalis fascia

Ext. obl.
Int. obl.
Transversus m.

RECTUS SHEATH

Rectus abd. m.

Ant. wall

Post.wall

Ext. obl.
Int. obl.
Transversus

LOWER 1/4 ABDOMINAL WALL

Transversalis fascia

Rectus abdominis m.

Arcuate line

Transversus m.

Int. oblique m.

Inf. epigastric a. & vv.

Interfoveolar lig.

Spermatic cord

Conjoined tendon (inguinal falx)

Pansky

174. NERVES OF THE ANTERIOR ABDOMINAL WALL

I. **Introduction:** the skin and muscles of the anterior abdominal wall are supplied almost entirely by the continuation of the thoracoabdominal (lower intercostal: T7–T11), subcostal (T12), iliohypogastric (L1), and ilioinguinal (L1) nerves

A. THE 7TH TO 11TH intercostal nerves supply the thoracic as well as abdominal walls

B. THE MAIN NERVE TRUNKS of the intercostal nerves pass first between the chondral attachments of the diaphragm and transversus abdominis muscles, then pass forward from the intercostal spaces, and run between the internal oblique and transversus abdominis mm. to supply the abdominal mm. and overlying skin

    1. A common nerve supply explains the contraction of the abdominal mm. when cold hands are applied to the abdomen

C. THE ANTERIOR CUTANEOUS BRANCHES of these nerves pierce the rectus sheath a short distance from the midline and supply the skin of the anterior abdominal wall

D. THE ANTERIOR CUTANEOUS BRANCHES of T7–T9 supply the skin above the umbilicus

E. T10 INNERVATES the skin at the umbilical level

F. T11, T12, AND L1 supply the skin below the umbilicus

    1. L1 appears just above the inguinal ligament as the iliohypogastric nerve

G. THE THORACOABDOMINAL NERVES supply intercostal, subcostal, serratus posterior inferior, transversus abdominis, external and internal oblique, and rectus abdominis muscles and give sensory twigs to the adjacent diaphragm, pleura, and peritoneum

II. **Thoracoabdominal nerves (T7–T11 intercostals)** leave the intercostal spaces and course downward and forward between the internal oblique and transversus mm., supplying them as well as the external oblique muscle

A. AS THEY ENTER THE RECTUS SHEATH, branches turn forward to supply the rectus abdominis muscle and overlying skin

    1. The anterior cutaneous branch pierces the rectus and anterior layer of the sheath to supply the overlying skin

    2. The lateral cutaneous branch pierces the external oblique and divides into anterior and posterior branches to supply the skin of back, side, and front of the abdominal wall

B. A VERTICAL INCISION along the linea semilunaris denervates the rectus abdominis m.

    1. An incision through the middle of the rectus only denervates its medial part

III. **Subcostal nerve (T12)** enters the abdomen behind the lateral arcuate ligament, courses downward and laterally behind the kidney, pierces the transversus abdominis, and passes between it and the internal oblique. It enters the rectus sheath, turns forward through its anterior layer, and becomes superficial about halfway between the umbilicus and pubic symphysis

A. LATERAL CUTANEOUS BRANCH supplies the skin and subcutaneous tissue of the gluteal region and lateral thigh as far as the greater trochanter

B. IT SUPPLIES PARTS of the transversus, oblique, and rectus muscles, pyramidalis, as well as the adjacent peritoneum

IV. **Iliohypogastric and ilioinguinal nerves (L1)** are predominantly cutaneous nerves

A. ILIOINGUINAL enters inguinal canal and accompanies the spermatic cord (or round ligament of uterus) to the scrotum (or labia majora)

    1. Cutaneous branches are sent to the thigh, and there are also anterior scrotal or anterior labial branches

B. ILIOHYPOGASTRIC divides into lateral and anterior cutaneous branches. The former supplies the skin over the side of the buttock; the latter supplies the skin above the pubis

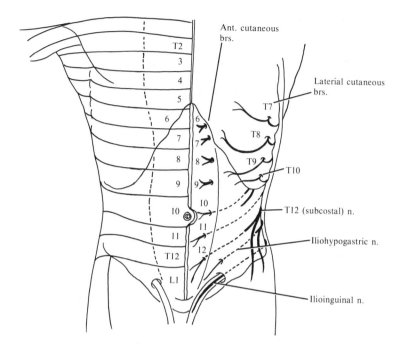

CUTANEOUS DISTRIBUTION OF THORACO-ABDOMINAL NERVES

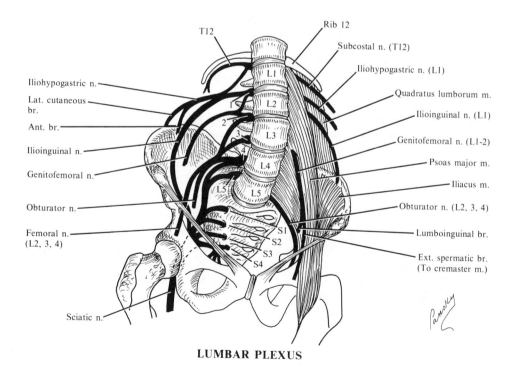

LUMBAR PLEXUS

175. SPERMATIC CORD, INGUINAL CANAL, AND HERNIA

I. Special features of the lower abdominal wall (see p. 384)

A. CONJOINED TENDON (INGUINAL FALX): fibers of the internal abdominal oblique muscle, which arise from the lateral end of the inguinal ligament, arch over the spermatic cord (or round ligament) and terminate in a tendinous band, common to it and the transversus abdominis. It inserts on the pubis and medial pectineal line dorsal to the lacunar ligament

B. CREMASTER MUSCLE arises from the inguinal ligament as the lowermost fibers of the internal oblique muscle. Sends long loops along the spermatic cord into the scrotum. Inserts on the pubis

C. TRANSVERSALIS FASCIA: the deep fascia that lines the entire abdomen deep to the transversus abdominis muscle
   1. Below, arcuate (semicircular) line is thickened and forms the posterior lamina of the rectus sheath

D. INTERFOVEOLAR LIGAMENT: a thickening of the transversalis fascia forming the medial boundary of the internal abdominal ring
   1. Divides fossa above the inguinal ligament into *medial* and *lateral inguinal fossae* (*fovea*)
   2. In the medial fovea is *Hesselbach's triangle,* bounded medially by the edge of the rectus abdominis, laterally by the inferior epigastric artery, and below by the inguinal ligament

II. Inguinal canal

A. EXTENT: 4 cm, from deep (abdominal) inguinal ring to subcutaneous ring

B. COURSE: oblique, running parallel to and just above the inguinal ligament

C. WALLS
   1. Anterior: aponeurosis of external oblique along entire length; internal oblique over lateral part
   2. Posterior: from medial to lateral—reflected inguinal ligament, inguinal falx, and transversalis fascia
   3. Cephalic: arched fibers of the internal oblique and transversus abdominis muscles
   4. Caudal: inguinal ligament and lacunar ligament

D. CONTENTS: spermatic cord (ductus deferens or round ligament, deferential vessels, testicular artery, pampiniform plexus of veins, lymphatics, and autonomic nerves), ilioinguinal and genital branch of genitofemoral nerve, cremasteric artery and muscle, and internal spermatic fascia

III. Coverings of spermatic cord below subcutaneous ring

A. EXTERNAL SPERMATIC FASCIA: prolongation of the deep abdominal fascia from the external oblique muscle

B. CREMASTERIC (MIDDLE SPERMATIC) FASCIA: derived from the internal oblique muscle and its fascia

C. INTERNAL SPERMATIC FASCIA: from the transversalis fascia at deep ring

IV. Hernia

A. INDIRECT (LATERAL): sac of hernia leaves abdomen in lateral fovea, through deep ring, lateral to the inferior epigastric vessels. Passes entire length of inguinal canal

B. DIRECT (MEDIAL): sac pushes forward from abdomen in inguinal (Hesselbach's) triangle, medial to the inferior epigastric vessels, and only passes through the medial end of the inguinal canal to the subcutaneous ring

C. AN INGUINAL HERNIA typically contains part of a viscus, most commonly a part of the small or large intestine, and an associated peritoneal diverticulum or sac

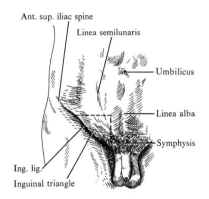

Ant. sup. iliac spine

Linea semilunaris

Umbilicus

Linea alba

Symphysis

Ing. lig.

Inguinal triangle

INGUINAL TOPOGRAPHY

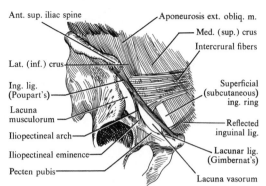

Ant. sup. iliac spine

Aponeurosis ext. obliq. m.

Med. (sup.) crus

Intercrural fibers

Lat. (inf.) crus

Ing. lig. (Poupart's)

Lacuna musculorum

Iliopectineal arch

Iliopectineal eminence

Pecten pubis

Superficial (subcutaneous) ing. ring

Reflected inguinal lig.

Lacunar lig. (Gimbernat's)

Lacuna vasorum

**INGUINAL LIG.
SUPERFICIAL INGUINAL RING**

LOWER ABDOMINAL WALL

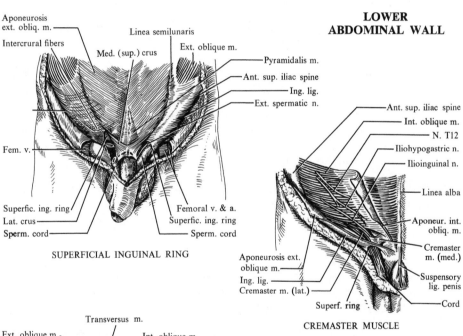

Aponeurosis ext. obliq. m.

Intercrural fibers

Med. (sup.) crus

Linea semilunaris

Ext. oblique m.

Pyramidalis m.

Ant. sup. iliac spine

Ing. lig.

Ext. spermatic n.

Fem. v.

Superfic. ing. ring

Lat. crus

Sperm. cord

Femoral v. & a.

Superfic. ing. ring

Sperm. cord

SUPERFICIAL INGUINAL RING

Ant. sup. iliac spine

Int. oblique m.

N. T12

Iliohypogastric n.

Ilioinguinal n.

Linea alba

Aponeur. int. obliq. m.

Cremaster m. (med.)

Suspensory lig. penis

Cord

Aponeurosis ext. oblique m.

Ing. lig.

Cremaster m. (lat.)

Superf. ring

CREMASTER MUSCLE

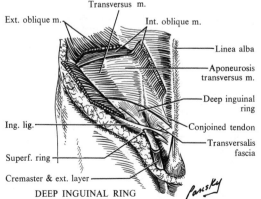

Transversus m.

Ext. oblique m.

Int. oblique m.

Linea alba

Aponeurosis transversus m.

Deep inguinal ring

Conjoined tendon

Transversalis fascia

Ing. lig.

Superf. ring

Cremaster & ext. layer

Pansky

DEEP INGUINAL RING

② Int. oblique m. (cremaster)

① Transversalis fascia

Cord

③ Ext. oblique m.

③ Ext. spermatic fascia

② Cremasteric (middle spermatic) fascia

Deep ring

Superf. ring

Cord

① Int. spermatic fascia

CORD LAYERS

176. PERITONEUM

I. Definition: the serous layer of the abdomen. Has a parietal and a visceral layer

II. Peritoneal cavity: potential space between the layers of the peritoneum

III. Ligaments and mesenteric folds of peritoneum
  A. FALCIFORM LIGAMENT AND LIGAMENTUM TERES OF LIVER (see p. 392)
  B. CORONARY LIGAMENTS OF LIVER
     1. Anterior layer: continuous with right side of falciform ligament, where peritoneum is reflected from the diaphragm to right lobe of liver
     2. Posterior layer: from the back of right lobe to right suprarenal and kidney
  C. LEFT AND RIGHT TRIANGULAR: where ant. and post. coronary ligs. meet
     1. Anterior layer: continuous with left side of falciform; the peritoneum is reflected from the diaphragm to left lobe of liver
     2. Posterior layer: posterior to above, where peritoneum is reflected from the left lobe of liver to diaphragm. The 2 layers join in a sharp fold at the left
  D. HEPATOGASTRIC AND HEPATODUODENAL: this is the *lesser omentum,* extending from the hepatic porta to the stomach and duodenum

IV. Bare area of liver: an area of the liver not covered by peritoneum, chiefly lies between the layers of the coronary ligaments

V. Great omentum: double layer of peritoneum, hangs from greater curvature of stomach, crosses transverse colon, and descends in front of abdominal viscera
  A. GASTROCOLIC LIGAMENT: the omentum between the stomach and transverse colon
  B. SPLENORENAL (PHRENICOLIENAL) LIGAMENT: dorsal mesentery from left kidney to spleen
  C. GASTROSPLENIC (GASTROLIENAL) LIGAMENT: dorsal mesentery joining spleen to stomach

VI. Transverse mesocolon: dorsal mesentery of transverse colon

VII. Phrenicocolic ligament: fold of peritoneum from the left colic flexure to diaphragm, helps to support spleen (sustentaculum lienis)

VIII. Mesentery proper: double sheet of peritoneum; broad, fanlike fold of peritoneum suspending jejunum and ileum from the posterior body wall

IX. Omental bursa (lesser sac): diverticulum from main cavity, behind stomach
  A. BOUNDARIES: anteriorly, caudate lobe of liver, lesser omentum, stomach, great omentum; posteriorly, great omentum, transverse colon and mesocolon, left suprarenal and left kidney; to the right, opens into the greater sac; to the left are the phrenicocolic ligament, hilum of spleen, and gastrosplenic (gastrolienal) ligament
  B. SUBDIVISIONS: vestibule, between the omental (epiploic) foramen and gastropancreatic fold; superior recess, between caudate lobe of liver and diaphragm; lienal recess, between spleen and stomach; inferior recess includes all the remaining lower part

X. Omental (epiploic) foramen: opening between the main cavity and lesser sac
  A. BOUNDARIES: anteriorly, edge of lesser omentum; posteriorly, inferior vena cava; cephalically, caudate lobe of liver; caudally, duodenum
     1. In free edge of lesser omentum are found: common bile duct to the right with the hepatic artery just to the left of the duct; the portal vein is behind both

XI. Clinical considerations
  A. PERITONITIS: an inflammation of the peritoneum as a result of infection. If patient survives, adhesions may result
  B. ASCITES: an accumulation of serous fluid in the peritoneal cavity

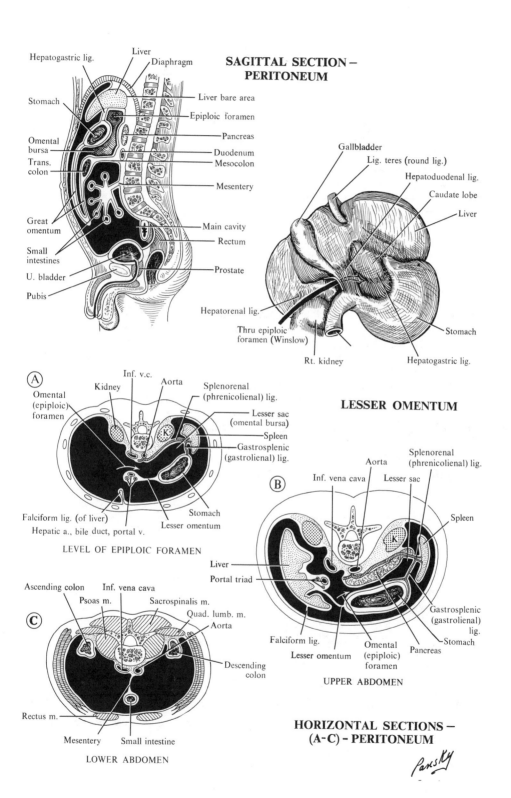

SAGITTAL SECTION – PERITONEUM

Hepatogastric lig.
Liver
Diaphragm
Stomach
Liver bare area
Epiploic foramen
Pancreas
Omental bursa
Duodenum
Trans. colon
Mesocolon
Mesentery
Great omentum
Main cavity
Small intestines
Rectum
U. bladder
Prostate
Pubis

Gallbladder
Lig. teres (round lig.)
Hepatoduodenal lig.
Caudate lobe
Liver

Hepatorenal lig.
Thru epiploic foramen (Winslow)
Rt. kidney
Stomach
Hepatogastric lig.

LESSER OMENTUM

Ⓐ
Inf. v.c.
Kidney
Aorta
Splenorenal (phrenicolienal) lig.
Omental (epiploic) foramen
Lesser sac (omental bursa)
Spleen
Gastrosplenic (gastrolienal) lig.
Falciform lig. (of liver)
Stomach
Hepatic a., bile duct, portal v.
Lesser omentum

LEVEL OF EPIPLOIC FORAMEN

Ⓑ
Aorta
Inf. vena cava
Lesser sac
Splenorenal (phrenicolienal) lig.
Spleen
Liver
Portal triad
Gastrosplenic (gastrolienal) lig.
Falciform lig.
Stomach
Lesser omentum
Omental (epiploic) foramen
Pancreas

UPPER ABDOMEN

Ⓒ
Ascending colon
Inf. vena cava
Psoas m.
Sacrospinalis m.
Quad. lumb. m.
Aorta
Descending colon
Rectus m.
Mesentery
Small intestine

LOWER ABDOMEN

HORIZONTAL SECTIONS – (A-C) – PERITONEUM

Pansky

177. PERITONEUM OF ABDOMINAL WALL AND PELVIS

I. Posterior abdominal wall. With all viscera cut away, the lines of reflection of peritoneum demonstrate the continuity of the peritoneum, both on the dorsal wall and also on the upper ventral wall and diaphragm. From above downward: the falciform ligament is continuous with the ventral layer of the coronary and triangular ligaments of liver; the dorsal layer of coronary and triangular ligaments is continuous with the lesser omentum at the abdominal end of the esophagus; the splitting of lesser omentum around the esophagus (and stomach) is reunited and becomes continuous with the dorsal mesentery as the splenorenal (phrenicolienal) ligament; there is a fusion of the dorsal mesentery (the great omentum) to the transverse mesocolon; the great omentum passes around the pylorus to become continuous with the free edge of the lesser omentum; the transverse mesocolon dorsally becomes continuous with the peritoneum covering the ascending and descending colons; the peritoneum of the descending colon, farther caudally, merges with the sigmoid mesocolon

II. Anterior abdominal wall
A. ABOVE UMBILICUS: starting at umbilicus and extending cephalically toward liver is the *falciform ligament containing the ligamentum teres*
B. BELOW UMBILICUS
    1. Median umbilical fold, in midline: a fold caused by the presence of the middle umbilical ligament (urachus), which extends from umbilicus to apex of the bladder
    2. Medial umbilical folds, 1 on either side: due to peritoneum covering lateral umbilical ligaments, which begin at umbilicus, diverge laterally and caudally to join superior vesical branch of internal iliac artery. The ligaments are remnants of the fetal umbilical arteries
    3. Lateral umbilical (epigastric) folds: folds of peritoneum covering inferior epigastric vessels

III. Fossae or fovea of anterior abdominal wall: produced by the folds described above
A. LATERAL AND MEDIAL INGUINAL FOSSAE (FOVEAE): depressions in peritoneum on either side of the lateral umbilical folds
B. SUPRAVESICAL FOSSAE (FOVEAE): between middle and medial umbilical folds

IV. Pelvic peritoneum: extending down from the abdominal walls, this peritoneum covers the pelvic viscera and related structures

V. Special features
A. DEEP INGUINAL RING: with peritoneum intact, a slight depression may be seen in lateral inguinal fossa just lateral to the lateral umbilical fold, above middle of inguinal ligament. Site of indirect (lateral) hernia
B. INGUINAL (HESSELBACH'S) TRIANGLE: an area bounded laterally by lateral umbilical fold, medially by the border of rectus abdominis, and below by the inguinal ligament. Most frequent site of direct (medial) hernia

VI. Clinical considerations
A. CULDOCENTESIS: aspiration of fluids accumulated in rectouterine pouch
B. CULDOSCOPY: introduction of an endoscope through the posterior vaginal fornix into the rectouterine pouch for viewing of the pelvic viscera
C. THE MAJOR FUNCTIONS of the peritoneum are to minimize friction and resist infection. It exudes fluid, and in response to injury (or infection) it tends to wall it off or localize it
D. PERITONEAL CAVITY in the male is a completely closed sac; in the female, the uterine tubes open into it

PERITONEUM

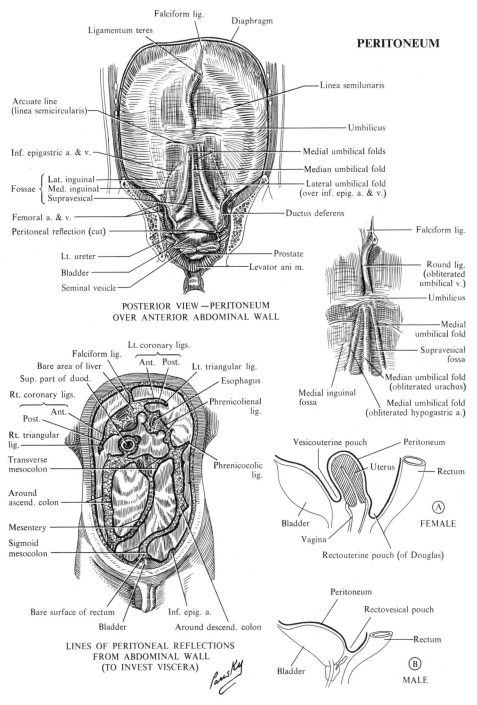

Falciform lig.

Ligamentum teres

Diaphragm

Arcuate line
(linea semicircularis)

Inf. epigastric a. & v.

Fossae { Lat. inguinal
Med. inguinal
Supravesical

Femoral a. & v.

Peritoneal reflection (cut)

Lt. ureter

Bladder

Seminal vesicle

Linea semilunaris

Umbilicus

Medial umbilical folds

Median umbilical fold

Lateral umbilical fold
(over inf. epig. a. & v.)

Ductus deferens

Prostate

Levator ani m.

**POSTERIOR VIEW—PERITONEUM
OVER ANTERIOR ABDOMINAL WALL**

Falciform lig.

Round lig.
(obliterated
umbilical v.)

Umbilicus

Medial
umbilical fold

Supravesical
fossa

Median umbilical fold
(obliterated urachus)

Medial inguinal
fossa

Medial umbilical fold
(obliterated hypogastric a.)

Lt. coronary ligs.
Falciform lig.
Bare area of liver   Ant. Post.
Sup. part of duod.   Lt. triangular lig.
Rt. coronary ligs.   Esophagus
Post.   Phrenicolienal
Ant.   lig.
Rt. triangular
lig.
Transverse
mesocolon
Around
ascend. colon   Phrenicocolic
                          lig.
Mesentery
Sigmoid
mesocolon

Bare surface of rectum
Bladder   Inf. epig. a.
                      Around descend. colon

**LINES OF PERITONEAL REFLECTIONS
FROM ABDOMINAL WALL
(TO INVEST VISCERA)**

Vesicouterine pouch   Peritoneum
                  Uterus
                           Rectum

Bladder

Vagina   Ⓐ
                           FEMALE
Rectouterine pouch (of Douglas)

Peritoneum
          Rectovesical pouch

                  Rectum
Bladder   Ⓑ
                  MALE

SAGITTAL VIEW—PELVIC PERITONEUM

178. STOMACH STRUCTURE: PARTS AND RELATIONS

I. Orifices (for surface projections, see p. 378)

(for surface projections, see p. 378)

A. CARDIAC: between abdominal end of esophagus and stomach and opens toward left. Right side of esophagus is continuous with the lesser curvature and left side with the greater curvature. The latter is indicated by an acute angle—*incisura cardiaca*

B. PYLORIC opens into duodenum toward the right and is indicated by the *duodenopyloric constriction*

II. Curvatures

A. LESSER: continuation of right side of esophagus, is the right or concave border. The *incisura angularis,* a notch in this border, divides the stomach into right and left portions
   1. Hepatogastric ligament (lesser omentum) attaches to it. Between the layers of this ligament run the right and left gastric arteries and the superior gastric (lt. gastric) lymph nodes

B. GREATER: a continuation of the left side of the esophagus, is the left or convex border. The *pyloric antrum,* a dilatation opposite the angular incisure, is limited on the right by the *sulcus intermedius.* The *pyloric canal* is that area between the *sulcus intermedius* and the *pylorus* (opening between stomach and duodenum surrounded by pyloric sphincter)
   1. Gastrosplenic (gastrolienal) ligament attached to its left portion
   2. Great omentum attached to its right portion. Between the layers of the omentum are the gastro-omental (epiploic) vessels and inferior gastric (rt. gastro-omental (epiploic)) lymph nodes

III. Parts of stomach

A. BODY lies to the left of a vertical line passing through the angular incisure
   1. Fundus: that part of body lying superior to a horizontal line through the cardiac opening

B. PYLORIC PORTION lies to right of a line through angular incisure. Is further subdivided into pyloric antrum and pyloric canal, by a line drawn through the sulcus intermedius

IV. Relations of stomach

A. ANTERIOR SURFACE: entire surface is covered with peritoneum. Left half is in contact with diaphragm; right half is in contact with left and quadrate lobes of liver and abdominal wall

B. POSTERIOR SURFACE: entire surface is covered with peritoneum except near the cardiac opening, where the *gastrophrenic ligament* is attached. Is in contact with the diaphragm, spleen, left suprarenal gland, part of left kidney, pancreas, left colic flexure, and upper surface of the transverse mesocolon

V. Special feature

A. MUSCULAR COAT made up of 3 layers of smooth muscle—an inner oblique, chiefly at cardia and spreading to anterior and posterior surfaces; circular, well developed over entire organ; and longitudinal, concentrated along curvatures
   1. The muscularis externis of the lower esophagus, although not very thick, is often referred to as the *esophageal* or *cardiac "sphincter."* Its functions are physiologic and not truly anatomic, but it prevents reflux into the esophagus
   2. The middle layer is greatly thickened at the pyloris to form the *pyloric sphincter* which controls rate of discharge of stomach contents into the duodenum

VI. Clinical considerations

A. STOMACH ULCER: due to an excess of acid secretion associated with vagal nerve involvement. The bleeding *peptic ulcer* is usually located posterior, and the perforating type is located anterior

B. PARTIAL GASTRECTOMY (STOMACH REMOVAL): most common operation performed on stomach in cases of duodenal ulcer, gastric ulcer, or malignancy

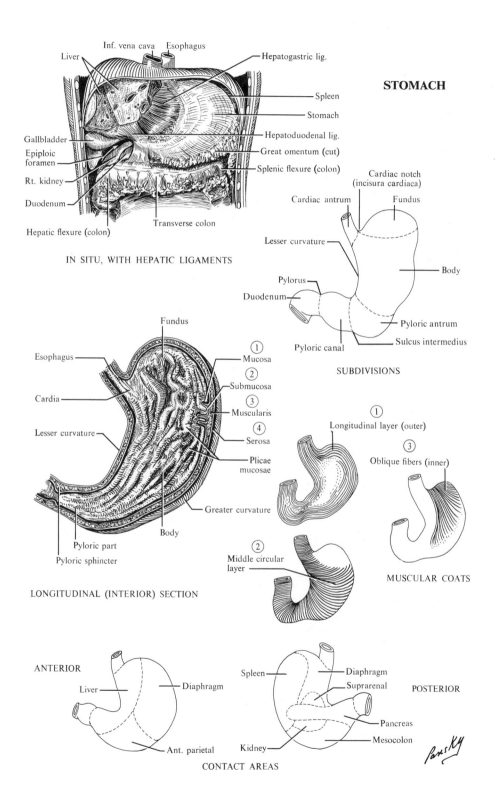

STOMACH

IN SITU, WITH HEPATIC LIGAMENTS

Inf. vena cava
Esophagus
Liver
Hepatogastric lig.
Spleen
Stomach
Hepatoduodenal lig.
Gallbladder
Epiploic foramen
Great omentum (cut)
Splenic flexure (colon)
Rt. kidney
Duodenum
Hepatic flexure (colon)
Transverse colon

SUBDIVISIONS

Cardiac notch (incisura cardiaca)
Cardiac antrum
Fundus
Lesser curvature
Body
Pylorus
Duodenum
Pyloric antrum
Sulcus intermedius
Pyloric canal

LONGITUDINAL (INTERIOR) SECTION

Fundus
Esophagus
① Mucosa
② Submucosa
③ Muscularis
④ Serosa
Cardia
Plicae mucosae
Lesser curvature
Greater curvature
Body
Pyloric part
Pyloric sphincter

MUSCULAR COATS

① Longitudinal layer (outer)
③ Oblique fibers (inner)
② Middle circular layer

CONTACT AREAS

ANTERIOR
Liver
Diaphragm
Ant. parietal

Spleen
Diaphragm
Suprarenal
POSTERIOR
Pancreas
Kidney
Mesocolon

179. BLOOD SUPPLY, LYMPH DRAINAGE, AND INNERVATION OF THE STOMACH

I. Blood supply

A. ARTERIES: all derived directly or indirectly from the celiac artery

1. Left gastric: directly from celiac, runs upward and to left across dorsal wall of omental bursa to cephalic end of lesser curvature, which it follows
2. Right gastric: branch from common hepatic artery, runs upward along the lesser curvature to anastomose with 1, above
3. Right gastro-omental (gastroepiploic): 1 of terminal branches of the gastroduodenal artery from the common hepatic artery, runs toward the left on the greater curvature. Has a large *pyloric branch*
4. Left gastro-omental (gastroepiploic): from splenic artery, through gastrosplenic (gastrolienal) ligament, runs from left to right along the greater curvature to meet 3, above
5. Short gastric arteries: 5 to 7 small branches from the splenic to that part of great curvature above the left gastro-omental (gastroepiploic) artery

B. VEINS: venous drainage directly or indirectly into portal vein

1. Short gastric veins from greater curvature and fundus to lienal vein
2. Left gastro-omental (gastroepiploic) along greater curvature to lienal vein
3. Right gastro-omental (gastroepiploic) from right end of greater curvature to superior mesenteric vein
4. Left gastric (coronary) vein runs length of lesser curvature from cardia to portal vein; accompanies left gastric artery
5. Right gastric is a small vein which accompanies the right gastric artery
6. Pyloric along pyloric part of lesser curvature to portal vein

II. Nerve supply

A. PARASYMPATHETIC: preganglionics from posterior vagal trunk (right vagus) posteriorly and anterior vagal trunk (left vagus) anteriorly. These synapse within the walls of the stomach. Thus, the postganglionics are very short

B. SYMPATHETIC: preganglionic fibers mainly in the thoracic splanchnic nerves; postganglionics arise in ganglia of celiac plexus

III. Lymphatic drainage

A. VISCERAL NODES

1. Gastric
   a. Superior (left gastric) along left gastric artery
   b. Inferior along right half of greater curvature
2. Hepatic. Subdivided into groups: along hepatic artery, near neck of gallbladder, and in the angle between superior and descending duodenum
3. Pancreaticosplenic (lienal) along splenic artery

B. VESSELS OF STOMACH may follow along the left gastric artery to superior gastric nodes, from fundus and body (to left of esophagus) along the left gastro-omental (gastroepiploic) artery to pancreaticosplenic (lienal) nodes, from right part of greater curvature to inferior gastric and hepatic nodes, and from pyloric region into hepatic and superior gastric nodes

IV. Clinical considerations

A. THE DIRECTION OF LYMPH FLOW and the position of the major lymph nodes are essential in understanding the possible spread of malignancy from stomach

B. THE VAGUS NERVES largely control the secretion of acid by the parietal cells of the stomach. Since excess acid secretion is associated with peptic ulcers (either in the stomach or duodenum), section of the vagus trunks as they enter the abdomen is often carried out to reduce acid production (often in conjunction with resection of the ulcerated area)

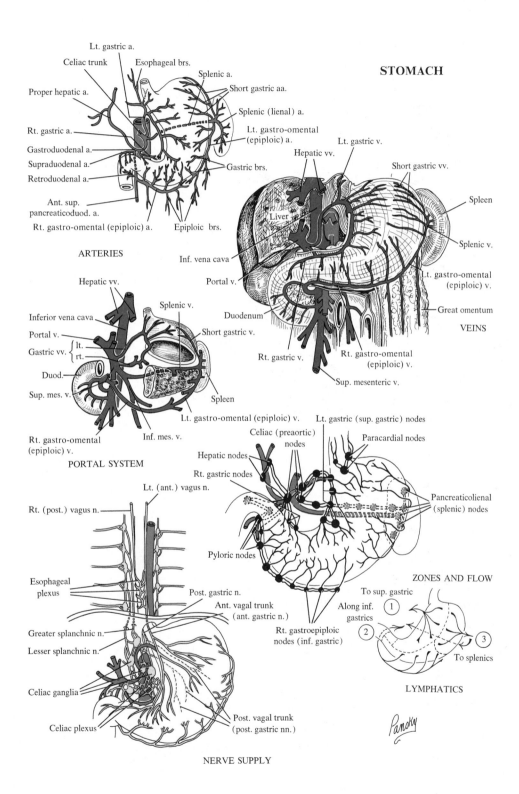

STOMACH

ARTERIES

Lt. gastric a.
Celiac trunk
Esophageal brs.
Splenic a.
Short gastric aa.
Proper hepatic a.
Splenic (lienal) a.
Rt. gastric a.
Lt. gastro-omental (epiploic) a.
Gastroduodenal a.
Supraduodenal a.
Gastric brs.
Retroduodenal a.
Ant. sup. pancreaticoduod. a.
Rt. gastro-omental (epiploic) a.
Epiploic brs.

VEINS

Lt. gastric v.
Hepatic vv.
Short gastric vv.
Spleen
Liver
Splenic v.
Lt. gastro-omental (epiploic) v.
Inf. vena cava
Portal v.
Great omentum
Duodenum
Rt. gastric v.
Rt. gastro-omental (epiploic) v.
Sup. mesenteric v.

PORTAL SYSTEM

Hepatic vv.
Splenic v.
Inferior vena cava
Portal v.
Short gastric v.
Gastric vv. { lt. rt.
Duod.
Sup. mes. v.
Spleen
Rt. gastro-omental (epiploic) v.
Inf. mes. v.
Lt. gastro-omental (epiploic) v.

NERVE SUPPLY

Lt. gastric (sup. gastric) nodes
Celiac (preaortic) nodes
Paracardial nodes
Hepatic nodes
Rt. gastric nodes
Pancreaticolienal (splenic) nodes
Lt. (ant.) vagus n.
Rt. (post.) vagus n.
Pyloric nodes
Esophageal plexus
Post. gastric n.
Ant. vagal trunk (ant. gastric n.)
Greater splanchnic n.
Rt. gastroepiploic nodes (inf. gastric)
Lesser splanchnic n.
Celiac ganglia
Celiac plexus
Post. vagal trunk (post. gastric nn.)

ZONES AND FLOW

To sup. gastric
Along inf. gastrics
① ② ③
To splenics

LYMPHATICS

Pansky

180. DUODENUM

I. Definition: first and shortest part of small intestine

II. Extent: from pylorus to duodenojejunal flexure, 25 cm long

III. Parts and relations
  A. SUPERIOR extends from pylorus to the right, under quadrate lobe of liver to neck of gallbladder, where it bends sharply caudad. Nearly completely covered by peritoneum except at neck of gallbladder. Hepatoduodenal ligament attached to upper border of same region. Related above and anteriorly to liver and gallbladder; posteriorly to gastroduodenal artery, common bile duct, and portal vein; below and posteriorly to pancreas
  B. DESCENDING from level of neck of gallbladder at first lumbar vertebra, along right side of vertebral column to upper body of L4. Covered over anteriorly by peritoneum, except where crossed by transverse mesocolon. Related posteriorly to medial side of right kidney and structures at its hilum (vessels, ureter), inferior vena cava, and psoas major muscle; anteriorly to liver, transverse colon, coils of jejunum; medially to head of pancreas and common duct; and laterally to right colic flexure
     1. Common duct and major pancreatic duct pierce wall about 7.0 cm below pylorus. Accessory duct is 2 cm cephalic to this
  C. INFERIOR (HORIZONTAL) passes from right to left, with slight cephalic deviation along upper border of L4. Covered anteriorly by peritoneum, except near midline, where crossed by vessels. Related anteriorly to superior mesenteric vessels, which cross it; posteriorly to right crus of diaphragm, inferior vena cava, and aorta; and cephalically to pancreas
  D. ASCENDING rises cephalically to left of aorta to upper border of L2, where it turns sharply to join jejunum. Covered anteriorly by peritoneum. Related posteriorly to left psoas major muscle and left renal vessels; and on the right to the uncinate process of pancreas

IV. Structural characteristics
  A. SEROSA ONLY PARTLY COMPLETE
  B. HIGH FOLDS OF THE CONNECTIVE TISSUE of the submucosa form numerous, nearly circular projections into lumen. These are the *circular folds* (plicae), which are numerous and well developed in duodenum, beginning about 2.5 cm from pylorus
  C. DUODENAL GLANDS (OF BRUNNER): compound tubuloalveolar glands of mucous type found in the submucosa
  D. VILLI are numerous and large
  E. MAJOR DUODENAL PAPILLA with *sphincter of Oddi* around common duct and major duct of pancreas (hepatopancreatic ampulla) within wall of duodenum

V. Clinical considerations
  A. MOST OF THE DUODENAL ULCERS OCCUR within 5 cm of the pylorus and more frequently on the anterior wall
  B. DUODENECTOMY: because attachment of the duodenum to the posterior body wall structures is secondary, both it and the associated head of the pancreas can be separated from underlying viscera (right kidney) without endangering the blood supply or ducts of the kidney
  C. THE FIRST PART of the duodenum has the poorest blood supply, for it is not supplied from the arcades but by small branches from the gastroduodenal artery
  D. ANOMALIES ARE RELATIVELY UNCOMMON and include atresia (discontinuity of lumen) and stenosis (complete or incomplete)

DUODENUM

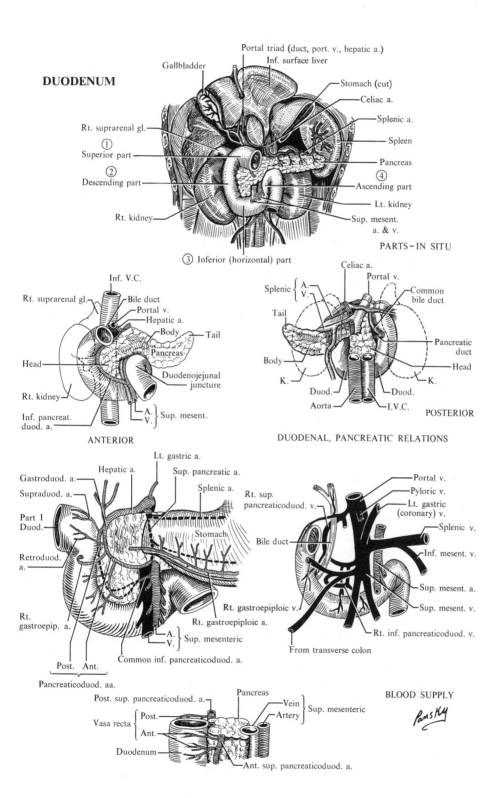

Portal triad (duct, port. v., hepatic a.)
Inf. surface liver
Gallbladder
Stomach (cut)
Celiac a.
Splenic a.
Spleen
Pancreas
Rt. suprarenal gl.
① Superior part
② Descending part
④ Ascending part
Lt. kidney
Rt. kidney
Sup. mesent. a. & v.
③ Inferior (horizontal) part

PARTS – IN SITU

Inf. V.C.
Rt. suprarenal gl.
Bile duct
Portal v.
Hepatic a.
Body
Tail
Pancreas
Head
Duodenojejunal juncture
Rt. kidney
Inf. pancreat. duod. a.
A. V. } Sup. mesent.

ANTERIOR

Celiac a.
Portal v.
Splenic { A. V.
Common bile duct
Tail
Pancreatic duct
Body
Head
K.
K.
Duod.
Duod.
Aorta
I.V.C.

POSTERIOR

DUODENAL, PANCREATIC RELATIONS

Lt. gastric a.
Hepatic a.
Sup. pancreatic a.
Splenic a.
Gastroduod. a.
Supraduod. a.
Rt. sup. pancreaticoduod. v.
Part I Duod.
Stomach
Retroduod. a.
Bile duct
Rt. gastroepip. a.
Rt. gastroepiploic v.
Rt. gastroepiploic a.
A. V. } Sup. mesenteric
Common inf. pancreaticoduod. a.
Post. Ant.
Pancreaticoduod. aa.

Portal v.
Pyloric v.
Lt. gastric (coronary) v.
Splenic v.
Inf. mesent. v.
Sup. mesent. a.
Sup. mesent. v.
Rt. inf. pancreaticoduod. v.
From transverse colon

BLOOD SUPPLY

Post. sup. pancreaticoduod. a.
Pancreas
Vasa recta { Post. Ant.
Vein Artery } Sup. mesenteric
Duodenum
Ant. sup. pancreaticoduod. a.

Pansky

181. DUODENUM: VESSELS, NERVES, AND LYMPHATIC DRAINAGE

I. **Arterial supply:** predominantly from the two arcades formed by the superior and inferior pancreaticoduodenal arteries and branches of the gastroduodenal and superior mesenteric arteries. A lesser supply, principally to the 1st part of the duodenum is derived from the supraduodenal and retroduodenal branches of the gastroduodenal artery

A. THE ARTERIES APPROACH the duodenum through its concavity. Thus, an incision along the right edge of the 2nd portion of the duodenum will free or mobilize the organ and head of the pancreas without endangering their blood supplies

B. PROXIMAL HALF: supplied by the superior pancreaticoduodenal artery

C. DISTAL HALF: supplied by the inferior pancreaticoduodenal artery

D. THE ABOVE VESSELS (A and B) anastomose to form the anterior and posterior arterial arcades which occupy the angle between the duodenum and the pancreas

E. SUPERIOR PART, in addition to above, is also supplied by supraduodenal (branch of hepatic proper), rt. gastric, rt. gastro-omental, and gastroduodenal (retroduodenal) arteries

II. **Venous drainage:** veins generally follow the arteries, but are more variable and drain into the portal venous system

A. MOST VEINS DRAIN into sup. mesenteric v. A few enter portal vein directly

B. A NUMBER OF SMALL VEINS on the anterior and posterior duodenal surfaces of the superior part of the duodenum drain into the superior pancreaticoduodenal vein

C. THE PREPYLORIC VEIN (of Mayo): an anterior vein which drains into the rt. gastric vein. Used surgically as a guide to the gastroduodenal junction

III. **Lymphatic drainage:** lymph vessels on the anterior and posterior surfaces of the duodenum drain into ant. and post. collecting trunks and nodes that lie in front of and behind the pancreas, anastomose freely with each other, and ultimately drain into the thoracic duct

A. THE ANT. EFFERENT VESSELS follow the arteries and drain upward via the pancreatico-duodenal nodes to the gastroduodenal nodes and finally into the celiac nodes

B. THE POST. EFFERENT VESSELS pass behind the head of the pancreas and drain downward into the sup. mesenteric nodes (found around origin of sup. mesenteric artery)

IV. **Nerve supply:** sympathetic and parasympathetic nerves, as well as sensory fibers from the celiac and sup. mesenteric plexuses supply the duodenum

A. THE SYMPATHETIC SYSTEM supplies its fibers to all organs that receive blood from the celiac axis via the celiac ganglia.

1. Preganglionic fibers begin in the 5th–9th segments of the cord, emerge via ventral roots, and proceed to the sympathetic chain ganglia via white rami communicantes. The majority of fibers do not synapse in the chain ganglia but continue via the greater splanchnic nerves to the celiac ganglion, where they synapse with postganglionic fibers. The latter pass to the organs with the respective arteries

B. THE PARASYMPATHETIC SYSTEM: derived entirely from the vagus nerves

1. Left vagus: continues from ant. surface of esophagus onto lesser curvature and anterior surface of stomach as *ant. gastric n.* It enters free edge of lesser omentum and gives fibers to the liver, gallbladder, and a few to duodenum

2. Right vagus: from post. surface of esophagus, it sends branches to post. surface of stomach (post. gastric n.), but also a large contribution to join the celiac ganglia. There is no synapse here. Fibers continue through and reach the organs to be innervated. Synapse occurs in the intrinsic ganglia in the walls of the organs innervated

C. SENSORY FIBERS are found in the greater splanchnic nerves. Pain is associated with distention or violent contraction

D. SYMPATHETICS DECREASE MOTILITY and secretion and cause vasoconstriction; parasympathetics increase peristalsis and gastric secretion and relate to vasodilation

DUODENUM

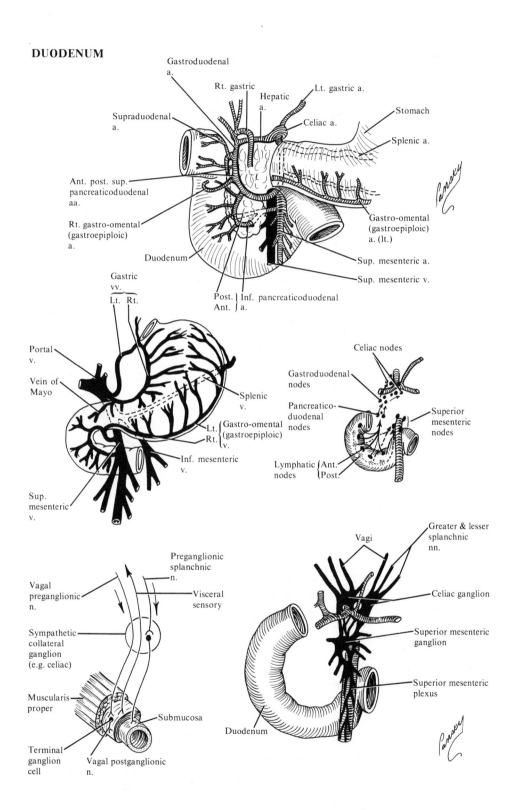

Gastroduodenal a.

Rt. gastric

Hepatic a.

Lt. gastric a.

Stomach

Supraduodenal a.

Celiac a.

Splenic a.

Ant. post. sup. pancreaticoduodenal aa.

Rt. gastro-omental (gastroepiploic) a.

Gastro-omental (gastroepiploic) a. (lt.)

Duodenum

Sup. mesenteric a.

Sup. mesenteric v.

Gastric vv.
Lt. Rt.

Post. | Inf. pancreaticoduodenal
Ant. | a.

Portal v.

Vein of Mayo

Splenic v.

Lt. | Gastro-omental
Rt. | (gastroepiploic) v.

Inf. mesenteric v.

Sup. mesenteric v.

Celiac nodes

Gastroduodenal nodes

Pancreatico-duodenal nodes

Superior mesenteric nodes

Lymphatic nodes | Ant. Post.

Preganglionic splanchnic n.

Visceral sensory

Vagal preganglionic n.

Sympathetic collateral ganglion (e.g. celiac)

Muscularis proper

Submucosa

Terminal ganglion cell

Vagal postganglionic n.

Vagi

Greater & lesser splanchnic nn.

Celiac ganglion

Superior mesenteric ganglion

Superior mesenteric plexus

Duodenum

-401-

182. PANCREAS

I. Type of gland: soft, fleshy organ with little connective tissue
A. ENDOCRINE: islets of Langerhans
B. EXOCRINE: compound tubuloalveolar, serous secretion

II. Parts and their relations
A. HEAD: the broad, right extremity. Lies within the curve of the duodenum
   1. Uncinate process: a prolongation of the left and caudal borders of the head
      a. Superior mesenteric artery with superior mesenteric vein on its right side crosses the uncinate process
   2. Anterior surface: most of the right side separated from the transverse colon by areolar tissue (no peritoneum); lower part of surface below transverse colon is covered by peritoneum; is in contact with coils of the small intestine
   3. Posterior surface: without peritoneum, is in contact with inferior vena cava, common bile duct, renal veins, right crus of diaphragm, and aorta
B. NECK: a constricted portion to the left of the head. Above, it adjoins the pylorus. Behind, it is related to the origin of the portal vein and the gastroduodenal artery
C. BODY
   1. Anterior surface: separated from the stomach by the omental bursa
   2. Posterior surface: nonperitoneal; related to the aorta, splenic vein, left kidney and vessels; left suprarenal; origin of superior mesenteric artery, and crura of diaphragm
   3. Inferior surface: peritoneal; related to duodenojejunal flexure, coils of jejunum, and left colic flexure
   4. Anterior border: layers of transverse mesocolon diverge along this
   5. Superior border: related to celiac artery, with hepatic artery to right and splenic artery to left
D. TAIL: left extremity, extending to surface of spleen, in splenorenal (phrenicolienal) ligament

III. Ducts
A. MAJOR (OF WIRSUNG) extends toward right, reaches neck of pancreas, where it turns caudally and dorsally and comes in contact with the common bile duct, forming the hepatopancreatic ampulla (of Vater). These ducts pass obliquely through the wall of the descending duodenum and open through a common orifice into its lumen
B. ACCESSORY (MINOR OF SANTORINI) drains part of head, enters the duodenum above the major

IV. Blood supply
A. ARTERIES
   1. Numerous small branches from the splenic artery
   2. Retroduodenal branch of gastroduodenal artery
   3. Superior pancreaticoduodenal from the gastroduodenal artery
   4. Inferior pancreaticoduodenal artery from the superior mesenteric artery
B. VEINS drain into both the splenic and superior mesenteric veins

V. Lymphatic drainage (see p. 432) follows the course of blood vessels to terminate in the pancreaticosplenic (lienal), pancreaticoduodenal, and celiac nodes

VI. Nerves: autonomics and sensory: via celiac and superior mesenteric plexuses

VII. Clinical considerations
A. HYPERTROPHY OF HEAD may cause portal or bile duct obstruction
B. DEGENERATION OF THE ISLETS OF LANGERHANS leads to diabetes mellitus
C. PANCREATITIS is a serious inflammatory condition of the exocrine pancreas

PANCREAS

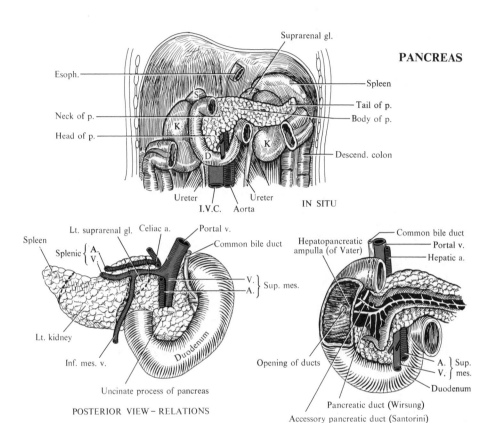

Suprarenal gl.

Esoph.

Spleen

Neck of p.

Tail of p.

Body of p.

Head of p.

K

K

D

Descend. colon

Ureter

Ureter

I.V.C.   Aorta

IN SITU

Spleen

Lt. suprarenal gl.   Celiac a.

Portal v.

Common bile duct

Splenic { A. V. }

V. } Sup. mes.
A. }

Lt. kidney

Inf. mes. v.

Duodenum

Uncinate process of pancreas

POSTERIOR VIEW – RELATIONS

Hepatopancreatic ampulla (of Vater)

Common bile duct

Portal v.

Hepatic a.

A. } Sup.
V. } mes.

Duodenum

Opening of ducts

Pancreatic duct (Wirsung)

Accessory pancreatic duct (Santorini)

PANCREATIC DUCTS

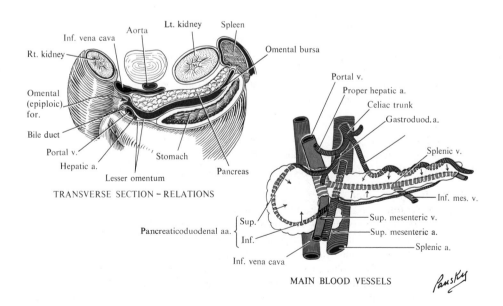

Inf. vena cava   Aorta   Lt. kidney   Spleen

Rt. kidney

Omental bursa

Omental (epiploic) for.

Portal v.

Proper hepatic a.

Celiac trunk

Gastroduod. a.

Bile duct

Splenic v.

Portal v.

Hepatic a.

Lesser omentum

Stomach

Pancreas

Inf. mes. v.

Sup. mesenteric v.

Sup. mesenteric a.

TRANSVERSE SECTION – RELATIONS

Pancreaticoduodenal aa. { Sup. Inf. }

Splenic a.

Inf. vena cava

MAIN BLOOD VESSELS

Pansky

183. Duodenum–Jejunum–Ileum, Duodenal Fossae, and Superior Mesenteric Artery

I. Duodenojejunal flexure: point where ascending part of duodenum turns sharply anteriorly and caudally to join jejunum
  A. LOCATION: upper border, second lumbar vertebra, on left side
  B. SUSPENSORY MUSCLE (LIGAMENT OF TREITZ): musculofibrous band, from tissue around celiac artery and right crus of diaphragm to flexure and ascending duodenum, which continues into mesentery. Acts as suspensory ligament

II. Duodenal recesses (fossae)
  A. INFERIOR at level of third lumbar vertebra, on left side of ascending duodenum, opens cephalically, and extends down behind ascending duodenum
  B. SUPERIOR ventral to body of second lumbar vertebra on left of ascending duodenum, opens caudally, lies behind peritoneal fold, superior duodenal (*duodenojejunal fold*)
  C. DUODENOJEJUNAL lies below pancreas, between the aorta and the left kidney
  D. PARADUODENAL (rarely found): small pocket seen behind ascending branch of lt. colic a.
  E. RETRODUODENAL (rarely found) lies behind transverse and ascending parts of duodenum, anterior to aorta

III. Intestinal arteries: usually 12–15
  A. ORIGIN: superior mesenteric artery (from abdominal aorta)
  B. DISTRIBUTION: jejunum and ileum
  C. COURSE: in mesentery, running parallel with each other
  D. BRANCHING: each divides into 2 branches, which unite with branches from adjoining arteries, forming arches (arterial arcades) with convexities toward the intestine
    1. In upper part of mesentery, from this arch, fairly long, straight arteries arise which go to the gut (*vaso recta*). The latter do not anastomose within the mesentery, but there are many anastomoses in the intestinal wall
    2. As the intestinal tract descends, branches arise from the 1st set of arches to unite with similar branches from the arch above or below, forming a 2nd set of arches, located nearer the intestine than the 1st. The straight arteries arising from these are shorter. Still farther caudad, 3, 4, or 5 more arches are added so last arch in lower ileum is close to the gut and the straight aa. are very short
  E. VENOUS DRAINAGE: superior mesenteric v. drains jejunum and ileum and unites with splenic vein, behind neck of pancreas, to form the portal vein
  F. LYMPHATICS: from lacteals in villi, lymph vessels in wall of jejunum, and ileum pass in mesentery to the mesenteric nodes
  G. NERVES: derived from vagus and splanchnic (visceral) nn. via celiac ganglion and plexuses around the superior mesenteric artery

IV. Meckel's diverticulum: present in a small percentage of cases
  A. LOCATION: usually about 1 m above ileocolic valve on the antimesenteric border of ileum. End may be free or attached to abdominal wall
  B. CAUSE: failure of entire omphalovitelline duct of fetus to atrophy

V. The jejunum–ileum
  A. SUPPORT: from dorsal body wall by the *mesentery*
  B. JEJUNUM, upper two fifths (2.4 m); ileum, lower three fifths (3.6 m)
  C. JEJUNUM: greater diameter, thicker wall, and more vascular
    1. Jejunum lies in umbilical region; ileum in hypogastric and pelvic regions
    2. The circular folds (plicae circulares) of the mucous membrane are large and well developed in upper jejunum, small in upper ileum, and absent in terminal ileum
  D. ILEUM: distal part always in pelvis. More fat in mesentery. Aggregated lymph nodules (Peyer's patches) seen in antimesenteric border
  E. THE VASCULARITY of the wall is greater in jejunum than in ileum, but the arterial arcades are shorter and more complex in the ileum

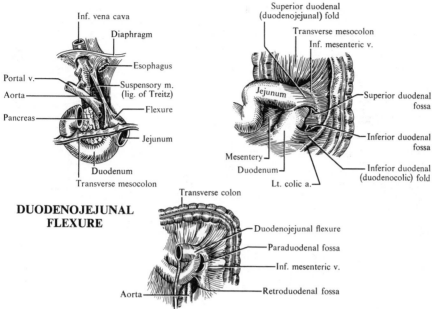

Inf. vena cava
Diaphragm
Portal v.
Aorta
Pancreas
Esophagus
Suspensory m.
(lig. of Treitz)
Flexure
Jejunum
Duodenum
Transverse mesocolon

**DUODENOJEJUNAL
FLEXURE**

Superior duodenal
(duodenojejunal) fold
Transverse mesocolon
Inf. mesenteric v.
Jejunum
Superior duodenal
fossa
Inferior duodenal
fossa
Mesentery
Duodenum
Inferior duodenal
(duodenocolic) fold
Lt. colic a.

Transverse colon
Duodenojejunal flexure
Paraduodenal fossa
Inf. mesenteric v.
Aorta
Retroduodenal fossa

**DUODENAL FOSSAE
AND FOLDS**

① ②

ABOUT 0.9 METER FROM
DUODENOJEJUNAL FLEXURE

ABOUT 2.1 METERS FROM
DUODENOJEJUNAL FLEXURE (Lunette's)

**VASCULAR CHANGES FROM
JEJUNUM TO ILEUM**
(After Thorek)

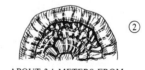

③ ④

ABOUT 3.0–3.6 METERS FROM FLEXURE
(TERTIARY LOOPS)

TERMINAL ILEUM (Fat)

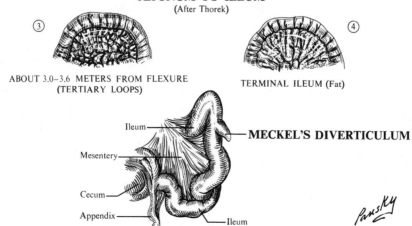

Ileum
Mesentery
Cecum
Appendix
Ileum

MECKEL'S DIVERTICULUM

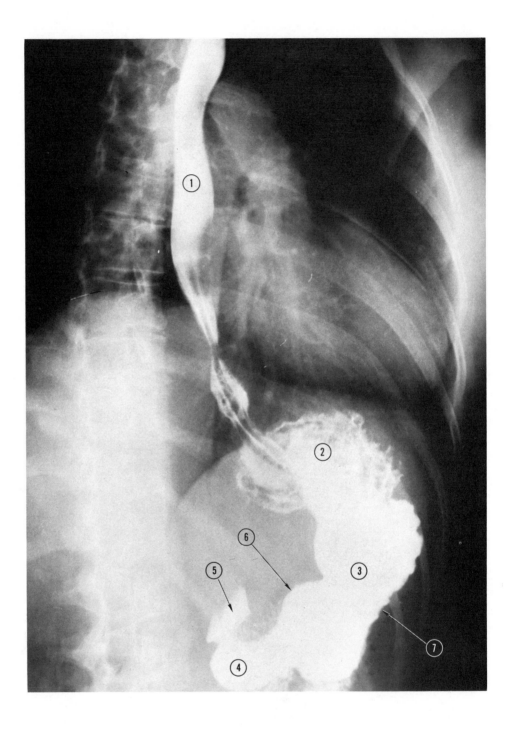

FIGURE 33. **Esophagus and stomach.** *1*, Esophagus; *2*, fundus of stomach; *3*, body of stomach; *4*, pylorus; *5*, duodenal cap; *6*, lesser curvature; *7*, greater curvature.

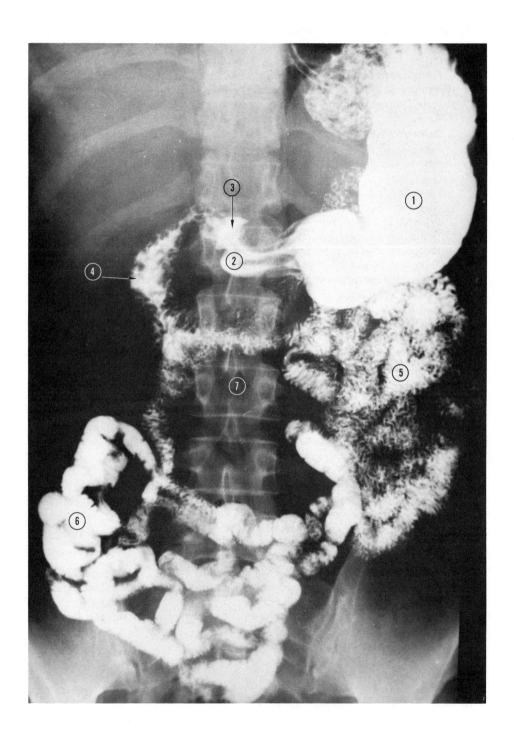

FIGURE 34. **Stomach and small intestine after barium meal.** *1*, Body of stomach; *2*, pylorus; *3*, duodenal cap; *4*, descending duodenum; *5*, jejunum; *6*, ileum; *7*, body of L3.

184. THE LARGE INTESTINE

I. Extent: from terminal ileum to anus, 1.5 m

II. Parts

A. CECUM: blind pouch that extends caudally below ileocolic valve, located in right iliac fossa above inguinal ligament. Its shape varies. Usually covered by peritoneum
   1. Appendix: long and narrow tube (8.2 cm), which begins at apex of cecum. Position varies. Fixed by *mesoappendix* containing the *appendicular artery*
   2. Ileocecal valve: ileum opens into junction between cecum and ascending colon. Opening is guarded by 2 lips, which project into lumen. The lips merge on either side of opening, forming membranous ridges, the *frenula of the valve*

B. COLON
   1. Ascending: begins at iliocecal valve and ascends through right lateral (lumbar) and hypochondriac regions to visceral surface of liver, to right of gallbladder. It then bends sharply to the left as the *right colic (hepatic) flexure*. Its ventral surface and sides are covered by peritoneum. Dorsally, it is separated from the iliacus, quadratus lumborum, transversus abdominis muscles, and lateral part of right kidney
   2. Transverse: longest and most movable. From right hypochondriac region arches to come across umbilical zone and then upward into left hypochondriac region, where it bends caudally at *left colic flexure (splenic)* below spleen. It is invested in peritoneum and is suspended from body wall by *transverse mesocolon*
   3. Descending: extends caudally through left hypochondriac and lateral (lumbar) regions along lateral border of left kidney. At caudal end of kidney bends medially and descends in groove between psoas and quadratus lumborum muscles to crest of ileum. Covered anteriorly and on sides with peritoneum, which helps to fix it
   4. Iliac colon: in left iliac fossa, from iliac crest to brim of true pelvis. Anterior to iliacus and psoas muscles. Covered anteriorly and at sides with peritoneum
   5. Sigmoid: begins at pelvic brim, crosses sacrum, and then curves to midline at third sacral segment, where it enters rectum. Usually, completely invested with peritoneum and has a mesocolon. Posteriorly are the left external iliac vessels, left piriformis muscle, and left sacral plexus. Anteriorly, coils of small intestine

III. Special structural characteristics

A. TAENIAE COLI: the longitudinal smooth muscle coat of the colon is incomplete, being collected in 3 bands, the taeniae. These create sacculations (haustra)
   1. Semilunar folds: the crescent-shaped folds between the haustra
B. EPIPLOIC APPENDAGES: small fat-filled sacs of peritoneum attached along taeniae

IV. Peritoneal recesses (fossae) (retroperitoneal fossae)

A. SUPERIOR ILEOCECAL: fold of peritoneum over the branch of the ileocolic artery
B. INFERIOR ILEOCECAL: behind the ileocecal junction produced by the ileocecal fold (bloodless fold of Treves) from the antimesenteric border of the ileum, crosses the ileocecal junction and joins the mesoappendix
C. RETROCECAL (CECAL): behind the cecum

V. Clinical considerations

A. TAENIAE converge at root of appendix and may be used to locate its root
B. McBURNEY'S POINT: located by drawing a line from the right anterior superior iliac spine to the umbilicus. The midpoint of this line locates root of the appendix
C. COLOSTOMY: an artificial opening between the colon and skin
D. DIVERTICULITIS: chiefly in sigmoid colon; inflammation of abnormal outpocketings

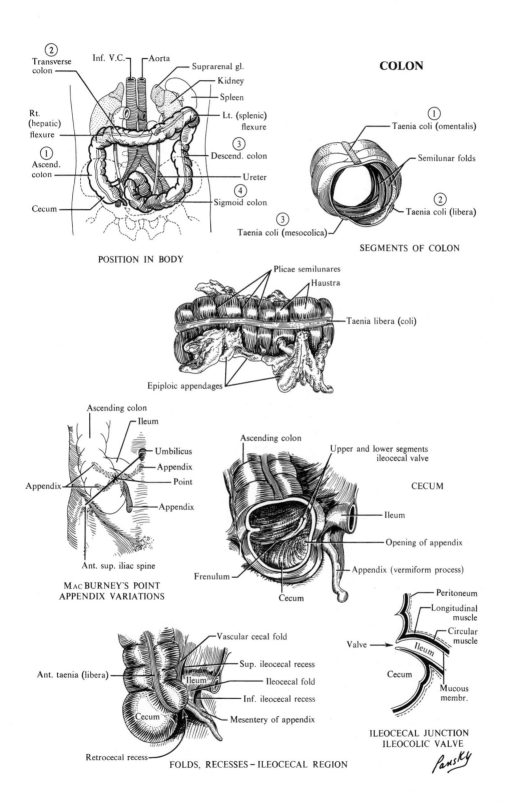

COLON

POSITION IN BODY

② Transverse colon — Inf. V.C. — Aorta — Suprarenal gl. — Kidney — Spleen — Lt. (splenic) flexure

Rt. (hepatic) flexure

① Ascend. colon

Cecum

③ Descend. colon — Ureter — ④ Sigmoid colon

SEGMENTS OF COLON

① Taenia coli (omentalis) — Semilunar folds — ② Taenia coli (libera) — ③ Taenia coli (mesocolica)

Plicae semilunares — Haustra — Taenia libera (coli) — Epiploic appendages

Ascending colon — Ileum — Umbilicus — Appendix — Point — Appendix — Appendix — Ant. sup. iliac spine

MACBURNEY'S POINT APPENDIX VARIATIONS

Ascending colon — Upper and lower segments ileocecal valve — Ileum — Opening of appendix — Appendix (vermiform process) — Frenulum — Cecum

CECUM

Vascular cecal fold — Sup. ileocecal recess — Ileocecal fold — Inf. ileocecal recess — Mesentery of appendix — Ant. taenia (libera) — Ileum — Cecum — Retrocecal recess

FOLDS, RECESSES – ILEOCECAL REGION

Peritoneum — Longitudinal muscle — Circular muscle — Valve — Ileum — Cecum — Mucous membr.

ILEOCECAL JUNCTION ILEOCOLIC VALVE

Pansky

185. BLOOD SUPPLY OF THE COLON AND APPENDIX

I. Arteries

A. ILEOCOLIC ARTERY
  1. Origin: lowest branch of superior mesenteric artery
  2. Course: caudally and to the right, into the right iliac fossa
  3. Branches
     a. Superior ascends along ascending colon to join right colic artery
     b. Inferior runs toward ileocolic junction. Branches: *ascending (colic)*, to ascending colon; *cecal*, anterior and posterior to cecum; *appendicular*, descends posterior to terminal ileum to mesoappendix of appendix; *ileal*, passes to left on ileum to anastomose with last intestinal branch of superior mesenteric artery

B. RIGHT COLIC
  1. Origin: from middle of superior mesenteric artery
  2. Course: to right, behind peritoneum, crossing anterior to internal spermatic or ovarian vessels, right ureter, and psoas major muscle
  3. Branches
     a. Descending: anastomoses with superior branch of ileocolic artery
     b. Ascending: course cephalically to join middle colic artery

C. MIDDLE COLIC
  1. Origin: from superior mesenteric artery, just below pancreas
  2. Course: caudally and anteriorly in transverse mesocolon
  3. Branches: near border of transverse colon
     a. Right: anastomoses with ascending branch of right colic artery
     b. Left: anastomoses with ascending branch of left colic artery

D. LEFT COLIC
  1. Origin: from inferior mesenteric artery
  2. Course: toward the left, behind the peritoneum, in front of left psoas major muscle, and crosses left ureter and internal spermatic vessels
  3. Branches
     a. Ascending ascends in front of left kidney to enter transverse mesocolon, and anastomoses with left branch of middle colic artery
     b. Descending descends to join highest sigmoid artery

E. SIGMOID ARTERIES: 2 or 3
  1. Origin: from inferior mesenteric artery
  2. Course and branches: runs caudally and laterally behind peritoneum, but anterior to psoas major muscle, ureter, and internal spermatic vessels. Enters sigmoid mesocolon and is distributed to that region. *Superior sigmoidal* anastomoses with left colic artery. *Inferior sigmoidal* anastomoses with superior rectal artery

F. SUPERIOR RECTAL
  1. Origin: continuation of inferior mesenteric artery
  2. Course: into pelvis in sigmoid mesocolon and crosses left common iliac vessels
  3. Branches: divides at third sacral segment giving 1 branch to either side of rectum

II. Veins

A. PORTAL VEIN formed by union of splenic and superior mesenteric veins
  1. Splenic vein receives inf. mesenteric and veins from pancreas and stomach
     a. Inferior mesenteric begins in rectum as superior rectal vein and drains sigmoid and descending colon
  2. Superior mesenteric vein receives veins from stomach, pancreas, duodenum, jejunum, ileum, cecum, appendix, ascending and transverse colon

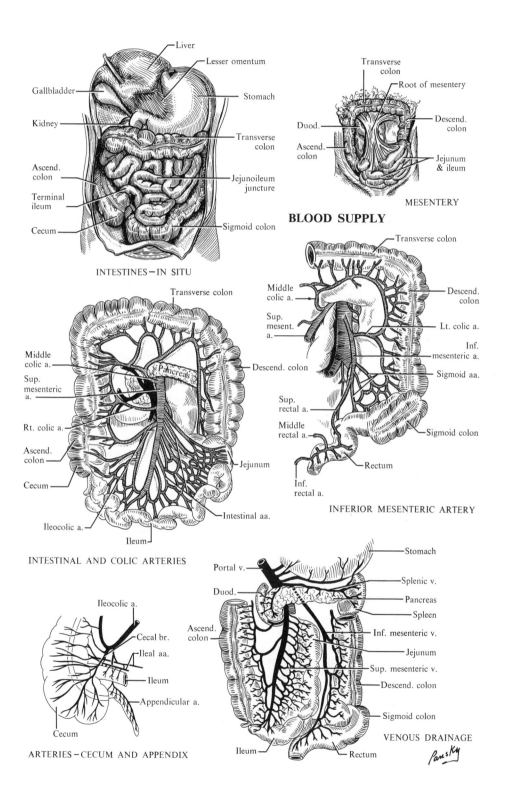

Liver

Lesser omentum

Gallbladder

Stomach

Kidney

Transverse
colon

Ascend.
colon

Jejunoileum
juncture

Terminal
ileum

Cecum

Sigmoid colon

INTESTINES—IN SITU

Transverse
colon

Root of mesentery

Duod.

Descend.
colon

Ascend.
colon

Jejunum
& ileum

MESENTERY

BLOOD SUPPLY

Transverse colon

Transverse colon

Middle
colic a.

Descend.
colon

Sup.
mesent.
a.

Lt. colic a.

Inf.
mesenteric a.

Sigmoid aa.

Middle
colic a.

Pancreas

Descend. colon

Sup.
mesenteric
a.

Rt. colic a.

Sup.
rectal a.

Ascend.
colon

Middle
rectal a.

Sigmoid colon

Cecum

Jejunum

Rectum

Ileocolic a.

Intestinal aa.

Inf.
rectal a.

Ileum

INFERIOR MESENTERIC ARTERY

INTESTINAL AND COLIC ARTERIES

Stomach

Portal v.

Splenic v.

Ileocolic a.

Duod.

Pancreas

Cecal br.

Ascend.
colon

Spleen

Ileal aa.

Inf. mesenteric v.

Ileum

Jejunum

Appendicular a.

Sup. mesenteric v.

Descend. colon

Cecum

Sigmoid colon

VENOUS DRAINAGE

ARTERIES—CECUM AND APPENDIX

Ileum

Rectum

Pansky

186. GALLBLADDER AND DUCTS

I. Surface projections: fundus lies at right border of rectus muscle at end of ninth costal cartilage

II. Location: in fossa for gallbladder on visceral surface of liver. Cephalically is joined to liver by connective tissue. Caudally is covered by peritoneum

III. Parts
A. FUNDUS directed caudally, anteriorly, and to right
B. BODY extends cephalically, posteriorly, and to left
C. NECK in form of S-shaped curve

IV. Relations
A. FUNDUS: abdominal wall
B. BODY: cephalically, the liver; caudally, transverse colon and descending duodenum

V. Duct system
A. CYSTIC: about 4 cm long, runs posteriorly, caudally, and toward left from neck of gallbladder and joins hepatic duct to form common bile duct. From neck to common duct contains *spiral folds (valves of Reister)*
B. HEPATIC DUCT arises from a major duct of the right and left lobes of liver, which leave through porta hepatis and join to form the main hepatic duct. Runs caudally and to the right in lesser omentum to join the cystic duct, forming common bile duct
C. COMMON DUCT: 7.5 cm in length. Runs caudally in free edge of hepatoduodenal ligament, with hepatic artery to its left and portal vein behind. Passes posterior to superior duodenum, then passes near right border of the head of pancreas. Major duct of pancreas approaches common duct, and the 2 run side by side in an oblique course through wall of duodenum
  1. Walls of terminal ends of 2 ducts are thickened by smooth muscle—*sphincter of ampulla (of Oddi)*
    a. Major duodenal papilla: an elevation into lumen of descending duodenum caused by above sphincter
    b. Common duct and duct of pancreas have common opening on major papilla in most cases

VI. Blood supply and innervation
A. ARTERIES: cystic from right hepatic artery; gives a superficial branch, which supplies the free, inferior surface, and a deep branch to cephalic surface. (The origin and number of the cystic artery[ies] are variable)
B. VEINS
  1. Into liver capillaries
  2. Cystic vein opens into the right branch of the portal vein
C. LYMPHATICS (see p. 432) drain into hepatic nodes from bladder and into hepatic and pancreaticoduodenal nodes from the common bile duct
D. NERVES: autonomics by way of celiac plexus
  1. Pain fibers from the bile passages reach the spinal cord via the splanchnic nerves. Pain is excruciating (due to distention or spasm) and felt in upper rt. quadrant or in epigastrium, but is often referred posteriorly to region of rt. scapula and is sometimes cardiac in type and distribution

VII. Clinical consideration
A. OBSTRUCTION OF THE DUCT SYSTEM as a result of "stone" formation (gallstone) leads to pain and jaundice

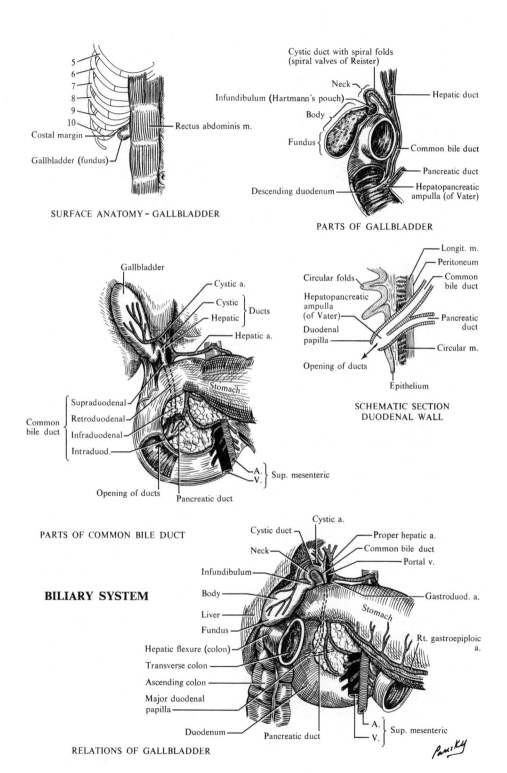

SURFACE ANATOMY - GALLBLADDER

5
6
7
8
9
10
Costal margin
Gallbladder (fundus)
Rectus abdominis m.

PARTS OF GALLBLADDER

Cystic duct with spiral folds
(spiral valves of Reister)
Neck
Infundibulum (Hartmann's pouch)
Body
Fundus
Descending duodenum
Hepatic duct
Common bile duct
Pancreatic duct
Hepatopancreatic
ampulla (of Vater)

PARTS OF COMMON BILE DUCT

Gallbladder
Cystic a.
Cystic
Hepatic
Ducts
Hepatic a.
Stomach
Supraduodenal
Retroduodenal
Infraduodenal
Intraduod.
Common
bile duct
Opening of ducts
Pancreatic duct
A.
V.
Sup. mesenteric

SCHEMATIC SECTION
DUODENAL WALL

Circular folds
Hepatopancreatic
ampulla
(of Vater)
Duodenal
papilla
Opening of ducts
Epithelium
Longit. m.
Peritoneum
Common
bile duct
Pancreatic
duct
Circular m.

BILIARY SYSTEM

RELATIONS OF GALLBLADDER

Cystic duct
Neck
Infundibulum
Body
Liver
Fundus
Hepatic flexure (colon)
Transverse colon
Ascending colon
Major duodenal
papilla
Duodenum
Pancreatic duct
Cystic a.
Proper hepatic a.
Common bile duct
Portal v.
Gastroduod. a.
Stomach
Rt. gastroepiploic
a.
A.
V.
Sup. mesenteric

Pansky

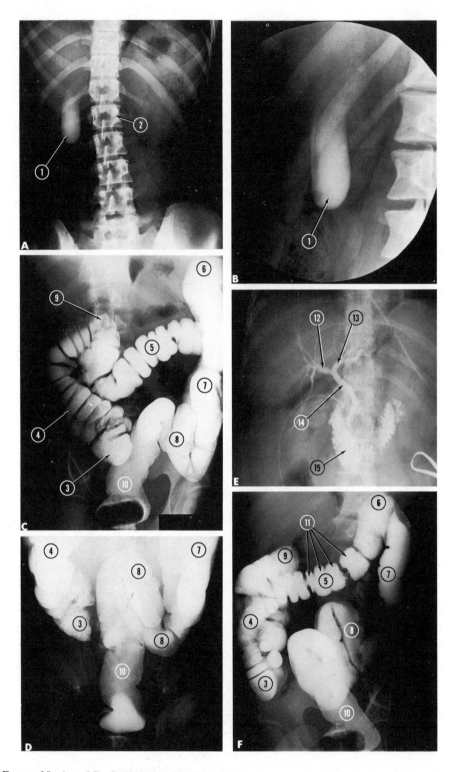

FIGURE 35. **A and B, Gallbladder; C, D, and F, large intestine; E, hepatic and common bile ducts.** *1*, Gallbladder; *2*, 1st lumbar vertebra; *3*, cecum; *4*, ascending colon; *5*, transverse colon; *6*, splenic (lt. colic) flexure; *7*, descending colon; *8*, sigmoid colon; *9*, hepatic (right colic) flexure; *10*, rectum; *11*, haustra; *12*, rt. hepatic duct; *13*, lt. hepatic duct; *14*, common bile duct; *15*, duodenum.

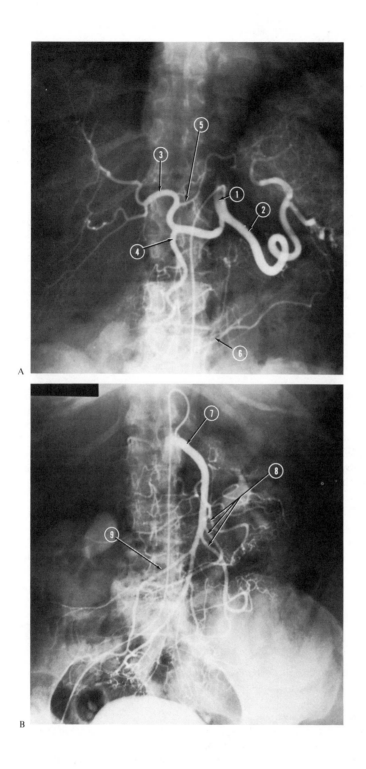

FIGURE 36. **A, Celiac artery arteriogram; B, superior mesenteric arteriogram.** *1*, Celiac; *2*, splenic; *3*, hepatic; *4*, gastroduodenal; *5*, right gastric; *6*, right gastro-omental (epiploic); *7*, superior mesenteric; *8*, intestinal; *9*, iliocolic.

187. LIVER

I. Surfaces
A. DIAPHRAGMATIC
   1. Anterior faces diaphragm, which separates it from sixth to tenth ribs and cartilages on right and from seventh and eighth cartilages on left. Covered by peritoneum except along attachment of falciform ligament
   2. Superior under dome of diaphragm, which separates it from lungs on right and heart on left. Covered by peritoneum except dorsally, at edge of bare area
   3. Posterior (dorsal) fitted against vertebral column and crura of diaphragm. Large area not covered by peritoneum—*bare area*
      a. Concavity for vertebral column
      b. Sulcus (fossa) for inferior vena cava to right of this
      c. Fossa for ductus venosus to left of vena cava
      d. Suprarenal impression to right of vena cava
      e. Esophageal impression to left of fossa for ductus venosus
B. VISCERAL faces posteriorly, caudally, and toward left. Covered by peritoneum except at gallbladder and porta
   1. Porta: fissure in left central part for blood vessels and bile duct
   2. Right portion shows colic, renal, and duodenal impressions
   3. Fossa for gallbladder and fossa for umbilical vein
   4. Left portion: gastric impression, tuber omentale, and caudate process

II. Division into lobes
A. ON VISCERAL SURFACE there is an H-shaped arrangement of fossae and fissures, which delimit the lobes. The left sagittal fossa is made up of the fossa for the umbilical vein (ligamentum teres) anteriorly and the fossa for the ductus venosus posteriorly. The right fossa is divided into an anterior part (fossa for the gallbladder) and a posterior part (fossa for the inferior vena cava) by the *caudate process,* a band of liver tissue joining the caudate and right lobes. The *porta* is the transverse part of the H

B. LOBES
   1. Right: largest and lies to the right of the right sagittal fossa
   2. Quadrate: lies between the gallbladder and the fossa for the umbilical vein. Is ventral to porta
   3. Caudate: lies between vena cava and fossa for the ductus venosus. Is posterior to the porta and is joined to the right lobe by the *caudate process*
   4. Left: to the left of the left sagittal fossa

III. Relations: the various fossae and impressions given above indicate the relations, except for the *tuber omentale,* which is directed toward the lesser curvature of the stomach and overlies the lesser omentum

IV. Liver segments: further subdivisions of the liver lobes into smaller segments based on the liver vascular pattern

V. Clinical considerations
A. JAUNDICE, in relation to the liver, is an accumulation of bile pigment in the blood stream, frequently as a result of obstruction of the duct system
B. BECAUSE OF ITS GREAT VASCULARITY, the liver is a good site for secondary carcinoma from almost any other body site (particularly GI tract)
C. CIRRHOSIS OF THE LIVER is due to an atrophy of the parenchyma and a hypertrophy of the connective tissue
D. HEPATOMEGALY: enlarged liver associated with carcinoma, heart failure, fatty infiltration, or Hodgkin's disease

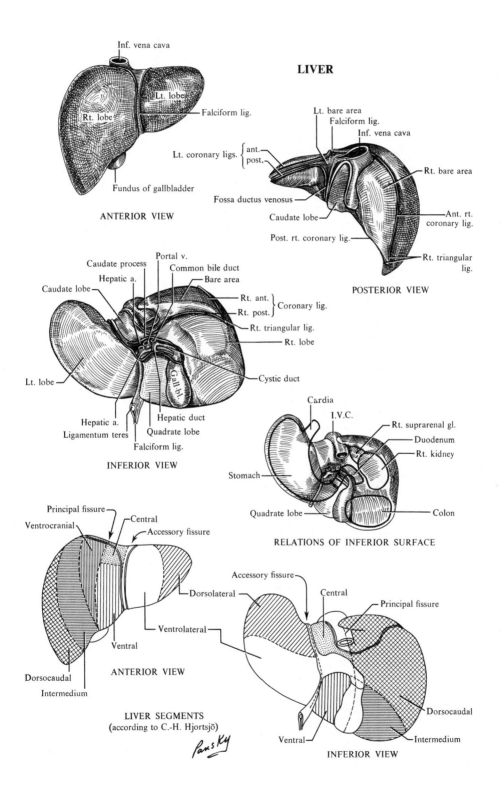

LIVER

Inf. vena cava

Lt. lobe

Rt. lobe

Falciform lig.

Lt. coronary ligs. { ant. / post.

Fundus of gallbladder

ANTERIOR VIEW

Lt. bare area

Falciform lig.

Inf. vena cava

Rt. bare area

Fossa ductus venosus

Caudate lobe

Ant. rt. coronary lig.

Post. rt. coronary lig.

Rt. triangular lig.

POSTERIOR VIEW

Caudate process

Portal v.

Common bile duct

Hepatic a.

Bare area

Caudate lobe

Rt. ant. } Coronary lig.
Rt. post.

Rt. triangular lig.

Rt. lobe

Lt. lobe

Gall bl.

Cystic duct

Hepatic a.

Hepatic duct

Ligamentum teres

Quadrate lobe

Falciform lig.

INFERIOR VIEW

Cardia

I.V.C.

Rt. suprarenal gl.

Duodenum

Rt. kidney

Stomach

Quadrate lobe

Colon

RELATIONS OF INFERIOR SURFACE

Principal fissure

Ventrocranial

Central

Accessory fissure

Dorsolateral

Ventrolateral

Ventral

Dorsocaudal

Intermedium

ANTERIOR VIEW

Accessory fissure

Central

Principal fissure

Dorsocaudal

Ventral

Intermedium

INFERIOR VIEW

LIVER SEGMENTS
(according to C.-H. Hjortsjö)

Pansky

188. LIVER LOBULE AND PORTAL CIRCULATION

I. Liver lobule: unit of structure
A. SHAPE: polygonal, with scant connective tissue between adjoining lobules
B. COMPOSITION
1. Hepatic cells: arranged in "cords" or "plates" with bile canaliculi compressed between 2 adjoining cells
2. Sinusoids: narrow, endothelial-lined channels between liver cords
3. Central vein: large venous channel running longitudinally, in center of lobule
4. Portal triad: in connective tissue, usually at one of the angles of the lobule, consisting of a group of 3 structures: bile duct, branch of portal vein, and branch of hepatic artery
C. BILE FLOW: bile is formed in hepatic cells, drains toward periphery of lobule through the canaliculi between the cells, and empties into small duct of triad
D. CIRCULATION: venous blood, carrying materials absorbed from alimentary canal, enters liver through portal vein and passes through branchings to reach the portal triad. From these branches, the portal blood enters sinusoids to reach central vein of lobule. Arterial blood enters through hepatic artery, carrying oxygenated blood through branches in portal triad, from which it enters sinusoids to reach central vein. The central veins are the actual beginnings of the hepatic vein system. Central veins from several lobules enter *sublobular* veins. Sublobular veins unite into increasingly larger trunks, which finally converge to form 3 hepatic veins that enter the vena cava

II. Collateral portal circulation: important clinically in case of obstruction of the portal vein. These are areas where portal and systemic systems anastomose
A. GASTROESOPHAGEAL REGION: gastric veins of portal and esophageal veins of azygos system
B. ANORECTAL REGION: inferior mesenteric of portal and rectal of internal iliac system
C. PARAUMBILICAL REGION: branches along falciform (paraumbilical) tributaries of the portal vein and superior and inferior epigastric veins
D. RETROPERITONEAL REGION: intestinal veins of portal (rt. colic, ileocolic, lt. colic) with retroperitoneal (lumbar) tributaries of the inferior vena cava

III. Nerves of liver are numerous and contain both sympathetic and parasympathetic (vagal) fibers
A. NERVES REACH THE LIVER via an extensive hepatic plexus, the largest derivative of the celiac plexus
1. Receives filaments from lt. and rt. vagus and rt. phrenic nerves
2. Hepatic plexus accompanies the hepatic artery and portal vein and their branches and enters liver at the porta hepatis

IV. Clinical considerations
A. PORTAL OBSTRUCTION: in an attempt to get portal blood back to the heart by bypassing the obstruction, the veins in the region of anastomoses may be engorged, creating esophageal varicosities, hemorrhoids, and varicosities on the abdominal wall around umbilicus (caput medusae)
B. PORTACAVAL SHUNT: a procedure for relief of portal vein obstruction. Basically, it consists of either anastomosing the portal vein directly with the inferior vena cava or anastomosing one of its tributaries into the vena cava or into another systemic vein, for example, the splenic vein into the left renal vein
C. PORTAL HYPERTENSION: abnormal increase in pressure in portal vein and its tributaries, often due to liver cirrhosis. With no functionally competent valves in portal venous system, increased pressure is reflected throughout system

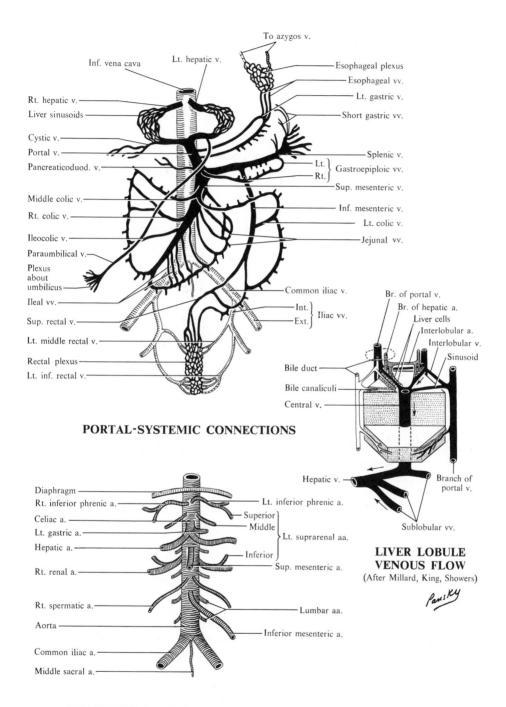

PORTAL-SYSTEMIC CONNECTIONS

**LIVER LOBULE
VENOUS FLOW**
(After Millard, King, Showers)

BRANCHES OF THE ABDOMINAL AORTA

189. SPLEEN

I. Type of structure: a lymphatic organ interposed in the blood stream

II. Surface projection
A. LONG AXIS: in line of left tenth rib
B. ENDS: medially 4 cm from dorsal midline, laterally in midaxillary line in ninth interspace (left side)
C. BORDERS: cephalic, upper left ninth rib; caudal, lower left eleventh rib

III. Relations
A. DIAPHRAGMATIC: convex, smooth, facing upward, backward, and to left. Related to diaphragm, which separates it from ninth, tenth, and eleventh ribs, and left lung and pleura
B. VISCERAL
   1. Gastric: concave; faces anteriorly, cephalically, and medially. Contacts posterior side of stomach and tail of pancreas
   2. Renal: flattened; faces medially and caudally. Related to upper anterior surface of left kidney
   3. Colic: small and slightly concave, at anterior extremity, related to left colic flexure
C. POSTERIOR (SUPERIOR) EXTREMITY: directed toward vertebral column
D. ANTERIOR (INFERIOR) EXTREMITY: rests on left colic flexure and phrenicocolic ligament

IV. Support
A. SPLENORENAL (PHRENICOLIENAL) LIGAMENT: reflection of peritoneum running from diaphragm and anterior aspect of left kidney to hilum of spleen
   1. Contents: splenic vessels
B. GASTROSPLENIC (GASTROLIENAL): dorsal mesentery between spleen and stomach
   1. Contents: short gastric and left gastro-omental (gastroepiploic) vessels
C. PHRENICOCOLIC LIGAMENT: beneath caudal end of spleen

V. Blood supply
A. SPLENIC ARTERY (from celiac artery): large; tortuous course across the posterior wall of the omental bursa and runs through the splenorenal (phrenicolienal) ligament to hilum of spleen. Divides into 6 or more branches in hilum
B. SPLENIC VEIN arises from the union of 6 or more veins that emerge from hilum. Vein runs in groove on back of pancreas, below the artery, and ends behind the neck of the pancreas by joining the superior mesenteric vein to form the *portal vein*

VI. Lymphatic drainage into pancreaticosplenic (lienal) nodes (see p. 432)

VII. Nerves: chiefly postganglionics of the sympathetic system from the celiac plexus to the blood vessels, capsule, and trabeculae of the organ

VIII. Functions: storage of red blood cells, which can be forced back into the circulation in a respiratory crisis by contraction of the smooth muscle in the capsule and trabeculae; destruction of worn-out red blood cells; removal of foreign material from the blood stream; and production of mononuclear leukocytes

IX. Clinical considerations
A. A HYPERTROPHIC SPLEEN due to overactivity of its macrophage system can be removed without any *apparent* ill effects
B. SPLENOMEGALY: any degree of enlargement is abnormal

SPLEEN

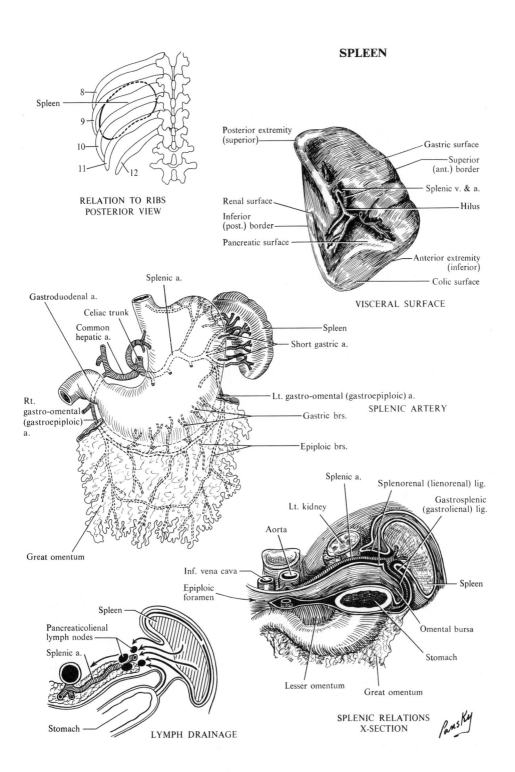

Spleen

8
9
10
11 12

RELATION TO RIBS
POSTERIOR VIEW

Posterior extremity
(superior)

Renal surface

Inferior
(post.) border

Pancreatic surface

Gastric surface

Superior
(ant.) border

Splenic v. & a.

Hilus

Anterior extremity
(inferior)

Colic surface

VISCERAL SURFACE

Splenic a.

Gastroduodenal a.

Celiac trunk

Common
hepatic a.

Rt.
gastro-omental
(gastroepiploic)
a.

Great omentum

Spleen

Short gastric a.

Lt. gastro-omental (gastroepiploic) a.

Gastric brs.

Epiploic brs.

SPLENIC ARTERY

Splenic a.

Lt. kidney

Splenorenal (lienorenal) lig.

Gastrosplenic
(gastrolienal) lig.

Aorta

Inf. vena cava

Epiploic
foramen

Spleen

Omental bursa

Stomach

Lesser omentum

Great omentum

SPLENIC RELATIONS
X-SECTION

Spleen

Pancreaticolienal
lymph nodes

Splenic a.

Stomach

LYMPH DRAINAGE

Pansky

190. THE KIDNEY AND ITS BLOOD VESSELS

I. Nephron: the functional unit of the kidney (a million or more per kidney)

A. PARTS: glomerulus, an arterial capillary net; Bowman's capsule surrounding the capillary net; proximal convoluted tubule; Henle's loop; distal convoluted tubule

II. Duct system: nephron joins the duct system, which eventually opens at the apex of the renal pyramid into the minor calyces; the minor calyces unite to form 2 or 3 major calyces, which, in turn, join to form the renal pelvis, the expanded cephalic end of the ureter

III. General structure

A. CAPSULE: tough, fibrous tissue

B. HILUM: medial fissure, which expands into a large central cavity, the *renal sinus*, which contains the proximal part of the renal pelvis, the calyces, branches of renal vessels, nerves, and fat

C. MEDULLA: inner part, 8 to 18 pyramidal masses (renal pyramids), striated in appearance; contains collecting ducts, portions of Henle's loop, and parts of the secretory tubules

D. CORTEX lies beneath capsule, overlies bases of medullary pyramids and dips down between them as the *renal columns*
1. Pars radiata: conical projections from bases of pyramids into cortex. Contains ducts and parts of Henle's loop
2. Pars convoluta: surrounds radiate part. Contains glomeruli, capsules, and convoluted parts of nephron

E. KIDNEY LOBE is a pyramid and its associated cortex. Five to six are seen in fetus and may persist in the adult, but this is unusual

IV. Blood vessels

A. ARTERIES: renal artery branches into *interlobar arteries* between pyramids. At bases of pyramids, the interlobar arteries form the *arcuate arteries*. Along their course, the arcuate arteries send branches into the cortex as the *interlobular arteries*. The latter give rise chiefly to *afferent glomerular arteries* and to some *nutrient* and *perforating capsular arteries*. The capillary net of the glomerulus coalesces to form the *efferent glomerular arteries*. These vessels break up into a true capillary net around the nephrons and also give rise to a few *arteriolae rectae*, which enter the medulla and run toward the pelvis

B. VEINS: begin in venous plexuses draining the capillary bed around the tubules. These open into *venae rectae*, then into *interlobular veins, arcuate veins, interlobar veins*, and finally the *renal vein*
1. Stellate veins lie beneath the capsule and drain part of the area supplied by perforating capsular arteries. These veins drain into the interlobular veins

V. Anastomoses between renal and systemic vessels occur in fat around kidney where the perforating capsular arteries join branches from suprarenal, spermatic (or ovarian), superior and inferior mesenteric arteries

VI. Nerve supply: extensive, from the extensions of the celiac (aorticorenal) and intermesenteric plexuses that accompany the renal artery, as well as from direct branches of the thoracic and lumbar splanchnic nerves

A. PAIN FIBERS, mainly from the renal pelvis and upper part of the ureter, enter the spinal cord via the splanchnic nerves

VII. Special features

A. WITHIN THE RENAL SINUS, each arterial ramus rebranches, and although the pattern is quite variable, it is said that the distribution is constant enough to allow the division of the kidney into vascular segments that correspond to the prevailing vascular pattern. *Five segmental arteries*, therefore, *five renal segments* are described

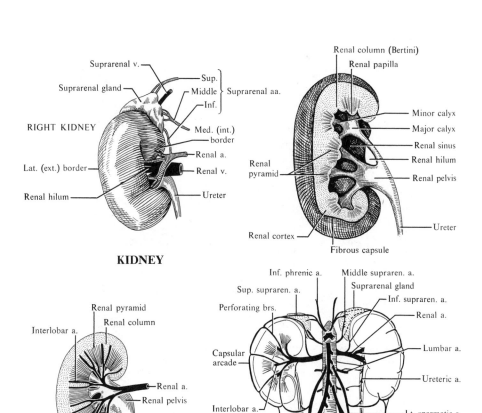

Suprarenal v.

Suprarenal gland

Sup. ⎫
Middle ⎬ Suprarenal aa.
Inf. ⎭

RIGHT KIDNEY

Med. (int.) border

Lat. (ext.) border

Renal a.

Renal v.

Renal hilum

Ureter

Renal column (Bertini)

Renal papilla

Minor calyx

Major calyx

Renal sinus

Renal hilum

Renal pelvis

Renal pyramid

Ureter

Renal cortex

Fibrous capsule

KIDNEY

Renal pyramid

Renal column

Interlobar a.

Renal a.

Renal pelvis

Ureter

Fibrous capsule

Renal cortex

Inf. phrenic a.   Middle supraren. a.

Sup. supraren. a.   Suprarenal gland

Perforating brs.   Inf. supraren. a.

Renal a.

Capsular arcade

Lumbar a.

Ureteric a.

Interlobar a.

Rt. spermatic a.   Lt. spermatic a.

Rt. colic a.   Lt. colic a.

SUPRARENAL AND RENAL CAPSULAR
ARTERIES (After Papin)

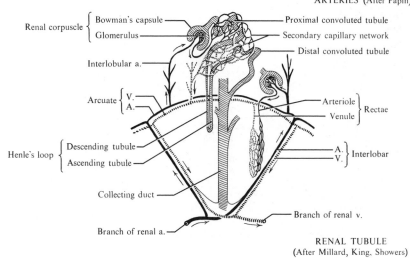

Renal corpuscle ⎰ Bowman's capsule
⎱ Glomerulus

Proximal convoluted tubule

Secondary capillary network

Distal convoluted tubule

Interlobular a.

Arcuate ⎰ V.
⎱ A.

Arteriole ⎱
Venule ⎰ Rectae

Henle's loop ⎰ Descending tubule
⎱ Ascending tubule

A. ⎱
V. ⎰ Interlobar

Collecting duct

Branch of renal v.

Branch of renal a.

RENAL TUBULE
(After Millard, King, Showers)

Pansky

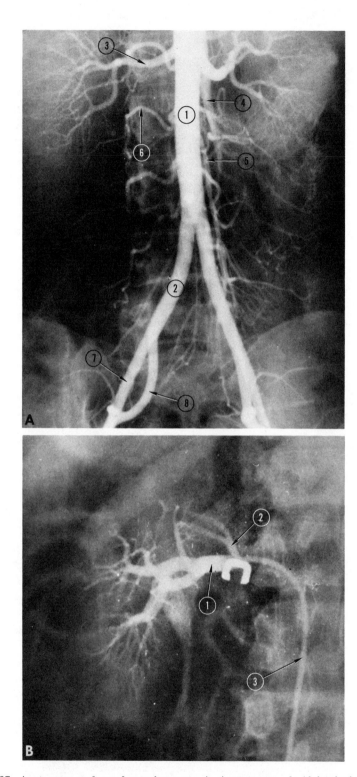

FIGURE 37. **Aortogram and renal arteriogram.** **A, Aortogram.** *1,* Abdominal aorta; *2,* right common iliac artery; *3,* right renal artery; *4,* superior mesenteric artery; *5,* inferior mesenteric artery; *6,* lumbar artery; *7,* right external iliac artery; *8,* right internal iliac artery. **B, Renal arteriogram.** *1,* Renal artery; *2,* suprarenal artery; *3,* catheter.

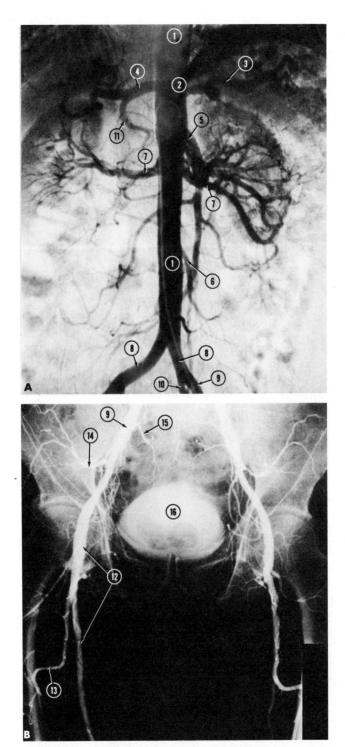

FIGURE 38. **A, Aortogram; B, Iliac arteriogram.** *1*, Abdominal aorta; *2*, celiac; *3*, splenic; *4*, hepatic; *5*, superior mesenteric; *6*, inferior mesenteric; *7*, renals; *8*, common iliac; *9*, external iliac; *10*, internal iliac; *11*, gastroduodenal; *12*, femoral; *13*, deep (profunda) femoral; *14*, iliolumbar; *15*, sacral; *16*, bladder.

191. RELATIONS AND FIXATION OF THE KIDNEY

I. Relations
A. RIGHT
   1. Anterior: convex, faces ventrally and to the right. Related to *suprarenal gland, visceral surface of the liver, colic flexure,* and the *small intestine*
   2. Posterior: more flattened. Embedded in fat with no peritoneum. Cephalic pole lies on twelfth rib; below this are the diaphragm, lumbocostal arches, psoas and quadratus lumborum muscles, and tendon of transversus abdominis muscle. Crossed by upper lumbar arteries, T12, iliohypogastric, and ilioinguinal nerves
   3. Lateral border: convex without important relations
   4. Medial border: *hilum* for renal vessels and ureter near center. Above the hilum, it is in contact with the suprarenal, and below the hilum, with the ureter
B. LEFT
   1. Anterior: convex; faces anteriorly and to the left. Related to *suprarenal, spleen, body of pancreas* (splenic vessels), *stomach, left colic flexure, small intestine*
   2. Posterior: less convex. Embedded in fat with no peritoneum. Superior pole rests on eleventh rib, the twelfth rib more caudally. In other respects, similar to right
   3. Left border: similar to right
   4. Medial: similar to right

II. Relations of structures at hilum: renal vein most anterior, artery intermediate, and ureter most posterior. Branches of arteries and veins may pass posterior to the ureter

III. Renal vessels
A. VEINS terminate in the vena cava
   1. Left: longer than right, crosses anterior side of aorta just below the superior mesenteric artery and opens into the inferior vena cava above the right vein
     a. Tributaries: left inferior phrenic, left internal spermatic, and left suprarenal
   2. Right: short, lies in front of renal artery. No extrarenal tributaries
B. ARTERIES arise from the aorta at right angles, at level of disk between L1 and L2
   1. Left: slightly above level of right renal artery. Lies posterior to renal vein, body of pancreas, and splenic vein. Inferior mesenteric vein crosses it anteriorly
   2. Right: longer than left. Passes behind inferior vena cava and right renal vein with head of pancreas and descending duodenum overlying the veins

IV. Renal fascia
A. FORMATION: from subserous extraperitoneal fascia; splits near the lateral border of the kidney
   1. Anterior layer over anterior surface and continues over renal vessels and aorta to join similar layer of the other side
   2. Posterior layer continues beneath the kidney, but passes posterior to the aorta to the other side, but is adherent to the deep fascia
B. CONNECTIONS TO KIDNEY: fibrous strands of renal fascia connect to capsule
C. ADIPOSE CAPSULE (PERIRENAL FAT): fatty tissue lying between the fascia and surface of the kidney
   1. Pararenal fat: fat behind renal fascia

V. Support: renal fascia and vessels, adipose capsule, and pararenal fat

VI. Clinical considerations
A. CONGENITAL ANOMALIES
   1. Unascended kidney: kidney may remain near original site of development and might be found near pelvic brim with common iliac vessels
   2. Horseshoe kidney: the caudal poles of the two kidneys are joined

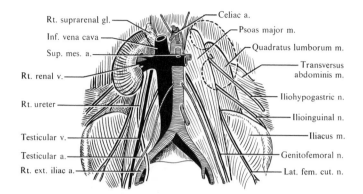

Rt. suprarenal gl. — Celiac a.
Inf. vena cava — Psoas major m.
Sup. mes. a. — Quadratus lumborum m.
Rt. renal v. — Transversus abdominis m.
Rt. ureter — Iliohypogastric n.
— Ilioinguinal n.
Testicular v. — Iliacus m.
Testicular a. — Genitofemoral n.
Rt. ext. iliac a. — Lat. fem. cut. n.

KIDNEY RELATIONS

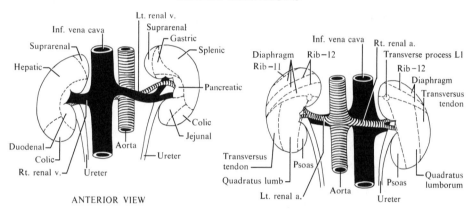

Lt. renal v.
Inf. vena cava — Suprarenal
Suprarenal — Gastric
Hepatic — Splenic
— Pancreatic
Duodenal — Colic
Colic — Jejunal
Rt. renal v. — Aorta — Ureter
Ureter

ANTERIOR VIEW

Inf. vena cava — Rt. renal a.
Diaphragm — Rib −12 — Transverse process L1
Rib −11 — Rib −12
— Diaphragm
— Transversus tendon
Transversus tendon — Psoas
Quadratus lumb — Quadratus lumborum
Lt. renal a. — Psoas
Aorta — Ureter

POSTERIOR VIEW

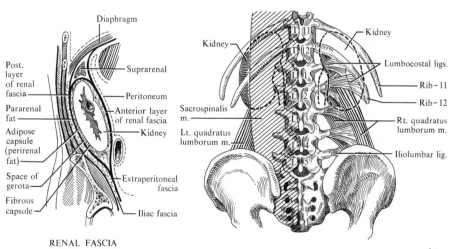

Diaphragm
Post. layer of renal fascia — Suprarenal
— Peritoneum
Pararenal fat — Anterior layer of renal fascia
Adipose capsule (perirenal fat) — Kidney
Space of gerota — Extraperitoneal fascia
Fibrous capsule — Iliac fascia

RENAL FASCIA LONGITUDINAL SECTION

Kidney — Kidney
— Lumbocostal ligs.
— Rib −11
— Rib −12
Sacrospinalis m. — Rt. quadratus lumborum m.
Lt. quadratus lumborum m. — Iliolumbar lig.

RELATIONS TO MUSCLES, LIGAMENTS

Pansky

192. URETERS

I. Origin: renal pelvis at the level of the spine of L1

II. Parts and relations
- A. ABDOMINAL: both lie beneath the peritoneum, embedded in subserous tissue on the medial part of psoas major muscle; both are crossed by the spermatic (ovarian) vessels. This part ends as it enters the true pelvis, crossing the bifurcation of the common iliac vessels
  1. Right: near origin, covered by descending duodenum, to the right of the inferior vena cava. Crossed by right colic and ileocolic vessels, the mesentery, and terminal ileum
  2. Left: crossed by left colic vessels and sigmoid mesocolon
- B. PELVIC
  1. Male: runs caudally on the lateral pelvic wall along the anterior border of the greater sciatic notch. Is anterior to internal iliac artery and medial to obturator, inferior vesical, and middle rectal arteries. At level of lower part of sciatic notch, it turns medially to reach the lateral angle of the bladder. Here it lies anterior to the seminal vesicle. The vas crosses over it as it approaches the bladder
  2. Female (see p. 452): forms the posterior boundary of the ovarian fossa. Runs medially and anteriorly on lateral aspect of cervix and upper vagina to fundus of the bladder. In part of its course is accompanied by uterine artery, which then crosses over the ureter
- C. INTRAMURAL (see p. 438): runs obliquely through the bladder wall for a distance of 2 cm to open at lateral angles of trigone of bladder

III. Constrictions: areas of diminished diameter
- A. URETEROPELVIC JUNCTION
- B. AT CROSSING OF ILIAC VESSELS
- C. AT JUNCTION WITH BLADDER

IV. Vessels and nerves
- A. ARTERIES: ureteric branches of renal, internal spermatic, superior and inferior vesical arteries
- B. VEINS follow correspondingly named arteries and terminate in correspondingly named veins
- C. LYMPHATICS pass to the lumbar and internal iliac nodes
- D. NERVES: spermatic, renal, and hypogastric plexuses

V. Clinical considerations
- A. KIDNEY STONES may descend in the ureter and become lodged, particularly in the areas of ureteric constriction, and result in a great deal of pain and urinary retention, which can damage the kidney structure
- B. CONGENITAL ANOMALIES: ureters may be double over part or all of extent
- C. SURGICAL APPROACH TO BOTH KIDNEY AND ADRENAL: from back and side, by an incision below and parallel to the 12th rib. If necessary, it can be extended to the front of the abdomen, paralleling the inguinal ligament. In this way, the entire procedure can remain retroperitoneal; it avoids cutting nerves since the incision is parallel to their course; and the kidney structures can be separated from overlying structures such as duodenum and pancreas, since kidney belongs to the dorsal body wall, whereas other organs have become only secondarily adherent
- D. OBSTRUCTION of the ureter at any level leads to dilation of the parts above, including the renal pelvis and calices, resulting in *hydronephrosis*

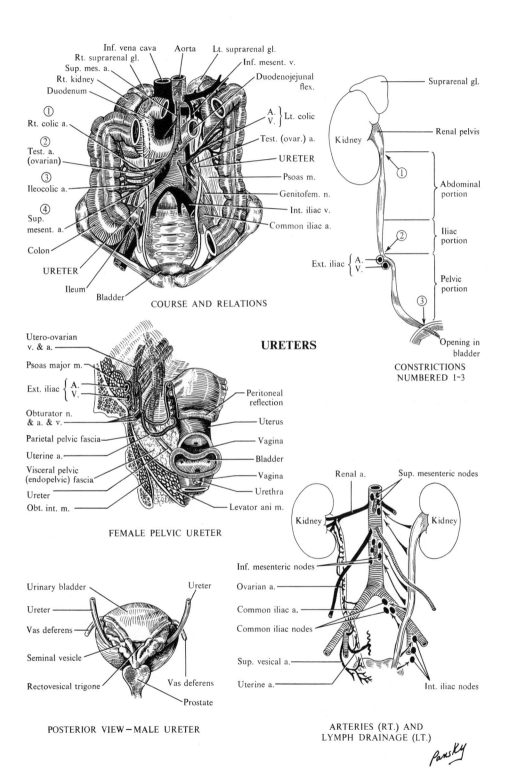

URETERS

Inf. vena cava — **Aorta** — **Lt. suprarenal gl.**

Rt. suprarenal gl.

Sup. mes. a.

Rt. kidney

Duodenum

① Rt. colic a.

② Test. a. (ovarian)

③ Ileocolic a.

④ Sup. mesent. a.

Colon

URETER

Ileum

Bladder

Inf. mesent. v.

Duodenojejunal flex.

A. } Lt. colic
V. }

Test. (ovar.) a.

URETER

Psoas m.

Genitofem. n.

Int. iliac v.

Common iliac a.

COURSE AND RELATIONS

Suprarenal gl.

Kidney

Renal pelvis

① Abdominal portion

Ext. iliac { A.
V. } ② Iliac portion

Pelvic portion

③ Opening in bladder

CONSTRICTIONS NUMBERED 1-3

Utero-ovarian v. & a.

Psoas major m.

Ext. iliac { A.
V. }

Obturator n. & a. & v.

Parietal pelvic fascia

Uterine a.

Visceral pelvic (endopelvic) fascia

Ureter

Obt. int. m.

Peritoneal reflection

Uterus

Vagina

Bladder

Vagina

Urethra

Levator ani m.

FEMALE PELVIC URETER

Urinary bladder

Ureter

Vas deferens

Seminal vesicle

Rectovesical trigone

Ureter

Vas deferens

Prostate

POSTERIOR VIEW — MALE URETER

Renal a.   Sup. mesenteric nodes

Kidney   Kidney

Inf. mesenteric nodes

Ovarian a.

Common iliac a.

Common iliac nodes

Sup. vesical a.

Uterine a.

Int. iliac nodes

ARTERIES (RT.) AND LYMPH DRAINAGE (LT.)

Pansky

193. SUPRARENAL GLAND

I. Position, size, shape, and relations
A. LOCATION: at cranial pole of kidney, at L1, within renal fascia
B. SIZE: length and width, 3.0–5.0 cm; thickness, 0.4–0.6 cm; weight, 3.5–6.0 g, heavier in male than female
C. RELATIONS AND SHAPE
   1. Right: pyramidal, with hilum below apex near anterior border
      a. Anterior: medially, inferior vena cava without peritoneum; laterally liver, without peritoneum above, with peritoneum below
      b. Posterior: diaphragm above; cranial pole and anterior surface of right kidney below
   2. Left: semilunar in shape; hilum near caudal end of anterior surface. Slightly larger than right
      a. Anterior: peritoneum of omental bursa above; pancreas and splenic vein, without peritoneum, below
      b. Posterior: medially, left crus of diaphragm; laterally, anterior surface of left kidney

II. Blood and nerve supply
A. ARTERIES
   1. Superior suprarenal artery from inferior phrenic artery (from aorta)
   2. Middle suprarenal artery, directly from aorta
   3. Inferior suprarenal artery from renal artery (from aorta)
B. VEINS
   1. Suprarenal vein receives blood from all parts of the gland and leaves through the hilum
      a. Right enters inferior vena cava
      b. Left enters the left renal vein
C. NERVE SUPPLY: via the celiac plexus and thoracic and lumbar splanchnic nerves. The fibers are mostly preganglionic sympathetic fibers that go directly to the cells of the medulla (apparently the cortex receives no nerve supply)

III. Parts
A. THICK CORTEX derived from the mesoderm
B. MEDULLA derived from the embryonic neural crest ectoderm
C. CAPSULE: tough, connective tissue

IV. Secretions
A. CORTEX: cortisol, corticosterone, aldosterone, 11-dehydroepiandrosterone, progesterone, estradiol, and estrone
B. MEDULLA: epinephrine and norepinephrine

V. General functions
A. CORTEX: gluconeogenesis; enhances water diuresis, probably by increasing glomerular filtration; maintains electrolyte balance, thus maintaining blood volume and pressure; fat deposition; has effect on lymphocytes (lympholytic); anti-inflammatory and antiallergic. Normally has little effect on sex. Cortex essential to life
B. MEDULLA: essential to the "fight-or-flight" mechanism; raises blood pressure; increases heart rate; dilates bronchi and breaks down glycogen, thus elevating blood sugar

VI. Clinical considerations (chief pathology involves cortex only)
A. HYPERACTIVITY: Cushing's disease, adrenogenital syndrome, primary or secondary aldosteronism
B. HYPOFUNCTION: acute adrenal insufficiency (Waterhouse-Friderichsen syndrome); chronic adrenal insufficiency (Addison's disease)

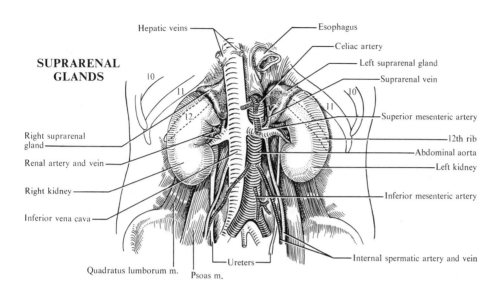

SUPRARENAL GLANDS

Hepatic veins

Esophagus

Celiac artery

Left suprarenal gland

Suprarenal vein

Superior mesenteric artery

10

11

12

10

11

12th rib

Abdominal aorta

Left kidney

Inferior mesenteric artery

Internal spermatic artery and vein

Right suprarenal gland

Renal artery and vein

Right kidney

Inferior vena cava

Quadratus lumborum m.

Psoas m.

Ureters

SUPRARENAL VESSELS

Esophagus

Phrenic a.

Sup.
Mid. } Suprarenal aa.
Inf.

Suprarenal vein

Suprarenal vein

Renal vein

Left kidney

Inferior vena cava

Int. sperm. a. and v.

SUPRARENAL ARTERIES

Phrenic a

Superior
Middle } Suprarenal aa.
Inf.

Renal a.

Aorta

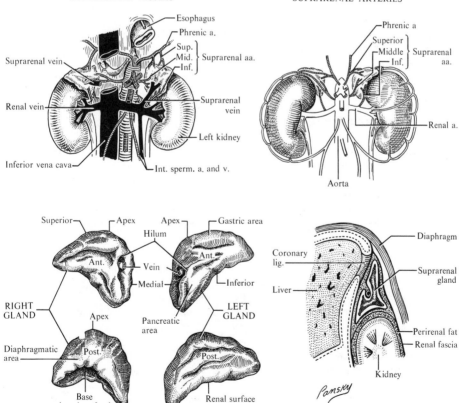

Superior

Apex

Apex

Gastric area

Hilum

Ant.

Ant.

Vein

Medial

Inferior

RIGHT GLAND

Apex

LEFT GLAND

Pancreatic area

Diaphragmatic area

Post.

Post.

Base (renal surface)

Renal surface

Coronary lig.

Liver

Diaphragm

Suprarenal gland

Perirenal fat

Renal fascia

Kidney

Pansky

– 431 –

194. LYMPHATICS OF THE ABDOMEN

I. Lymph nodes
A. PARIETAL
1. Epigastric: along inferior epigastric vessels
2. Lumbar
   a. Right lateral aortic: anterior to inferior vena cava at level of renal vessels, posterior to inferior vena cava, on the origins of the right psoas muscle and right crus of diaphragm. Afferents: from common iliac nodes; ovary, testis, uterus, kidney, suprarenal, and abdominal muscles. Efferents: chiefly form the right lumbar trunk, but some pass to pre- and retroaortic nodes or thoracic duct
   b. Left lateral aortic: on left side of aorta, on origin of left psoas muscle and left crus of diaphragm. Afferents and efferents similar to above but for left side
   c. Preaortic: in front of aorta around origin of 3 major arterial branches and named accordingly: celiac, superior mesenteric, and inferior mesenteric. Afferents: few from lateral aortics, mainly from viscera supplied by the related arteries. Efferents: few to retroaortics, mainly to intestinal trunk
   d. Retroaortic: on bodies of third and fourth lumbar vertebrae behind aorta. Afferents: from lateral and preaortic nodes. Efferents to cisterna chyli

II. Lymphatic drainage of viscera
A. STOMACH: see p. 396
B. LIVER
1. Convex surface: to posterior mediastinal nodes, superior gastric nodes, celiac group of preaortic nodes, and hepatic nodes
2. Visceral surface: hepatic and posterior mediastinal nodes
C. GALLBLADDER: to hepatic and pancreaticoduodenal nodes; to hepatic nodes from common duct
D. DUODENUM: to pancreaticoduodenal nodes and thence to hepatic and preaortic (superior mesenteric) nodes
E. JEJUNUM AND ILEUM: vessels are the lacteals to mesenteric nodes in mesentery and then to superior mesenteric group of preaortic nodes
F. COLON
1. Ascending and transverse: through right colic and middle colic nodes to mesenteric nodes to superior mesenteric group of preaortics
2. Descending and sigmoid; left colic nodes along left colic and sigmoid arteries to inferior mesenteric group of preaortics
G. PANCREAS: to pancreaticosplenic (lienal) nodes (thence to celiac group of preaortics), to pancreaticoduodenal nodes, and to superior mesenteric group of preaortic nodes

III. Cisterna chyli
A. LOCATION: in front of second lumbar vertebra, behind and to right of aorta beside right crus of diaphragm
B. FORMATION: right and left lumbar trunks and intestinal trunk
C. TERMINATION: narrows down and passes through aortic hiatus of diaphragm to become thoracic duct

IV. Clinical consideration
A. REGIONAL LYMPH NODES: a surgeon may judge the extent of metastases from a malignancy by examining nodes draining the area; for example, from the sigmoid colon, first check for nodes in sigmoid mesocolon, and then examine the inferior mesenteric group found at the origin of the inferior mesenteric artery

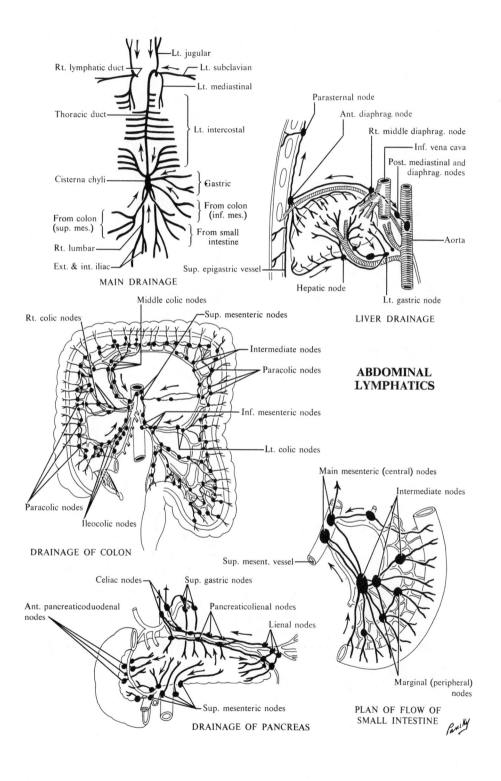

MAIN DRAINAGE

Lt. jugular
Rt. lymphatic duct
Lt. subclavian
Lt. mediastinal
Thoracic duct
Lt. intercostal
Cisterna chyli
Gastric
From colon (inf. mes.)
From colon (sup. mes.)
From small intestine
Rt. lumbar
Ext. & int. iliac
Sup. epigastric vessel

LIVER DRAINAGE

Parasternal node
Ant. diaphrag. node
Rt. middle diaphrag. node
Inf. vena cava
Post. mediastinal and diaphrag. nodes
Aorta
Hepatic node
Lt. gastric node

ABDOMINAL LYMPHATICS

DRAINAGE OF COLON

Middle colic nodes
Rt. colic nodes
Sup. mesenteric nodes
Intermediate nodes
Paracolic nodes
Inf. mesenteric nodes
Lt. colic nodes
Paracolic nodes
Ileocolic nodes

PLAN OF FLOW OF SMALL INTESTINE

Main mesenteric (central) nodes
Intermediate nodes
Sup. mesent. vessel
Marginal (peripheral) nodes

DRAINAGE OF PANCREAS

Celiac nodes
Sup. gastric nodes
Pancreaticolienal nodes
Ant. pancreaticoduodenal nodes
Lienal nodes
Sup. mesenteric nodes

Pansky

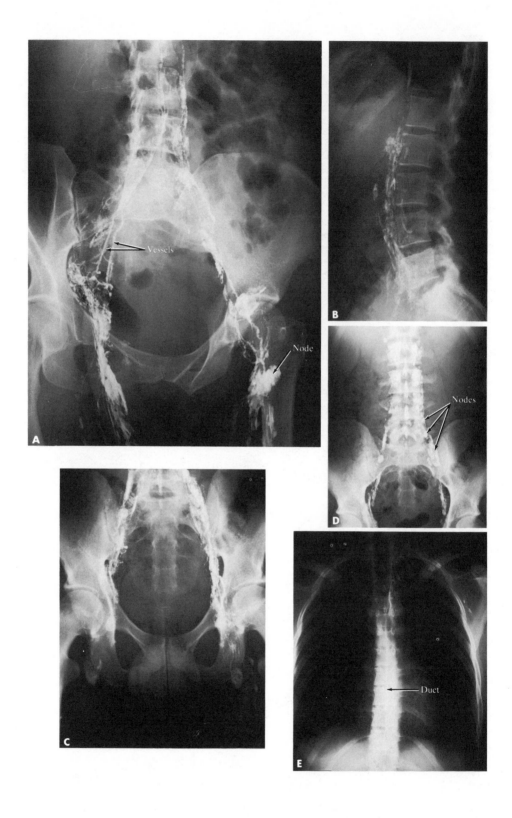

FIGURE 39. **Lower lumbar lymphangiograms. A, C, and D, Anteroposterior views of nodes and vessels; B, lateral view of nodes and vessels; E, thoracic duct seen lying on thoracic vertebrae.** Inguinal and iliac nodes are also shown.

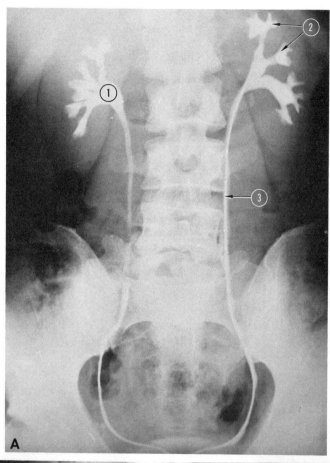

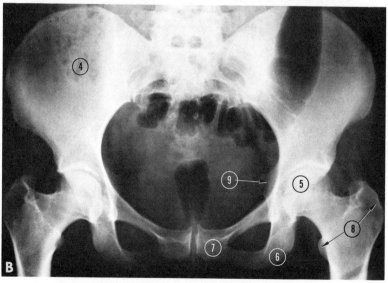

FIGURE 40. **Retrograde pyelogram and normal pelvis.   A, Retrograde pyelogram; B, normal pelvis, adult.**   *1,* Pelvis of kidney; *2,* major and minor calyces; *3,* ureter; *4,* ilium; *5,* head of femur; *6,* ischium; *7,* pubis; *8,* greater and lesser trochanters of femur; *9,* spine of ischium.

195. THE DIAPHRAGM

I. Location, configuration, and composition
A. SERVES AS A SEPTUM between the thoracic and abdominal cavities
B. DOME-SHAPED, with concavity facing caudally
C. COMPOSED OF SKELETAL MUSCLE and dense collagenous connective tissue

II. Origin of muscular fibers
A. STERNAL PART: 2 muscular bands from the dorsal side of the xiphoid process
B. COSTAL PART: from costal cartilages and bone of ribs 7–12
C. LUMBAR PART: from the lumbocostal arches and crura
   1. Lumbocostal ligaments (arches)
      a. Medial arcuate ligament: tendinous arch crossing the psoas muscle. It is attached medially to the body of first (and second) lumbar vertebrae and laterally to the anterior transverse process of first (and second) lumbar vertebrae
      b. Lateral arcuate ligament: tendinous arch crossing the quadratus lumborum muscle. It is attached medially to the anterior transverse process of the first lumbar vertebra and laterally to the tip of rib 12
   2. Crura
      a. Right: larger than left and arises from the bodies and disks of lumbar vertebrae 1–3. The most medial fibers cross in front of the aorta to the left side
      b. Left: arises from the bodies and disks of lumbar vertebrae 1 and 2. Some of its medial fibers cross to other side

III. Insertion: central tendon, where muscular fibers converge and become tendinous near the center of diaphragm

IV. Innervation and action
A. INNERVATION: phrenic nerve of cervical plexus, C4 (also C3 and C5)
B. ACTION: contraction of the muscle causes descent of the central tendon. This decreases pressure and increases the volume of the thoracic cavity, resulting in air being "pushed" into the lungs

V. Orifices in the diaphragm, with the major structures passing through
A. ESOPHAGEAL HIATUS: at the level of the tenth thoracic vertebra. Transmits esophagus and right and left vagus nerves
B. AORTIC HIATUS: at the level of the twelfth thoracic vertebra, just to the left of the midline. Transmits aorta, azygos vein, and thoracic duct. (Some anatomists consider this hiatus as being behind diaphragm)
C. VENA CAVAL FORAMEN (HIATUS): about at the level of the disk between the eighth and ninth thoracic vertebrae, to the right of the midline. Transmits inferior vena cava and small branches of right phrenic nerve
D. MINOR OPENINGS
   1. Right crus: 2, for the right greater and lesser splanchnic nerves
   2. Left crus: 3, for the left greater and lesser splanchnic nerves and the hemiazygos vein
   3. Anteriorly: between sternal and costal parts of the diaphragm for the passage of the superior epigastric artery

VI. Clinical consideration
A. HIATUS HERNIA: an opening, usually on the left side, of the diaphragm near the esophageal hiatus, permitting the abdominal viscera to ascend into the thorax

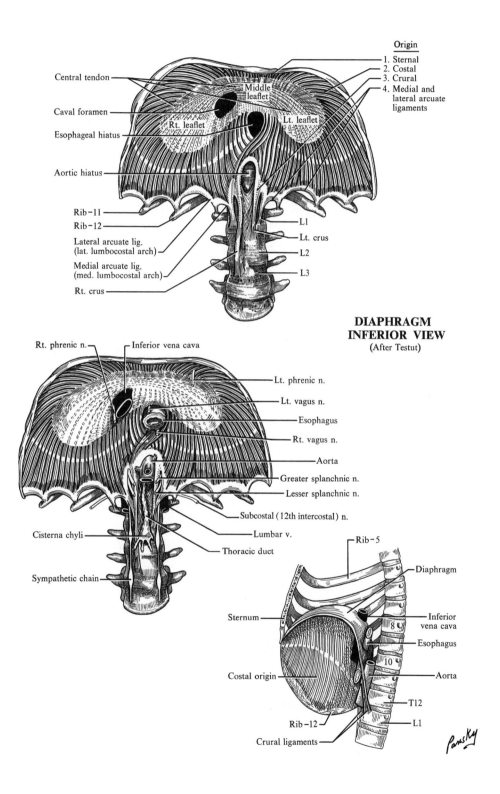

Origin
1. Sternal
2. Costal
3. Crural
4. Medial and lateral arcuate ligaments

Central tendon

Middle leaflet

Caval foramen

Rt. leaflet Lt. leaflet

Esophageal hiatus

Aortic hiatus

Rib-11

Rib-12

Lateral arcuate lig. (lat. lumbocostal arch)

Medial arcuate lig. (med. lumbocostal arch)

Rt. crus

L1

Lt. crus

L2

L3

**DIAPHRAGM
INFERIOR VIEW**
(After Testut)

Rt. phrenic n. Inferior vena cava

Lt. phrenic n.

Lt. vagus n.

Esophagus

Rt. vagus n.

Aorta

Greater splanchnic n.

Lesser splanchnic n.

Subcostal (12th intercostal) n.

Cisterna chyli

Lumbar v.

Thoracic duct

Sympathetic chain

Rib-5

Diaphragm

Sternum

Inferior vena cava

8

Esophagus

10

Aorta

Costal origin

T12

Rib-12

L1

Crural ligaments

196. BLADDER AND MALE URETHRA

I. Bladder, male: surfaces and relations
A. FUNDUS (posterior): triangular, directed caudally and posteriorly. Separated from rectum by rectovesical septum, seminal vesicles, and vas deferens
B. APEX: directed toward pubic symphysis
   1. Median umbilical ligament (urachus) continued up abdominal wall from the apex to the umbilicus
C. SUPERIOR SURFACE: bounded laterally by lateral borders that delimit it from inferior surface, bounded posteriorly by a line connecting the ureters. Is covered by peritoneum and is related to the sigmoid colon and coils of ileum
D. INFEROLATERAL SURFACE: directed caudally and is without peritoneum; separated from pubis by prevesical cleft (space of Retzius)
E. NECK: triangular, in contact with base of prostate, contains urethral orifice

II. Bladder, female: surfaces and relations
A. FUNDUS separated above from anterior surface of uterus by vesicouterine pouch; below and behind it is related to cervix and upper vaginal wall
B. SUPERIOR SURFACE: uterus rests on this when bladder is empty
C. INFERIOR SURFACE rests on pelvic and urogenital diaphragms

III. Fixation (see p. 441): held in place by ligaments attached to its inferior surface
A. TRUE LIGAMENTS
   1. Pubovesicals: between pubis and bladder, directly in the female; in the male are attached to the prostate as the medial and lateral puboprostatic ligaments
   2. Rectovesical: from bladder to sides of rectum and sacrum
   3. Median umbilical: from apex of bladder to abdominal wall
B. FALSE LIGAMENTS: a group of peritoneal folds from the bladder to the abdominal or pelvic walls: 1 median, 2 medial, 2 lateral, and 2 sacrogenital (posterior false ligaments)

IV. Special features of interior
A. TRIGONE: a smooth triangular area above the urethral orifice. The posterolateral angles are formed by the ureteric orifices; the base is formed by the interureteric ridge, between the orifices; and the anterior angle is at the internal urethral orifice

V. Vessels and nerves (see p. 440)
A. ARTERIES: superior and inferior vesical, middle rectal
B. VEINS: vesical plexus to vesical veins to internal iliac veins, communications with prostatic plexus
C. LYMPHATICS: to external, internal, sacral, and median common iliac nodes
D. NERVES: via inferior hypogastric and vesical plexuses

VI. Distended bladder, male: surface and relations
A. FUNDUS: little change
B. SUMMIT directed anteriorly and cephalically above attachment of median umbilical ligament to apex, resulting in a peritoneal pouch between the anterior body wall and summit
C. POSTEROSUPERIOR SURFACE directed cephalically and posteriorly, covered by peritoneum, separated from rectum by rectovesical pouch
D. ANTEROINFERIOR SURFACE: nonperitoneal, below, related to pubic bones; above, with anterior abdominal wall
E. LATERAL SURFACES: below, nonperitoneal; related to lateral walls of pelvis

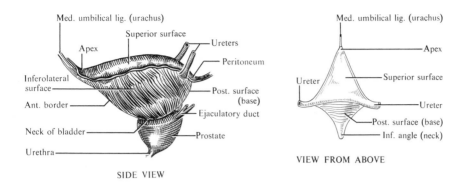

Med. umbilical lig. (urachus)

Superior surface

Apex

Ureters

Peritoneum

Inferolateral surface

Ant. border

Post. surface (base)

Ejaculatory duct

Neck of bladder

Prostate

Urethra

SIDE VIEW

Med. umbilical lig. (urachus)

Apex

Superior surface

Ureter

Ureter

Post. surface (base)

Inf. angle (neck)

VIEW FROM ABOVE

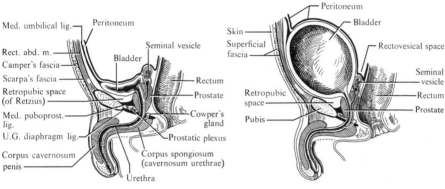

Med. umbilical lig.

Peritoneum

Rect. abd. m.

Seminal vesicle

Camper's fascia

Bladder

Scarpa's fascia

Rectum

Retropubic space (of Retzius)

Prostate

Med. puboprost. lig.

Cowper's gland

U.G. diaphragm lig.

Prostatic plexus

Corpus cavernosum penis

Corpus spongiosum (cavernosum urethrae)

Urethra

MIDSAGITTAL SECTION – NORMAL

Peritoneum

Bladder

Skin

Superficial fascia

Rectovesical space

Seminal vesicle

Retropubic space

Rectum

Pubis

Prostate

MIDSAGITTAL SECTION – DISTENDED

BLADDER

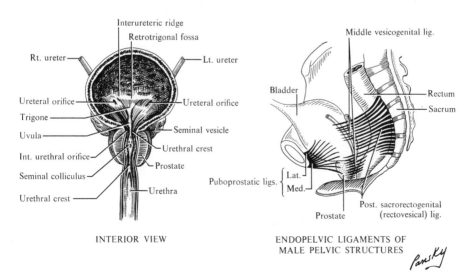

Interureteric ridge

Retrotrigonal fossa

Rt. ureter

Lt. ureter

Ureteral orifice

Ureteral orifice

Trigone

Uvula

Seminal vesicle

Int. urethral orifice

Urethral crest

Seminal colliculus

Prostate

Urethral crest

Urethra

INTERIOR VIEW

Middle vesicogenital lig.

Bladder

Rectum

Sacrum

Puboprostatic ligs. { Lat.
Med.

Post. sacrorectogenital (rectovesical) lig.

Prostate

ENDOPELVIC LIGAMENTS OF
MALE PELVIC STRUCTURES

Pansky

197. LYMPHATICS, NERVES, AND FASCIA OF BLADDER AND PROSTATE

I. Lymphatics
A. BLADDER: arise from the submucous plexus and form 3 groups of vessels
   1. From superior and inferolateral surfaces to external iliac nodes
   2. From fundus to external and internal iliac nodes
   3. From neck, with vessels of prostate, to sacral and medial common iliac nodes
B. PROSTATE, with those of 3, above, and from seminal vesicle to sacral, internal iliac, and common iliac nodes

II. Nerves
A. AUTONOMIC
   1. Sympathetic: preganglionics from lower thoracic and upper lumbar levels. Postganglionics from superior hypogastric plexus, which is a caudal continuation of the aortic and inferior mesenteric plexuses via the hypogastric nerves into the inferior hypogastric plexus. Some postganglionics are derived from the sacral trunk
   2. Parasympathetic: preganglionics run from S2, S3, and S4 as white rami communicantes, through the vesical division of pelvic plexus, with ganglia located on or in wall of bladder
B. SOMATIC MOTOR to external sphincter via branches from pudendal nerve arising from S2, S3, and S4
C. AFFERENT (visceral) travel with both sympathetic and parasympathetic fibers. Those associated with emptying the bladder travel with the parasympathetics

III. Micturition (emptying of bladder)
A. AFFERENT LIMB begins in stretch receptors in muscular wall and travels as IIC, above
B. EFFERENT LIMB: mainly the result of excitation brought to involuntary muscle through the pelvic splanchnics. This not only compresses the bladder but opens its internal sphincter. There is a simultaneous relaxation of the external sphincter.

IV. Urinary retention: mainly through sympathetic excitation to internal sphincter and through voluntary control via the pudendal nerve to the external sphincter

V. Endopelvic fascia (subserous). This is the fascia located between the peritoneum (above) and the fascia of the pelvic wall and floor. It invests the pelvic viscera and their vascular pedicles. Condensations of this fascia acquire special terminology
A. HYPOGASTRIC SHEATH (stalk): around hypogastric vessels
B. SUPERIOR AND INFERIOR ALAR OR VESICAL WINGS: around superior and inferior vesical vessels
C. MACKENRODT'S LIGAMENT (cardinal ligament): in the female, around uterine vessels
D. MIDDLE RECTAL LIGAMENT: around middle rectal vessels
E. SACROGENITAL LIGAMENT: around the nerve plexuses from the sacral region to viscera
F. RECTOVESICAL (MALE) AND RECTOVAGINAL (FEMALE) FASCIA: between respective organs
G. PUBOPROSTATIC LIGAMENTS: between pubis and the bladder and prostate
H. UTEROVESICAL AND VESICOVAGINAL LIGAMENTS: between uterus, vagina, and bladder
 I. PRESACRAL LIGAMENT: around the superior rectal vessels, hypogastric (presacral) nerves, and sympathetic trunk

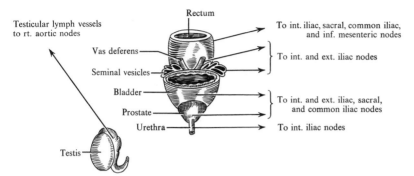

Testicular lymph vessels to rt. aortic nodes

Rectum

To int. iliac, sacral, common iliac, and inf. mesenteric nodes

Vas deferens

To int. and ext. iliac nodes

Seminal vesicles

Bladder

To int. and ext. iliac, sacral, and common iliac nodes

Prostate

Urethra

To int. iliac nodes

Testis

LYMPH DRAINAGE OF MALE PELVIS

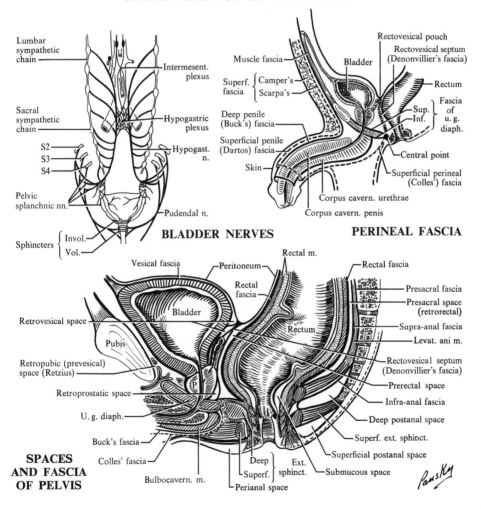

Lumbar sympathetic chain

Intermesent. plexus

Sacral sympathetic chain

Hypogastric plexus

S2
S3
S4

Hypogast. n.

Pelvic splanchnic nn.

Pudendal n.

Sphincters { Invol.
Vol.

BLADDER NERVES

Muscle fascia

Rectovesical pouch

Bladder

Rectovesical septum (Denonvillier's fascia)

Superf. fascia { Camper's
Scarpa's

Rectum

Deep penile (Buck's) fascia

Sup.
Inf.

Fascia of u. g. diaph.

Superficial penile (Dartos) fascia

Central point

Skin

Superficial perineal (Colles') fascia

Corpus cavern. urethrae

Corpus cavern. penis

PERINEAL FASCIA

Vesical fascia

Rectal m.

Peritoneum

Rectal fascia

Rectal fascia

Presacral fascia

Bladder

Presacral space (retrorectal)

Retrovesical space

Rectum

Supra-anal fascia

Pubis

Levat. ani m.

Retropubic (prevesical) space (Retzius)

Rectovesical septum (Denonvillier's fascia)

Retroprostatic space

Prerectal space

U. g. diaph.

Infra-anal fascia

Deep postanal space

Buck's fascia

Superf. ext. sphinct.

SPACES AND FASCIA OF PELVIS

Colles' fascia

Superficial postanal space

Bulbocavern. m.

Deep } Superf.

Ext. sphinct.

Submucous space

Perianal space

Pansky

198. VAS DEFERENS, SEMINAL VESICLES, AND PROSTATE

I. Ductus (vas) deferens: the excretory duct of the testis
A. COURSE: ascends along posterior border of the testis; in spermatic cord, passes through the inguinal canal to the deep inguinal ring; bends around inferior epigastric artery; curves posteriorly and caudally; crosses anterior to the external iliac vessels and enters pelvis; descends on medial side of lateral umbilical ligament and obturator vessels and nerve; crosses to medial side of ureter to run between fundus of bladder and seminal vesicles; to base of prostate, where its terminal part widens into an *ampulla*. The ampulla is joined by the duct of the seminal vesicle to form the *ejaculatory duct,* which runs through the prostate to open into the urethra

II. Seminal vesicle: bilateral lobulated sacs consisting of irregular pouches
A. RELATIONS: anterior surface against fundus of bladder; posterior surface separated from rectum by rectovesical fascia; superiorly, related to vas deferens and ureters; inferiorly, joins vas deferens at the posterior surface of the prostate

III. Prostate: a cone-shaped glandular body, the size of a chestnut, containing much connective tissue and smooth muscle
A. PARTS AND RELATIONS
  1. Base faces cephalad against neck of bladder
  2. Apex directed caudad against urogenital diaphragm
  3. Posterior surface separated from rectum by rectovesical septum
  4. Anterior surface separated from symphysis pubis by a venous plexus and fat. Urethra opens through this surface just above apex
  5. Inferolateral surfaces separated from levator ani muscles by venous plexus
  6. Lobes
    a. Anterior: small, nonglandular area in front of urethra
    b. Posterior: posterior to urethra and ejaculatory ducts and behind middle lobe
    c. Two lateral (right and left) occupy almost entire base of gland, lateral and anterior to urethra
    d. Median (middle): glandular, posterior to urethra but anterior to ejaculatory ducts; here are found the subtrigonal and cervical glands (Albarran's)

IV. Special features of the prostatic urethra
A. URETHRAL CREST: a longitudinal ridge on the posterior wall
B. PROSTATIC SINUS: depression on sides of crest into which the prostatic ducts open
C. SEMINAL COLLICULUS: summit of the urethral crest on which open the ejaculatory ducts and a median blind-ending sac, the *prostatic utricle*

V. Fixation: by puboprostatic ligaments, urogenital diaphragm, and levator ani mm.

VI. Vessels and nerves
A. ARTERIES: middle rectal and inferior vesical
B. VEINS: prostatic plexus to internal iliac vein
C. LYMPHATICS terminate in the internal iliac and sacral nodes
D. NERVES: prostatic plexus from inferior hypogastric plexus

VII. Clinical considerations
A. ANTERIOR LOBE: adenomas rare; no urethral encroachment
B. POSTERIOR LOBE: adenomas rare; lobe encountered in digital examination
C. LATERAL LOBE: hypertrophy causes urinary obstruction
D. MEDIAN LOBE: important clinically; enlargement of mucous glands leads to obstruction; adenomas frequent, encroaching into urethra, blocking internal orifice

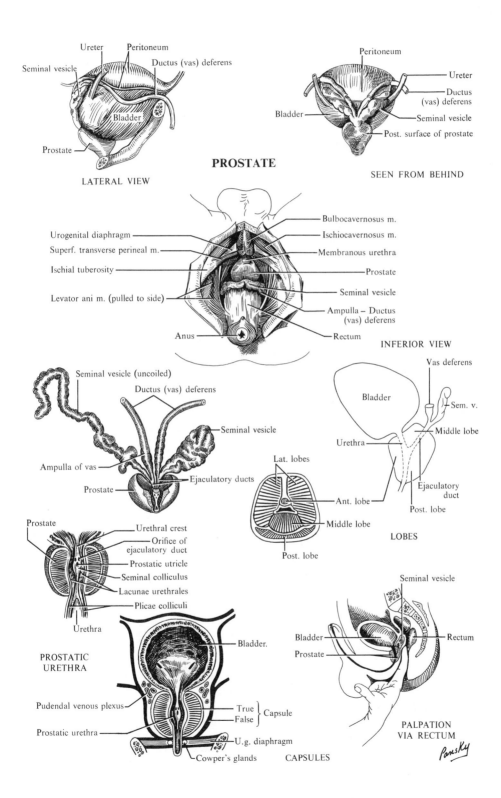

Ureter   Peritoneum

Seminal vesicle   Ductus (vas) deferens

Bladder

Prostate

LATERAL VIEW

Peritoneum

Ureter

Ductus (vas) deferens

Seminal vesicle

Bladder

Post. surface of prostate

SEEN FROM BEHIND

PROSTATE

Urogenital diaphragm

Superf. transverse perineal m.

Ischial tuberosity

Levator ani m. (pulled to side)

Anus

Bulbocavernosus m.

Ischiocavernosus m.

Membranous urethra

Prostate

Seminal vesicle

Ampulla – Ductus (vas) deferens

Rectum

INFERIOR VIEW

Seminal vesicle (uncoiled)

Ductus (vas) deferens

Seminal vesicle

Ampulla of vas

Ejaculatory ducts

Prostate

Vas deferens

Bladder

Sem. v.

Urethra

Middle lobe

Lat. lobes

Ant. lobe

Ejaculatory duct

Post. lobe

Middle lobe

Post. lobe

LOBES

Prostate

Urethral crest

Orifice of ejaculatory duct

Prostatic utricle

Seminal colliculus

Lacunae urethrales

Plicae colliculi

Urethra

PROSTATIC URETHRA

Bladder.

Pudendal venous plexus

True
False
} Capsule

Prostatic urethra

U.g. diaphragm

Cowper's glands

CAPSULES

Seminal vesicle

Bladder

Rectum

Prostate

PALPATION VIA RECTUM

Pansky

– 443 –

199. RECTUM AND ANAL CANAL

I. Rectum
A. COURSE AND EXTENT: from 3rd sacral segment to slightly below tip of coccyx. Here bends abruptly posteriorly to anal canal. Total length, 12 cm
B. CURVATURES
1. Two posteroanterior: upper convex posteriorly; lower convex anteriorly
2. Two lateral: to right, at junction of third and fourth sacral segments; to left, at sacrococcygeal articulation
C. DIAMETER
1. Cephalic end similar to colon; caudal end dilated to form *rectal ampulla*
D. SPECIAL STRUCTURAL CHARACTERISTICS
1. Transverse rectal folds (Houston's valves): permanent, 3 transverse folds, which project into lumen
a. Upper, from right side, near cephalic end
b. Middle, 3 cm below first, extends inward from left
c. Lower, largest, opposite bladder and extends posteriorly from anterior wall
2. Unlike colon, the rectum has a complete outer longitudinal muscle coat
E. PERITONEAL RELATIONSHIP: upper two thirds has some peritoneum: most cephalic portion has peritoneum anteriorly and laterally; lower down it has peritoneum anteriorly. The lower third has no peritoneum
F. RELATIONS
1. Posteriorly: superior rectal vessels, left piriformis muscle, left sacral plexus of nerves, and the fascia covering the sacrum, coccyx, and levator ani muscle
2. Anteriorly
a. Male: separated by coils of intestine in rectovesical fossa from fundus of bladder, and by rectovesical septum from the triangular area of the fundus of bladder, seminal vesicles, ductus deferens, and prostate
b. Female: separated by intestinal coils in rectouterine fossa from the uterus and by rectovaginal septum from the posterior wall of the vagina

II. Anal canal
A. COURSE AND EXTENT: origin at level of apex of prostate, directed posteriorly and caudally. Length, 2.5–4.0 cm
B. SURROUNDED BY INTERNAL AND EXTERNAL ANAL SPHINCTERS: supported by levator ani muscle
C. SPECIAL FEATURES
1. Anal columns: vertical folds due to dilated veins of rectal plexus
2. Anal sinuses: furrows between columns
3. Anal valves: folds joining columns at caudal end of anal sinuses
D. RELATIONS
1. Posteriorly: anococcygeal body (musculofibrous tissue)
2. Anteriorly
a. Male: separated by central tendon of perineum from membranous and bulbar urethra
b. Female: central tendon of perineum separates it from vagina

III. Anal sphincters
A. INTERNAL: a thickening of the intrinsic smooth muscle coat of the intestinal wall
B. EXTERNAL: composed of striated, voluntary muscle
1. Subcutaneous: portion immediately around anal orifice
2. Superficial part (main portion) arises from the anococcygeal raphé, splits to encircle canal, and inserts in central tendon of perineum
3. Deep part (a true sphincter) encircles the anus, the fibers of the 2 sides decussating ventral and dorsal to anus

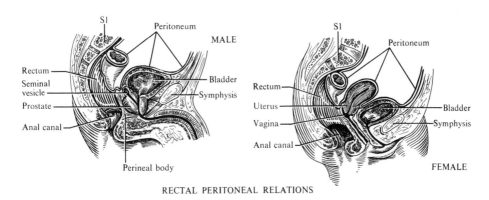

RECTAL PERITONEAL RELATIONS

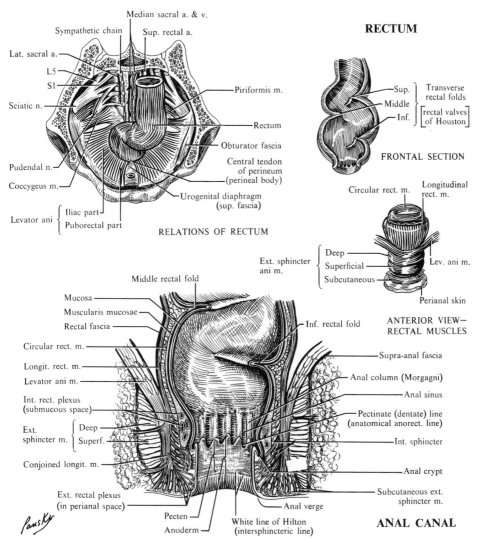

RECTUM

RELATIONS OF RECTUM

FRONTAL SECTION

ANTERIOR VIEW— RECTAL MUSCLES

ANAL CANAL

200. BLOOD SUPPLY AND LYMPH DRAINAGE OF RECTUM

I. Blood supply
A. ARTERIES
1. Superior rectal arises from the inferior mesenteric artery (see p. 410). Two branches descend on either side of rectum. Above anus, each gives rise to several small branches, which pass straight caudally, regularly spaced, to the level of the internal sphincter, at which point they form loops around the caudal rectum and anastomose with other vessels
2. Middle rectal arises either directly from the internal iliac artery or by a common stem with the inferior vesical artery, which approaches rectum from side and joins loop at caudal end of rectum and upper anal canal
3. Inferior rectal arises from internal pudendal artery, pierces wall of the pudendal (Alcock's) canal, and gives 2 or 3 branches, which pass medially through the ischiorectal fossa to muscle and skin around anus
B. VEINS
1. Rectal plexus: network of vessels around anal canal. Usually, at cephalic border of canal, some veins are dilated or sacculated. This plexus drains
   a. Chiefly through 6 ascending vessels, which lie between muscularis and mucosa for about 12.0 cm. At this level, they unite to form the superior rectal vein, a tributary of the inferior mesenteric vein
   b. Middle rectal: from plexus, with tributaries from bladder, prostate, and seminal vesicle, swings laterally on pelvic surface of levator ani to internal iliac vein
   c. Inferior rectal: from lower plexus into internal pudendal vein and then into internal iliac vein

II. Lymphatic drainage
A. FROM ANUS: with lymphatics of superficial perineum and scrotum into superficial inguinal nodes
B. FROM ANAL CANAL: accompany middle rectal artery and end in internal iliac nodes around internal iliac vessels; from these to common iliac nodes and then to lateral aortic group
C. FROM RECTUM: through pararectal nodes which lie on rectal muscles and sigmoid mesocolon to the inferior mesenteric group of preaortic nodes

III. Nerves (see p. 458) derived from the inferior mesenteric and hypogastric plexuses

IV. Clinical considerations
A. THE ANAL CANAL presents 4 landmarks. The *anocutaneous line* marks the lower end of the gastrointestinal tract. *Hilton's white line* marks the interval between the external and internal anal sphincters. The *pectin* is in the mucocutaneous junction, internal hemorrhoids developing above and external hemorrhoids below this line; this line is also a lymphatic dividing line between the flow of lymph upward into the pelvis, primarily to the internal iliac nodes, or downward to the subinguinal nodes. The *anorectal line* is the line above the anal crypts and sinuses, marking the beginning of the anal canal
B. THE LINE around the anal canal that can be traced by following the anal valves and the bases of the anal columns is usually referred to by clinicians as the *pectinate, dentate,* or *mucocutaneous line*. It is an important landmark
1. The change between columnar or cuboidal epithelium in the upper part of the canal and the stratified epithelium in the lower part occurs at or close to this level and is important because carcinomas from the two types of epithelium differ
2. The line is only about 0.25 to 1.0 cm below a divide in the nerve supply, with the afferent innervation about it through fibers of the pelvic plexus (visceral type) and that below of somatic nerve fibers in the pudendal nerve

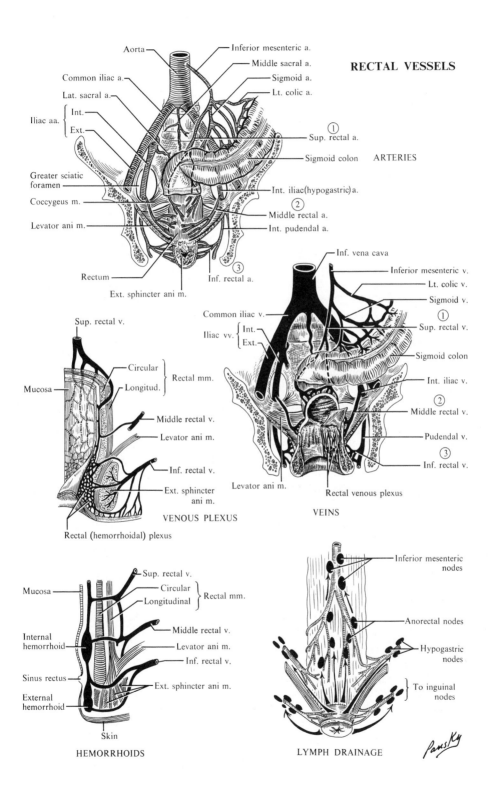

RECTAL VESSELS

Aorta — Inferior mesenteric a.

Common iliac a. — Middle sacral a.

Lat. sacral a. — Sigmoid a.

— Lt. colic a.

Iliac aa. { Int. / Ext.

Sup. rectal a. ①

Sigmoid colon   ARTERIES

Greater sciatic foramen

Int. iliac (hypogastric) a.

Coccygeus m.

② Middle rectal a.

Levator ani m.

Int. pudendal a.

Rectum

Ext. sphincter ani m.   Inf. rectal a. ③

Inf. vena cava

Inferior mesenteric v.

Lt. colic v.

Sigmoid v.

Common iliac v.

Iliac vv. { Int. / Ext.

① Sup. rectal v.

Sigmoid colon

Int. iliac v.

② Middle rectal v.

Pudendal v.

③ Inf. rectal v.

Levator ani m.

Rectal venous plexus

VEINS

Sup. rectal v.

Circular | Rectal mm.
Longitud. |

Mucosa

Middle rectal v.

Levator ani m.

Inf. rectal v.

Ext. sphincter ani m.

VENOUS PLEXUS

Rectal (hemorrhoidal) plexus

Sup. rectal v.

Circular | Rectal mm.
Longitudinal |

Mucosa

Middle rectal v.

Internal hemorrhoid

Levator ani m.

Inf. rectal v.

Sinus rectus

Ext. sphincter ani m.

External hemorrhoid

Skin

HEMORRHOIDS

Inferior mesenteric nodes

Anorectal nodes

Hypogastric nodes

To inguinal nodes

LYMPH DRAINAGE

Pansky

201. THE PELVIC DIAPHRAGM

I. Composition: the levator ani and coccygeus muscles with their superior and inferior fasciae

II. Location: forms a sling across pelvic cavity

III. Function: it is the pelvic floor that holds and supports the viscera

IV. Muscles
A. LEVATOR ANI, innervated by S4 (sometimes S3 or S5) through the pudendal plexus, consists of the following parts:
  1. Pubococcygeus arises from pubis and its superior ramus to insert into anococcygeal raphé and coccyx
    a. Puborectalis: most medial portion of the pubococcygeus originating at pubis and the superior layer of the urogenital diaphragm to pass along the side of the rectum to meet its counterpart behind that organ, forming a sling. Some of its fibers continue to the tip of the coccyx
  2. Iliococcygeus arises from the arcus tendineus and the ischial spine. Inserts on last segments of coccyx and anococcygeal raphé.
B. COCCYGEUS innervated by S4 and S5. Arises from the spine of the ischium and sacrospinous ligament. Inserts into the slides of lower sacrum and coccyx

V. Parietal pelvic fascia (fascia lining walls of pelvis) is continuous with transversalis fascia over brim of the pelvis and is attached to bones along rim of true pelvis. Has 2 primary regional divisions
A. PIRIFORM covers piriformis muscle and leaves pelvis with that muscle through greater sciatic foramen. Sacral nerves are behind this layer; internal iliac vessels are in front of it
B. OBTURATOR: divisible into intrapelvic and extrapelvic portions by the arcus tendineus of pelvic fascia (a thickening of fascia from ischial spine to pubic bone near obturator membrane). Ensheaths obturator vessels and nerve. At arcus, splits into 3 layers
  1. Supra-anal covers inner (superior) surface of levator ani and coccygeus muscles
  2. Infra-anal covers inferior surface of levator ani and coccygeus muscles
  3. Extrapelvic continuation of obturator covers obturator muscle below arcus and lines the lateral wall of ischiorectal fossa
    a. Pudendal (Alcock's) canal: a split in the fascia to invest the pudendal nerve and internal pudendal vessels

VI. Visceral pelvic (subserous endopelvic) fascia (see p. 440) invests pelvic viscera and their vessels

VII. Ischiorectal fossa: a wedge-shaped space at either side of anal canal. Its base is directed caudally, and its apex is cephalic, along the arcus tendineus. It contains fat, connective tissue, rectal vessels, and nerves. It is bounded medially by the infra-anal fascia over the levator ani and external anal sphincter muscles, laterally by extrapelvic obturator fascia over the obturator internus muscle and the ischial tuberosities, posteriorly by the gluteus maximus muscle and sacrotuberous ligament, anteriorly by the posterior edge of the urogenital diaphragm, inferiorly by fascia and skin
A. ANTERIOR RECESS: a continuation of fossa anteriorly, between urogenital diaphragm below and the levator and obturator internus muscles above and lateral

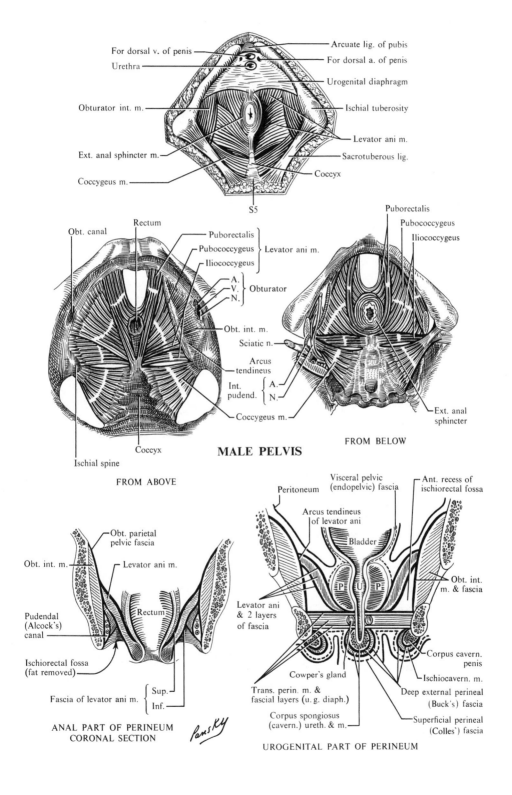

For dorsal v. of penis — Arcuate lig. of pubis

Urethra — For dorsal a. of penis

— Urogenital diaphragm

Obturator int. m. — Ischial tuberosity

— Levator ani m.

Ext. anal sphincter m. — Sacrotuberous lig.

Coccygeus m. — Coccyx

S5

Obt. canal   Rectum

Puborectalis   Puborectalis
Pubococcygeus   Pubococcygeus
Iliococcygeus   Iliococcygeus
} Levator ani m.

A.
V. } Obturator
N.

Obt. int. m.

Sciatic n.

Arcus
tendineus

Int.
pudend. { A.
N.

Coccygeus m.

Ischial spine

Coccyx

MALE PELVIS

FROM BELOW

Ext. anal
sphincter

FROM ABOVE

Obt. parietal
pelvic fascia

Obt. int. m.   Levator ani m.

Pudendal
(Alcock's)
canal

Rectum

Ischiorectal fossa
(fat removed)

Fascia of levator ani m. { Sup.
Inf.

ANAL PART OF PERINEUM
CORONAL SECTION

Peritoneum

Visceral pelvic
(endopelvic) fascia

Ant. recess of
ischiorectal fossa

Arcus tendineus
of levator ani

Bladder

P U P

Obt. int.
m. & fascia

Levator ani
& 2 layers
of fascia

Cowper's gland

Trans. perin. m. &
fascial layers (u. g. diaph.)

Corpus spongiosus
(cavern.) ureth. & m.

Corpus cavern.
penis

Ischiocavern. m.

Deep external perineal
(Buck's) fascia

Superficial perineal
(Colles') fascia

Pansky

UROGENITAL PART OF PERINEUM

202. RELATIONS OF THE UTERUS, TUBES, AND OVARY— PART I

I. Ovary: the germinal and endocrine gland of the female

A. LOCATION AND RELATIONS: in the ovarian fossa on the lateral wall of the pelvis, behind the broad ligament, with the external iliac vessels above, the ureter posteroinferiorly, and covered by the fimbria of the uterine tube medially

B. SURFACES AND BORDERS: lateral and medial surfaces, upper (tubal) extremity, lower (uterine) extremity, anterior (mesovarian) border, posterior (free) border

C. FIXATION: *suspensory ligament,* a peritoneal fold running from the upper extremity to the iliac vessels; *proper ligament of the ovary,* from the lower extremity through the mesometrium (broad ligament) to the lateral angle of the uterus; *mesovarium,* a mesentery joining the anterior border to the posterior side of the broad ligament

II. Uterine tube; duct to carry ova from ovaries to uterus

A. LOCATION AND RELATIONS: lies in the free, cephalic border of the broad ligament—*the mesosalpinx*

B. PARTS: *infundibulum,* with abdominal ostium surrounded by *fimbria; ampulla,* the middle, wide part, which curves over ovary; *isthmus,* the constricted medial part, which enters the uterus; and *interstitial* or *intrauterine* part

III. Uterus: organ adapted for the development of the fertilized ovum

A. PARTS
  1. Body
     a. Vesical (anterior) surface: lies on the superior surface of the bladder, covered with peritoneum, which is reflected onto bladder forming the vesicouterine pouch
     b. Intestinal (posterior) surface: related to the sigmoid colon and coils of small intestine, covered with peritoneum
     c. Fundus: directed anteriorly and cephalically, related to the coils of the small intestine
     d. Lateral margins: mesometrium (broad), round, with ovarian ligaments attached here, receives uterine tubes
  2. Cervix: the constricted part of the uterus demarcated from the body by the *isthmus,* which indicates the position of the internal os. The axis is a curve with the concavity directed anteriorly. Divided into 2 parts by the vagina
     a. Supravaginal: separated from the bladder anteriorly by the parametrium (fatty, fibrous tissue, which is continuous with the tissue between the layers of the broad ligament); posteriorly it is covered with peritoneum and is separated from the rectum by coils of small intestine
     b. Vaginal: protrudes into the vagina
        i. Ostium (external os): the opening of cervix into vagina
        ii. Fornices (anterior, posterior, and lateral): the groove that lies between the walls of the vagina and cervix. The posterior fornix is deepest

IV. Clinical considerations

A. PROLAPSE: when supports of the uterus become stretched and very lax, the cervix may descend, for varying degrees, into the vagina or even out into the vestibule

B. THE RELATIONS of the body of the uterus are markedly changed by pregnancy, progressive enlargement bringing the uterus high into the abdominal cavity

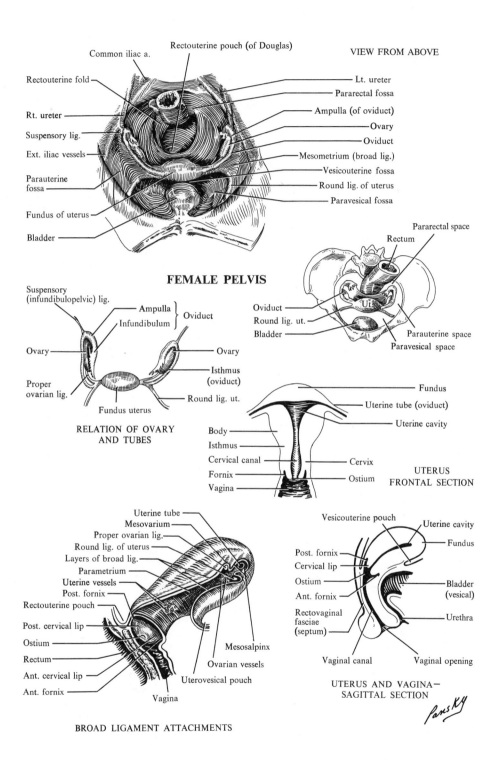

Rectouterine pouch (of Douglas)

Common iliac a.

VIEW FROM ABOVE

Rectouterine fold

Rt. ureter

Suspensory lig.

Ext. iliac vessels

Parauterine fossa

Fundus of uterus

Bladder

Lt. ureter

Pararectal fossa

Ampulla (of oviduct)

Ovary

Oviduct

Mesometrium (broad lig.)

Vesicouterine fossa

Round lig. of uterus

Paravesical fossa

FEMALE PELVIS

Pararectal space

Rectum

Oviduct

Round lig. ut.

Bladder

Ut.

Parauterine space

Paravesical space

Suspensory (infundibulopelvic) lig.

Ampulla

Infundibulum

Oviduct

Ovary

Ovary

Isthmus (oviduct)

Proper ovarian lig.

Round lig. ut.

Fundus uterus

RELATION OF OVARY AND TUBES

Fundus

Uterine tube (oviduct)

Uterine cavity

Body

Isthmus

Cervical canal

Fornix

Vagina

Cervix

Ostium

UTERUS FRONTAL SECTION

Uterine tube

Mesovarium

Proper ovarian lig.

Round lig. of uterus

Layers of broad lig.

Parametrium

Uterine vessels

Post. fornix

Rectouterine pouch

Post. cervical lip

Ostium

Rectum

Ant. cervical lip

Ant. fornix

Mesosalpinx

Ovarian vessels

Uterovesical pouch

Vagina

BROAD LIGAMENT ATTACHMENTS

Vesicouterine pouch

Uterine cavity

Post. fornix

Cervical lip

Ostium

Ant. fornix

Rectovaginal fasciae (septum)

Fundus

Bladder (vesical)

Urethra

Vaginal canal

Vaginal opening

UTERUS AND VAGINA— SAGITTAL SECTION

Pansky

203. RELATIONS OF THE UTERUS, TUBES, AND OVARY— PART II

III.
B. FIXATION AND SUPPORT
1. Mesometrium (broad ligaments): folds of peritoneum extending across the pelvis containing uterus, parametrium, vessels, nerves, uterine tubes, ureters, round ligaments, paroophoron, and epoophoron
2. Round ligaments: fibrous cords attached to superior lateral borders of uterus, pass over external iliac vessels and inguinal ligaments to leave the abdomen through the deep inguinal rings. They traverse the inguinal canal and are anchored in the labia majora
3. Cardinal ligaments: subserous tissue (endopelvic fascia) around vagina and cervix, which extends laterally across the pelvis within the lowest part of the broad ligaments to attach to the deep fascia covering the levator ani muscles. Vaginal and uterine vessels lie in it
4. Uterosacral ligaments: subserous tissue (endopelvic tissue) attached to the cervix, running posteriorly to join the deep fascia over the sacrum
5. Levator ani muscles and fascia support it from below
6. Peritoneal ligaments (false): these are similar to the false ligaments of the bladder and are reflections of the peritoneum from uterus to other organs or are folds covering deeper fibrous bands: vesicouterine (anterior ligament) to bladder, rectovaginal (posterior ligament) to rectum, sacrogenital covering uterosacral ligaments.

V. Vagina: that part of the birth canal extending from the uterus to the vestibule
A. INCLINATION: cephalically and posteriorly, forms an angle of about 90° with uterus. The posterior wall is 2.5–3.0 cm longer than the anterior
B. RELATIONS: upper end embraces cervix; anteriorly are the fundus of bladder and urethra; posteriorly lie the rectouterine peritoneal pouch, rectovaginal fascia, and perineal body. The pelvic ureters lie close to the lateral fornices and, as these reach the bladder, lie anterior to the anterior fornix

VI. Vessels, nerves, and lymphatics (see p. 454)

VII. Clinical considerations
A. THE UTERUS, normally anteflexed and anteverted, may assume a variety of abnormal positions, for example, retroversion, retroflexion, or prolapse (a falling or sinking into the vagina)
B. IN HYSTERECTOMY (removal of uterus), the relations of the ureters to the cervix, vagina, and uterine vessels must be recalled
C. THE RELATIONS OF THE POSTERIOR FORNIX to the bottom of the rectouterine pouch makes possible easier palpation of the pelvic viscera, drainage of abdominal fluids, and inspection of pelvic viscera without abdominal incisions
D. HYSTERECTOMY IS EITHER PARTIAL, in which at least a part of the cervix is left, or complete (panhysterectomy)
E. SALPINGITIS is an inflammation of the uterine tube(s)
F. SALPINGECTOMY is a removal of the uterine tube(s)
G. TUBAL PREGNANCY is a situation in which the fertilized ovum implants in the uterine tube
H. OVARIAN CYST FORMATION (single or multiple) may occur as a result of lack of ovulation of the follicle and its continued growth
I. A BULGING OF THE BLADDER into the anterior vaginal wall is known as *cystocele,* and a bulging into the posterior vaginal wall is known as *rectocele*. Each involves laceration of the intervening connective tissue and adjacent vaginal wall

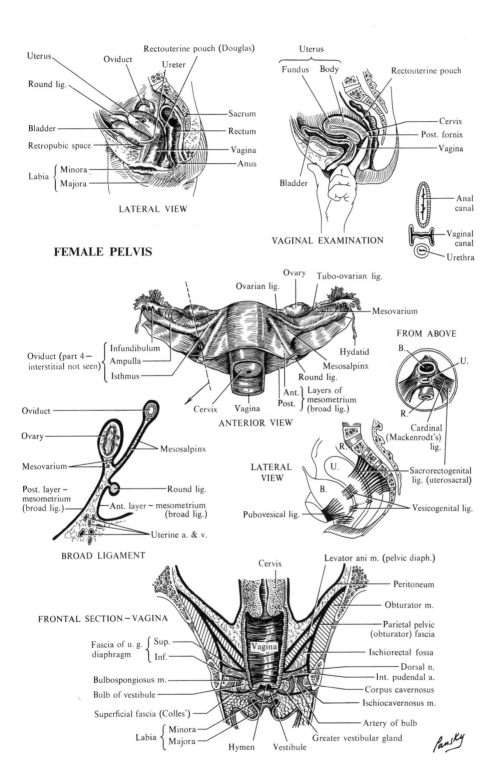

Rectouterine pouch (Douglas)

Uterus
Oviduct
Round lig.
Ureter

Bladder
Retropubic space
Labia { Minora / Majora

Sacrum
Rectum
Vagina
Anus

LATERAL VIEW

FEMALE PELVIS

Uterus
Fundus   Body

Rectouterine pouch

Cervix
Post. fornix
Vagina

Bladder

VAGINAL EXAMINATION

Anal canal
Vaginal canal
Urethra

Ovary
Ovarian lig.   Tubo-ovarian lig.

Mesovarium

FROM ABOVE

B.
U.
R.

Oviduct (part 4 — interstitial not seen) { Infundibulum / Ampulla / Isthmus

Hydatid
Mesosalpinx
Round lig.

Cervix   Vagina

Ant. / Post. } Layers of mesometrium (broad lig.)

ANTERIOR VIEW

Cardinal (Mackenrodt's) lig.

Oviduct
Ovary
Mesovarium

Mesosalpinx

Post. layer — mesometrium (broad lig.)

Round lig.

Ant. layer — mesometrium (broad lig.)

Uterine a. & v.

BROAD LIGAMENT

LATERAL VIEW

R.
U.
B.

Pubovesical lig.

Sacrorectogenital lig. (uterosacral)

Vesicogenital lig.

Cervix

Levator ani m. (pelvic diaph.)

Peritoneum
Obturator m.

FRONTAL SECTION — VAGINA

Parietal pelvic (obturator) fascia

Fascia of u. g. diaphragm { Sup. / Inf.

Vagina

Ischiorectal fossa
Dorsal n.
Int. pudendal a.

Bulbospongiosus m.
Bulb of vestibule
Superficial fascia (Colles')

Corpus cavernosus
Ischiocavernosus m.
Artery of bulb

Labia { Minora / Majora

Greater vestibular gland

Hymen   Vestibule

Pansky

– 453 –

204. BLOOD SUPPLY, INNERVATION, AND LYMPH DRAINAGE OF FEMALE GENITAL SYSTEM

I. Blood supply

A. OVARY

1. Artery: ovarian artery from aorta descends in the suspensory ligament of the ovary to broad ligament, sending branches to ovary and uterine tube. Anastomoses with uterine artery
2. Vein: pampiniform plexus to ovarian vein, which travels with the ovarian artery to terminate on the right in the vena cava, on the left in the renal vein

B. UTERINE TUBE

1. Artery: similar to that of the ovary
2. Vein: similar to that of the ovary, with some flow to uterine plexus

C. UTERUS

1. Arteries: ovarian; uterine from anterior division of internal iliac crosses ureter to reach side of uterus through broad ligament, where it ascends to level of uterine tube. Supplies cervix, upper vagina, body of uterus, uterine tube, and round ligament
2. Veins: from uterine plexus on the sides and superior angles of uterus. Communicates with vaginal plexus but is drained chiefly by uterine veins, which end in internal iliac veins

D. VAGINA

1. Artery: vaginal artery from internal iliac (comparable to inferior vesical in the male). Sends branches to uterus and joins branches from uterine artery to form the azygos artery of the vagina
2. Vein: from vaginal plexus, which has communications with vesical, rectal, and uterine plexuses, through vaginal vein to internal iliac vein

II. Lymphatic drainage

A. OVARY: vessels following ovarian artery to enter lateral and preaortic nodes
B. UTERINE TUBE: follows ovarian and uterine drainage
C. UTERUS

1. Cervix: to external, internal, and common iliac nodes
2. Body and fundus: mostly follow ovarian drainage to lateral and preaortic nodes, some to external iliac and superficial inguinal nodes (along round lig. to labia)

D. VAGINA: upper, middle, and lower portions drain to external, internal, and common iliac nodes; vulvar drainage ends in superficial inguinal nodes

III. Lymph nodes concerned in drainage of female genital system

A. COMMON ILIAC: on sides of common iliac artery at bifurcation of aorta. Most afferents from internal and external iliac nodes; efferents to lateral aortic nodes
B. EXTERNAL ILIAC: along external iliac artery. Afferents mainly from inguinal nodes but also from glans clitoridis and urethra, bladder, cervix, and upper vagina. Efferents to lateral aortic nodes
C. INTERNAL ILIAC: along corresponding artery. Afferents follow all branches of internal iliac artery from all pelvic viscera. Efferents to common iliac nodes

IV. Nerves

A. OVARY receives fibers from hypogastric and ovarian plexuses
B. UTERINE TUBE: fibers from hypogastric and ovarian plexuses
C. UTERUS: uterovaginal portion of hypogastric plexus
D. VAGINA: uterovaginal portion of hypogastric plexus
E. MOST OF THE PAIN FIBERS FROM THE VAGINA travel with the sacral parasympathetic fibers, as do those from the cervix uteri, and enter the cord via nerves S2, S3, and S4
F. THE LOWER 2.5 CM OF THE VAGINA receives its innervation from the pudendal nerves (also S2, S3, and S4)

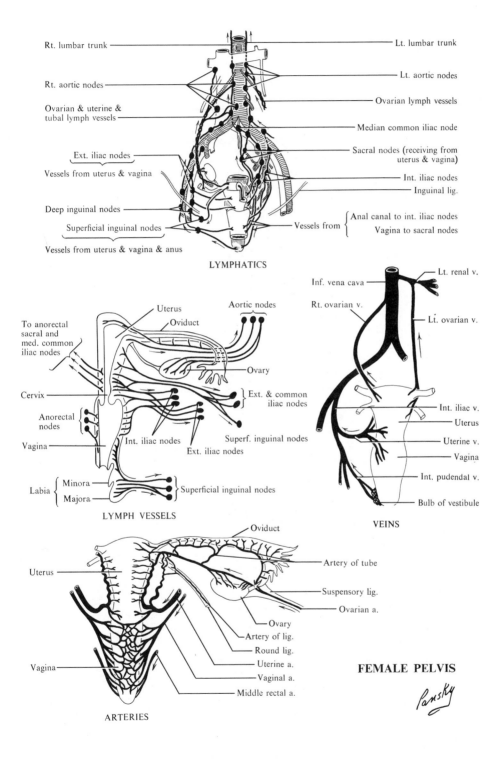

Rt. lumbar trunk
Lt. lumbar trunk

Rt. aortic nodes
Lt. aortic nodes

Ovarian & uterine &
tubal lymph vessels
Ovarian lymph vessels

Median common iliac node

Ext. iliac nodes
Sacral nodes (receiving from
uterus & vagina)

Vessels from uterus & vagina
Int. iliac nodes

Inguinal lig.

Deep inguinal nodes

Superficial inguinal nodes
Vessels from { Anal canal to int. iliac nodes
Vagina to sacral nodes

Vessels from uterus & vagina & anus

LYMPHATICS

Uterus
Aortic nodes
Oviduct

To anorectal
sacral and
med. common
iliac nodes

Ovary

Inf. vena cava
Lt. renal v.
Rt. ovarian v.
Lt. ovarian v.

Ext. & common
iliac nodes

Cervix

Anorectal
nodes

Int. iliac nodes
Superf. inguinal nodes
Int. iliac v.
Uterus

Vagina
Ext. iliac nodes
Uterine v.
Vagina

Labia { Minora
Majora
Superficial inguinal nodes
Int. pudendal v.

Bulb of vestibule

LYMPH VESSELS
VEINS

Oviduct

Uterus
Artery of tube

Suspensory lig.
Ovarian a.

Ovary
Artery of lig.
Round lig.

Vagina
Uterine a.
Vaginal a.
Middle rectal a.

FEMALE PELVIS

Pansky

ARTERIES

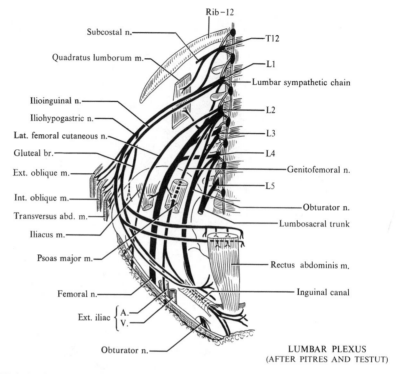

LUMBAR PLEXUS
(AFTER PITRES AND TESTUT)

REVIEW OF LUMBAR AND SACRAL PLEXUSES

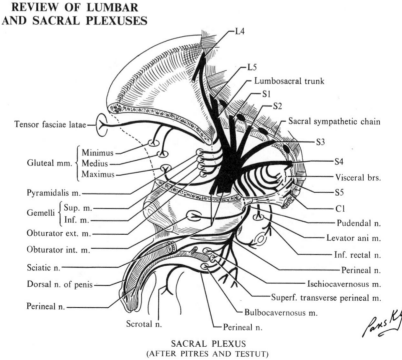

SACRAL PLEXUS
(AFTER PITRES AND TESTUT)

FIGURE 41. **Lumbosacral plexus.**

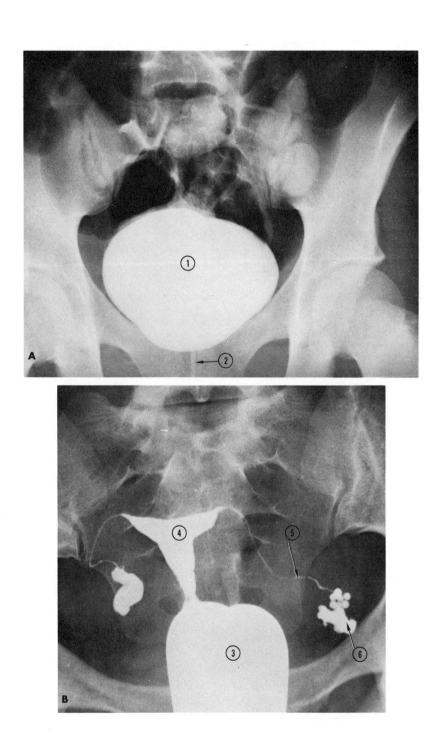

FIGURE 42. **Cystogram and female genital tract. A, Cystogram; B, female genital tract after barium sulfate injection.** *1,* Bladder; *2,* catheter in urethra; *3,* vagina; *4,* uterus; *5,* uterine tube; *6,* infundibulum of tube.

205. ABDOMINAL AND PELVIC AUTONOMICS

I. **Lumbar part of sympathetic trunk** continuous with the thoracic trunk under medial lumbocostal arch; lies anterior to bodies of lumbar vertebrae; on right, partly behind inferior vena cava; on left, behind aorta

II. **Lumbar sympathetic ganglia:** number 2–6 with a variety of fusions
 A. ROOTS
  1. White rami, as far as L2 to each of upper 3 ganglia
 B. BRANCHES
  1. Gray rami to each of the lumbar nerves
  2. Lumbar splanchnics to aortic network from upper 3 lumbar nerves
  3. Visceral branches

III. **Pelvic sympathetic trunk** is a continuation of the lumbar trunk, lies on sacrum medial to sacral foramina, with communications across midline
 A. GANGLIA: number 4 or 5, with frequent fusions; lowest ganglion, the coccygeal, of each side usually fuses in midline as *ganglion impar*
 B. BRANCHES: gray rami to all sacral and coccygeal nerves, visceral branches to hypogastric and pelvic plexuses

IV. **Celiac plexus:** a network of fibers and ganglia surrounding the aorta in the region of origin of the celiac artery and its branches at level of L1
 A. GANGLIA: 2 large groups of interconnected ganglia on either side of the aorta. The preganglionics of the thoracic splanchnic nerves synapse here. The vagal fibers merely pass through
 B. ASSOCIATED GANGLIA: aorticorenal at the origin of the renal artery, and the superior mesenteric at the origin of the superior mesenteric artery
 C. ROOTS OF PLEXUS (see pp. 366, 369): preganglionics of the great splanchnic nerve go to the celiac ganglion, and those of the lesser splanchnic nerve to the aorticorenal ganglion. Postganglionics are distributed with the celiac plexus along the major branches of the abdominal aorta

V. **Intermesenteric (aorta) plexus** surrounds aorta between the mesenteric vessels
 A. INFERIOR MESENTERIC GANGLION: around origin of the inferior mesenteric artery
 B. ROOTS: celiac, from celiac plexus; lumbar splanchnics from L1, L2
 C. BRANCHES: to kidney, testis or ovary, ureter, colon, and rectum

VI. **Superior hypogastric plexus:** caudal prolongation of the intermesenteric plexus in region of aortic bifurcation
 A. ROOTS: last 3 lumbar splanchnics and intermesenteric branches
 B. BRANCHES: hypogastric nerves, which cross pelvic brim and spread out as a plexus in the pelvis, the inferior hypogastric or pelvic plexus, for the pelvic viscera

VII. **Clinical considerations**
 A. LUMBAR SYMPATHETIC GANGLIONECTOMY: a trunk resection to reduce peripheral arterial deficiency resulting from arteriosclerosis, thromboangiitis obliterans, or femoral embolus
 B. RESECTION OF SUPERIOR HYPOGASTRIC PLEXUS to restore normal bladder function in cases of retention resulting from lesions of the spinal cord

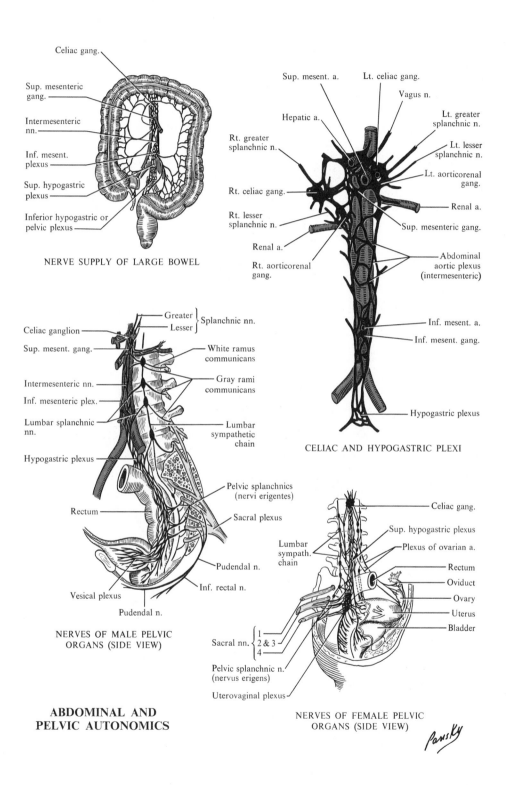

Celiac gang.

Sup. mesenteric gang.

Intermesenteric nn.

Inf. mesent. plexus

Sup. hypogastric plexus

Inferior hypogastric or pelvic plexus

NERVE SUPPLY OF LARGE BOWEL

Sup. mesent. a.

Hepatic a.

Rt. greater splanchnic n.

Rt. celiac gang.

Rt. lesser splanchnic n.

Renal a.

Rt. aorticorenal gang.

Lt. celiac gang.

Vagus n.

Lt. greater splanchnic n.

Lt. lesser splanchnic n.

Lt. aorticorenal gang.

Renal a.

Sup. mesenteric gang.

Abdominal aortic plexus (intermesenteric)

Inf. mesent. a.

Inf. mesent. gang.

Hypogastric plexus

CELIAC AND HYPOGASTRIC PLEXI

Greater \
Lesser / Splanchnic nn.

Celiac ganglion

Sup. mesent. gang.

White ramus communicans

Gray rami communicans

Intermesenteric nn.

Inf. mesenteric plex.

Lumbar splanchnic nn.

Lumbar sympathetic chain

Hypogastric plexus

Rectum

Pelvic splanchnics (nervi erigentes)

Sacral plexus

Pudendal n.

Inf. rectal n.

Vesical plexus

Pudendal n.

NERVES OF MALE PELVIC ORGANS (SIDE VIEW)

Celiac gang.

Sup. hypogastric plexus

Plexus of ovarian a.

Lumbar sympath. chain

Rectum

Oviduct

Ovary

Uterus

Bladder

Sacral nn. { 1
2 & 3
4

Pelvic splanchnic n. (nervus erigens)

Uterovaginal plexus

ABDOMINAL AND PELVIC AUTONOMICS

NERVES OF FEMALE PELVIC ORGANS (SIDE VIEW)

Pansky

ABDOMINOPELVIC NERVES

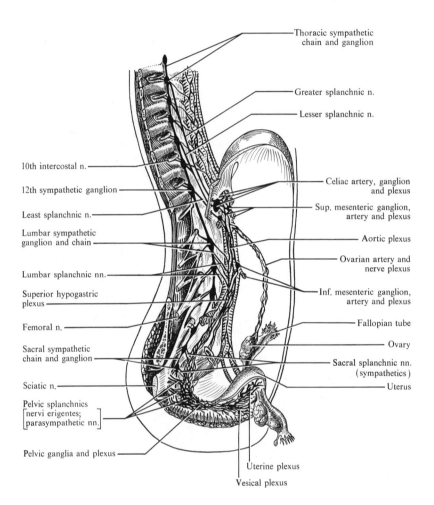

Thoracic sympathetic
chain and ganglion

Greater splanchnic n.

Lesser splanchnic n.

10th intercostal n.

12th sympathetic ganglion

Least splanchnic n.

Lumbar sympathetic
ganglion and chain

Lumbar splanchnic nn.

Superior hypogastric
plexus

Femoral n.

Sacral sympathetic
chain and ganglion

Sciatic n.

Pelvic splanchnics
[nervi erigentes;
parasympathetic nn.]

Pelvic ganglia and plexus

Celiac artery, ganglion
and plexus

Sup. mesenteric ganglion,
artery and plexus

Aortic plexus

Ovarian artery and
nerve plexus

Inf. mesenteric ganglion,
artery and plexus

Fallopian tube

Ovary

Sacral splanchnic nn.
(sympathetics)

Uterus

Uterine plexus

Vesical plexus

SAGITTAL VIEW

FIGURE 43. **Abdominal pelvic nerves.**

– 460 –

NERVES OF UTERUS AND PERINEUM

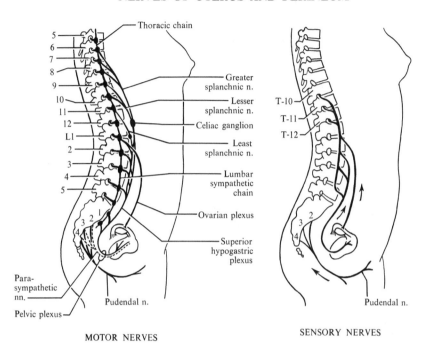

MOTOR NERVES

SENSORY NERVES

SPINAL ANALGESIA

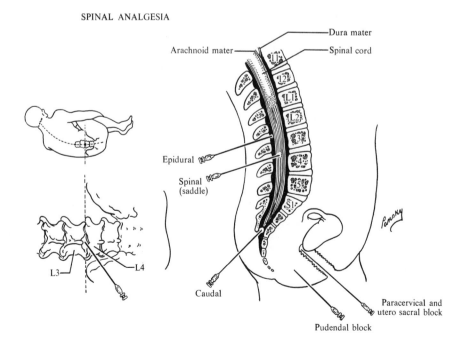

FIGURE 44. **Nerves of uterus and perineum.**

206. PELVIC VESSELS; PERINEAL VESSELS AND NERVES

I. **Arteries of pelvis** originate from internal iliac artery, which descends from the region of the lumbosacral articulation to the greater sciatic notch, and divides into

A. ANTERIOR BRANCH, from which arise

1. Superior vesical to upper bladder and vas deferens (*artery to vas deferens*)
2. Middle vesical to fundus of bladder and seminal vesicle
3. Inferior vesical to fundus of bladder, prostate, and seminal vesicles
4. Middle rectal to rectum
5. Obturator leaves pelvis to medial thigh via obturator canal
6. Internal pudendal: 1 of terminal branches, leaves pelvis through greater sciatic foramen, below piriformis muscle
7. Inferior gluteal: larger of terminal branches, curves posteriorly between first and second or second and third sacral nerves, then runs between piriformis and coccygeus muscles, through greater sciatic foramen into gluteal region
8. Uterine (see p. 454) to uterus and cervix
9. Vaginal (see p. 454) to cervix and vagina

B. POSTERIOR BRANCHES

1. Iliolumbar ascends posterior to obturator nerve and ext. iliac vessels to medial border of psoas m.; divides into a lumbar branch to psoas and quadratus lumborum mm. and to spinal cord; and an iliac branch to iliac, gluteal, abdominal mm.
2. Lateral sacral
   a. Superior: large, runs medially and enters first or second sacral foramen
   b. Inferior: crosses to sacral foramina; descends sends branches through foramina
3. Superior gluteal runs posteriorly between lumbosacral trunk and first sacral nerve, and leaves pelvis through greater sciatic foramen above the piriformis muscle

II. **Veins of pelvis** correspond to the named arteries; originate in venous plexuses

III. **Vessels of the female perineum** (for male, see p. 470)

A. ARTERIES: internal pudendal artery crosses spine of ischium and enters ischiorectal fossa through lesser sciatic foramen. Lies in pudendal (Alcock's) canal. Gives *inferior rectal artery,* which crosses ischiorectal fossa to lower rectum and anal canal; *perineal artery,* which enters superficial perineal compartment to supply ischiocavernosus and bulbospongiosus and superficial transverse perineal muscles to end as *posterior labial arteries.* From Alcock's canal, the internal pudendal artery enters the deep perineal compartment and gives branches to the bulb, urethra, and greater vestibular glands as well as the *deep* and *dorsal arteries* to the *clitoris*

B. VEINS originate in deep veins of clitoris, receive tributaries from bulb, and run through the perineum as the internal pudendal veins, to end in the int. iliac veins

IV. **Nerves of the female perineum**

A. PERINEAL BRANCH OF POSTERIOR FEMORAL CUTANEOUS to labia

B. PUDENDAL: from S2–4, leaves pelvis through greater sciatic foramen, passes over spine of ischium, and enters ischiorectal fossa through lesser sciatic foramen

1. Inferior rectal nerve: crosses ischiorectal fossa to external sphincter and skin
2. Perineal nerve enters deep compartment and divides into
   a. Superficial branches pierce inferior layer of UG diaphragm to labium majus
   b. Deep branches mainly to muscles of perineum including sphincter urethrae. Nerve to bulb comes from the branch to the bulbospongiosus muscle and also supplies the mucous membrane of urethra
3. Dorsal nerve of clitoris: terminal branch, which runs with dorsal artery

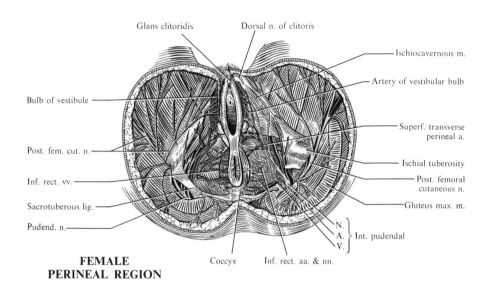

Glans clitoridis — Dorsal n. of clitoris

Ischiocavernous m.

Artery of vestibular bulb

Bulb of vestibule

Superf. transverse perineal a.

Post. fem. cut. n.

Ischial tuberosity

Post. femoral cutaneous n.

Inf. rect. vv.

Sacrotuberous lig.

Gluteus max. m.

Pudend. n.

N.
A. } Int. pudendal
V.

**FEMALE
PERINEAL REGION**

Coccyx   Inf. rect. aa. & nn.

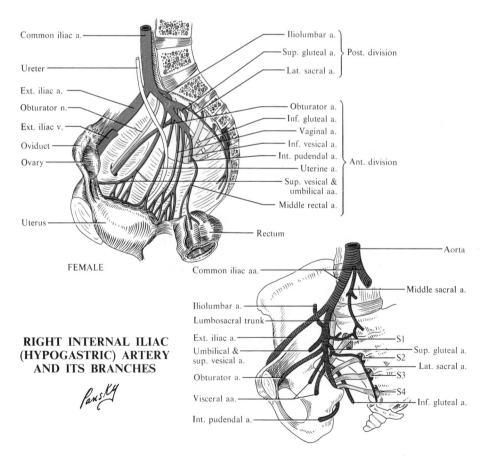

Common iliac a.

Iliolumbar a.
Sup. gluteal a. } Post. division
Lat. sacral a.

Ureter

Ext. iliac a.

Obturator n.

Obturator a.
Inf. gluteal a.
Vaginal a.
Inf. vesical a.
Int. pudendal a. } Ant. division
Uterine a.
Sup. vesical & umbilical aa.
Middle rectal a.

Ext. iliac v.

Oviduct

Ovary

Uterus

Rectum

FEMALE

Aorta

Common iliac aa.

Middle sacral a.

Iliolumbar a.

Lumbosacral trunk

**RIGHT INTERNAL ILIAC
(HYPOGASTRIC) ARTERY
AND ITS BRANCHES**

Ext. iliac a.

Umbilical & sup. vesical a.

Obturator a.

Visceral aa.

Int. pudendal a.

S1
S2   Sup. gluteal a.
S3   Lat. sacral a.
S4   Inf. gluteal a.

Pansky

207. THE FEMALE PERINEUM

I. **Boundaries:** superficial and deep fascia, superficial and deep compartments, similar to male (see p. 466)

II. **Central tendon of perineum** (perineal body, central point, in the female "the perineum"): a mass of fibrous and muscular tissue in midline between anus and vagina (anus and bulb in the male), may be 4 cm in diameter (2 cm in the male). In this center, fasciae of perineum fuse; muscles also arise, interdigitate, or insert here

III. **Muscles:** except for the differences listed below, are similar to the male (see p. 466)
   A. SUPERFICIAL TRANSVERSE PERINEAL similar to, but smaller than, male
   B. BULBOSPONGIOSUS (SPHINCTER VAGINAE) surrounds orifice of vagina and covers vestibular bulb. Arises in central tendon and inserts on corpora cavernosa of clitoris. It narrows vaginal orifice and maintains erection of clitoris
   C. DEEP TRANSVERSE PERINEAL inserts into sides of vagina

IV. **External genitalia**
   A. MONS PUBIS: eminence overlying symphysis pubis, composed mainly of fat, covered with hair
   B. LABIA MAJORA: longitudinal folds of skin that extend caudally and posteriorly from mons. Skin of outer surface is pigmented and set with hair. Inner layer is smooth with large sebaceous glands. Between layers are loose connective tissue, fat, blood vessels, nerves, and glands
      1. Pudendal cleft (rima): between labia
      2. Anterior labial commissure: where labia join anteriorly
      3. Posterior labial commissure: mainly skin connecting the posterior ends of labia
   C. LABIA MINORA: 2 small folds between labia majora, surrounding vaginal orifice
      1. Frenulum of labia (fourchette): where these folds join posteriorly
      2. Prepuce: most anterior part of labia, which have split to pass above clitoris
      3. Frenulum of clitoris: parts of labia that pass below clitoris
   D. CLITORIS: located beneath anterior labial commissure and partly hidden by prepuce and frenulum of labia minora. Has 2 corpora and a glans but no urethra
   E. VESTIBULE: area posterior to clitoris and between labia minora. Has several openings: external urethral; vaginal, behind the urethra; ducts of major vestibular glands
      1. Hymen: membrane of variable size and form partly blocking the vaginal orifice in the virgin
   F. BULB OF VESTIBULE: 2 masses of erectile tissue on either side of vaginal orifice
   G. GREATER VESTIBULAR GLANDS lie beneath posterior ends of bulb. Each has duct that opens into the vestibule in a groove between hymen and labia minora

V. **Clinical considerations**
   A. VAGINAL HYSTERECTOMY: for repair of procidentia and prolapse associated with cystocele and rectocele
   B. POSTERIOR COLPOTOMY: in treatment of pelvic abscess
   C. RECTO- AND VESICOVAGINAL FISTULAE (abnormal connections between bladder and rectum and vagina): from injuries during childbirth or from faulty development
   D. CYSTS OF THE MAJOR VESTIBULAR GLANDS (Bartholin) can occur
   E. IMPERFORATE HYMEN: a failure of an opening to occur in the hymen, which may not be recognized until puberty when menstrual fluid dilates the genital canal, creating pressure (hematocolpos)
   F. EPISIOTOMY is an incision in the vulva which is made to direct the tear in the perineum to the side during labor

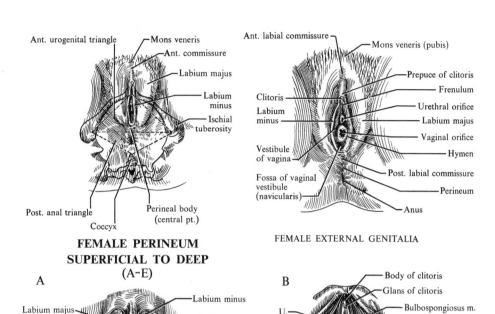

Ant. urogenital triangle — Mons veneris
Ant. commissure
Labium majus
Labium minus
Ischial tuberosity

Post. anal triangle
Coccyx
Perineal body (central pt.)

FEMALE PERINEUM
SUPERFICIAL TO DEEP
(A-E)

Ant. labial commissure — Mons veneris (pubis)
Prepuce of clitoris
Frenulum
Clitoris
Labium minus
Urethral orifice
Labium majus
Vaginal orifice
Hymen
Vestibule of vagina
Post. labial commissure
Fossa of vaginal vestibule (navicularis)
Perineum
Anus

FEMALE EXTERNAL GENITALIA

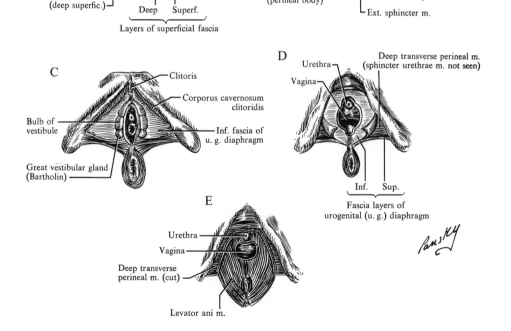

A
Labium majus
Labium minus
Ischiocav. m.
Bulbospon. m.
Ischial tuberosity
Ischial tuberosity
Colles' fascia (deep superfic.)
Deep   Superf.
Layers of superficial fascia

B
Body of clitoris
Glans of clitoris
U.
V.
Bulbospongiosus m.
Ischiocavernosus m.
Ischial tuberosity
Inf. fascia of u. g. diaphragm
Labium minus
Central tendon of perineum (perineal body)
Superficial transverse perineal m.
Ext. sphincter m.

C
Clitoris
Corporus cavernosum clitoridis
Bulb of vestibule
Inf. fascia of u. g. diaphragm
Great vestibular gland (Bartholin)

D
Urethra
Vagina
Deep transverse perineal m. (sphincter urethrae m. not seen)
Inf.   Sup.
Fascia layers of urogenital (u. g.) diaphragm

Pansky

E
Urethra
Vagina
Deep transverse perineal m. (cut)
Levator ani m.

208. THE MALE PERINEUM

I. **Boundaries:** ventrally, pubic arch and arcuate ligaments; dorsally, tip of the coccyx; laterally, inferior rami of pubis and ischium and sacrotuberous ligaments

II. **Divisions:** dorsal, anal region; ventral, urogenital region

III. **Superficial fascia:** 2-layered as on abdomen
   A. SUPERFICIAL LAYER: in UG region, continuous with Camper's fascia, superficial fascia of thigh, and superficial anal fascia; medially, continuous with the dartos of scrotum. In anal region, it is fatty and fills ischiorectal fossa
   B. DEEP LAYER (COLLES' FASCIA) continues with Scarpa's fascia of abdomen, attached to ischiopubic rami, curves around superficial transverse perineal muscle to fuse with deep fascia and central tendon. In anal region, adherent to superficial layer

IV. **Muscles:** all innervated by the perineal branch of pudendal nerve

Name	Origin	Insertion	Action
A. IN SUPERFICIAL PERINEAL POUCH			
Superfic. trans. perineal	Ischial tuberos.	Central tendon	Fixes central tendon
Bulbospongiosus	Central tend., med. raphé	Side and dorsum of penis	Compresses urethra, helps in erection
Ischiocavernosus	Ischial tuberos.	Crura of penis	Maintains erection
B. IN DEEP PERINEAL COMPARTMENT			
Deep trans. perineal	Ramus of ischium	Joins muscle of other side	Compresses memb. urethra
Sphinct. ureth.	Ramus of ischium	Same as above	Same as above

V. **Deep fascia:** obturator and infra-anal
   A. OBTURATOR (see p. 448): over obturator m. on lat. wall of ischiorectal fossa
   B. INFRA-ANAL covers inferior surface of levator ani and coccygeus muscles, joins obturator along arcus tendineus. At border of urogenital region splits into 3 sheets
      1. Deep (external) perineal attached to ischiopubic rami and urogenital diaphragm. Is continuous with deep abdominal fascia, fascia lata, deep penile (Buck's) fascia. Covers superficial perineal muscles
      2. Inferior (superficial) layer of urogenital diaphragm. Attached to ischiopubic rami, leaving a slight space between the ventral edge of this layer and the arcuate ligament to permit the passage of the dorsal vein of penis
         a. Pierced by urethra, ducts of bulbourethral glands, arteries to bulb, deep arteries to penis, dorsal arteries, and nerves of penis
      3. Superior (deep) layer of urogenital diaphragm. Similar to above, with deeper attachments. Prostate rests on it and capsule attached to it. Pierced by urethra

VI. **Perineal compartments**
   A. SUPERFICIAL: space between deep external perineal fascia (Gallaudet's fascia) and the inferior layer of the urogenital diaphragm. It contains crura and bulb of penis, ischio- and bulbospongiosus muscles, superficial transverse perineal muscle, deep artery of penis, vessels and nerves of all the muscles, major vestibular glands of female
   B. DEEP: space between the inferior and superior fascia of the urogenital diaphragm. Contains deep transverse perineal and sphincter urethrae muscles, bulbourethral glands, internal pudendal vessels, nerves, and their branches
   C. SUPERFICIAL PERINEAL FASCIAL CLEFT: between deep external perineal fascia (Gallaudet's) and deep layer of superficial fascia of perineum (Colles')

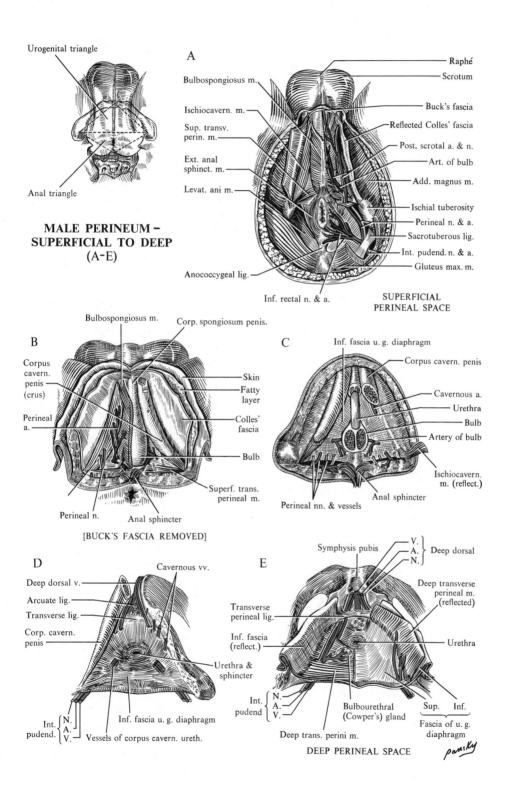

Urogenital triangle

Anal triangle

**MALE PERINEUM –
SUPERFICIAL TO DEEP
(A-E)**

A

Bulbospongiosus m.

Ischiocavern. m.

Sup. transv. perin. m.

Ext. anal sphinct. m.

Levat. ani m.

Anococcygeal lig.

Inf. rectal n. & a.

Raphé

Scrotum

Buck's fascia

Reflected Colles' fascia

Post. scrotal a. & n.

Art. of bulb

Add. magnus m.

Ischial tuberosity

Perineal n. & a.

Sacrotuberous lig.

Int. pudend. n. & a.

Gluteus max. m.

SUPERFICIAL
PERINEAL SPACE

B

Bulbospongiosus m.

Corp. spongiosum penis.

Corpus cavern. penis (crus)

Perineal a.

Perineal n.

Anal sphincter

Skin

Fatty layer

Colles' fascia

Bulb

Superf. trans. perineal m.

[BUCK'S FASCIA REMOVED]

C

Inf. fascia u. g. diaphragm

Corpus cavern. penis

Cavernous a.

Urethra

Bulb

Artery of bulb

Ischiocavern. m. (reflect.)

Perineal nn. & vessels

Anal sphincter

D

Cavernous vv.

Deep dorsal v.

Arcuate lig.

Transverse lig.

Corp. cavern. penis

Int. pudend. { N. A. V.

Vessels of corpus cavern. ureth.

Urethra & sphincter

Inf. fascia u. g. diaphragm

E

Symphysis pubis

V. A. N. } Deep dorsal

Deep transverse perineal m. (reflected)

Transverse perineal lig.

Inf. fascia (reflect.)

Urethra

Int. pudend { N. A. V.

Bulbourethral (Cowper's) gland

Deep trans. perini m.

Sup. Inf.

Fascia of u. g. diaphragm

DEEP PERINEAL SPACE

pansky

209. SCROTUM AND TESTIS

I. Scrotum: cutaneous sac to contain testis. Divided by a median raphé, which is continuous on penis and along midline to anus. Composed of 2 layers

A. SKIN: thin and pigmented, with hair and sebaceous glands

B. DARTOS: superficial fascia containing scattered smooth muscle. Is closely bound to skin but only loosely attached to deeper layers

II. Coverings of testis

A. EXTERNAL SPERMATIC FASCIA (intercrural, intercolumnar): at subcutaneous inguinal ring is continuous with deep abdominal fascia of external oblique muscle which is attached to crura of ring

B. CREMASTERIC LAYER (middle spermatic fascia): this is composed of cremasteric muscle fibers interconnected by fascia. It is derived from the internal abdominal oblique muscle and its fascia

C. INTERNAL SPERMATIC FASCIA is derived from and continuous with the transversalis fascia at the deep inguinal ring

D. TUNICA VAGINALIS: serous membrane derived from the peritoneum of the abdomen. Has 2 layers
   1. Visceral covers testis and epididymis. Posteriorly is reflected onto scrotal lining
   2. Parietal covers more area than visceral layer, for it lines scrotum and extends a short distance into the spermatic cord
   3. Cavity of tunica vaginalis: space between the visceral and parietal layers

III. Testis

A. STRUCTURE
   1. Capsule, the tunica albuginea: a dense, fibrous membrane
   2. Mediastinum testis: reflection of capsule along posterior border into the interior of gland, forming a partial septum
   3. Trabeculae (septa): partial partitions radiating from the front and sides of mediastinum. These divide testis into numerous conical lobules
   4. Lobules: each contains several highly convoluted seminiferous tubules, which give rise to sperm
   5. Tubuli recti: at apex of lobule, next to mediastinum, seminiferous tubules straighten out, carry sperm to rete testis
   6. Rete testis: network of irregular spaces that transmit sperm across mediastinum to efferent ducts (ductules)
   7. Efferent ductules: lead from rete, penetrate tunica albuginea to the head of the epididymis, where each of the 12–15 ductules becomes highly convoluted
   8. Duct of epididymis: the efferent ducts open into this. In the head and body of the epididymis, this duct is convoluted. In the tail of the epididymis, it becomes straighter where it leads into the vas deferens

IV. Epididymis: an important storehouse for sperm. In general, all parts are covered by the visceral layer of serous membrane, which helps to bind this structure to the testis. As noted above, it contains efferent ductules and the duct of the epididymis

V. Clinical considerations

A. HYDROCELE: an accumulation of fluid in the cavity of tunica vaginalis

B. ORCHITIS: an inflammation of the glandular structure of the testis; may occur as a complication of mumps and can lead to sterility

C. VARICOCELE is a condition of enlargement (varicosity) of the veins of the spermatic cord

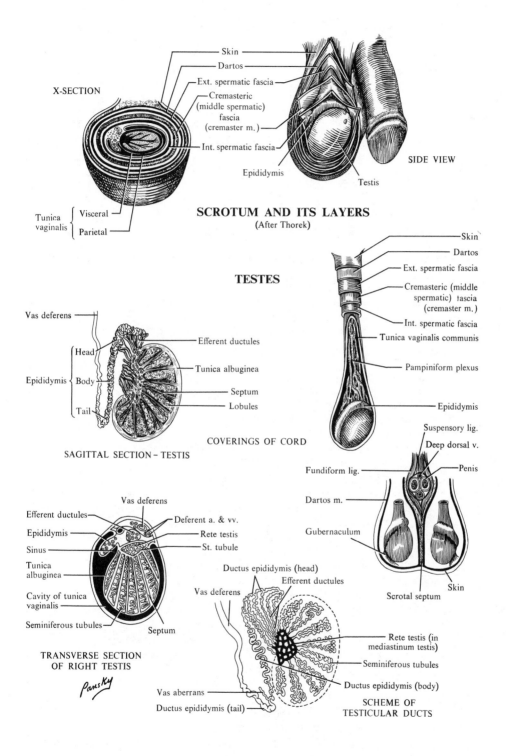

X-SECTION

Skin
Dartos
Ext. spermatic fascia
Cremasteric (middle spermatic) fascia (cremaster m.)
Int. spermatic fascia
Epididymis
Testis
SIDE VIEW

Tunica vaginalis { Visceral / Parietal

SCROTUM AND ITS LAYERS
(After Thorek)

TESTES

Vas deferens
Head
Epididymis { Body
Tail

Efferent ductules
Tunica albuginea
Septum
Lobules

SAGITTAL SECTION – TESTIS

COVERINGS OF CORD

Skin
Dartos
Ext. spermatic fascia
Cremasteric (middle spermatic) fascia (cremaster m.)
Int. spermatic fascia
Tunica vaginalis communis
Pampiniform plexus
Epididymis

Suspensory lig.
Deep dorsal v.
Fundiform lig.
Penis
Dartos m.
Gubernaculum
Scrotal septum
Skin

Vas deferens
Efferent ductules
Epididymis
Sinus
Tunica albuginea
Cavity of tunica vaginalis
Seminiferous tubules
Deferent a. & vv.
Rete testis
St. tubule
Septum

TRANSVERSE SECTION
OF RIGHT TESTIS

Ductus epididymis (head)
Efferent ductules
Vas deferens
Rete testis (in mediastinum testis)
Seminiferous tubules
Ductus epididymis (body)
Vas aberrans
Ductus epididymis (tail)
SCHEME OF
TESTICULAR DUCTS

210. THE PENIS

I. Composition: 2 dorsal cylindrical masses, the corpora cavernosa penis, and a single ventral mass, the corpus spongiosum (corpus cavernosum urethrae)

A. CORPORA CAVERNOSA PENIS: anterior three fourths composed of erectile tissue contained in a strong fibrous capsule. Proximally, the capsules of the adjacent sides of the corpora form a complete *septum penis;* distally, the septum is incomplete. The posterior one fourth of each corpus diverges as the *crura of the penis.* The latter consists of tendonlike tissue; serves as attachments of the penis to the ischiopubic rami

B. CORPUS SPONGIOSUM: erectile tissue containing penile urethra. 2 expansions
   1. Glans: conical tip that forms a cap over the other corpora. The *corona glandis* is the widest, proximal portion
   2. Bulbus penis. The proximal enlargement is entered by the urethra

II. Ligaments

A. FUNDIFORM: sling of Scarpa's fascia extending to the sides and dorsum of the penis

B. SUSPENSORY: deep fascia from linea alba, symphysis pubis, and arcuate ligament

III. Skin: thin, hairless, and dark. Forms fold over glans called the prepuce (foreskin)

IV. Fascia

A. SUPERFICIAL continuous with Scarpa's fascia and the dartos of the scrotum

B. DEEP (FASCIA PENIS, BUCK'S) starts from deep (external) perineal fascia and fascia investing muscles of penis, which extend distally as a single sheet covering the corpora to the glans

V. Vessels and nerves (see also p. 467)

A. ARTERIES: from the internal pudendal artery in the deep perineal compartment
   1. Dorsal arteries pierce urogenital diaphragm, pass through suspensory ligament to lie on either side of the deep dorsal vein beneath deep (Buck's) fascia
   2. Deep arteries pierce urogenital diaphragm and enter crura to run in center of corpora cavernosa penis
   3. Artery of bulb pierces urogenital diaphragm to bulb and corpus spongiosum
   4. Urethral artery pierces UG diaphragm to glans in corpus spongiosum penis

B. VEINS: all begin in cavernous spaces of erectile tissue
   1. Superficial dorsal vein drains prepuce and skin, runs in subcutaneous tissue, and ends in superficial external pudendal vein (of saphenous)
   2. Deep vein drains erectile tissue, runs under deep fascia between dorsal arteries, passes through suspensory lig. and into pelvis between arcuate lig. and urogenital diaphragm and drains into int. pudendal v. and then to int. iliac v.

C. LYMPHATICS: superficial and deep drain to superficial inguinal nodes, glans drains to deep inguinal and external iliac nodes

D. NERVES: anterior scrotal of ilioinguinal, perineal and dorsal nerves from pudendal nerve. Autonomics from hypogastric and pelvic plexuses

VI. Erection caused by contraction of ischio- and bulbospongiosus muscles, which compress veins; at the same time, elevated blood pressure and dilation of arteries increase flow of blood into the cavernous spaces of the erectile tissue

VII. Ejaculation consists of

A. EMISSION: reflex wave of muscle contraction in ductus deferens with contraction of seminal vesicle and prostate. Under sympathetic control

B. EJACULATION PROPER: rhythmic contraction of bulbocavernosus (and ischiocavernosus) muscles. Ejaculation and erection are under parasympathetic control

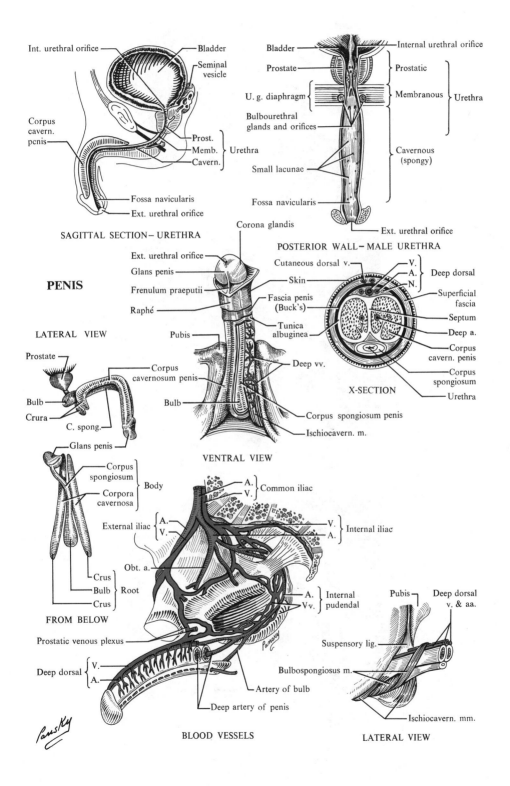

Int. urethral orifice

Bladder

Seminal vesicle

Corpus cavern. penis

Prost.
Memb. } Urethra
Cavern.

Fossa navicularis

Ext. urethral orifice

SAGITTAL SECTION– URETHRA

Bladder

Internal urethral orifice

Prostate

Prostatic

U. g. diaphragm

Membranous } Urethra

Bulbourethral glands and orifices

Small lacunae

Cavernous (spongy)

Fossa navicularis

Ext. urethral orifice

POSTERIOR WALL – MALE URETHRA

PENIS

Corona glandis

Ext. urethral orifice

Glans penis

Frenulum praeputii

Raphé

Cutaneous dorsal v.

V.
A. } Deep dorsal
N.

Skin

Superficial fascia

Fascia penis (Buck's)

Septum

Tunica albuginea

Deep a.

LATERAL VIEW

Pubis

Corpus cavern. penis

Prostate

Corpus cavernosum penis

Deep vv.

Corpus spongiosus

Bulb

Urethra

Bulb

Crura

C. spong.

Corpus spongiosum penis

X-SECTION

Ischiocavern. m.

Glans penis

VENTRAL VIEW

Corpus spongiosum

Corpora cavernosa } Body

A.
V. } Common iliac

External iliac {
A.
V.

V.
A. } Internal iliac

Obt. a.

Crus }
Bulb } Root
Crus }

A.
Vv. } Internal pudendal

Pubis

Deep dorsal v. & aa.

FROM BELOW

Prostatic venous plexus

Suspensory lig.

Deep dorsal {
V.
A.

Bulbospongiosus m.

Artery of bulb

Deep artery of penis

Ischiocavern. mm.

BLOOD VESSELS

LATERAL VIEW

– 471 –

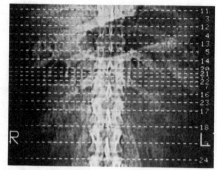

ABDOMEN KEY—LEVELS

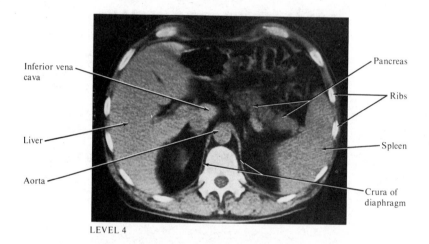

Inferior vena cava

Liver

Aorta

Pancreas

Ribs

Spleen

Crura of diaphragm

LEVEL 4

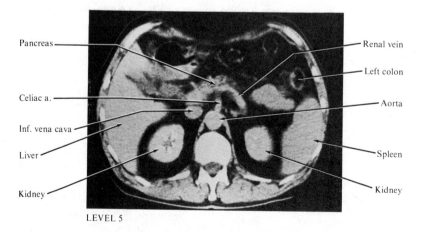

Pancreas

Celiac a.

Inf. vena cava

Liver

Kidney

Renal vein

Left colon

Aorta

Spleen

Kidney

LEVEL 5

FIGURE 45. **Abdominal CT scans. Key levels and levels 4 and 5.**

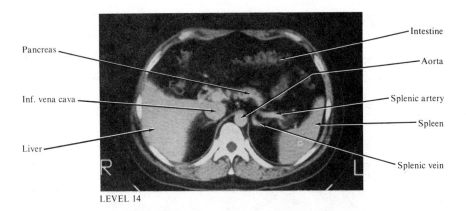

Pancreas

Inf. vena cava

Liver

Intestine

Aorta

Splenic artery

Spleen

Splenic vein

R   L

LEVEL 14

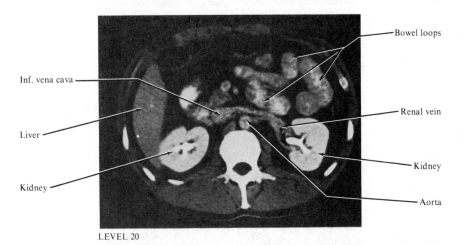

Inf. vena cava

Liver

Kidney

Bowel loops

Renal vein

Kidney

Aorta

LEVEL 20

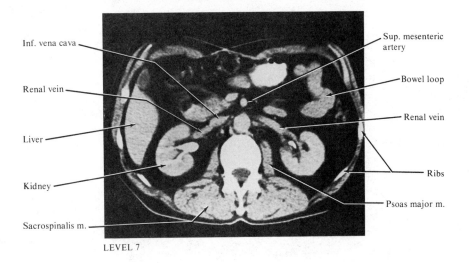

Inf. vena cava

Renal vein

Liver

Kidney

Sacrospinalis m.

Sup. mesenteric artery

Bowel loop

Renal vein

Ribs

Psoas major m.

LEVEL 7

FIGURE 46. **Abdominal CT scans.   Levels 14, 20, and 7.**

– 473 –

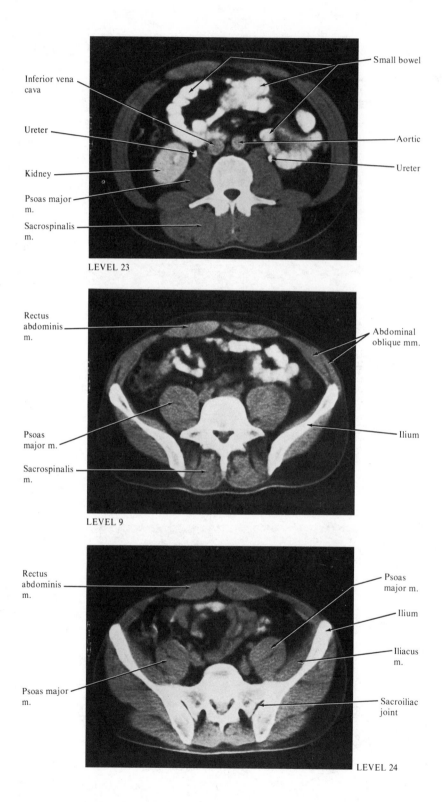

FIGURE 47. **Abdominal CT scans. Levels 23, 9, and 24.**

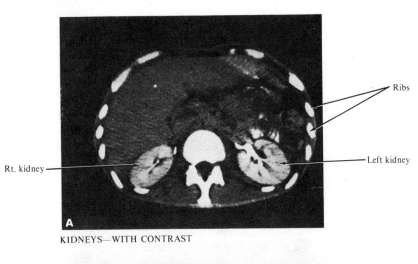

Ribs

Rt. kidney

Left kidney

KIDNEYS—WITH CONTRAST

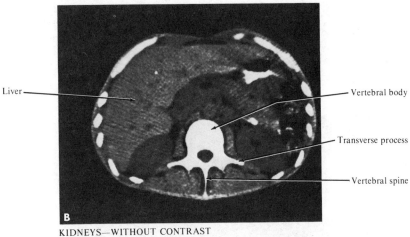

Liver

Vertebral body

Transverse process

Vertebral spine

KIDNEYS—WITHOUT CONTRAST

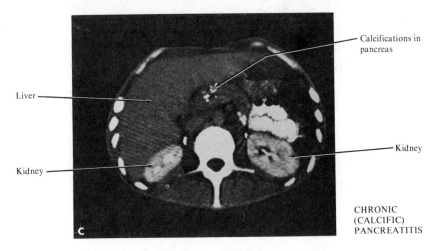

Calcifications in
pancreas

Liver

Kidney

Kidney

CHRONIC
(CALCIFIC)
PANCREATITIS

FIGURE 48. **Abdominal CT scans. A, Kidneys with contrast; B, kidneys without contrast; C, pancreatitis.**

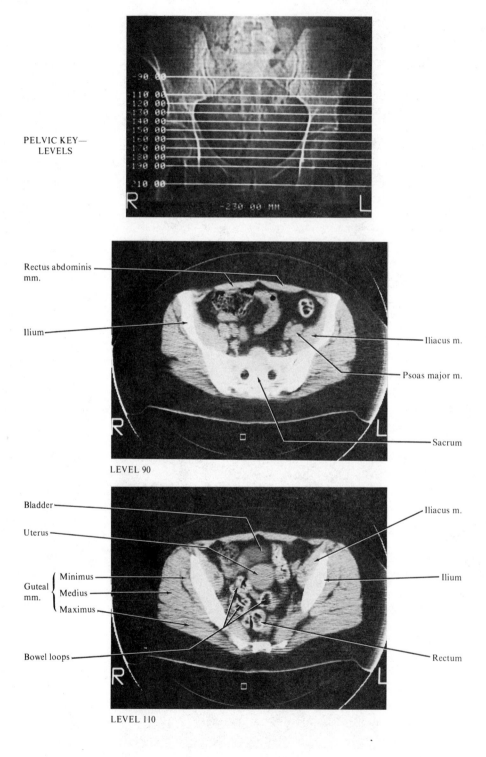

PELVIC KEY—
LEVELS

-90 00
110 00
120 00
130 00
140 00
150 00
160 00
170 00
180 00
190 00
210 00

R   -230 00 MM   L

Rectus abdominis
mm.

Ilium

Iliacus m.

Psoas major m.

Sacrum

R   L

LEVEL 90

Bladder

Uterus

Iliacus m.

Guteal
mm.

Minimus

Medius

Maximus

Ilium

Bowel loops

Rectum

R   L

LEVEL 110

FIGURE 49. **Pelvic CT scans.   Key levels and level 90 and 110.**

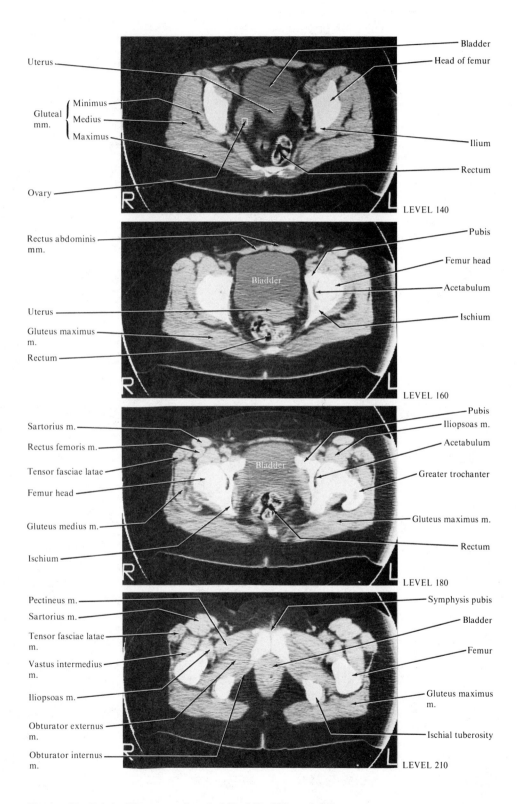

Uterus

Gluteal { Minimus
mm. { Medius
{ Maximus

Ovary

Bladder

Head of femur

Ilium

Rectum

R   L   LEVEL 140

Rectus abdominis
mm.

Uterus

Gluteus maximus
m.

Rectum

Pubis

Femur head

Acetabulum

Ischium

Bladder

R   L   LEVEL 160

Sartorius m.

Rectus femoris m.

Tensor fasciae latae

Femur head

Gluteus medius m.

Ischium

Pubis

Iliopsoas m.

Acetabulum

Greater trochanter

Gluteus maximus m.

Rectum

Bladder

R   L   LEVEL 180

Pectineus m.

Sartorius m.

Tensor fasciae latae
m.

Vastus intermedius
m.

Iliopsoas m.

Obturator externus
m.

Obturator internus
m.

Symphysis pubis

Bladder

Femur

Gluteus maximus
m.

Ischial tuberosity

R   L   LEVEL 210

FIGURE 50. **Pelvic CT scans. Levels 140, 160, 180, and 210.**

– 477 –

UNIT SIX

Lower
Extremity

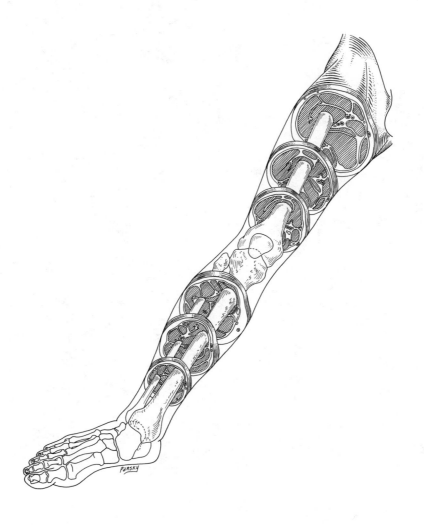

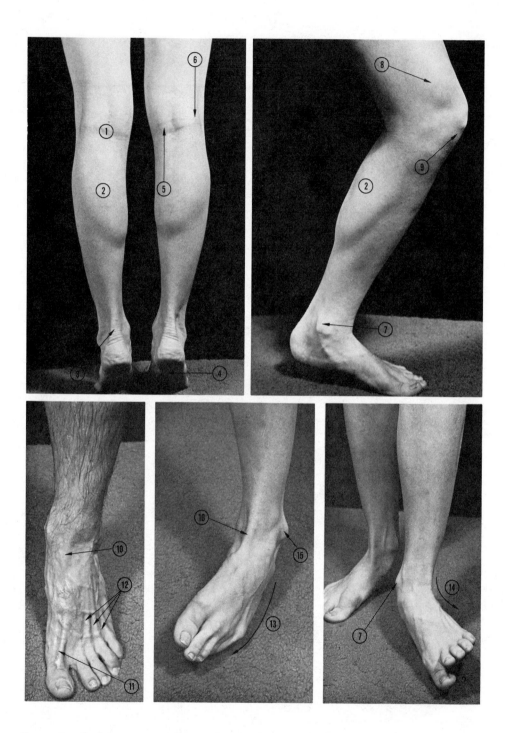

FIGURE 51. **Surface anatomy of lower extremity.** *1*, Popliteal fossa; *2*, gastrocnemius m.; *3*, tendo calcaneus (Achilles); *4*, plantar surface of foot; *5*, semitendinosus m.; *6*, biceps femoris m.; *7*, medial malleolus; *8*, vastus medialis m.; *9*, patella; *10*, tibialis anterior m.; *11*, extensor hallucis longus tendon; *12*, extensor digitorum longus tendons; *13*, inversion; *14*, eversion; *15*, lateral malleolus.

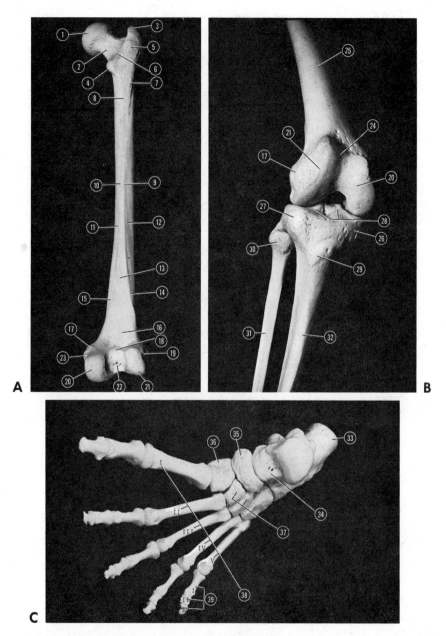

FIGURE 52. **A, Right femur, posterior view; B, Right femur, tibia, fibula, anterior view; C, Left foot.** *1,* head; *2,* neck; *3,* greater trochanter; *4,* lesser trochanter; *5,* quadrate tubercle; *6,* intertrochanteric line; *7,* gluteal tuberosity, *8,* pectineal line; *9 and 10,* lateral and medial lip of linea aspera; *11,* medial surface; *12,* lateral surface; *13,* posterior surface; *14 and 15,* lateral and medial supracondylar lines; *16,* popliteal surface; *17,* adductor tubercle; *18,* intercondylar line; *19,* lateral epicondyle; *20,* medial condyle; *21,* lateral condyle; *22,* intercondylar fossa; *23,* medial epicondyle; *24,* patellar surface; *25,* anterior surface of shaft; *26,* medial condyle; *27,* lateral condyle; *28,* intercondylar tubercles; *29,* tuberosity; *30,* head of fibula; *31,* shaft of fibula; *32,* shaft of tibia; *33,* calcaneus; *34,* talus; *35,* navicular; *36,* cuboid; *37,* lateral intermediate, and medial cuneiforms; *38,* metatarsals; *39,* phalanges of little toe.

– 481 –

211. CUTANEOUS NERVES AND DERMATOMES OF THE LOWER EXTREMITY

I. Source of cutaneous nerves: lumbar plexus, sacral plexus, posterior primary divisions of lumbar and sacral nerves

II. Origin and distribution

Name	Spinal Component	Distribution
A. FROM LUMBAR PLEXUS (L1–L4)		
1. Iliohypogastric	L1	
Lat. branch		Lat. thigh, ant. gluteal region
Ant. branch		Abdomen
2. Ilioinguinal	L1	Upper med. thigh, ext. genitalia
3. Genitofemoral	L1, 2	
Genital br.		Upper med. thigh, scrotum
Femoral br.		Upper ant. thigh
4. Lat. fem. cutaneous	L2, 3	Lat. ant. thigh, post. lat. thigh
5. Obturator	L2, 3, 4	Med. thigh, just above knee
6. Femoral		
Ant. cutaneous	L2, 3	Ant. and anteromed. thigh
Saphenous	L3, 4	Med. side knee, leg, ankle
B. FROM SACRAL PLEXUS (L4–S3)		
1. Post. fem. cutaneous	S1, 2, 3	Lower buttock, post. thigh, middle post. leg
2. Sural	S1, 2	Post. leg, lat. side and dorsum of foot
3. Lat. sural cutan.	S1, 2	Lateral side of leg
4. Medial calcaneal	S1, 2	Heel
5. Medial plantar	L4, 5	Sole of foot, adjacent sides first 4 toes
6. Lateral plantar	S1, 2	Lat. sole of foot, sides of toes 4 and 5
7. Deep peroneal (fibular)	L4, 5	Adjacent sides of dorsum, toes 1 and 2
8. Superfic. peroneal (fibular)	L4, 5, S1	Ant. lower leg and ankle, dorsum of foot, adjacent sides toes 1–4
C. FROM POSTERIOR PRIMARY DIVISIONS OF LUMBAR AND SACRAL NERVES		
1. Superior cluneal	L1, 2, 3	Upper med. buttock
2. Middle cluneal	S1, 2, 3	Med. aspect of buttock

III. Dermatomes

A. DEFINITION: these represent *specific skin areas* supplied by *specific spinal nerves,* regardless of the named cutaneous nerves that are distributed to the same general territory

B. ARRANGEMENT: the dermatomes of the lower extremity are related to the development of the limb. Their spiral course in the lower extremity is an expression of the rotation of the limb as an adaptation to the erect position

C. CLINICAL SIGNIFICANCE: the skin is easily tested for various sensory modalities (for example, touch or temperature). If certain dermatomes are found insensitive, this can be related to definite spinal cord levels, thus, a lesion in the central nervous system can be located more readily

IV. Clinical considerations

A. SPINAL-VERTEBRAL SEGMENTS: a patient with numbness in middle of dorsum of foot and anterolateral leg may have a lesion in the 5th lumbar cord segment, which lies at the level of 11th thoracic vertebra

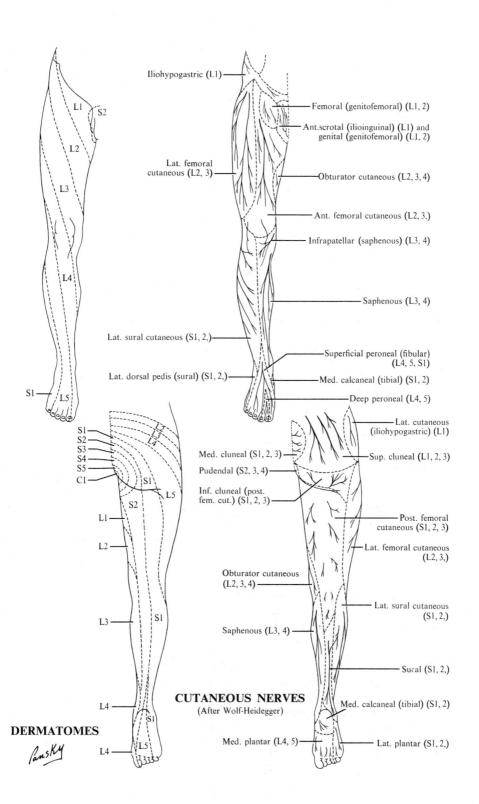

Iliohypogastric (L1)

Femoral (genitofemoral) (L1, 2)

Ant. scrotal (ilioinguinal) (L1) and
genital (genitofemoral) (L1, 2)

Lat. femoral
cutaneous (L2, 3)

Obturator cutaneous (L2, 3, 4)

Ant. femoral cutaneous (L2, 3,)

Infrapatellar (saphenous) (L3, 4)

Saphenous (L3, 4)

Lat. sural cutaneous (S1, 2,)

Superficial peroneal (fibular)
(L4, 5, S1)

Lat. dorsal pedis (sural) (S1, 2,)

Med. calcaneal (tibial) (S1, 2)

Deep peroneal (L4, 5)

L1
S2
L2
L3
L4
S1
L5

S1
S2
S3
S4
S5
C1

L3
L4
S1
L5
S2
L1
L2
L3
S1
L4
S1
L5
L4

Lat. cutaneous
(iliohypogastric) (L1)

Med. cluneal (S1, 2, 3)

Sup. cluneal (L1, 2, 3)

Pudendal (S2, 3, 4)

Inf. cluneal (post.
fem. cut.) (S1, 2, 3)

Post. femoral
cutaneous (S1, 2, 3)

Lat. femoral cutaneous
(L2, 3,)

Obturator cutaneous
(L2, 3, 4)

Lat. sural cutaneous
(S1, 2,)

Saphenous (L3, 4)

Sural (S1, 2,)

CUTANEOUS NERVES
(After Wolf-Heidegger)

Med. calcaneal (tibial) (S1, 2)

Med. plantar (L4, 5)

Lat. plantar (S1, 2,)

DERMATOMES

Pansky

212. SUPERFICIAL VEINS OF THE LOWER EXTREMITY

I. **General:** lie between the layers of the superficial fascia. Valves are more numerous in the veins of the lower extremity than they are in the upper

A. DORSAL DIGITAL VEINS run along the two dorsal margins of each toe and unite at the webs of the toes to form *dorsal metatarsal veins*. The latter empty into the *dorsal venous arch* (overlies metatarsal bones and lies in the subcutaneous tissue). The arch receives communications from the plantar venous arch. Proximally, it is connected with the irregular *dorsal venous network* of the foot

II. **Specific veins**

A. GREAT (LONG) SAPHENOUS VEIN: longest vein in body
   1. Origin: medial side of dorsal venous arch of foot
   2. Course: ascends in front of medial malleolus, along medial side of leg, behind the medial condyles of tibia and femur, along medial side of thigh to the saphenous hiatus, where it pierces cribriform fascia, and the femoral sheath
   3. Termination: in femoral vein about 3.75 cm below inguinal ligament
   4. Tributaries
      a. External pudendal from genital area
      b. Superficial circumflex iliac from upper lateral thigh
      c. Superficial epigastric from abdomen, has communication with lateral thoracic vein by way of *thoracoepigastric vein*
      d. Variable number of other unnamed tributaries from foot, leg, and thigh

B. SMALL (SHORT) SAPHENOUS VEIN
   1. Origin: lateral side of dorsal venous arch of foot
   2. Course: from behind lateral malleolus along lateral side of tendo calcaneus (Achilles), crosses the latter to middle of back of leg, runs straight upward to pierce the deep fascia in lower popliteal space
   3. Termination: in popliteal vein between heads of gastrocnemius muscle
   4. Tributaries: especially from lateral foot and leg, send communication upward to join great saphenous vein

III. **Clinical considerations**

A. VARICOSE VEINS: abnormal dilation and loss of valvular competence in superficial veins of lower extremity, especially the great saphenous system
   1. Prevalent in lower limb because of great weight of the long column of blood (heart to foot). Since venous blood flow depends on muscular movement, long-standing or increased pressure on the more caudal part of the system (as in pregnancy) causes venous stasis. This causes walls to dilate in regions of the numerous valves
   2. To relieve this condition, superficial veins can be ligated, stripped, etc., without interference with venous return because of the numerous communications with the deep veins

B. VENIPUNCTURE: the great saphenous vein is most frequently used, especially its lower end. A cutdown is made about 1.25 cm anterior to the medial malleolus. When necessary, the cephalic portion may be used, although here it may be embedded in fat. The site used is often described as 2 fingerbreadths medial and 1 fingerbreadth below the pubic tubercle. Another method of finding the vein is given as 1 fingerbreadth below the inguinal ligament just medial to the point where the pulsating femoral artery can be palpated

C. VEIN GRAFTS: great saphenous vein has been used to bypass obstructions

D. THE SAPHENOUS NERVE accompanies the greater vein anterior to the medial malleolus and, if it is trapped by a ligature in a saphenous cutdown, the patient may complain of pain along the medial side of the foot

E. THROMBOPHLEBITIS: marked inflammation of a vein with secondary thrombus or clot formation. May destroy valves of the deep veins, resulting in varicosities. May lead to a traveling thrombus and eventual pulmonary embolus

F. PHLEBOTHROMBOSIS: inflammatory element is mild, and the thrombosis takes the form of a loosely attached clot, originating peripherally and propagating centrally

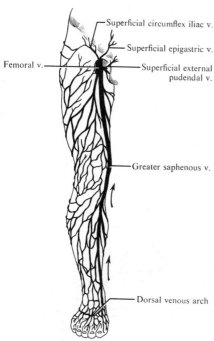

Superficial circumflex iliac v.

Superficial epigastric v.

Femoral v.

Superficial external pudendal v.

Greater saphenous v.

Dorsal venous arch

ANTERIOR VIEW

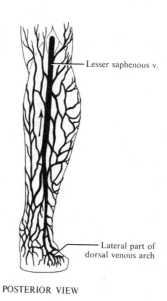

Lesser saphenous v.

Lateral part of dorsal venous arch

POSTERIOR VIEW

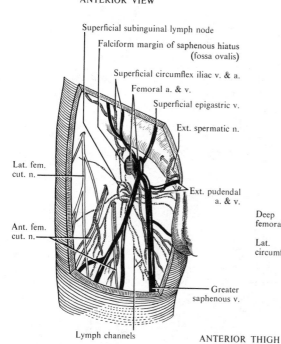

Superficial subinguinal lymph node

Falciform margin of saphenous hiatus (fossa ovalis)

Superficial circumflex iliac v. & a.

Femoral a. & v.

Superficial epigastric v.

Ext. spermatic n.

Lat. fem. cut. n.

Ext. pudendal a. & v.

Ant. fem. cut. n.

Greater saphenous v.

Lymph channels

ANTERIOR THIGH

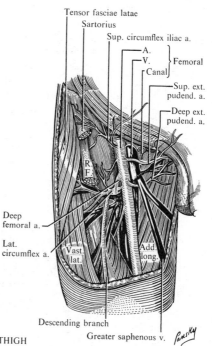

Tensor fasciae latae

Sartorius

Sup. circumflex iliac a.

A.
V. } Femoral
Canal

Sup. ext. pudend. a.

Deep ext. pudend. a.

R. F.

Deep femoral a.

Lat. circumflex a.

Vast. lat.

Add. long.

Descending branch

Greater saphenous v.

Pansky

– 485 –

213. LYMPHATIC DRAINAGE OF THE LOWER EXTREMITY

I. Superficial vessels lie in superficial fascia

A. MEDIAL GROUP follow great saphenous vein (three to seven trunks)
   1. Origin and course: medial side of dorsum of foot, pass on both sides of medial malleolus, run behind medial femoral condyle
   2. Termination: superficial subinguinal nodes

B. LATERAL GROUP (one to two main trunks)
   1. Origin and course: lateral side of foot, ascend leg on both posterior and anterior sides; those on anterior side join medial group (see above)
   2. Termination: for anterior vessels, see A2, above; the posterior group end in the popliteal nodes

II. Deep vessels accompany deep blood vessels; thus, these are anterior and posterior tibial and peroneal sets, 2 or 3 lymphatics with each artery. They terminate in the popliteal nodes

III. Lymph nodes: 3 sets

A. ANTERIOR TIBIAL (inconstant): when present, receive vessels along anterior tibial artery; their efferents go to popliteal nodes

B. POPLITEAL NODES: usually 6 or 7 scattered in fat of popliteal fossa
   1. Afferents: vessels running with small saphenous vein, vessels from knee joint running with genicular arteries, and vessels running with anterior and posterior tibial arteries
   2. Efferents: some follow great saphenous vein to superficial subinguinal nodes; most follow femoral vessels to deep inguinal nodes

C. INGUINAL: vary from 12 to over 20, located in proximal part of femoral trigone. Divided into 2 sets by a transverse line drawn across thigh at the level of the end of the great saphenous vein
   1. Superficial inguinal lie above the line
      a. Afferents from penis, scrotum, perineum, buttock, and abdominal wall
      b. Efferents to external iliac nodes
   2. Subinguinal: below line. Consist of 2 groups
      a. Superficial, on either side of great saphenous vein
         i. Afferents follow superficial vessels of lower limb, especially those with great saphenous vein; a few from external genitalia and buttock
         ii. Efferents to deep subinguinal and external iliac nodes
      b. Deep, under deep fascia on medial side of femoral vein; 1–3 in number. One of these may lie in femoral canal (node of Cloquet)
         i. Afferents: deep lymphatics of lower extremity; from superficial subinguinal nodes
         ii. Efferents: to external iliac nodes

IV. Clinical considerations

A. DRAINAGE of the external genitalia, perineum, buttock, and lower anal canal into the inguinal nodes should be noted

B. THE UTERUS adjoining the attachment of the round ligament sends its lymphatics along the ligament to drain into the inguinal nodes

C. AS THEY COURSE UPWARD in the thigh, the superficial lymphatics from the lateral side run anteriorly and then medially and those from the medial side run anteriorly and then lateral, to converge toward the path of the great saphenous vein. This creates a sort of "lymphatic divide" on the back of the thigh

D. INJECTION OF RADIOPAQUE MATERIAL into the lymphatics of the lower limb can demonstrate the lymphatics as well as lymph nodes

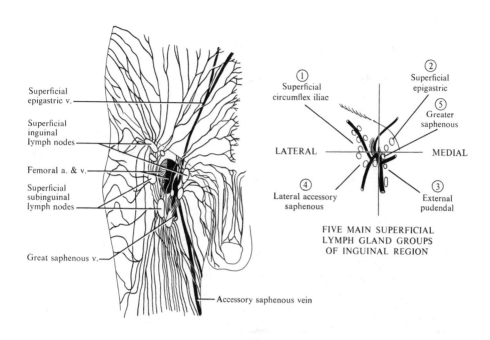

Superficial
epigastric v.

Superficial
inguinal
lymph nodes

Femoral a. & v.

Superficial
subinguinal
lymph nodes

Great saphenous v.

Accessory saphenous vein

① Superficial
circumflex iliac

② Superficial
epigastric

⑤ Greater
saphenous

LATERAL

MEDIAL

④ Lateral accessory
saphenous

③ External
pudendal

FIVE MAIN SUPERFICIAL
LYMPH GLAND GROUPS
OF INGUINAL REGION

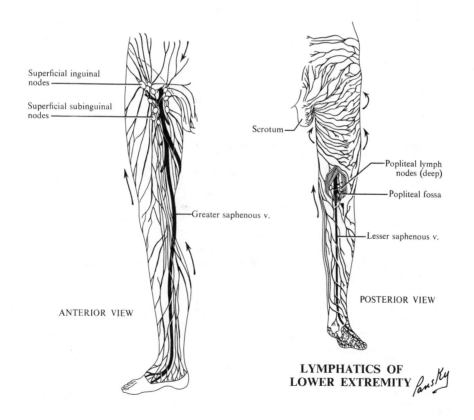

Superficial inguinal
nodes

Superficial subinguinal
nodes

Greater saphenous v.

ANTERIOR VIEW

Scrotum

Popliteal lymph
nodes (deep)

Popliteal fossa

Lesser saphenous v.

POSTERIOR VIEW

**LYMPHATICS OF
LOWER EXTREMITY** *Pansky*

214. BONES OF THE HIP

I. Hip bone
A. PARTS: ilium, ischium, and pubis
   1. Ilium: ala showing an arched crest with a tubercle; anterior, posterior, and inferior gluteal lines on the lateral surface; anterior superior and inferior spines projecting ventrally; posterior superior and inferior spines projecting dorsally; and the iliac fossa, arcuate line, auricular surface, and tuberosity, on its medial side
   2. Ischium shows a spine, tuberosity, and rami
   3. Pubis shows rami and crest, tubercle, pectin, iliopubic eminence, and symphysis
B. SPECIAL FEATURES
   1. Acetabulum formed superiorly by ilium, posteroinferiorly by ischium, and anteroinferiorly by pubis
   2. Obturator foramen and groove
   3. Greater and lesser sciatic notches

II. Clinical features
A. FRACTURES: most commonly due to crushing forces, occurring at thinnest places—either ala of ilium or ischiopubic rami. Usually little displacement
B. BURSITIS: inflammation, especially of bursa, over ischial tuberosity, from sitting for long hours on hard seat, "tailor's or weaver's bottom"
C. OSSIFICATION: from 8 centers

Location	When Appears	When Closes
Lower ilium	8th–9th fetal week	18th year
Superior ramus, ischium	3rd fetal month	18th year
Superior ramus, pubis	4th–5th fetal month	18th year
Acetabulum (1 or more)	12th year	18th year
Iliac crest	Puberty	20th–25th year
Tuberosity of ischium	Puberty	20th–25th year
Symphysis pubis	Puberty	20th–25th year

III. Special features
A. THE PELVIS serves two purposes: it transmits the weight of the body to the lower limbs, and it also forms the lower part of the abdominal cavity (pelvic cavity) and houses part of the abdominopelvic viscera
B. PELVIC INLET divides the pelvis into an "open" anterosuperior region and a confined posteroinferior tunnel, the *true pelvis*. The inlet extends from the sacral promontory on to each ala of the sacrum, across a sacroiliac joint, and along each iliopectineal line on to the superior aspect of each pubic bone beside the centrally located symphysis
C. PELVIC OUTLET: from the tip of the 5th piece of the sacrum (not the tip of coccyx), then on each side the sacrotuberous ligament (between sacrum and tuberosity of ischium), to the ischial tuberosity, then the ischiopubic ramus, and finally, anteriorly, the anteroinferior aspect of the symphysis pubis (a diamond-shaped area)
D. THE PELVIC CAVITY lies between the inlet and outlet
E. PELVIS MAJOR (greater or false) lies above superior pelvic aperture. Formed by ala of sacrum, iliac fossa, L5 and S1 vertebrae, and abdominal wall. Its cavity is part of abdominal cavity and contains abdominal viscera
F. PELVIS MINOR (lesser or true) lies below superior pelvic aperture. Limited by pelvic inlet (above) and pelvic outlet (below); sacrum and coccyx, posterior; symphysis pubis, body of pubis, and pubic rami, anterior; pelvic aspects of ilium and ischium, lateral. Contains pelvic viscera (bladder, rectum, and parts of urogenital organs), blood vessels, nerves, and lymphatics

HIP BONE

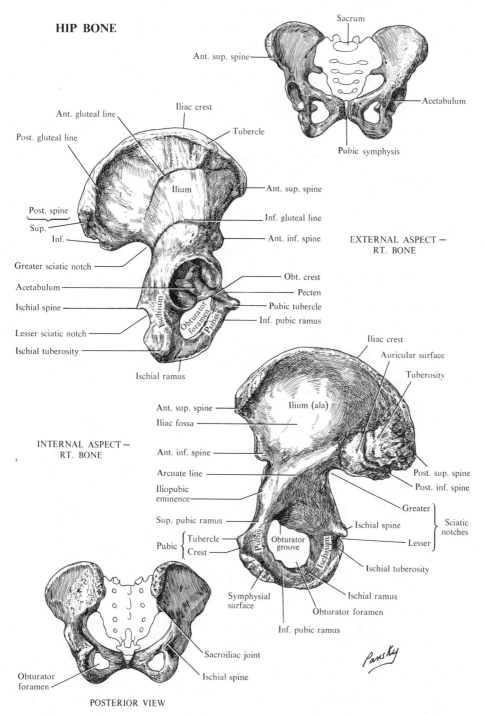

ANTERIOR VIEW

Sacrum

Ant. sup. spine

Acetabulum

Pubic symphysis

Ant. gluteal line

Iliac crest

Post. gluteal line

Tubercle

Ilium

Ant. sup. spine

Inf. gluteal line

Post. spine

Sup.

Ant. inf. spine

Inf.

EXTERNAL ASPECT — RT. BONE

Greater sciatic notch

Acetabulum

Obt. crest

Ischium

Pecten

Ischial spine

Pubic tubercle

Obturator foramen

Inf. pubic ramus

Lesser sciatic notch

Pubis

Ischial tuberosity

Ischial ramus

Iliac crest

Auricular surface

Tuberosity

Ant. sup. spine

Ilium (ala)

Iliac fossa

INTERNAL ASPECT — RT. BONE

Ant. inf. spine

Arcuate line

Post. sup. spine

Post. inf. spine

Iliopubic eminence

Greater

Sup. pubic ramus

Ischial spine

Sciatic notches

Pubic { Tubercle / Crest

Pubis

Obturator groove

Lesser

Ischium

Ischial tuberosity

Symphysial surface

Ischial ramus

Obturator foramen

Inf. pubic ramus

Pansky

Obturator foramen

Sacroiliac joint

Ischial spine

POSTERIOR VIEW

215. FUNCTIONAL BONY PELVIS

I. Introduction: the bony pelvis may be a reliable indicator of sex, particularly in forensic medicine. Certain measured areas are described

A. INLET
  1. Anteroposterior (AP) or conjugate diameter: passes from the upper margin of the pubic symphysis to the middle of the sacral promontory
  2. Obstetric conjugate diameter: measured from the back of the pubic symphysis to the sacral promontory. It is shorter than the AP diameter and is the minimal distance between the promontory and the symphysis
  3. Diagonal conjugate diameter: the only one measured per vaginum. It is the distance between the lower pubic symphysis margin and the sacral promontory
     a. When the promontory cannot be reached via the vagina, the AP inlet diameter is said to be adequate for parturition; if it can be felt, the pelvis is regarded as contracted
  4. Transverse diameter: passes across the widest part of the inlet (side to side)
  5. Oblique diameter: extends from the sacroiliac joint of one side to the iliopubic eminence of the opposite side (or center of obturator foramen)

B. OUTLET
  1. AP or conjugate diameter: measured from lower margin of pubic symphysis to tip of coccyx
  2. Transverse diameter: measured between the ischial tuberosities
  3. Oblique diameter: measured from junction of ischial and pubic rami of one side to the point of crossing of the sacrotuberous and sacrospinous ligaments of the opposite side

II. Pelvic classification: one depends on the shape of the pelvic inlet; the other on measurements of diameters. These are especially applicable in the female where size and shape of inlet affect parturition. The inlet shape is most used. Types described are

A. ANTHROPOID (apelike): 23%; oval sides are long and narrow with AP diameter greater than transverse; long sacrum with deep pelvic cavity; prominent ischial spines and narrow subpubic angle
  1. May encounter difficulty in delivery due to narrow transverse diameter

B. GYNECOID: about 50%; wide, circular inlet; wide subpubic arch and widely spaced ischial spines
  1. With such a pelvis, one can expect a reasonably uneventful delivery

C. ANDROID: twice as common in white as in nonwhite females; heart-shaped inlet; narrow subpubic angle; prominent ischial spines; resembles the male pelvis

D. PLATYPELLOID: 2.5%; flattened type; short sacrum; AP diameter is short, and transverse is long
  1. May be difficult for fetal head to engage inlet, and cesarean section may be necessary

III. Axis of birth canal: path followed by the fetal head in its course through the pelvis. Extends downward and backward in the axis of the inlet as far as the ischial spines (level of uterovaginal angle), and here it turns forward and downward, at almost a right angle and continues in the axis of the vagina (approximately parallel to inlet)

IV. Male pelvis vs. female pelvis

A. FEMALE: bones are thinner and lighter; muscle markings are not as prominent; cavity is less funnel-shaped; distance between ischial spines and tuberosities are greater; sciatic notch is wider; surfaces of sacrum for articulation with ilium and with the 5th lumbar vertebra are smaller; and the subpubic angle is close to a right angle in the female and is more acute in the male

B. MALE: in addition to above, the ischial spines in the male are heavier and project farther into the pelvic cavity

FUNCTIONAL BONY PELVIS

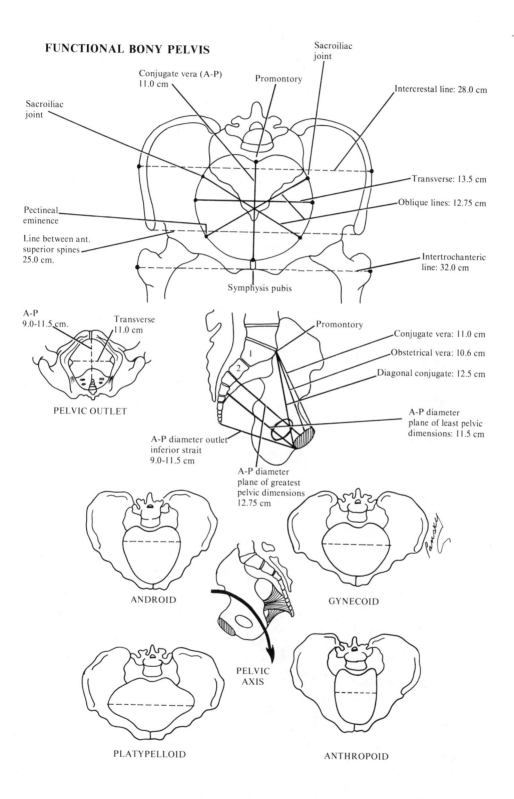

Sacroiliac joint

Conjugate vera (A-P) 11.0 cm

Promontory

Sacroiliac joint

Intercrestal line: 28.0 cm

Sacroiliac joint

Pectineal eminence

Line between ant. superior spines 25.0 cm.

Transverse: 13.5 cm

Oblique lines: 12.75 cm

Intertrochanteric line: 32.0 cm

Symphysis pubis

A-P 9.0-11.5 cm.

Transverse 11.0 cm

PELVIC OUTLET

Promontory

Conjugate vera: 11.0 cm

Obstetrical vera: 10.6 cm

Diagonal conjugate: 12.5 cm

A-P diameter plane of least pelvic dimensions: 11.5 cm

A-P diameter outlet inferior strait 9.0-11.5 cm

A-P diameter plane of greatest pelvic dimensions 12.75 cm

ANDROID

GYNECOID

PELVIC AXIS

PLATYPELLOID

ANTHROPOID

216. THE PELVIC GIRDLE

I. Definition
A. BONES THROUGH WHICH LOWER EXTREMITY IS ATTACHED TO THE TRUNK

II. Bones (see p. 488)

III. Articulations
A. SACROILIAC, between auricular surface of sacrum and ilium
   1. Type: amphiarthrodial (syndesmosis)
   2. Ligaments
      a. Anterior from anterolateral sacrum to auricular and preauricular sulcus of ilium
      b. Posterior from first 3 transverse tubercles of sacrum to tuberosity and posterior superior spine of ilium
      c. Interosseous: short fibers between tuberosities of sacrum and ilium
B. SACROCOCCYGEAL, between apex of sacrum and base of coccyx
   1. Type: amphiarthrodial (syndesmosis)
   2. Ligaments: anterior, posterior, lateral, interarticular, and fibrocartilage disk
C. INTERPUBIC, between the pubic bones of opposite sides
   1. Type: amphiarthrodial (symphysis)
   2. Ligaments
      a. Superior pubic connects bones superiorly
      b. Arcuate connects bones inferiorly
      c. Interpubic: fibrocartilage disk
D. LIGAMENTS between sacrum and ischium
   1. Sacrotuberous from posterior inferior iliac spine, fourth and fifth transverse tubercles and sides of sacrum, and side of coccyx to the inner ischial tuberosity
   2. Sacrospinous from side of sacrum and coccyx to spine of ischium

IV. Special features
A. OBTURATOR MEMBRANE: interlacing fibers, attached to bony margins of foramen, which the membrane nearly fills. Obturator internus and externus muscles arise from inner and outer surfaces
B. GREATER SCIATIC FORAMEN
   1. Boundaries: in front and above, ilium and rim of great sciatic notch; behind, sacrotuberous ligament; below, sacrospinous ligament
   2. Structures passing through or lying in it: piriformis muscle, superior gluteal vessels and nerve, inferior gluteal vessels and nerve, internal pudendal vessels and nerve, sciatic nerve, posterior femoral cutaneous nerve, nerves to obturator internus and quadratus femoris muscles
C. LESSER SCIATIC FORAMEN
   1. Boundaries: in front, tuberosity of ischium; above, spine of ischium and sacrospinous ligament; behind, sacrotuberous ligament
   2. Structures passing through: tendon of obturator internus muscle, nerve to obturator internus muscle, internal pudendal vessels and nerve

V. Clinical considerations
A. SACROILIAC STRAIN is probably quite rare, although formerly often diagnosed, due to the great strength of the supporting ligaments
B. DURING PREGNANCY, the vertebropelvic ligaments progressively relax, and movements between the vertebral column and pelvis become freer. The symphysis pubis relaxes (due to hormone *relaxin*), and the distance between pubic bones increases considerably. The above facilitates passage of fetus through birth canal during parturition

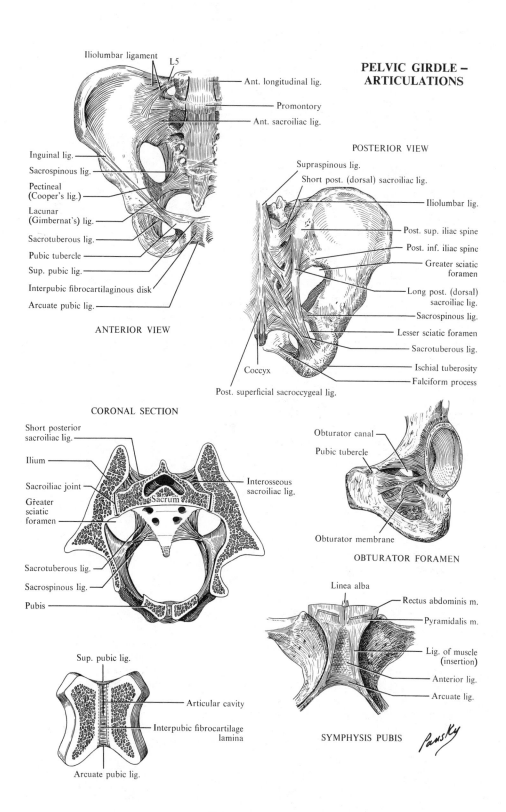

PELVIC GIRDLE –
ARTICULATIONS

Anterior view (top left):

Iliolumbar ligament

L5

Ant. longitudinal lig.

Promontory

Ant. sacroiliac lig.

Inguinal lig.

Sacrospinous lig.

Pectineal
(Cooper's lig.)

Lacunar
(Gimbernat's) lig.

Sacrotuberous lig.

Pubic tubercle

Sup. pubic lig.

Interpubic fibrocartilaginous disk

Arcuate pubic lig.

ANTERIOR VIEW

Posterior view:

POSTERIOR VIEW

Supraspinous lig.

Short post. (dorsal) sacroiliac lig.

Iliolumbar lig.

Post. sup. iliac spine

Post. inf. iliac spine

Greater sciatic foramen

Long post. (dorsal) sacroiliac lig.

Sacrospinous lig.

Lesser sciatic foramen

Sacrotuberous lig.

Ischial tuberosity

Falciform process

Coccyx

Post. superficial sacrococcygeal lig.

Coronal section:

CORONAL SECTION

Short posterior sacroiliac lig.

Ilium

Sacroiliac joint

Greater sciatic foramen

Sacrum

Interosseous sacroiliac lig.

Sacrotuberous lig.

Sacrospinous lig.

Pubis

Obturator foramen:

Obturator canal

Pubic tubercle

Obturator membrane

OBTURATOR FORAMEN

Symphysis pubis:

Linea alba

Rectus abdominis m.

Pyramidalis m.

Lig. of muscle (insertion)

Anterior lig.

Arcuate lig.

SYMPHYSIS PUBIS

Sup. pubic lig.

Articular cavity

Interpubic fibrocartilage lamina

Arcuate pubic lig.

Pansky

217. BONE OF THE THIGH

I. Femur
A. PARTS: proximal extremity, shaft, distal extremity
1. Proximal: head with fovea; neck, constricted portion between head and intertrochanteric crest and line; lesser trochanter, intertrochanteric crest with quadrate tubercle, intertrochanteric line, gluteal tuberosity, and greater trochanter
2. Shaft: linea aspera with a medial and lateral lip; pectineal line, an upward continuation of medial lip; continuation of lateral lip to gluteal tuberosity; inferior continuation of lips of linea aspera as supracondylar lines; nutrient foramen
3. Distal: popliteal plane (planum popliteum), flat space between supracondylar lines; medial and lateral condyles and epicondyles; intercondylar fossa; adductor tubercle
B. SPECIAL FEATURES
1. Longest, largest, heaviest bone of body
2. Angle between shaft and neck: 120–125°, smaller in female, greater in children
3. In erect posture, shaft runs obliquely, the medial distal ends of femora in contact
4. Anatomical neck: epiphyseal line between head and neck
5. Surgical neck: junction of shaft and proximal extremity, just below lesser trochanter
6. In standing position, femur transmits weight from hip bone to tibia
7. In the living state, the tibia is covered with muscles so that only its upper and lower ends are palpable
C. OSSIFICATION, FROM 5 CENTERS

Location	When Appears	When Closes
Body	7th fetal week	
Distal end	9th fetal week	20th year
Head	1st year	After puberty, last to close
Greater trochanter	4th year	After puberty, 2nd to close
Lesser trochanter	13th–14th year	After puberty, 1st to close

II. Clinical features
A. FRACTURE
1. Subtrochanteric, a break just below lesser trochanter. Due to power of iliopsoas, proximal fragment of bone is flexed, adducted, and laterally rotated. The distal part will be shortened by pull of hamstrings, rectus femoris, adductors, and sartorius muscles; abducted by gluteal muscles
2. Fractures in the lower third. Proximal part is fairly stable. Distal part will override (shorten) by pull of the hamstrings, rectus femoris, sartorius, and adductor magnus muscles and will be pulled backward by attachments of the gastrocnemius and plantaris muscles. This could lead to damage of the popliteal artery, which lies close to the bone in that position
3. Fractures of the neck, intertrochanteric (between trochanters) or pertrochanteric (through the trochanters), are common over 60 years of age. They are more common in older women than men, because women's bones are weakened due to senile and postmenopausal osteoporosis
4. The usual "broken hip" more often than not is a fracture of the femoral neck
B. THE ANGLE between the neck and body of the femur is called the *angle of inclination*. Marked decrease in this angle, that is, a more transverse direction of the neck, can result from weight-bearing on a bone that is not capable of standing it. Such an abnormal decrease is known as *coxa vara* (bent hip). It is usually due to rickets, but the same effect can result from disease of the bone or a fracture of the neck that is improperly set
1. When the angle is increased, we speak of *coxa valga*
C. THE BLOOD SUPPLY OF THE FEMUR consists of multiple arteries entering at each end and one or two *nutrient arteries* entering the body. The latter are derived from upper perforating branches of the profunda femoris (deep femoral)

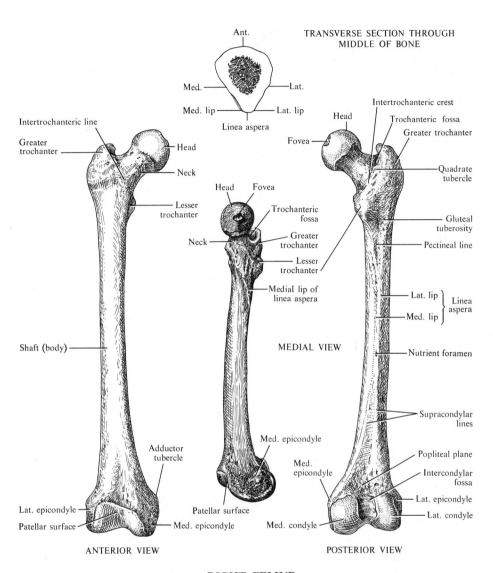

TRANSVERSE SECTION THROUGH
MIDDLE OF BONE

Ant.

Med. — — Lat.

Med. lip — — Lat. lip

Linea aspera

Intertrochanteric line

Greater trochanter

Head

Neck

Lesser trochanter

Shaft (body)

Adductor tubercle

Lat. epicondyle

Patellar surface

Med. epicondyle

ANTERIOR VIEW

Head   Fovea

Trochanteric fossa

Neck

Greater trochanter

Lesser trochanter

Medial lip of linea aspera

Med. epicondyle

Patellar surface

MEDIAL VIEW

Head

Fovea

Intertrochanteric crest

Trochanteric fossa

Greater trochanter

Quadrate tubercle

Gluteal tuberosity

Pectineal line

Lat. lip

Med. lip

Linea aspera

Nutrient foramen

Supracondylar lines

Med. epicondyle

Popliteal plane

Intercondylar fossa

Lat. epicondyle

Lat. condyle

Med. condyle

POSTERIOR VIEW

RIGHT FEMUR

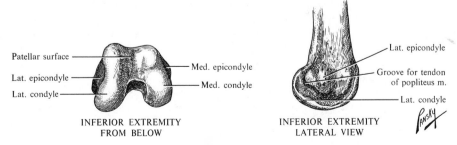

Patellar surface

Lat. epicondyle

Lat. condyle

Med. epicondyle

Med. condyle

INFERIOR EXTREMITY
FROM BELOW

Lat. epicondyle

Groove for tendon of popliteus m.

Lat. condyle

INFERIOR EXTREMITY
LATERAL VIEW

PANSKY

218. MUSCLES OF THE ANTERIOR THIGH

I.

Name	Origin	Insertion	Action	Nerve
Sartorius	Anterior superior iliac spine	Upper medial tibia	Flexes and rotates thigh laterally Flexes leg, rotates it medially	Femoral
Quadriceps Rectus femoris	Ant. inf. iliac spine Rim of acetabulum	Base of patella	Flexes thigh Extends leg	Femoral
Vastus lateralis	Intertrochanteric line Great trochanter Glut. tuber. Lat. linea aspera	Lateral patella	Extends leg	Femoral
Vastus medius	Intertrochanteric line Med. linea aspera Supracondylar line	Medial patella	Extends leg	Femoral
Vastus intermedius	Upper ant. shaft of femur	Base of patella	Extends leg	Femoral
Iliopsoas	Out of field, see p. 382	Lesser trochanter	Flexes thigh Lat. rotates and adducts thigh	Nn. to iliopsoas (L2, 3, 4)

II. Special features
- A. ILIOPSOAS flexes thigh. When extremity is free, it rotates thigh laterally. When extremity is fixed, as when foot is on ground, it rotates pelvis

III. Clinical considerations (see p. 494)
- A. THE MEMBERS OF THE ANTERIOR GROUP of muscles differ considerably in their actions. In general, the iliopsoas and pectineus act only at the hip joint; the sartorius and rectus femoris act both at the hip and knee joints (but with opposite actions at the knee joint); and the vastus muscles act only at the knee joint
- B. WITH PARALYSIS of the quadriceps femoris muscle, the knee cannot be extended, but the patient can stand erect since the body weight tends to overextend the knee. The patient also may be able to walk with short steps if the pelvis is rotated to prevent overextension of the hip which would flex the knee
- C. HIP POINTER: a bruise or contusion over a bony prominence, particularly the anterior superior iliac spine of the iliac crest from which the sartorius muscle and inguinal ligament arise

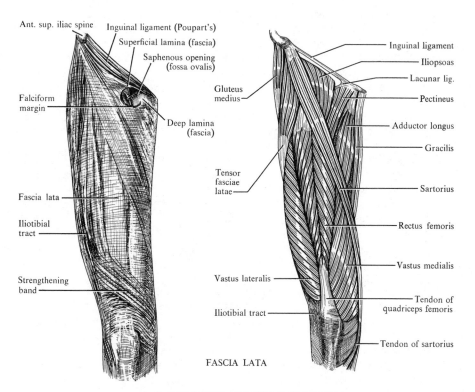

Ant. sup. iliac spine

Inguinal ligament (Poupart's)

Superficial lamina (fascia)

Saphenous opening
(fossa ovalis)

Gluteus
medius

Falciform
margin

Deep lamina
(fascia)

Tensor
fasciae
latae

Fascia lata

Iliotibial
tract

Strengthening
band

Vastus lateralis

Iliotibial tract

Inguinal ligament

Iliopsoas

Lacunar lig.

Pectineus

Adductor longus

Gracilis

Sartorius

Rectus femoris

Vastus medialis

Tendon of
quadriceps femoris

Tendon of sartorius

FASCIA LATA

ANTERIOR THIGH MUSCLES

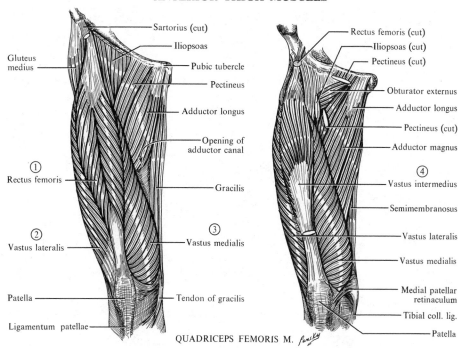

Gluteus
medius

Sartorius (cut)

Iliopsoas

Pubic tubercle

Pectineus

Adductor longus

Opening of
adductor canal

Rectus femoris

① Rectus femoris

Gracilis

② Vastus lateralis

③ Vastus medialis

Patella

Tendon of gracilis

Ligamentum patellae

Rectus femoris (cut)

Iliopsoas (cut)

Pectineus (cut)

Obturator externus

Adductor longus

Pectineus (cut)

Adductor magnus

④ Vastus intermedius

Semimembranosus

Vastus lateralis

Vastus medialis

Medial patellar
retinaculum

Tibial coll. lig.

Patella

QUADRICEPS FEMORIS M. *Pansky*

219. FASCIA, FEMORAL TRIANGLE, AND FEMORAL SHEATH

I. Fascia of the anterior and anteromedial thigh
A. SUPERFICIAL: continuous with that of the abdomen, gluteal region, and leg. Consists of 2 layers, with vessels and nerves between them. The more superficial layer is fatty. The deep layer is adherent to the deep fascia below the inguinal ligament along the medial thigh and at the edge of the saphenous hiatus (fossa ovalis)
   1. Cribriform fascia: the deep layer of the superficial fascia over the saphenous hiatus (fossa ovalis)
B. DEEP FASCIA (fascia lata) is attached above to the os coxa and inguinal ligament, below to the tibia, and is continuous with the crural fascia
   1. Part of fascia lata overlying vastus lateralis forms the *iliotibial tract*
C. INTERMUSCULAR SEPTA: inward extensions of deep fascia
   1. Medial divides the extensors from the adductors, extends inward to the bone between the quadriceps and adductor longus and brevis muscles
   2. Lateral extends inward from the fascia to the linea aspera and separates the vastus lateralis from the biceps femoris muscle
D. SAPHENOUS HIATUS (FOSSA OVALIS): gap in the fascia lata, where it is split into a superficial layer that curves laterally and downward from the pubic tubercle and inguinal ligament along the saphenous vein, and then under the vein to fuse to the deep layer medial to the vein. The deep layer is part of the iliopectineal fascia
   1. Structures piercing the cribriform fascia to enter or leave the fossa: great saphenous vein, superficial circumflex iliac vessels, superficial epigastric vessels, and external pudendal vessels

II. Femoral triangle
A. BOUNDED above by the inguinal ligament, laterally by the sartorius muscle, and medial by the adductor longus muscle
B. FLOOR, from lateral to medial: iliacus, psoas major, and pectineus muscles

III. Adductor (Hunter's) canal: a musculofibrous canal in the middle of the thigh, beginning at the apex of the femoral triangle and ending in the tendinous hiatus of the adductor magnus muscle
A. BOUNDED: anterolaterally, vastus medialis; behind, adductor longus and magnus muscles; and covered (roofed) anteromedially by fibrous tissue under the sartorius muscle (vastoadductor fascia)

IV. Interval behind inguinal ligament
A. ILIOPECTINEAL ARCH: a septum of iliopectineal fascia extends between the iliopubic eminence to the inguinal ligament, dividing the area into 2 compartments
   1. Muscular contains iliacus and psoas major muscles, femoral nerve, and lateral femoral cutaneous nerve
   2. Vascular contains femoral sheath and its contents (see below)

V. Femoral sheath: a prolongation of the transversalis fascia from the abdomen, fusing to the iliac fascia dorsally. Funnel-shaped, narrow end fuses with the intrinsic fascia of the vessels contained about 5 cm below the inguinal ligament. Two vertical septa divide the sheath into 3 compartments: lateral, for artery and femoral branch of genitofemoral nerve; intermediate, for vein; and medial, for femoral canal, containing a few small lymphatics and a lymph node (node of Cloquet)
A. THE FEMORAL RING is the cephalic end of the femoral canal

VI. Clinical consideration
A. THE FEMORAL CANAL is the most common site for femoral herniations

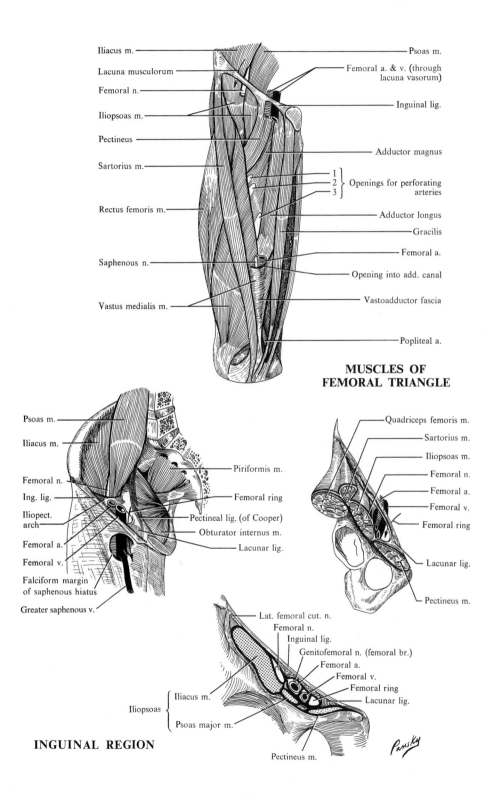

Iliacus m.

Lacuna musculorum

Femoral n.

Iliopsoas m.

Pectineus

Sartorius m.

Rectus femoris m.

Saphenous n.

Vastus medialis m.

Psoas m.

Femoral a. & v. (through lacuna vasorum)

Inguinal lig.

Adductor magnus

1
2 } Openings for perforating
3   arteries

Adductor longus

Gracilis

Femoral a.

Opening into add. canal

Vastoadductor fascia

Popliteal a.

**MUSCLES OF
FEMORAL TRIANGLE**

Psoas m.

Iliacus m.

Femoral n.

Ing. lig.

Iliopect. arch

Femoral a.

Femoral v.

Falciform margin of saphenous hiatus

Greater saphenous v.

Piriformis m.

Femoral ring

Pectineal lig. (of Cooper)

Obturator internus m.

Lacunar lig.

Quadriceps femoris m.

Sartorius m.

Iliopsoas m.

Femoral n.

Femoral a.

Femoral v.

Femoral ring

Lacunar lig.

Pectineus m.

Lat. femoral cut. n.

Femoral n.

Inguinal lig.

Genitofemoral n. (femoral br.)

Femoral a.

Femoral v.

Femoral ring

Lacunar lig.

Iliopsoas { Iliacus m.

Psoas major m.

Pectineus m.

INGUINAL REGION

Pansky

220. MUSCLES OF THE MEDIAL THIGH

I.

Name	Origin	Insertion	Action	Nerve
Pectineus	Pectineal line Pubis, between iliopubic eminence and tubercle	Pectineal line of femur	Flexes thigh Adducts and lat. rotates thigh	Femoral
Adductor longus	Front of pubis	Med. linea aspera of femur	Adducts, flexes, and lat. rotates thigh	Obturator
Gracilis	Lower symphysis Pubic arch	Upper med. tibia, below condyle	Adducts thigh Flexes, med. rotates leg	Obturator
Adductor brevis	Inferior pubic ramus	Pectineal line and upper med. linea aspera of femur	Adducts and flexes thigh	Obturator
Adductor magnus	Inf. ischiopubic rami Outer inf. ischial tuberosity	Med. gluteal tuberosity Med. linea aspera Supracondylar line and adductor tubercle	Adducts thigh Upper part flexes thigh Lower part extends and rotates thigh laterally	Obturator

II. Special features (see p. 513)

A. AT THE INSERTION OF THE ADDUCTOR MAGNUS MUSCLE along and close to the linea aspera are a series of small tendinous arches attached to the bone. The upper 4 are small apertures transmitting the perforating branches of the deep femoral artery, which supply the hamstring muscles at the back of the thigh as well as a nutrient vessel to the femur. The fifth and lowest of the openings is by far the largest, the *tendinous hiatus* (*adductor hiatus*), which represents the caudal end of the adductor canal through which the femoral vessels enter the popliteal space

B. MOST OF THESE MUSCLES are supplied by the obturator nerve except for the pectineus which is usually supplied by the femoral (or both) or by the accessory obturator nerve. The extensor part of the adductor magnus muscle is supplied, in addition, by the tibial part of the sciatic nerve

C. THE 3 ADDUCTORS (helped by the pectineus) are powerful muscles and are used in all movements where the thighs are approximated
   1. They are stabilizers during flexion and extension
   2. The longus and magnus are active during medial rotation (importance uncertain)
   3. The extensor portion of the magnus helps the hamstrings in extending the thigh

D. THE GRACILIS acts at both the hip and knee, but is chiefly an adductor, flexor, and medial rotator, especially during the swing phase of walking. It has no postural function

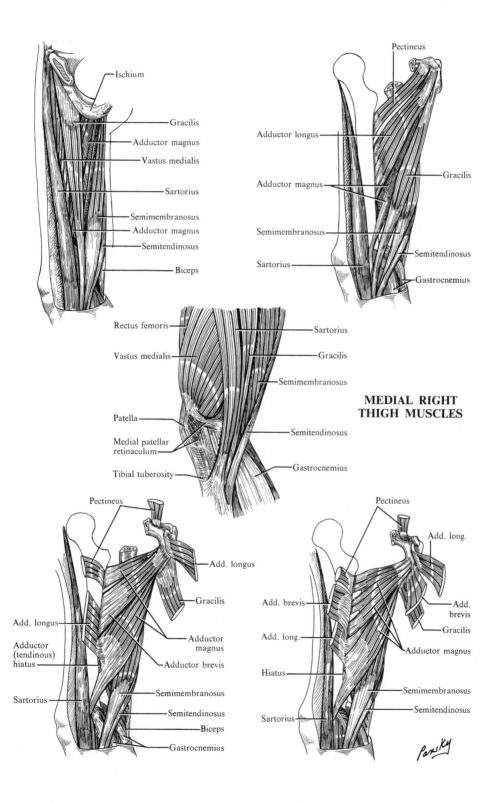

Ischium

Gracilis

Adductor magnus

Vastus medialis

Sartorius

Semimembranosus

Adductor magnus

Semitendinosus

Biceps

Pectineus

Adductor longus

Adductor magnus

Semimembranosus

Sartorius

Gracilis

Semitendinosus

Gastrocnemius

Rectus femoris

Vastus medialis

Patella

Medial patellar retinaculum

Tibial tuberosity

Sartorius

Gracilis

Semimembranosus

Semitendinosus

Gastrocnemius

MEDIAL RIGHT THIGH MUSCLES

Pectineus

Add. longus

Gracilis

Adductor magnus

Adductor brevis

Semimembranosus

Semitendinosus

Biceps

Gastrocnemius

Add. longus

Adductor (tendinous) hiatus

Sartorius

Pectineus

Add. long.

Add. brevis

Add. long.

Hiatus

Sartorius

Add. brevis

Gracilis

Adductor magnus

Semimembranosus

Semitendinosus

Pansky

221. FEMORAL AND OBTURATOR VESSELS AND NERVES

I. Femoral artery (see p. 499)

A. RELATIONS

Anterior
Fascia lata, nerve to vastus medialis muscle,
saphenous nerve, sartorius muscle, roof of
adductor canal

Lateral	**Femoral**	*Medial*
Vastus intermedius	**artery**	Adductor longus and
muscle		sartorius muscles, femoral vein

Posterior
Psoas major, pectineus, adductor longus
and adductor magnus muscles, deep
femoral and femoral veins

B. BRANCHES: superficial epigastric, superficial circumflex iliac, external pudendal (with scrotal and inguinal branches), deep femoral, and descending genicular arteries

II. Deep femoral artery arises from the back of the femoral artery. At first it lies lateral to, then behind, the artery as far as the medial side of the femur. It then runs behind the adductor longus muscle to end as the fourth perforating artery. Branches:

A. MEDIAL CIRCUMFLEX FEMORAL winds around the medial side of the femur between the pectineus and psoas major muscles, then between the obturator externus and adductor brevis muscles to the back of the thigh, where it anastomoses with the inferior gluteal, lateral circumflex femoral, and first perforating arteries

B. LATERAL CIRCUMFLEX FEMORAL arises from the lateral side, behind the sartorius and rectus femoris muscles. Runs in front of the femur to anastomose with the superior gluteal, deep circumflex iliac, and superior lateral genicular arteries

C. THREE PERFORATING ARTERIES pierce adductor magnus muscle to reach the back of thigh, forming anastomotic loops

D. THE TERMINATION OF THE ARTERY is the fourth perforating artery

III. Femoral nerve: distribution

A. MUSCULAR to pectineus, sartorius, and quadriceps muscles

B. CUTANEOUS: anterior branches to anterior and medial thigh
   1. Saphenous runs with the femoral artery, crossing the artery from lateral to medial in the adductor canal. It becomes superficial at the medial side of the knee to supply the skin on the medial side of the leg

C. NERVES TO HIP JOINT AND KNEE JOINT

IV. Femoral vein begins at the tendinous hiatus of the adductor magnus muscle. Receives muscular tributaries, the deep femoral vein, and the great saphenous vein

V. Obturator vessels and nerve

A. ARTERY is distributed to obturator externus, pectineus, adductors, and gracilis muscles, hip joint, and anastomoses with the inf. gluteal and med. circumflex arteries

B. NERVE: generally, if a nerve crosses or is in vicinity of a joint, it supplies the joint
   1. Anterior branch: lies deep to the pectineus and adductor longus muscles and superficial to the adductor brevis muscle. It supplies the adductor longus, brevis, and gracilis muscles, cutaneous areas on the medial thigh, and hip joint
   2. Posterior branch: as it enters thigh, lies posterior to the adductor brevis muscle but superficial to the adductor magnus muscle. It supplies the obturator externus, adductor magnus, and brevis muscles and also the knee joint

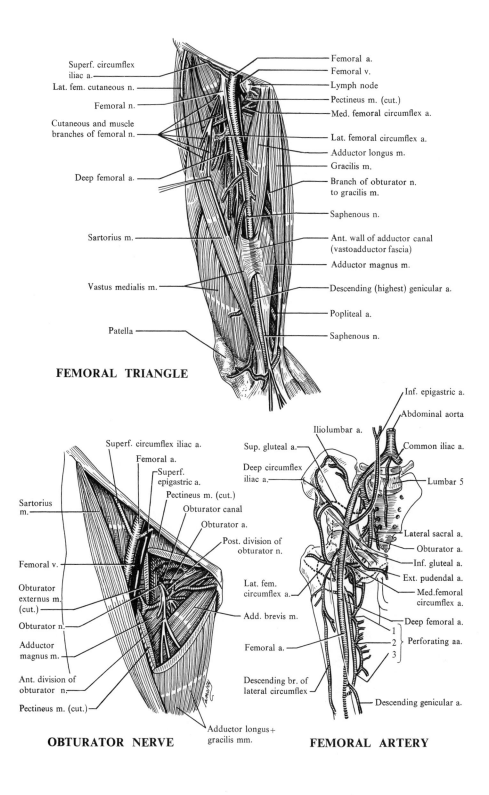

Superf. circumflex iliac a.
Lat. fem. cutaneous n.
Femoral n.
Cutaneous and muscle branches of femoral n.
Deep femoral a.
Sartorius m.
Vastus medialis m.
Patella

Femoral a.
Femoral v.
Lymph node
Pectineus m. (cut.)
Med. femoral circumflex a.
Lat. femoral circumflex a.
Adductor longus m.
Gracilis m.
Branch of obturator n. to gracilis m.
Saphenous n.
Ant. wall of adductor canal (vastoadductor fascia)
Adductor magnus m.
Descending (highest) genicular a.
Popliteal a.
Saphenous n.

FEMORAL TRIANGLE

Superf. circumflex iliac a.
Femoral a.
Superf. epigastric a.
Pectineus m. (cut.)
Obturator canal
Obturator a.
Post. division of obturator n.
Lat. fem. circumflex a.
Add. brevis m.
Femoral a.
Descending br. of lateral circumflex

Sartorius m.
Femoral v.
Obturator externus m. (cut.)
Obturator n.
Adductor magnus m.
Ant. division of obturator n.
Pectineus m. (cut.)
Adductor longus + gracilis mm.

OBTURATOR NERVE

Inf. epigastric a.
Abdominal aorta
Iliolumbar a.
Sup. gluteal a.
Common iliac a.
Deep circumflex iliac a.
Lumbar 5
Lateral sacral a.
Obturator a.
Inf. gluteal a.
Ext. pudendal a.
Med. femoral circumflex a.
Deep femoral a.
1
2 Perforating aa.
3
Descending genicular a.

FEMORAL ARTERY

– 503 –

222. MUSCLES OF THE GLUTEAL REGION

I.

Name	Origin	Insertion	Action	Nerve
Gluteus maximus	Ilium behind post. glut. line Sacrotuberous lig.	Iliotibial band Glut. tuberosity of femur	Extends and lat. rotates thigh Braces knee	Inf. glut.
Gluteus medius	Outer ilium and crest between post. and inf. glut. lines	Ridge, lat. side great. trochanter	Abducts thigh Ant. part rotates medially Post. part rotates laterally	Sup. glut.
Gluteus minimus	Outer ilium, between ant. and inf. glut. lines Sciatic notch	Ant. border, great. trochanter	Abducts thigh, rotates medially Weakly flexes	Sup. glut.
Tensor fasciae latae	Ant. iliac crest Notch between ant. sup. and ant. inf. iliac spines	Iliotibial band	Flexes, abducts, and rotates thigh med.	Sup. glut.
Piriformis	Ant. sacrum Great. sciatic notch	Upper border great. trochanter	Abducts and rotates thigh laterally	N. to piriformis
Obturator internus	Ischiopubic rami Inner obt. memb.	Med. surface, great. trochanter	Rotates laterally and abducts thigh	N. to obturator int.
Gemellus superior	Outer surface, ischial spine	Tendon obturator int. m.	See obt. int.	N. to obturator int.
Gemellus inferior	Ischial tuberosity	Tendon obturator int. m.	See obt. int.	N. to quadratus femoris
Obturator externus	Ischiopubic rami Outer obt. memb.	Trochanteric fossa	Rotates thigh lat.	Obturator
Quadratus femoris	Ischial tuberosity	Quadrate line	Rotates thigh lat.	N. to quadratus femoris

II. Special features
- A. PIRIFORMIS arises inside pelvis and divides the greater sciatic foramen into superior and inferior portions
- B. OBTURATOR INTERNUS MUSCLE arises inside pelvis
- C. GEMELLI insert into tendon of obturator internus muscle

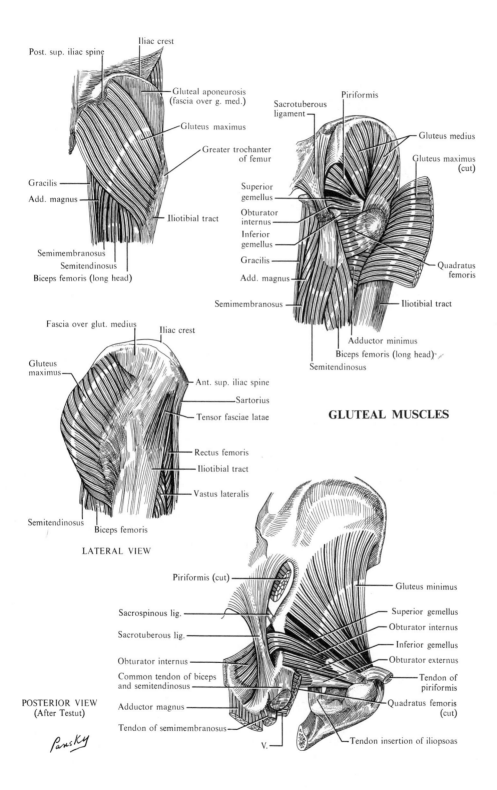

Post. sup. iliac spine
Iliac crest
Gluteal aponeurosis (fascia over g. med.)
Gluteus maximus
Greater trochanter of femur
Gracilis
Add. magnus
Iliotibial tract
Semimembranosus
Semitendinosus
Biceps femoris (long head)

Piriformis
Sacrotuberous ligament
Gluteus medius
Gluteus maximus (cut)
Superior gemellus
Obturator internus
Inferior gemellus
Gracilis
Quadratus femoris
Add. magnus
Semimembranosus
Iliotibial tract
Adductor minimus
Biceps femoris (long head)
Semitendinosus

GLUTEAL MUSCLES

Fascia over glut. medius
Iliac crest
Gluteus maximus
Ant. sup. iliac spine
Sartorius
Tensor fasciae latae
Rectus femoris
Iliotibial tract
Vastus lateralis
Semitendinosus
Biceps femoris

LATERAL VIEW

Piriformis (cut)
Gluteus minimus
Sacrospinous lig.
Superior gemellus
Sacrotuberous lig.
Obturator internus
Obturator internus
Inferior gemellus
Common tendon of biceps and semitendinosus
Obturator externus
Tendon of piriformis
POSTERIOR VIEW (After Testut)
Adductor magnus
Quadratus femoris (cut)
Tendon of semimembranosus
V.
Tendon insertion of iliopsoas

Pansky

– 505 –

223. VESSELS AND NERVES OF THE GLUTEAL REGION

I. Superficial nerves (see p. 482)

II. Arteries
A. SUPERIOR GLUTEAL: largest branch of internal iliac artery; leaves pelvis through greater sciatic foramen, above piriformis muscle. Branches:
   1. Superficial branch enters gluteus maximus muscle and anastomoses with inferior gluteal artery
   2. Deep branch runs forward under gluteus medius muscle; superior division continues along upper border of the gluteus minimus muscle to anterior superior iliac spine to anastomose with the deep circumflex iliac and lateral femoral circumflex arteries. The inferior division passes downward over the surface of the gluteus minimus muscle to the greater trochanter and anastomoses with the lateral femoral circumflex artery
B. INFERIOR GLUTEAL: from internal iliac artery, leaves pelvis through greater sciatic foramen below piriformis muscle. Branches:
   1. Ischiadic (sciatic), which runs with the sciatic nerve
   2. Anastomotic: forms the crucial anastomosis by uniting with the lateral and medial circumflex femoral and the first perforating artery from the profunda
C. INTERNAL PUDENDAL: from internal iliac artery, leaves pelvis through greater sciatic foramen, below piriformis muscle, crosses spine of ischium, and enters ischiorectal fossa through the lesser sciatic foramen

III. Nerves
A. SCIATIC: largest nerve of body; from sacral plexus, leaves pelvis through greater sciatic foramen below the piriformis muscle and is accompanied by the posterior femoral cutaneous nerve, inferior gluteal artery, pudendal vessels, and nerves. No gluteal branches except a branch to hip joint
B. PUDENDAL NERVE follows pudendal vessels
C. POSTERIOR FEMORAL CUTANEOUS NERVE enters region below piriformis muscle with inferior gluteal artery and gives inf. cluneal branches to skin at lower border of gluteus maximus muscle
D. SUPERIOR GLUTEAL NERVE runs with superior gluteal vessels to supply gluteus medius, gluteus minimus, and tensor fasciae latae muscles
E. INFERIOR GLUTEAL NERVE follows inferior gluteal vessels to gluteus maximus muscle
F. OTHER MUSCULAR BRANCHES OF PLEXUS
   1. Nerve to quadratus femoris and inferior gemellus muscles leaves pelvis below piriformis muscle
   2. Nerve to obturator internus and superior gemellus muscles leaves pelvis below piriformis muscle and enters ischiorectal fossa through lesser sciatic foramen
   3. Nerve to piriformis muscle does not leave pelvis

IV. Clinical considerations
A. SCIATICA is a neuritis (inflammation of the nerve) of the sciatic nerve characterized by intense pain at back of thigh
B. LESIONS OF THE SCIATIC NERVE lead to paralysis of all of the muscles below the knee. Sensation on the lateral side of the leg and both surfaces of the foot is also lost
C. THE SCIATIC NERVE leaves the buttock by passing just lateral to the ischial tuberosity, and thereafter runs downward in the posterior midline, behind the adductor magnus but deep to the other muscles arising from the tuberosity

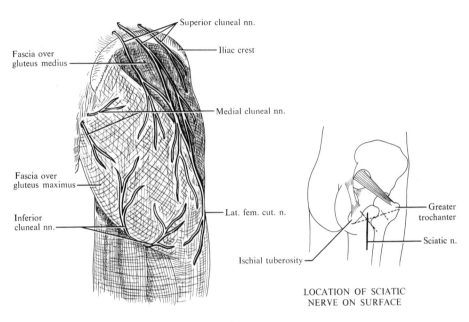

SUPERFICIAL NERVES – GLUTEAL REGION

Superior cluneal nn.

Iliac crest

Fascia over gluteus medius

Medial cluneal nn.

Fascia over gluteus maximus

Inferior cluneal nn.

Lat. fem. cut. n.

Greater trochanter

Sciatic n.

Ischial tuberosity

LOCATION OF SCIATIC
NERVE ON SURFACE

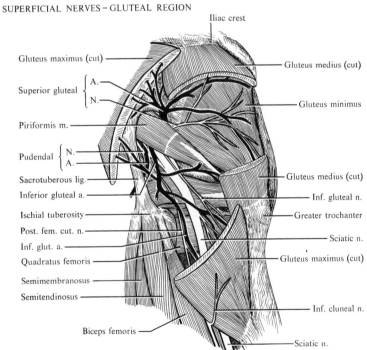

Iliac crest

Gluteus maximus (cut)

Gluteus medius (cut)

Superior gluteal { A. N.

Gluteus minimus

Piriformis m.

Pudendal { N. A.

Sacrotuberous lig.

Gluteus medius (cut)

Inferior gluteal a.

Inf. gluteal n.

Ischial tuberosity

Greater trochanter

Post. fem. cut. n.

Sciatic n.

Inf. glut. a.

Gluteus maximus (cut)

Quadratus femoris

Semimembranosus

Semitendinosus

Inf. cluneal n.

Biceps femoris

Sciatic n.

DEEP GLUTEAL STRUCTURES

Pansky

224. MUSCLES OF THE POSTERIOR THIGH

I.

Name	Origin	Insertion	Action	Nerve
Biceps femoris				
Long head	Inner medial ischial tuberosity Sacrotuberous lig.	Head of fibula Lat. condyle of tibia	Extends thigh Flexes leg	Tibial
Short head	Lat. linea aspera Lat. intermuscular septum	Lat. condyle of tibia	Flexes leg	Comm. peroneal
Semitendinosus	Lower medial ischial tuberosity	Medial body of tibia nearly to anterior crest	Extends thigh Flexes leg	Tibial
Semimembranosus	Upper outer ischial tuberosity	Post. medial aspect of tibial condyle	Extends thigh Flexes leg	Tibial

II. Special features
A. Popliteal fossa (space): a diamond-shaped space behind the knee
   1. Boundaries: laterally and above by biceps femoris muscle, medially and above by the semimembranosus and semitendinosus muscles, laterally and below by the lateral head of the gastrocnemius muscle, and medially and below by the medial head of the gastrocnemius muscle
   2. The floor, from above downward, is formed by popliteal surface of the femur, oblique popliteal ligament, and the popliteus muscle
   3. The roof: covered by fascia lata, superficial fascia, and skin
   4. Contents: popliteal vessels, tibial and common peroneal (fibular) nerves, termination of the small saphenous vein, lower end of the posterior femoral cutaneous nerve, articular branch of the obturator nerve, and a few small lymph nodes and fat

III. Clinical considerations
A. Charley horse: this may be expressed as pain or muscle stiffness after a bruise or excessive athletic activity and frequently occurs in the hamstring muscles (see table above)
B. In their actions, all the muscles arising from the ischial tuberosity assist the gluteus maximus to extend the hip joint, and since they normally maintain more tone than does the gluteus maximus, patients with paralyzed posterior hamstrings tend to fall forward
C. The short head of the biceps can act only at the knee. It is a more efficient flexor at the knee than is the long head, for the long head apparently relaxes during the last half of flexion, while the short head continues to contract
D. Pulled hamstrings are common sports injuries in people who run hard. There may be a "tearing off" or avulsion of part of the tendinous origin of the hamstrings from the ischial tuberosity. One usually also sees some contusion (bruising) and tearing of some muscle fibers, resulting in blood vessel rupture that creates a hematoma held in the dense fascia lata
E. Pain from an abscess or tumor in the popliteal region is usually severe due to the tenseness of the deep popliteal fascia not allowing for expansion
F. Injury to the common peroneal nerve results in loss of eversion and dorsiflexion of the foot (footdrop) and sensory loss to lateral leg and dorsum of foot

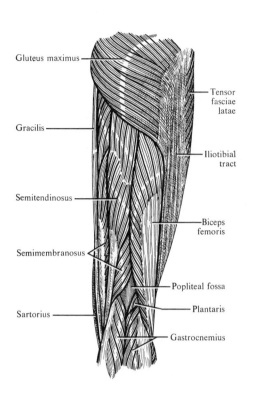

Gluteus maximus

Gracilis

Semitendinosus

Semimembranosus

Sartorius

Tensor fasciae latae

Iliotibial tract

Biceps femoris

Popliteal fossa

Plantaris

Gastrocnemius

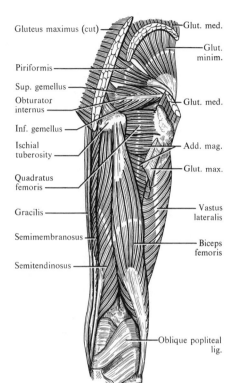

Gluteus maximus (cut)

Piriformis

Sup. gemellus

Obturator internus

Inf. gemellus

Ischial tuberosity

Quadratus femoris

Gracilis

Semimembranosus

Semitendinosus

Glut. med.

Glut. minim.

Glut. med.

Add. mag.

Glut. max.

Vastus lateralis

Biceps femoris

Oblique popliteal lig.

POSTERIOR THIGH MUSCLES

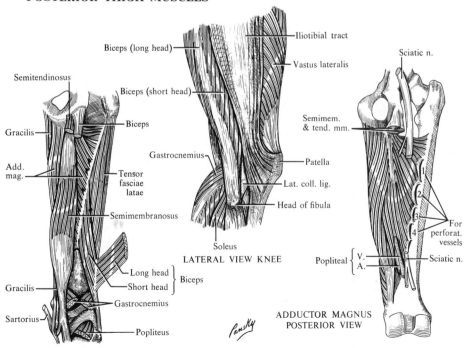

Semitendinosus

Gracilis

Add. mag.

Gracilis

Sartorius

Biceps

Tensor fasciae latae

Semimembranosus

Long head
Short head } Biceps

Gastrocnemius

Popliteus

Biceps (long head)

Biceps (short head)

Gastrocnemius

Soleus

Iliotibial tract

Vastus lateralis

Patella

Lat. coll. lig.

Head of fibula

LATERAL VIEW KNEE

Sciatic n.

Semimem. & tend. mm.

1

2

3

4

} For perforat. vessels

Popliteal { V.
A.

Sciatic n.

ADDUCTOR MAGNUS POSTERIOR VIEW

Pansky

225. SACRAL PLEXUS AND DERIVATIVES

I. **Formation:** lumbosacral trunk (L4 and L5) plus anterior primary divisions of sacral nerves 1–3 inclusive

A. LOCATION: posterolateral pelvis, between internal iliac vessels and piriformis muscle

B. MANNER OF FORMATION: nerves that enter plexus, except S3, have posterior and anterior branches that converge toward greater sciatic foramen

C. BRANCHES

Anterior Division	Posterior Division
N. to quadratus femoris and inf. gemellus mm. (L4, L5, S1)	N. to piriformis m. (S1, S2)
N. to obturator internus and sup. gemellus mm. (L5, S1, S2)	Superior gluteal (L4, L5, S1)
Post. fem. cutaneous (S2, S3)	Inferior gluteal (L5, S1, S2)
Sciatic	Post. fem. cutaneous (S1, S2)
Tibial (L4, L5, S1, S2, S3)	Perforating cutaneous (S2, S3)
	Sciatic
Pudendal (S2, S3, S4)	Common peroneal (fibular) (L4, L5, S1, S2)

D. DISTRIBUTION
1. Nerves to individually named muscles need no further comment
2. Posterior femoral cutaneous receives posterior divisions from S1 and S2 and anterior divisions from S2 and S3. Leaves pelvis with inferior gluteal artery to lower part of gluteus maximus muscle (see p. 507)
3. Perforating cutaneous nerve pierces sacrotuberous ligament, supplies lower and medial buttock
4. Superior gluteal follows superior gluteal artery, supplies gluteus medius, gluteus minimus, and tensor fasciae latae muscles
5. Inferior gluteal follows inferior gluteal vessels, supplies gluteus maximus muscle
6. Sciatic, actually 2 nerves, tibial and common peroneal (fibular) within the same sheath
    a. Relations: in front, lies on posterior ischium, obturator internus, gemelli, and quadratus femoris muscles; behind lies the gluteus maximus muscle; farther down it runs on the adductor magnus muscle and is crossed by the long head of the biceps muscle
    b. Splits just above popliteal space into its 2 parts (see p. 517)
        i. Tibial supplies all hamstrings except short head of biceps, and then enters leg (see p. 531)
        ii. Common peroneal (fibular) supplies short head of biceps and enters leg (see p. 531)
7. Pudendal leaves pelvis through greater sciatic foramen, crosses ischial spine, and enters ischiorectal fossa (see p. 467)

II. **Clinical considerations** (see p. 506)

A. LESIONS OF THE SUPERIOR GLUTEAL NERVE cause a drastic sag of the pelvis toward the side of the "swing phase" in walking

B. THE GLUTEAL REGION is a common site of intramuscular injection of drugs. Injection can only be made safely into the superolateral quadrant just inferior to the iliac crest where it penetrates the belly of the gluteus medius and possibly the gluteus minimus. Injections into either of the inferior two quadrants will endanger and possibly injure the sciatic or other nerves and vessels which emerge inferior to the piriform muscle. Injections into the superomedial quadrant may injure the superior gluteal nerve and/or vessels

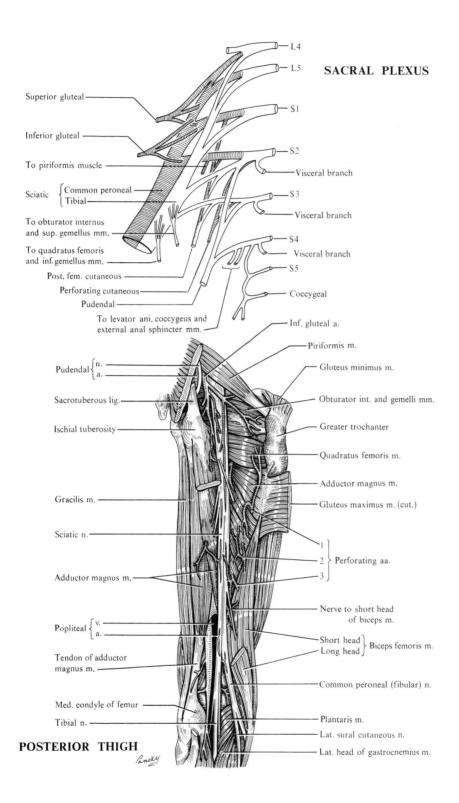

SACRAL PLEXUS

L4
L5
S1
Superior gluteal
Inferior gluteal
To piriformis muscle
S2
Visceral branch
Sciatic { Common peroneal
Tibial
S3
Visceral branch
To obturator internus
and sup. gemellus mm.
To quadratus femoris
and inf. gemellus mm.
S4
Visceral branch
Post. fem. cutaneous
S5
Perforating cutaneous
Pudendal
Coccygeal
To levator ani, coccygeus and
external anal sphincter mm.

Inf. gluteal a.
Piriformis m.
Pudendal { n.
a.
Gluteus minimus m.
Sacrotuberous lig.
Obturator int. and gemelli mm.
Ischial tuberosity
Greater trochanter
Quadratus femoris m.
Adductor magnus m.
Gracilis m.
Gluteus maximus m. (cut.)
Sciatic n.
1
2 } Perforating aa.
3
Adductor magnus m.
Nerve to short head
of biceps m.
Popliteal { v.
a.
Short head }
Long head } Biceps femoris m.
Tendon of adductor
magnus m.
Common peroneal (fibular) n.
Med. condyle of femur
Plantaris m.
Tibial n.
Lat. sural cutaneous n.

POSTERIOR THIGH
Pansky
Lat. head of gastrocnemius m.

– 511 –

226. THE HIP JOINT

I. Type: enarthrosis (ball and socket)

II. Bones: head of femur and acetabulum of hip bone

III. Movements: flexion, extension, abduction, adduction, lateral (external) rotation, and medial (internal) rotation

IV. Ligaments
   A. ACETABULAR LABRUM (glenoid labrum) deepens acetabulum, helps to hold head of femur
   B. TRANSVERSE ACETABULAR completes rim of acetabulum
   C. ARTICULAR CAPSULE extends from rim of acetabulum to the intertrochanteric crest and line
      1. Zona orbicularis: band of circularly arranged fibers in capsule
   D. ILIOFEMORAL (Y-shaped ligament of Bigelow): from anterior inferior iliac spine to intertrochanteric line, prevents overextension, abduction, and lateral rotation
   E. ISCHIOFEMORAL: from ischium behind the acetabulum to blend with the capsule, checks medial rotation
   F. PUBOFEMORAL: from superior pubic ramus, joins iliofemoral ligament, checks abduction
   G. LIGAMENTUM CAPITIS FEMORIS (TERES) (round ligament): from acetabular notch and transverse ligament to fovea of femur; little function as ligament but guides artery to head of femur

V. Synovial membrane lines articular capsule. Covers labrum, ligamentum teres, neck of femur, from the attachment of capsule, below, to the articular cartilage of head of femur

VI. Muscles acting on hip joint

Flexion	Extension	Abduction	Adduction	Med. Rotate	Lat. Rotate
Iliopsoas	Glut. max.	Glut. med.	Add. mag.	Glut. med.	Piriformis
Sartorius	Semitend.	Glut. min.	Add. long.	Glut. min.	Obturator int.
Pectineus	Semimemb.	Tensor f.l.	Add. brev.	Tensor f.l.	Gemelli
Rectus fem.	Biceps fem.	Piriformis	Gracilis	Add. mag.	Obturator ext.
Add. long.	Add. mag.	Sartorius	Pectineus	(post. part)	Quadrat. fem.
Add. brev.	(post. part)		Obturator ext.		Glut. max.
Add. mag.			Quadrat. fem.		Adductors (all)
(ant. part)					
Tensor f.l.					

VII. Clinical considerations
   A. DISLOCATIONS: since circulation to the head of femur over the ligamentum teres may be disrupted, reduction within 12 to 24 hours is mandatory
      1. Congenital: due to failure of acetabulum to deepen. More common in women
      2. Traumatic: usually caused by a blow on the knee when the thigh is flexed, tearing capsule
         a. The head of the femur is usually dislocated posteriorly, with a tearing of the posterior part of the capsule and frequently fracture of the acetabulum
         b. In anterior dislocation, much rarer than posterior, the head of the femur passes around the medial edge of the iliofemoral ligament and lodges against the body of the pubic bone or obturator foramen

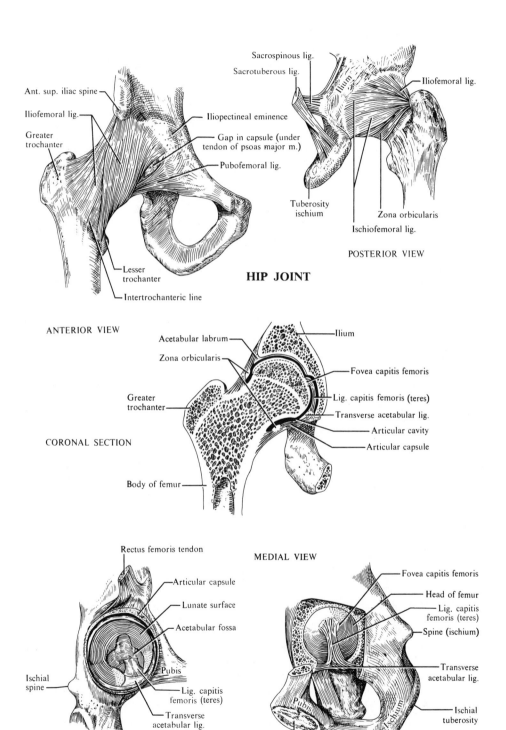

Ant. sup. iliac spine

Iliofemoral lig.

Greater trochanter

Iliopectineal eminence

Gap in capsule (under tendon of psoas major m.)

Pubofemoral lig.

Lesser trochanter

Intertrochanteric line

ANTERIOR VIEW

Sacrospinous lig.

Sacrotuberous lig.

Ilium

Iliofemoral lig.

Tuberosity ischium

Zona orbicularis

Ischiofemoral lig.

POSTERIOR VIEW

HIP JOINT

Acetabular labrum

Zona orbicularis

Ilium

Fovea capitis femoris

Greater trochanter

Lig. capitis femoris (teres)

Transverse acetabular lig.

Articular cavity

Articular capsule

CORONAL SECTION

Body of femur

MEDIAL VIEW

Rectus femoris tendon

Articular capsule

Lunate surface

Acetabular fossa

Pubis

Lig. capitis femoris (teres)

Ischial spine

Transverse acetabular lig.

INTERIOR OF RIGHT SOCKET

Fovea capitis femoris

Head of femur

Lig. capitis femoris (teres)

Spine (ischium)

Transverse acetabular lig.

Ischial tuberosity

Pubis

Ischium

Symphysial surface

Femur

Pansky

227. BONES OF THE LEG

I. Patella (knee cap) (see p. 521): a sesamoid bone in tendon of quadriceps femoris muscle
A. FLAT AND TRIANGULAR: superior border convex upward, apex downward. Anterior surface convex and rough. Posterior or articular surface is smooth and divided by a ridge into a larger lateral and smaller medial facet

II. Tibia (shin bone): second longest bone of body
A. PARTS: proximal (superior) extremity, shaft (body), and distal (inferior) extremity
B. PROXIMAL EXTREMITY shows medial and lateral condyles, anterior and posterior intercondylar areas (fossae), an intercondylar eminence, a tuberosity, and an articular facet for head of fibula
C. SHAFT shows an anterior border (crest); broad, convex, smooth medial (subcutaneous) surface; and a lateral (interosseous) border
D. LOWER EXTREMITY shows a medial malleolus; smooth, concave inferior articular surface for talus; fibular notch; posterior groove for flexor hallucis longus muscle; and a medial malleolar sulcus for the tibialis posterior and flexor digitorum longus muscles

III. Fibula (calf bone): most slender of long bones
A. PARTS: body and 2 extremities
B. PROXIMAL EXTREMITY shows a head, with superior articular surfaces, and an apex (styloid process)
C. SHAFT is described as having 3 borders and 3 surfaces
D. DISTAL EXTREMITY shows a lateral malleolus, with a smooth articular and inferior extremity for the talus and a roughened medial area for the tibia

IV. Ossification

Location	When Appears	When Closes
A. TIBIA, from 3 centers		
Body	7th fetal week	——
Prox. end	At time of birth	20th year
Dist. end	2nd year	18th year
B. FIBULA, from 3 centers		
Body	8th fetal week	——
Prox. end	4th year	25th year
Dist. end	2nd year	20th year
C. PATELLA, from 1 center	2nd or 3rd year	At puberty

V. Clinical considerations
A. FRACTURE OF PATELLA can be due to muscle pull. Fragments are usually drawn apart by the muscle pull, thus, closed reduction is often difficult due to tendons being caught in the break
B. FRACTURES OF FIBULAR NECK may damage common peroneal nerve
C. IN CASE OF FRACTURE of either fibula or tibia individually, the unbroken bone helps to splint the other with little displacement. If tibia is broken, look for break in fibula at another level. Fracture of both bones is referred to as Pott's fracture
D. LATERAL MALLEOLUS is often snapped off in overinversion of foot (ankle turn). Overeversion may break medial malleolus, and if force is strong enough, lateral may also break
E. THE BODY OF THE TIBIA is narrowest at junction of middle and lower thirds and is most frequent site of fracture and region where rickets affects bone during childhood
F. SPIRAL FRACTURE of tibia is caused by severe torsion in skiing, at junction of middle and lower thirds (usually see fractures of fibular neck, in addition)

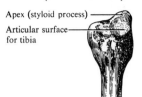

SUPERIOR EXTREMITY—FIBULA
(INTERNAL VIEW)

Apex (styloid process)

Articular surface for tibia

SUPERIOR EXTREMITY—TIBIA

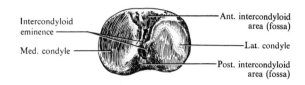

Intercondyloid eminence

Med. condyle

Ant. intercondyloid area (fossa)

Lat. condyle

Post. intercondyloid area (fossa)

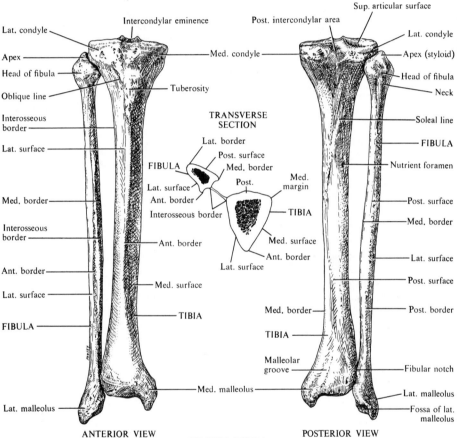

Lat. condyle

Apex

Head of fibula

Oblique line

Interosseous border

Lat. surface

Med. border

Interosseous border

Ant. border

Lat. surface

FIBULA

Lat. malleolus

Intercondylar eminence

Med. condyle

Tuberosity

TRANSVERSE
SECTION

Lat. border

Post. surface

Med. border

Post.

FIBULA

Lat. surface

Ant. border

Interosseous border

Ant. border

Lat. surface

TIBIA

Med. surface

TIBIA

Med. malleolus

ANTERIOR VIEW

Post. intercondylar area

Sup. articular surface

Lat. condyle

Apex (styloid)

Head of fibula

Neck

Soleal line

FIBULA

Nutrient foramen

Post. surface

Med. border

Lat. surface

Post. surface

Post. border

Med. margin

Med. border

TIBIA

Malleolar groove

Fibular notch

Lat. malleolus

Fossa of lat. malleolus

POSTERIOR VIEW

**RIGHT TIBIA
AND FIBULA**

For talus

Lat. malleolus

For post. talofibular ligament

INFERIOR EXTREMITY—FIBULA

POSTERIOR

Med. malleolus

Inf. articular surface

ANTERIOR

INFERIOR EXTREMITY—TIBIA
(FROM BELOW)

228. POPLITEAL REGION AND ANASTOMOSIS AROUND THE KNEE

I. Popliteal fossa (see p. 508)

II. Popliteal artery: a continuation of the femoral artery as it passes through tendinous hiatus in adductor magnus muscle

A. RELATIONS

	Anterior	
	Femur, oblique popliteal ligament, popliteus muscle	
Lateral		*Medial*
Biceps and lateral head gastrocnemius muscles; lateral condyle of femur, popliteal vein, tibial nerve	**Popliteal artery**	Semimembranosus, medial head gastrocnemius and plantaris muscles, medial condyle femur, tibial nerve, popliteal vein
	Posterior	
	Semimembranosus, gastrocnemius and plantaris muscles, popliteal vein, tibial nerve	

B. BRANCHES: medial and lateral superior genicular, middle genicular, medial and lateral inferior genicular

III. Tibial nerve
A. RELATIONS: enters fossa from beneath biceps muscle, lateral to popliteal vessels, and crosses superficial to vessels to reach medial side
B. BRANCHES follow the sup., inf. medial, and middle genicular aa. to joint

IV. Common peroneal (fibular) nerve
A. RELATIONS: along lateral side of fossa at border of biceps muscle
B. BRANCHES follow the superior and inferior lateral genicular arteries to joint

V. Popliteal vein formed by anterior and posterior tibial veins at lower border of popliteus muscle. First medial and then crosses superficial to the popliteal artery to the lateral side; becomes femoral vein as it passes through tendinous hiatus

VI. Anastomosis (in superficial and deep plexuses around patella) in case of ligation of

Above Ligation	*Below Ligation*
A. FEMORAL ARTERY, between origins of deep femoral and descending genicular arteries	
1. Desc. br. of lat. circumflex of profunda to	Lat. sup. and inf. genic. of poplit. Ant. tibial recur. of ant. tibial
	Through deep plexus to med. sup. and inf. genic. brs. of popliteal
B. FEMORAL, at tendinous hiatus	
1. Same as 1, above	
2. Descending (highest) genicular of femoral	
a. Musculoarticular brs. to	Lat. and med. sup. genic. of poplit.
b. Through saphenous br. to	Med. inf. genic. of poplit.
C. POPLITEAL, between origins of superior and inferior genicular arteries	
1. Lat. superior genicular to	Lat. inf. genic. of poplit. Fibular and recurrent tibial brs. of ant. tibial a.
2. Med. superior genicular to	Med. inf. genicular a.

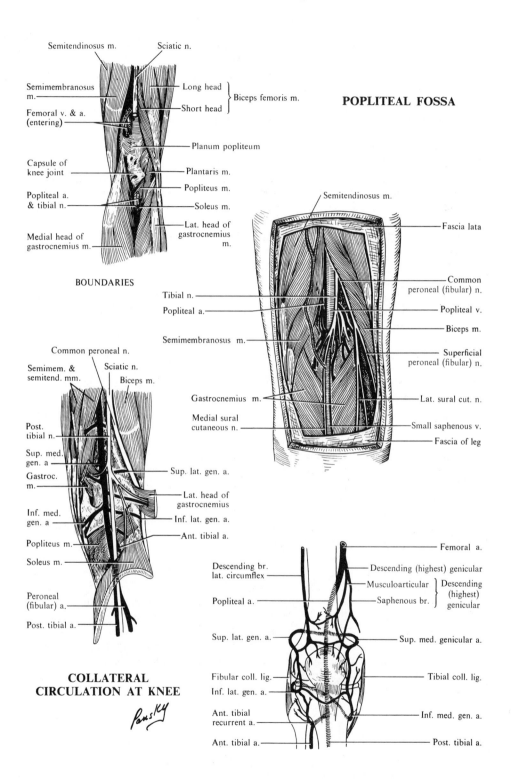

Semitendinosus m.

Sciatic n.

Semimembranosus m.

Long head
Short head } Biceps femoris m.

Femoral v. & a. (entering)

POPLITEAL FOSSA

Planum popliteum

Capsule of knee joint

Plantaris m.

Popliteus m.

Popliteal a. & tibial n.

Soleus m.

Medial head of gastrocnemius m.

Lat. head of gastrocnemius m.

Semitendinosus m.

Fascia lata

BOUNDARIES

Tibial n.

Popliteal a.

Common peroneal (fibular) n.

Popliteal v.

Biceps m.

Semimembranosus m.

Superficial peroneal (fibular) n.

Common peroneal n.

Semimem. & semitend. mm.

Sciatic n.

Biceps m.

Gastrocnemius m.

Lat. sural cut. n.

Medial sural cutaneous n.

Small saphenous v.

Post. tibial n.

Sup. med. gen. a

Fascia of leg

Gastroc. m.

Sup. lat. gen. a.

Inf. med. gen. a

Lat. head of gastrocnemius

Popliteus m.

Inf. lat. gen. a.

Soleus m.

Ant. tibial a.

Femoral a.

Descending br. lat. circumflex

Descending (highest) genicular

Peroneal (fibular) a.

Musculoarticular } Descending (highest) genicular

Popliteal a.

Saphenous br.

Post. tibial a.

Sup. lat. gen. a.

Sup. med. genicular a.

COLLATERAL CIRCULATION AT KNEE

Fibular coll. lig.

Tibial coll. lig.

Inf. lat. gen. a.

Inf. med. gen. a.

Ant. tibial recurrent a.

Ant. tibial a.

Post. tibial a.

229. KNEE JOINT—PART I

I. Type: ginglymus (hinge) between femur and tibia, arthrodial between femur and patella

II. Divisions: 3 joints in 1: 2 between medial and lateral condyles of femur and tibia, 1 between patella and femur

III. Movements: femur-tibia permits flexion and extension with slight rotation when leg is flexed; femur-patella, sliding up and down

IV. Ligaments

A. ARTICULAR CAPSULE: from femur to tibia, strengthened by fibers from fascia lata, iliotibial tract, and tendons of the vasti, hamstrings, and sartorius muscles

B. LIGAMENTUM PATELLAE: from apex of patella to tuberosity of tibia. Helps hold patella in place; serves as part of tendon of quadriceps muscle

C. OBLIQUE POPLITEAL: from lateral femur, over condyles to posterior head of tibia. Checks extension.

D. ARCUATE POPLITEAL: from lateral condyle of femur to styloid process of fibula. May check medial rotation of leg

E. TIBIAL COLLATERAL (medial ligament): from medial side of medial femoral condyle to medial condyle and body of tibia. Prevents lateral bending; checks extension, hyperflexion, and lateral rotation

F. FIBULAR COLLATERAL (lateral ligament): from back of lateral femoral condyle to lateral side of head and styloid process of fibula; checks hyperextension; is relaxed in flexion

G. CORONARY: from capsule to periphery of menisci and tibia. Helps hold menisci in place

H. ANTERIOR CRUCIATE: from medial back of lateral femoral condyle to front of tibial intercondylar eminence. Checks extension, lateral rotation, and anterior slipping of tibia on femur

I. POSTERIOR CRUCIATE: from front and lateral side of medial femoral condyle to posterior intercondylar fossa and posterior end of lateral meniscus. Checks extension, lateral rotation, and posterior slipping of tibia on femur

J. MEDIAL MENISCUS: crescent-shaped (oval); attached to tibia in front of anterior cruciate ligament and in posterior intercondylar fossa. Deepens medial tibial condyle

K. LATERAL MENISCUS: nearly circular; attached to tibia in front of anterior cruciate ligament, blending with latter; posteriorly attached behind intercondylar eminence in front of medial meniscus. Through the anterior meniscofemoral and posterior meniscofemoral (ligament of Wrisberg) ligaments, it is attached to medial femoral condyle. Deepens lateral tibial condyle

L. TRANSVERSE: interconnects anterior parts of 2 menisci

V. Clinical considerations

A. LESIONS OF THE MENISCI: displacements and tears are the most frequent type of damage, the medial being involved 5 to 15 times as often as the lateral. This is true not only because it is longer and less securely attached, but because the abnormal forces that cause the injury are most frequently applied to the lateral side of the knee and violently separate the medial tibial and femoral condyles. The meniscus may be torn or loosened from attachment to the tibial collateral ligament. In either case, all or part becomes jammed between the articular surfaces of the condyles, "locking" the joint

B. PAIN may be referred to the knee from hip disease

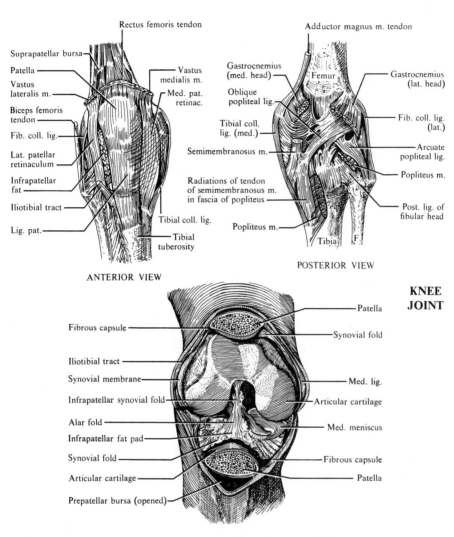

Rectus femoris tendon

Suprapatellar bursa
Patella
Vastus lateralis m.
Biceps femoris tendon
Fib. coll. lig.
Lat. patellar retinaculum
Infrapatellar fat
Iliotibial tract
Lig. pat.

Vastus medialis m.
Med. pat. retinac.

Tibial coll. lig.
Tibial tuberosity

ANTERIOR VIEW

Adductor magnus m. tendon

Gastrocnemius (med. head)
Oblique popliteal lig.
Tibial coll. lig. (med.)
Semimembranosus m.
Radiations of tendon of semimembranosus m. in fascia of popliteus
Popliteus m.

Femur

Gastrocnemius (lat. head)
Fib. coll. lig. (lat.)
Arcuate popliteal lig.
Popliteus m.
Post. lig. of fibular head

Tibia   F

POSTERIOR VIEW

KNEE JOINT

Fibrous capsule
Iliotibial tract
Synovial membrane
Infrapatellar synovial fold
Alar fold
Infrapatellar fat pad
Synovial fold
Articular cartilage
Prepatellar bursa (opened)

Patella
Synovial fold
Med. lig.
Articular cartilage
Med. meniscus
Fibrous capsule
Patella

OPENED ANTERIORLY (FLEXED JOINT)

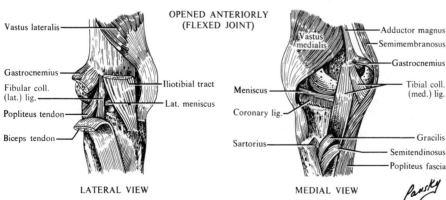

Vastus lateralis
Gastrocnemius
Fibular coll. (lat.) lig.
Popliteus tendon
Biceps tendon

Iliotibial tract
Lat. meniscus

LATERAL VIEW

Vastus medialis
Meniscus
Coronary lig.
Sartorius

Adductor magnus
Semimembranosus
Gastrocnemius
Tibial coll. (med.) lig.
Gracilis
Semitendinosus
Popliteus fascia

MEDIAL VIEW

Pansky

230. KNEE JOINT—PART II

VI. Synovial membrane: largest and most extensive in body

   A. EXTENDS ABOVE PATELLA and laterally and medially beneath vasti muscles

   B. EXTENDS DOWNWARD under patellar ligament; separated from latter by the infrapatellar pad (fat)

   C. SENDS FOLDS, *alar,* into joint cavity, which converge to form the *patellar fold*

   D. LINES CAPSULE, extends over menisci to free border, then under these to tibia

   E. BEHIND LATERAL MENISCUS forms blind sac between latter and tendon of popliteus muscle

   F. REFLECTED IN FRONT OF CRUCIATE LIGAMENTS

VII. Muscles acting on the joint

Flexion	*Extension*	*Medial Rotation*	*Lateral Rotation*
Semimembranosus	Quadriceps femoris	Popliteus	Biceps femoris
Semitendinosus	Tensor fasciae latae	Semimembranosus	
Biceps femoris		Semitendinosus	
Sartorius		Sartorius	
Gracilis		Gracilis	
Popliteus			
Gastrocnemius			
Plantaris			

VIII. Clinical considerations

   A. TEARS OF THE ANTERIOR CRUCIATE LIGAMENT are fairly frequently associated with tears of other ligaments, particularly the tibial collateral

      1. All the ligaments of the knee contribute to its stability, but the tibial collateral is especially important, for the major part of this strong ligament is tense in all positions of the joint and is thus a very valuable stabilizer

   B. THE CAPSULE OF THE KNEE JOINT is supplied by twigs from all the vessels that enter into anastomosis around the joint. In addition, the middle genicular artery penetrates the capsule posteriorly and is distributed especially to the tissue of the intercondylar region

   C. THE NERVES TO THE KNEE JOINT are typically derived from the femoral, the obturator, and both parts of the sciatic nerve. Many follow the arteries but some run directly to the capsule

   D. DISLOCATION OF PATELLA: uncommon; more often seen in female due to angle of femur

   E. FIBULAR COLLATERAL LIGAMENT is not commonly torn, being very strong; however, rupture of the tibial collateral ligament, often associated with tearing of the medial meniscus and anterior cruciate ligament, is a common football injury caused by a blow to the lateral side of the knee

   F. INJECTIONS into the synovial cavity of the knee joint are made for diagnostic and therapeutic reasons. In addition, aspiration of synovial fluid may be necessary to relieve pressure (due to increased secretion), to sample fluid diagnostically, or to remove blood in joint cavity as a result of soft tissue injury or fracture of a bone. *All procedures must be sterile to avoid infections*

      1. Injections are usually given laterally, with the knee flexed and the leg hanging. Inject into the center of a triangle marked by the apex of the patella, lateral plateau of the lateral tibial condyle, and the anterior prominence of the lateral femoral condyle. The needle passes posteromedially

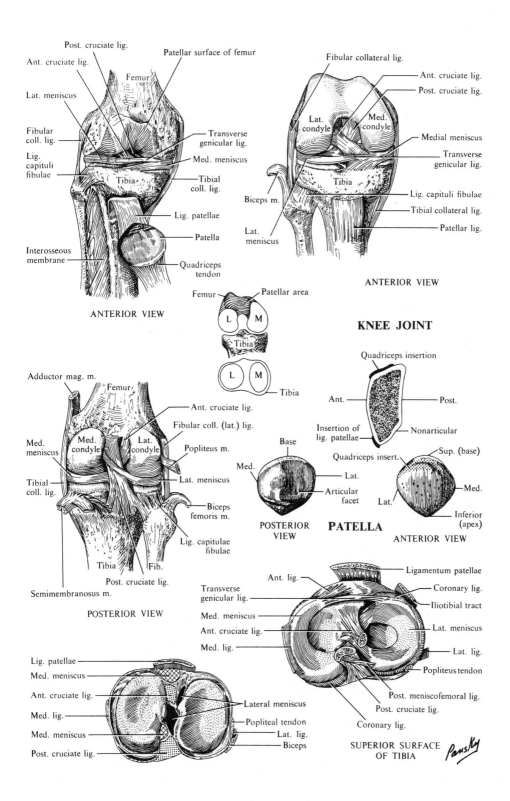

Post. cruciate lig.

Ant. cruciate lig.

Femur

Patellar surface of femur

Lat. meniscus

Fibular coll. lig.

Lig. capituli fibulae

Tibia

Transverse genicular lig.

Med. meniscus

Tibial coll. lig.

Lig. patellae

Patella

Interosseous membrane

Quadriceps tendon

ANTERIOR VIEW

Fibular collateral lig.

Ant. cruciate lig.

Post. cruciate lig.

Lat. condyle

Med. condyle

Medial meniscus

Transverse genicular lig.

Tibia

Biceps m.

Lig. capituli fibulae

Tibial collateral lig.

Lat. meniscus

Patellar lig.

ANTERIOR VIEW

Femur

Patellar area

L

M

Tibia

L

M

Tibia

KNEE JOINT

Quadriceps insertion

Ant.

Post.

Insertion of lig. patellae

Nonarticular

Adductor mag. m.

Femur

Ant. cruciate lig.

Fibular coll. (lat.) lig.

Med. condyle

Lat. condyle

Popliteus m.

Med. meniscus

Lat. meniscus

Tibial coll. lig.

Biceps femoris m.

Lig. capitulae fibulae

Tibia

Fib.

Post. cruciate lig.

Semimembranosus m.

POSTERIOR VIEW

Base

Med.

Lat.

Articular facet

POSTERIOR VIEW

Quadriceps insert.

Sup. (base)

Med.

Lat.

Inferior (apex)

PATELLA

ANTERIOR VIEW

Lig. patellae

Med. meniscus

Ant. cruciate lig.

Med. lig.

Med. meniscus

Post. cruciate lig.

Lateral meniscus

Popliteal tendon

Lat. lig.

Biceps

Ant. lig.

Transverse genicular lig.

Med. meniscus

Ant. cruciate lig.

Med. lig.

Ligamentum patellae

Coronary lig.

Iliotibial tract

Lat. meniscus

Lat. lig.

Popliteus tendon

Post. meniscofemoral lig.

Post. cruciate lig.

Coronary lig.

SUPERIOR SURFACE OF TIBIA

Pansky

– 521 –

231. KNEE JOINT—PART III, TIBIOFIBULAR ARTICULATION

I. **Bursae of knee:** usually named according to immediately adjacent structures (as tendons, etc.)

A. IN FRONT: 4. (1) Between front of patella and skin, (2) between front of femur and quadriceps femoris muscle (generally communicates with joint cavity), (3) between patellar ligament and upper tibia, (4) between tibial tuberosity and skin

B. LATERALLY: 3. (1) Between fibular collateral ligament and biceps muscle, (2) between fibular collateral ligament and popliteus muscle, (3) between popliteus muscle and lateral condyle of femur (generally communicates with joint cavity)

C. MEDIALLY: 3. (1) Between tibial collateral ligament and tendons of semitendinosus, sartorius, and gracilis muscles, (2) between tibial collateral ligament and tendon of semimembranosus muscle, (3) between tendon of semimembranosus muscle and tibia

D. POSTERIOR: 2. (1) Between lateral head of gastrocnemius muscle and capsule (sometimes communicates with joint), (2) between medial head of gastrocnemius muscle and capsule, extending under tendon of semimembranosus muscle (generally communicates with joint cavity)

II. **Tibiofibular joint** (proximal)

A. TYPE: diarthrodial

B. BONES: head of fibula and fibular facet, below lateral condyle of tibia

C. MOVEMENTS: gliding up and down

D. LIGAMENTS
1. Articular capsule
2. Anterior capitular
3. Posterior capitular

III. **Interosseous membrane** (intermediate)

A. TYPE: synarthrosis

B. FUNCTION: holds shafts together, strengthens fibula, for muscle attachments

C. LIGAMENT: the interosseous membrane. Fibers inclined downward and lateralward from lateral border of tibia to anteromedial border of fibula

IV. **Tibiofibular joint** (distal)

A. TYPE: syndesmosis

B. MOVEMENTS: slight up and down

C. LIGAMENTS
1. Anterior tibiofibular (lateral malleolar)
2. Posterior tibiofibular (lateral malleolar)
3. Interosseous
4. Inferior transverse

V. **Clinical considerations**

A. BURSITIS: bursa most frequently involved
1. Prepatellar: most common; becomes irritated by repeated or prolonged kneeling, thus, "housemaid's knee" or "nun's knee"
2. Gastrocnemius-semimembranosus (popliteal): may be caused by abnormal stress or strain; when distended called "Baker's cyst"
3. Between semimembranosus tendon and tibial collateral ligament and the tibia: when swollen and painful, may be mistaken for lesion of medial meniscus

B. GENU VALGUM: this is "knockknee" with medial condyle and shaft of femur protruding too far medially

C. GENU VARUM: this is "bowleg" with outward curving of the lower limb

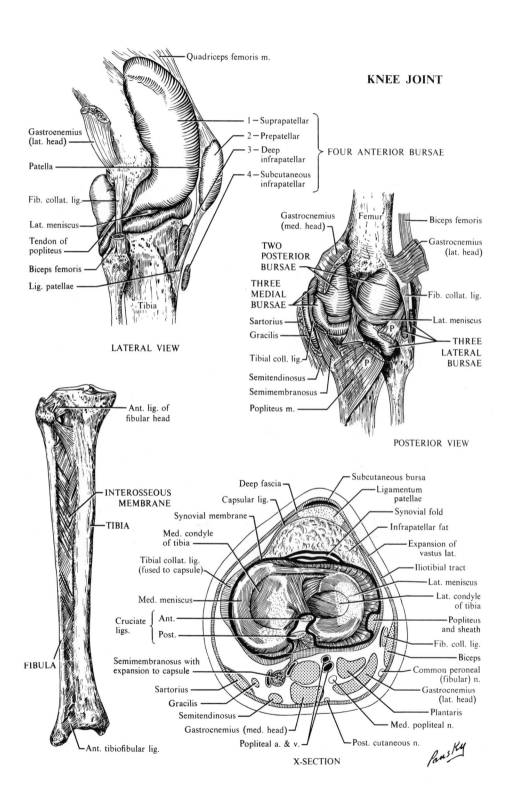

Quadriceps femoris m.

KNEE JOINT

Gastrocnemius
(lat. head)

Patella

1 — Suprapatellar

2 — Prepatellar

3 — Deep
    infrapatellar

4 — Subcutaneous
    infrapatellar

} FOUR ANTERIOR BURSAE

Fib. collat. lig.

Lat. meniscus

Tendon of
popliteus

Biceps femoris

Lig. patellae

Tibia

LATERAL VIEW

Gastrocnemius
(med. head)

Femur

Biceps femoris

Gastrocnemius
(lat. head)

TWO
POSTERIOR
BURSAE

THREE
MEDIAL
BURSAE

Fib. collat. lig.

Lat. meniscus

Sartorius

Gracilis

P

THREE
LATERAL
BURSAE

Tibial coll. lig.

Semitendinosus

P

Semimembranosus

Popliteus m.

POSTERIOR VIEW

Ant. lig. of
fibular head

INTEROSSEOUS
MEMBRANE

TIBIA

Deep fascia

Subcutaneous bursa

Capsular lig.

Ligamentum
patellae

Synovial membrane

Synovial fold

Med. condyle
of tibia

Infrapatellar fat

Tibial collat. lig.
(fused to capsule)

Expansion of
vastus lat.

Iliotibial tract

Med. meniscus

Lat. meniscus

Cruciate
ligs. { Ant.

Lat. condyle
of tibia

Post.

Popliteus
and sheath

FIBULA

Semimembranosus with
expansion to capsule

Fib. coll. lig.

Biceps

Sartorius

Common peroneal
(fibular) n.

Gracilis

Gastrocnemius
(lat. head)

Semitendinosus

Plantaris

Gastrocnemius (med. head)

Med. popliteal n.

Popliteal a. & v.

Post. cutaneous n.

Ant. tibiofibular lig.

Pansky

X-SECTION

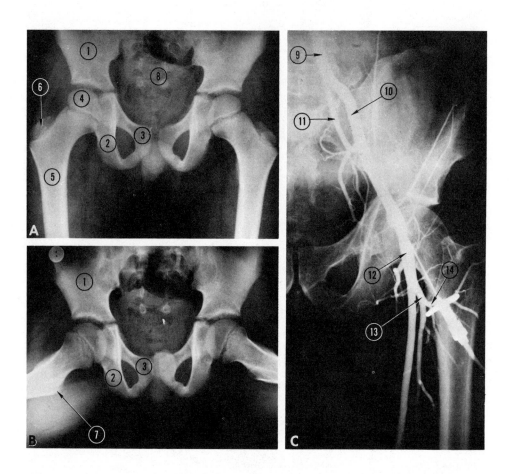

FIGURE 53. **Immature pelvis and arteriograms around hip. A, Immature pelvis, normal position; B, Immature pelvis, thighs abducted; C, arteriogram at hip.** *1,* Ilium; *2,* ischium; *3,* pubis; *4,* head of femur; *5,* shaft of femur; *6,* greater trochanter of femur; *7,* lesser trochanter of femur; *8,* sacrum; *9,* common iliac artery; *10,* external iliac artery; *11,* internal iliac artery; *12,* femoral artery; *13,* deep femoral artery; *14,* lateral circumflex artery.

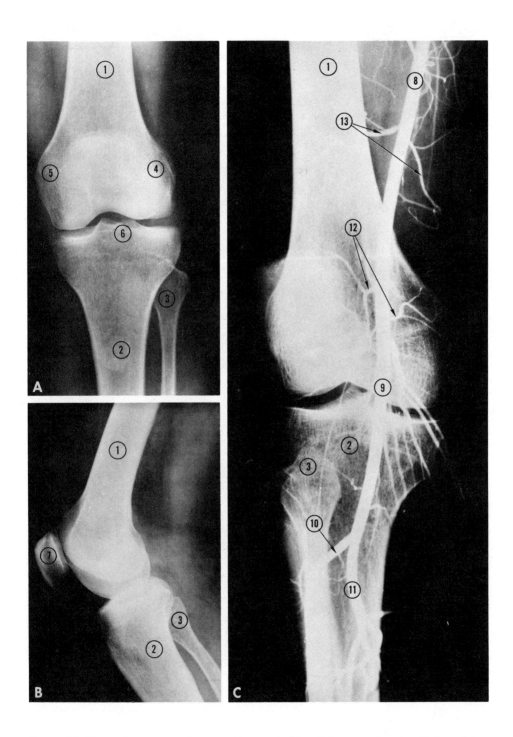

FIGURE 54. **Knee joint and arteriogram at knee. A, Knee joint, anterior view; B, knee joint, lateral view; C, arteriogram around knee.** *1*, Shaft of femur; *2*, tibia; *3*, head of fibula; *4*, lateral condyle; *5*, medial condyle of femur; *6*, intercondylar eminence; *7*, patella; *8*, femoral artery; *9*, popliteal artery; *10*, peroneal artery; *11*, posterior tibial artery; *12*, superior genicular arteries; *13*, muscular arteries.

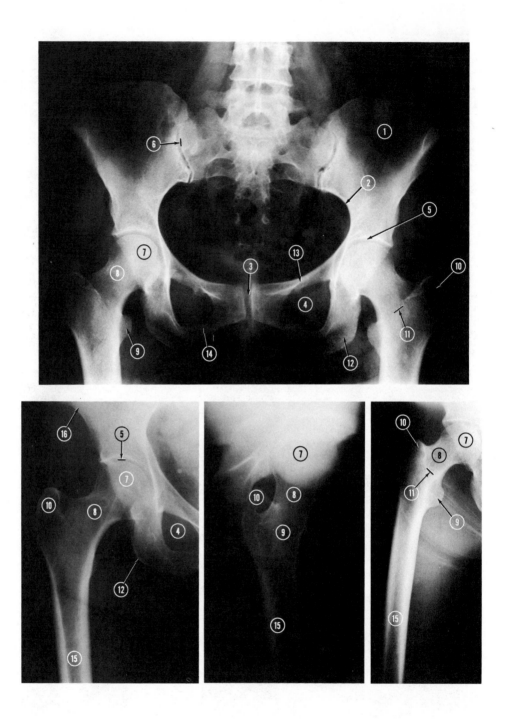

FIGURE 55. **Hip joint.** *1,* Iliac fossa; *2,* iliopectineal line; *3,* pubic symphysis; *4,* obturator foramen; *5,* acetabulum; *6,* sacroiliac joint; *7,* head of femur; *8,* neck of femur; *9,* lesser trochanter; *10,* greater trochanter; *11,* intertrochanteric crest; *12,* ischial tuberosity; *13,* superior pubic ramus; *14,* inferior pubic ramus; *15,* shaft of femur; *16,* anterior inferior iliac spine.

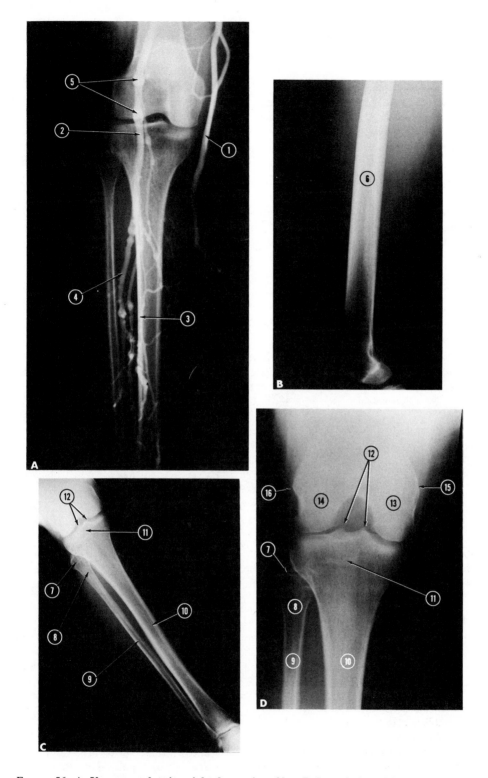

FIGURE 56. **A, Venogram showing right deep veins of leg; B, lateral view of femur; C, antero-posterior view of tibia and fibula; D, anteroposterior view of knee.** *1,* Great saphenous vein; *2,* popliteal vein; *3,* anterior tibial vein; *4,* posterior tibial vein; *5,* bulges of venous valves; *6,* shaft of femur; *7,* head of fibula; *8,* neck of fibula; *9,* shaft of fibula; *10,* shaft of tibia; *11,* tuberosity of tibia; *12,* intercondylar tubercles; *13,* medial condyle of femur; *14,* lateral condyle; *15,* medial epicondyle; *16,* lateral epicondyle.

Name	Origin	Insertion	Action	Nerve
I. Anterior				
Tibialis anterior	Lower lateral tibial condyle Lateral tibia Interosseous membrane	Medial side, first cuneiform Base, first metatarsal	Dorsiflexes foot Inverts foot	Deep peroneal (fibular)
Extensor digitorum longus	Lateral tibial condyle Anterior crest fibula Interosseous membrane	Bases of second and terminal phalanges of lateral 4 toes	Extends toes Dorsiflexes foot Everts foot	Deep peroneal (fibular)
Peroneus tertius	Distal fibula Interosseous membrane	Base fifth metatarsal	Dorsiflexes foot Everts foot	Deep peroneal (fibular)
Extensor hallucis longus	Middle anterior fibula Interosseous membrane	Base, distal phalanx of big toe	Extends big toe Everts foot	Deep peroneal (fibular)
II. Lateral				
Peroneus (fibularis) longus	Lateral tibial condyle Head of fibula Anterior capitular ligament Middle lateral fibula	Inferior, first cuneiform Lateral first metatarsal	Plantar flexes foot Everts foot Supports arch	Superficial peroneal (fibular)
Peroneus (fibularis) brevis	Middle lateral fibula	Dorsal surface of tuberosity, fifth metatarsal	Everts foot Plantar flexes foot	Superficial peroneal (fibular)

III. Special features

A. SUPERIOR AND INFERIOR EXTENSOR RETINACULA, flexor retinaculum, and peroneal retinacula (see p. 542)

B. COURSE OF TENDON OF PERONEUS LONGUS: passes behind lateral malleolus with tendon of peroneus brevis muscle but lies posterior to this, under peroneal retinacula. Along lateral side of calcaneus, it passes over lateral side of cuboid bone and then in a groove on plantar surface of this bone. It is held in the groove by the *long plantar ligament*. It then crosses foot obliquely to first cuneiform and metatarsal bones

C. THE FOUR ANTERIOR MUSCLES of the leg lie in a fascial compartment between the anterior intermuscular septum and the tibia. All 4 are innervated by the deep peroneal nerve. All have one action in common, for they all dorsiflex the foot, although with varying strength

D. SHIN SPLINTS: painful condition of anterior compartment of leg after vigorous and/or lengthy exercise. Ant. tibial muscles swell from overuse, reduce blood flow to muscles, and cramps develop

E. IF TIBIALIS ANTERIOR is paralyzed due to injury of common peroneal nerve or its deep branch, the *foot drops*

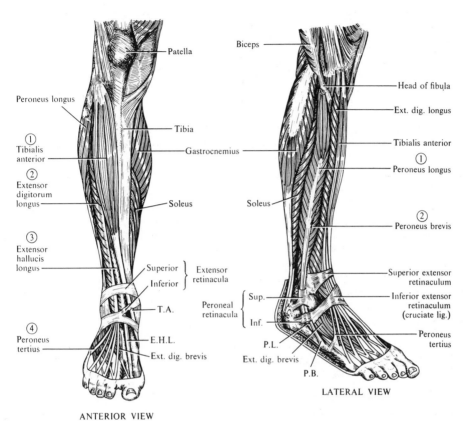

Patella

Biceps

Head of fibula

Ext. dig. longus

Peroneus longus

Tibialis anterior

① Peroneus longus

Tibia

② Extensor digitorum longus

Gastrocnemius

② Peroneus brevis

Soleus

Soleus

③ Extensor hallucis longus

Superior } Extensor
Inferior } retinacula

Superior extensor retinaculum

Inferior extensor retinaculum (cruciate lig.)

Peroneal { Sup.
retinacula { Inf.

T.A.

④ Peroneus tertius

Peroneus tertius

P.L.

E.H.L.

Ext. dig. brevis

Ext. dig. brevis

P.B.

ANTERIOR VIEW

LATERAL VIEW

ANTEROLATERAL MUSCLES

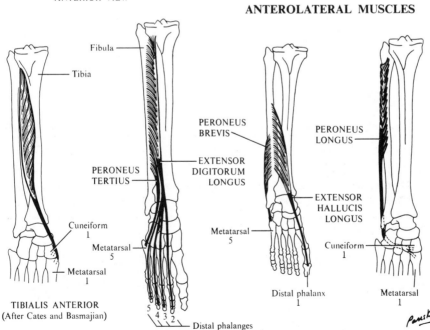

Fibula

Tibia

PERONEUS BREVIS

PERONEUS LONGUS

PERONEUS TERTIUS

EXTENSOR DIGITORUM LONGUS

EXTENSOR HALLUCIS LONGUS

Cuneiform 1

Metatarsal 5

Metatarsal 5

Cuneiform 1

Metatarsal 1

Distal phalanx 1

Metatarsal 1

TIBIALIS ANTERIOR
(After Cates and Basmajian)

5 4 3 2
Distal phalanges

233. COMPARTMENTS, DEEP VESSELS, AND NERVES OF THE LEG

I. Intermuscular septa: projections of fascia to anterior and lateral crests of fibula

A. ANTERIOR: between peronei and extensor digitorum longus muscles

B. POSTERIOR: between peronei and soleus muscles

II. Compartments: these septa, plus tibia, interosseous membrane, and fibula, divide the leg into anterior, lateral, and posterior compartments

A. DEEP TRANSVERSE FASCIA crosses posterior compartment, separating the soleus muscle from the other, deeper muscles

III. Anterior tibial artery begins at lower border of popliteus muscle. It passes between 2 heads of tibialis posterior muscle and above interosseous membrane to the front of the leg, where it is joined by the deep peroneal (anterior tibial) nerve

A. RELATIONS

	Anterior	
	Skin, superficial and deep fascia;	
	tibialis anterior, extensor digitorum longus, and	
	extensor hallucis longus muscles; deep peroneal nerve	
Lateral		*Medial*
Extensor digitorum longus and	**Anterior tibial**	Deep peroneal nerve and
extensor hallucis longus muscles;	**artery**	extensor hallucis longus muscle
deep peroneal nerve		
	Posterior	
	Interosseous membrane, tibia, ankle joint	

B. BRANCHES: anterior and posterior tibial recurrent; fibular; anterior medial and anterior lateral malleolar; and muscular

IV. Posterior tibial artery is a direct continuation of the popliteal artery

A. RELATIONS

	Anterior	
	Tibialis posterior and flexor digitorum	
	longus muscles, tibia, ankle joint	
Lateral	**Posterior tibial**	*Medial*
Tibial nerve	**artery**	Tibial nerve
	Posterior	
	Skin and fascia, gastrocnemius and soleus	
	muscles, tibial nerve, and deep transverse fascia	

B. BRANCHES: peroneal (fibular), posterior medial malleolar, communicating, medial calcaneal, muscular, and nutrient (tibial)

V. Peroneal artery arises from the posterior tibial artery, approaches fibula, and lies in a fibrous canal between tibialis posterior and flexor hallucis longus muscles

A. RELATIONS

	Anterior	
	Tibialis posterior muscle, interosseous membrane	
Lateral	**Peroneal (fibular)**	*Medial*
Flexor hallucis longus, fibula	**artery**	Flexor hallucis longus muscle
	Posterior	
	Soleus, flexor hallucis longus muscle,	
	deep transverse fascia	

B. BRANCHES: perforating, communicating, lateral calcaneal, muscular, and nutrient

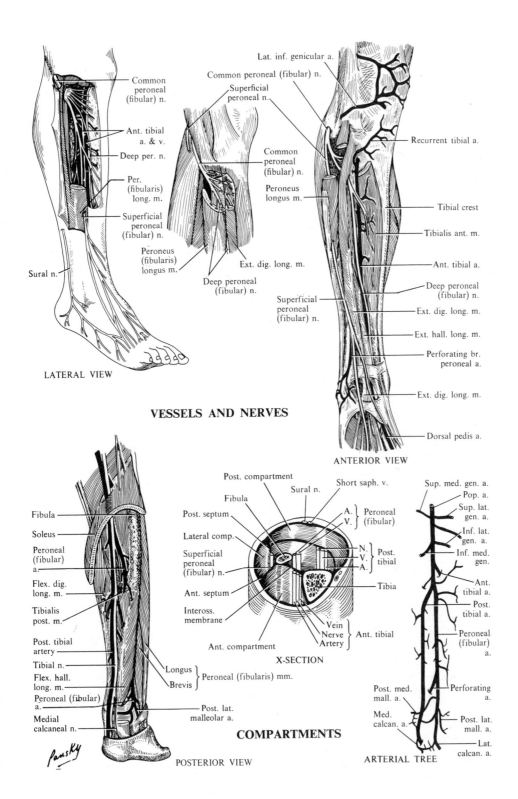

LATERAL VIEW

VESSELS AND NERVES

ANTERIOR VIEW

COMPARTMENTS

X-SECTION

POSTERIOR VIEW

ARTERIAL TREE

Common peroneal (fibular) n.

Ant. tibial a. & v.

Deep per. n.

Per. (fibularis) long. m.

Superficial peroneal (fibular) n.

Peroneus (fibularis) longus m.

Sural n.

Lat. inf. genicular a.

Common peroneal (fibular) n.

Superficial peroneal n.

Common peroneal (fibular) n.

Peroneus longus m.

Ext. dig. long. m.

Deep peroneal (fibular) n.

Superficial peroneal (fibular) n.

Recurrent tibial a.

Tibial crest

Tibialis ant. m.

Ant. tibial a.

Deep peroneal (fibular) n.

Ext. dig. long. m.

Ext. hall. long. m.

Perforating br. peroneal a.

Ext. dig. long. m.

Dorsal pedis a.

Fibula

Soleus

Peroneal (fibular) a.

Flex. dig. long. m.

Tibialis post. m.

Post. tibial artery

Tibial n.

Flex. hall. long. m.

Peroneal (fibular) a.

Medial calcaneal n.

Longus
Brevis } Peroneal (fibularis) mm.

Post. lat. malleolar a.

Post. compartment

Fibula

Post. septum

Lateral comp.

Superficial peroneal (fibular) n.

Ant. septum

Inteross. membrane

Ant. compartment

Sural n.

Short saph. v.

A.
V. } Peroneal (fibular)

N.
V.
A. } Post. tibial

Tibia

Vein
Nerve
Artery } Ant. tibial

Sup. med. gen. a.

Pop. a.

Sup. lat. gen. a.

Inf. lat. gen. a.

Inf. med. gen.

Ant. tibial a.

Post. tibial a.

Peroneal (fibular) a.

Perforating a.

Post. lat. mall. a.

Lat. calcan. a.

Post. med. mall. a.

Med. calcan. a.

−531−

234. MUSCLES OF THE POSTERIOR LEG

Name	Origin	Insertion	Action	Nerve
I. Superficial group				
Gastroc-nemius				
Medial head	Upper posterior medial condyle and femur above this	Through tendo calcan. into mid-post. calcan.	Flexes leg Plantar flexes foot	Tibial
Lateral head	Upper posterior lateral condyle and femur above this	See above	See above	Tibial
Soleus	Head and upper fibula Soleal line and upper med. tibia	See above	Plantar flexes foot	Tibial
Plantaris	Lower lateral supracondylar line	Posterior calcaneus	Flexes leg Plantar flexes foot	Tibial
II. Deep group				
Popliteus	Lateral condyle of femur Popliteal lig.	Posterior tibia above soleal line	Flexes leg Rotates leg medially	Tibial
Flexor hallucis longus	Inferior 2/3 posterior fibula Interosseous membrane	Base, distal phalanx, big toe	Flexes distal phalanx Plantar flexes and inverts foot	Tibial
Flexor digitorum longus	Posterior tibia below soleal line	Base, last phalanx, lateral 4 toes	Flexes toes Plantar flexes and inverts foot	Tibial
Tibialis posterior	Posterior surface, interosseous membrane Posterior shaft, tibia Upper shaft, fibula	Tuberosity navicular Sustentac. tali 3 cuneiforms, Cuboid Bases metatarsals 2, 3, 4	Plantar flexes, adducts, and inverts foot Supports arch	Tibial

III. Special features

  A. NOTE THE LONG, SLENDER PLANTARIS TENDON running on the medial border of the tendo calcaneus. May rupture

  B. THE FLEXOR HALLUCIS LONGUS TENDON runs in a groove on the posterior inferior tibia, posterior talus, and under sustentaculum tali

  C. THE FLEXOR DIGITORUM LONGUS TENDON runs in a groove behind medial malleolus with the tibialis posterior muscle, obliquely forward and lateralward superficial to the deltoid lig. into the sole of foot, where it crosses below flexor hallucis longus

  D. THE TIBIALIS POSTERIOR TENDON runs in front of the flexor digitorum longus in a groove with this muscle as they lie behind the medial malleolus, under the flexor retinaculum, but superficial to deltoid lig. It passes under calcaneonavicular lig.

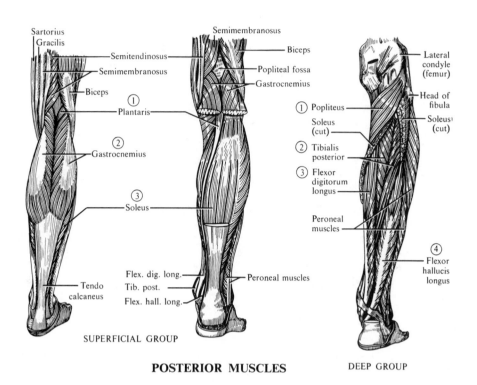

SUPERFICIAL GROUP

POSTERIOR MUSCLES

DEEP GROUP

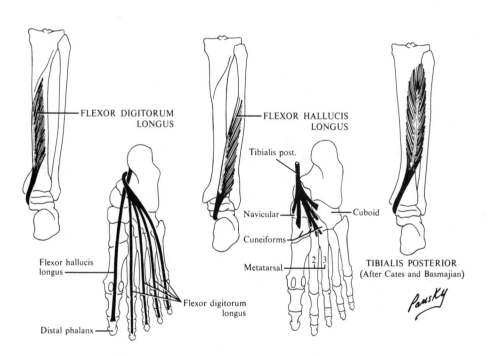

FLEXOR DIGITORUM LONGUS

FLEXOR HALLUCIS LONGUS

TIBIALIS POSTERIOR
(After Cates and Basmajian)

235. BONES OF ANKLE AND FOOT

I. Tarsals: talus, calcaneus, navicular, cuboid, 3 cuneiforms
A. TALUS (astragalus, ankle bone): 3 parts
    1. Body: behind, rests on the anterior part of calcaneus; above, lies under the tibia and is gripped by both malleoli; posteriorly has a medial and lateral tubercle with intervening groove for flexor hallucis longus muscle tendon; inferiorly has deep groove—*sulcus tali*
    2. Head: in front articulates with navicular and calcaneus
    3. Neck: constriction between 1 and 2, above
B. CALCANEUS (heel bone) has the following features:
    1. Sustentaculum tali: medial shelf to hold head of talus
    2. Groove for flexor hallucis longus muscle tendon runs below sustentaculum
    3. Posterior third ("heel") projects behind ankle joint
    4. Tuberosity: part in contact with ground; plantar surface has medial and lateral tubercles (processes)
    5. Peroneal trochlea on lateral aspect
    6. Groove corresponding to sulcus tali with which it forms *sinus tarsi*
C. NAVICULAR (boat-shaped)
    1. Tuberosity on medial side
D. CUBOID
    1. Groove on plantar surface for peroneus longus muscle tendon
E. CUNEIFORMS (wedge-shaped), 3: (1) medial, (2) intermediate, (3) lateral

II. Metatarsals: 5, numbered from medial to lateral
A. EACH CONSISTS OF
    1. Head (distal end): articulates with proximal phalanx
    2. Body (midportion)
    3. Base (proximal end): articulates with tarsals and bases of other metatarsals
B. FIRST: shortest, stoutest; 2 sesamoid bones at plantar surface of distal end
C. FIFTH has a tuberosity projecting posteriorly

III. Phalanges: 14 in number
A. THREE FOR TOES 2–5: (1) proximal, (2) middle, (3) distal
B. TWO FOR LARGE TOE (hallux)
C. EACH PHALANX consists of base (proximal end), body, and head (distal end)

IV. Ossification
A. TARSALS: all from 1 center except calcaneus, which has a second for heel
    1. Centers appear: calcaneus sixth, talus seventh, cuboid ninth fetal months; third cuneiform, first year; first cuneiform, third year; second cuneiform and navicular, fourth year; heel, tenth year. Latter joins body at puberty
B. METATARSALS: from 2 centers—body and head for 2–5; body and base for first
    1. Centers appear: body, ninth fetal week; base of first in third year; heads of 2–5, 5–8 years. All join body between 18–20 years
C. PHALANGES: from 2 centers—body and base
    1. Centers appear: body, tenth fetal week; base, 4–10 years. Join body at 18 years

V. Clinical considerations
A. FRACTURES OF THE BONES of the foot are common
    1. Jumping from a height, landing on the heels, fractures the calcanei (very disabling due to disruption of subtalar joint)
B. FRACTURES OF NECK OF TALUS occur during severe dorsiflexion of the ankle
C. FRACTURES OF METATARSALS AND PHALANGES usually occur when a heavy object falls on the foot. Phalangeal fracture may occur when stubbing the bare toes

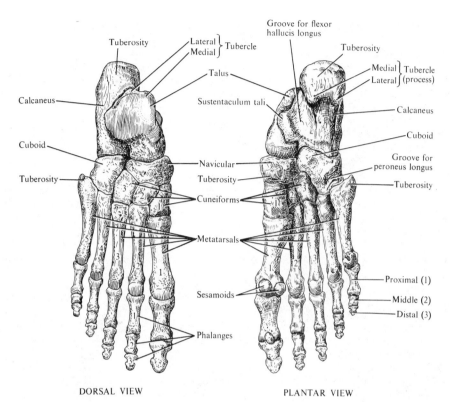

Groove for flexor hallucis longus

Tuberosity

Lateral ⎱ Tubercle
Medial ⎰

Talus

Medial ⎱ Tubercle
Lateral ⎰ (process)

Sustentaculum tali

Calcaneus

Cuboid

Navicular

Groove for peroneus longus

Cuboid

Tuberosity

Tuberosity

Tuberosity

Cuneiforms

Metatarsals

Proximal (1)

Middle (2)

Distal (3)

Sesamoids

Phalanges

DORSAL VIEW

PLANTAR VIEW

BONES OF ANKLE AND FOOT

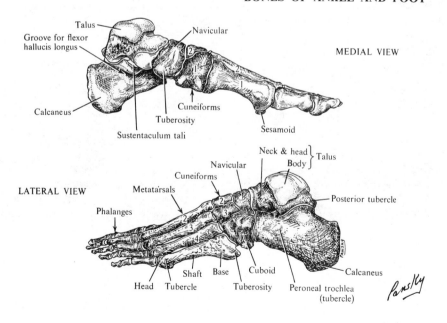

Talus

Navicular

Groove for flexor hallucis longus

MEDIAL VIEW

Calcaneus

Cuneiforms

Tuberosity

Sustentaculum tali

Sesamoid

Neck & head ⎱ Talus
Body ⎰

Navicular

Cuneiforms

Metatarsals

LATERAL VIEW

Phalanges

Posterior tubercle

Head   Tubercle   Base   Cuboid   Peroneal trochlea
Shaft   Tuberosity   (tubercle)

Calcaneus

Parsky

– 535 –

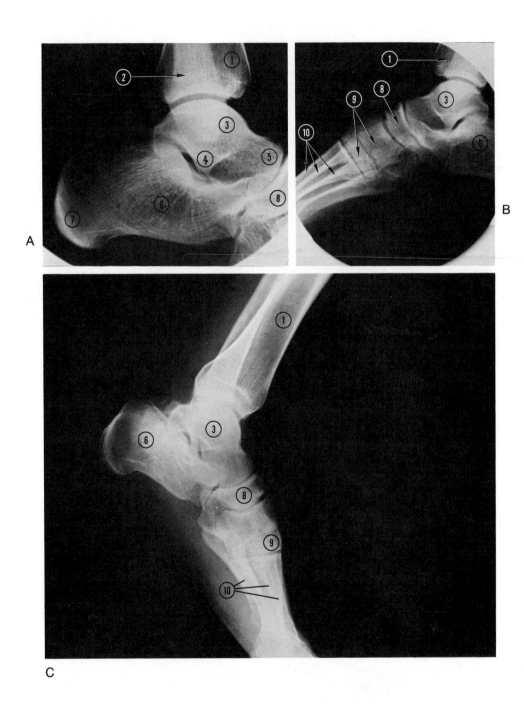

FIGURE 57. **Ankle joint. A, Left ankle joint; B, right ankle joint and foot, medial view; C, left ankle joint and foot.** *1*, Tibia; *2*, fibula; *3*, talus; *4*, sustentaculum tali; *5*, head of talus; *6*, calcaneus; *7*, tuberosity of calcaneus; *8*, navicular; *9*, cuneiforms; *10*, metatarsals.

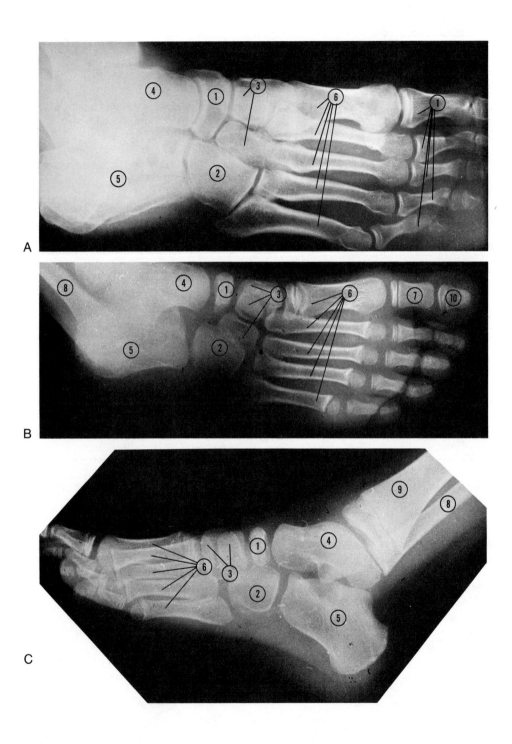

FIGURE 58. **Feet. A, Plantar view of left foot; B, plantar view of left foot; C, medial view of right foot.** *1*, Navicular; *2*, cuboid; *3*, cuneiforms; *4*, talus; *5*, calcaneus; *6*, metatarsals; *7*, proximal phalanges; *8*, fibula; *9*, tibia; *10*, distal phalanx (toe).

236. FLEXOR RETINACULUM, TENDON SHEATHS, AND VESSELS OF MEDIAL SIDE OF ANKLE

I. Flexor retinaculum holds the long flexor tendons at the medial side of ankle

A. EXTENDS FROM MARGINS OF MEDIAL MALLEOLUS TO CALCANEUS, passes over tendon of tibialis posterior muscle, attaches to bone behind this, and then forms superficial and deep layers

1. Superficial extends to tuberosity of calcaneus
2. Deep passes over flexor digitorum longus and flexor hallucis longus muscles and is attached to bone on either side of these, forming separate osseofibrous canals

B. THIS RETINACULUM converts the bony grooves into 4 osseofibrous canals for the following structures numbered from medial to lateral: tendon of tibialis posterior muscle, tendon of flexor digitorum longus muscle, posterior tibial artery and vein and tibial nerve, and the tendon of the flexor hallucis longus muscle

II. Tendon sheaths

A. BEHIND THE MEDIAL MALLEOLUS, the tendon of the tibialis posterior muscle lies in front of, but in the same groove with, the flexor digitorum longus muscle. Both have separate sheaths: that for the tibialis posterior starts 5 cm above malleolus and extends to tuberosity of navicular; that for flexor digitorum longus begins just above tip of malleolus and ends opposite first cuneiform. The tendon of the flexor hallucis longus muscle lies dorsal to that of the flexor digitorum longus muscle and crosses it from lateral to medial. Its sheath starts at the tip of the malleolus and extends to base of first metatarsal

III. Anastomoses about the ankle

Medial Malleolar Net	Lateral Malleolar Net
Ant. med. malleolar from ant. tibial a.	Ant. lat. malleolar from ant. tibial a.
Med. tarsal from dorsalis pedis a.	Lat. tarsal from dorsalis pedis a.
Post. med. malleolar from post. tibial a.	Perf. peroneal from peroneal (fibular) a.
Med. calcaneal from post. tibial a.	Lat. calcaneal from peroneal (fibular) a.
	Arcuate from dorsalis pedis a.

IV. The deep transverse fascial septum separates the deep and superficial groups of the posterior crural muscles. At the ankle it is continuous with the flexor retinaculum

V. Relations of vessels

A. DORSALIS PEDIS ARTERY is a continuation of the anterior tibial artery in front of the ankle, has deep peroneal nerve on its lateral side and tendon of flexor hallucis longus muscle on its medial side as they pass under the extensor retinacula

B. THE TIBIAL NERVE lies lateral to the posterior tibial artery as they pass in the third compartment beneath the flexor retinaculum

C. THE PERONEAL VESSELS lie behind the inferior tibiofibular articulation

VI. Clinical considerations

A. THE PULSE of the posterior tibial artery is usually palpated about halfway between the posterior surface of the medial malleolus and medial border of the tendo calcaneus

B. THE DORSALIS PEDIS PULSE can be felt where it passes over the navicular and cuneiform bones lateral to the extensor hallucis longus tendon or distal to this at the proximal end of the 1st interosseous space (10–20% may be too small to palpate or are found in unusual positions)

MEDIAL ANKLE

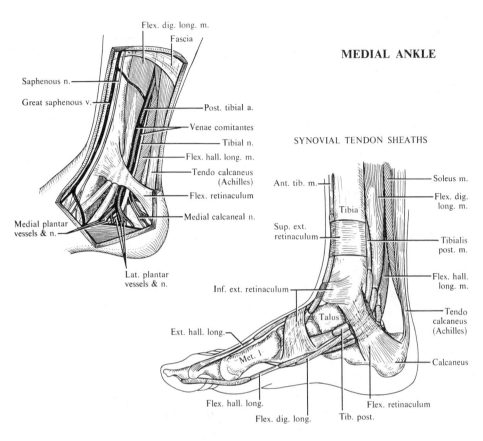

Flex. dig. long. m.

Fascia

Saphenous n.

Great saphenous v.

Post. tibial a.

Venae comitantes

Tibial n.

Flex. hall. long. m.

Tendo calcaneus (Achilles)

Flex. retinaculum

Medial calcaneal n.

Medial plantar vessels & n.

Lat. plantar vessels & n.

SYNOVIAL TENDON SHEATHS

Ant. tib. m.

Tibia

Sup. ext. retinaculum

Inf. ext. retinaculum

Talus

Ext. hall. long.

Met. 1

Flex. hall. long.

Flex. dig. long.

Tib. post.

Flex. retinaculum

Soleus m.

Flex. dig. long. m.

Tibialis post. m.

Flex. hall. long. m.

Tendo calcaneus (Achilles)

Calcaneus

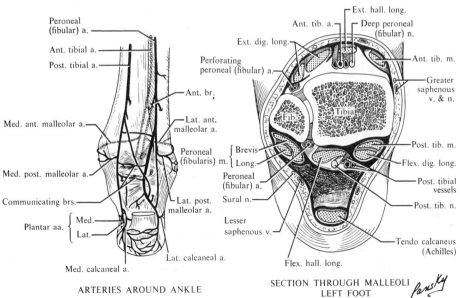

Peroneal (fibular) a.

Ant. tibial a.

Post. tibial a.

Ant. br.

Med. ant. malleolar a.

Lat. ant. malleolar a.

Peroneal (fibularis) m.

Med. post. malleolar a.

Communicating brs.

Lat. post. malleolar a.

Plantar aa. { Med.
              Lat.

Lat. calcaneal a.

Med. calcaneal a.

ARTERIES AROUND ANKLE

Ext. hall. long.

Ant. tib. a.

Deep peroneal (fibular) n.

Ext. dig. long.

Ant. tib. m.

Perforating peroneal (fibular) a.

Greater saphenous v. & n.

Fib.

Tibia

Brevis
Long.

Post. tib. m.

Peroneal (fibular) a.

Flex. dig. long.

Sural n.

Post. tibial vessels

Lesser saphenous v.

Post. tib. n.

Tendo calcaneus (Achilles)

Flex. hall. long.

SECTION THROUGH MALLEOLI LEFT FOOT

Pansky

237. MUSCLES OF DORSUM OF FOOT

I.

Name	Origin	Insertion	Action	Nerve
Extensor hallucis longus	Middle anterior fibula Interosseous membrane	Base, distal phalanx of big toe	Extends big toe Everts foot	Deep peroneal (fibular)
Extensor digitorum longus	Lateral tibial condyle Anterior crest fibula Interosseous membrane	Bases of second and terminal phalanges of lateral 4 toes	Extends toes Dorsiflexes foot Everts foot	Deep peroneal (fibular)
Peroneus (fibularis) tertius	Distal fibula Inteross. memb.	Base, fifth metatarsal	Dorsiflexes foot Everts foot	Deep peroneal (fibular)
Tibialis anterior	Lower lateral tibial condyle Lateral tibia Inteross. memb.	Medial side, first cuneiform Base, first metatarsal	Dorsiflexes foot Inverts foot	Deep peroneal (fibular)
Extensor digitorum brevis	Distal lateral and superior surface, calcaneus	Lateral side, long extensor tendons; slips to base of 1st phalanx of medial 4 toes	Extends medial 4 toes	Deep peroneal (fibular)
Dorsal interossei (4)	Adjacent sides and bases of metatarsals 1–4	1st and 2nd on either side, base of 1st phalanx of 2nd toe; 3 and 4 on lateral side, proximal phalanx, toes 3 and 4	Abduct toes Flex proximal phalanx Extend distal 2 phalanges	Deep branch of the lateral plantar

II. Special features

A. RETINACULA AROUND ANKLE (see p. 542)

B. SYNOVIAL TENDON SHEATH OF ANTERIOR MUSCLES

1. Tibialis anterior: from above superior extensor retinaculum, under inferior extensor retinaculum to level of talonavicular joint
2. Extensor hallucis longus: from just above superior limb of inferior extensor retinaculum, beneath both limbs of retinaculum to level of first tarsometatarsal joint
3. Tendon of extensor digitorum longus envelops this and peroneus tertius muscle from above the inferior extensor retinaculum to middle of cuboid bone

C. INSERTIONS OF EXTENSOR TENDONS

1. Extensor hallucis has medial and lateral expansions covering metatarsophalangeal joint and a prolongation of tendon to base of first phalanx
2. Extensor digitorum longus: tendons to toes 2, 3, 4 are joined by extensor brevis; each receives fibers from lumbricales and interossei, spreads out on dorsum of first phalanx of each toe, splits into 3 slips, the intermediate to base of second phalanx and the 2 collateral to base of third phalanx

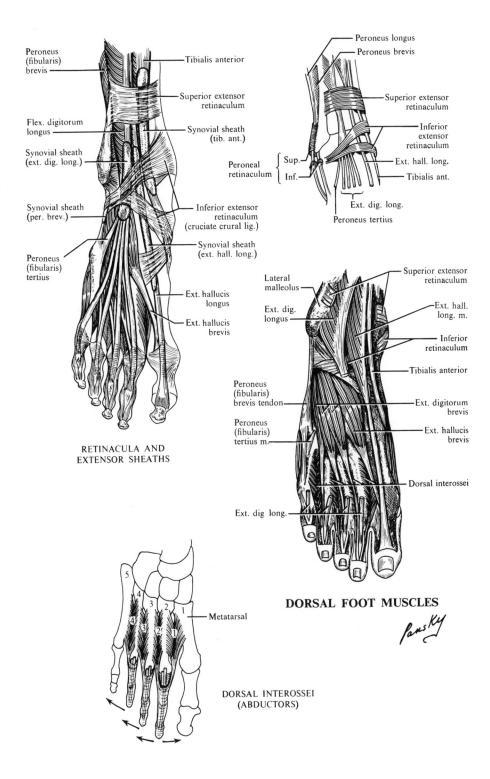

Peroneus (fibularis) brevis

Tibialis anterior

Superior extensor retinaculum

Flex. digitorum longus

Synovial sheath (tib. ant.)

Synovial sheath (ext. dig. long.)

Synovial sheath (per. brev.)

Inferior extensor retinaculum (cruciate crural lig.)

Synovial sheath (ext. hall. long.)

Peroneus (fibularis) tertius

Ext. hallucis longus

Ext. hallucis brevis

RETINACULA AND EXTENSOR SHEATHS

Peroneus longus

Peroneus brevis

Superior extensor retinaculum

Inferior extensor retinaculum

Ext. hall. long.

Tibialis ant.

Peroneal retinaculum { Sup. / Inf. }

Ext. dig. long.

Peroneus tertius

Lateral malleolus

Ext. dig. longus

Superior extensor retinaculum

Ext. hall. long. m.

Inferior retinaculum

Tibialis anterior

Peroneus (fibularis) brevis tendon

Ext. digitorum brevis

Peroneus (fibularis) tertius m.

Ext. hallucis brevis

Dorsal interossei

Ext. dig long.

DORSAL FOOT MUSCLES

Pansky

Metatarsal

DORSAL INTEROSSEI (ABDUCTORS)

– 541 –

238. EXTENSOR AND PERONEAL RETINACULA, SHEATHS, AND ARTERIES ON DORSUM OF FOOT

I. Extensor retinacula hold extensor tendons in place

A. SUPERIOR (transverse crural) across leg above ankle. Consists of transverse fibers from the medial side of tibia to the anterior side of fibula

B. INFERIOR (cruciate crural) has 2 parts
1. Superficial: Y-shaped, stem attached to lateral side of calcaneus, splits after passing over the extensor digitorum longus muscle
   a. Upper limb passes upward and medialward to attach to the medial malleolus
   b. Lower limb passes medially and downward over foot to insert on first cuneiform and sole of foot
2. Deep: beneath the stem of the Y and passes deep to the peroneus tertius and extensor longus muscles

II. Peroneal (fibular) retinacula hold peroneal muscles in place

A. SUPERIOR from lateral malleolus to fascia of back of leg and lateral side of calcaneus

B. INFERIOR overlies tendons on lateral calcaneus and is attached to the bone on either side of the tendons. Joins superficial part of the inferior extensor retinaculum

III. Synovial sheaths: for tibialis anterior, extends from above superior extensor retinaculum to interval between the limbs of the inferior extensor retinaculum. For extensor digitorum longus and extensor hallucis longus muscles, the sheath begins just below the superior extensor retinaculum, the former extending to the level of the base of the fifth metatarsal, the latter to the base of the first metatarsal. Both peronei lie in a common sheath extending 4 cm above and below the tip of the malleolus

IV. Arteries

A. DORSALIS PEDIS passes down dorsum of foot to the proximal end of the first intermetatarsal space and terminates as the *first dorsal metatarsal* and *deep plantar arteries*. Branches:
1. Lateral tarsal
2. Medial tarsal
3. Arcuate, which gives rise to *second, third,* and *fourth metatarsal arteries*. These subsequently split to form the *dorsal digital arteries* to the adjoining sides of the toes
4. Terminal branches
   a. First dorsal metatarsal to both sides of big toe and adjoining sides of big and second toes
   b. Deep plantar (see p. 548)

V. Veins (see p. 484)

VI. Nerves

A. MOST OF THE DORSUM OF THE FOOT is supplied by the superficial peroneal nerve. The sural contributes to the lateral side and the saphenous to the medial side of the foot

B. ON THE TOES, the deep peroneal nerve supplies adjoining areas on the first and second toes; the superficial peroneal nerve covers the remainder of the first, second, and third toes to the second joint. Part of the fourth toe is supplied by the medial plantar nerve. Adjoining parts of the fourth and fifth, plus the terminal ends of the first 3 toes, receive lateral plantar branches

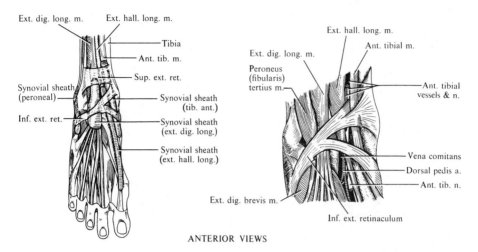

ANTERIOR VIEWS

DORSUM OF FOOT

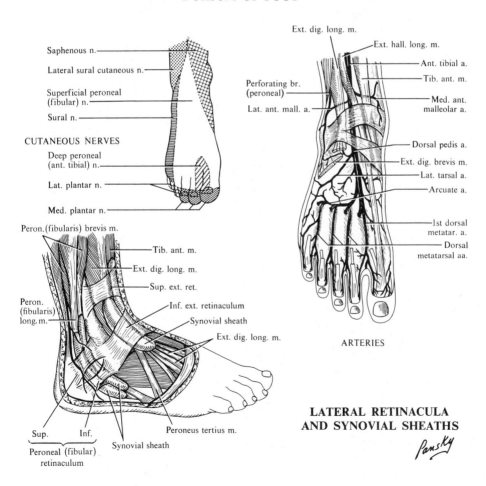

CUTANEOUS NERVES

ARTERIES

**LATERAL RETINACULA
AND SYNOVIAL SHEATHS**

Pansky

SURFACE ANATOMY-
LOWER EXTREMITY

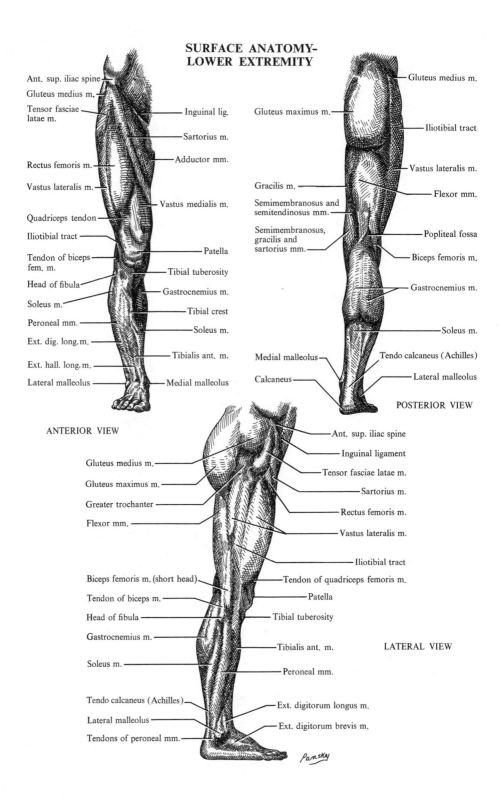

Ant. sup. iliac spine

Gluteus medius m.

Tensor fasciae latae m.

Inguinal lig.

Sartorius m.

Adductor mm.

Rectus femoris m.

Vastus lateralis m.

Vastus medialis m.

Quadriceps tendon

Iliotibial tract

Tendon of biceps fem. m.

Patella

Tibial tuberosity

Head of fibula

Gastrocnemius m.

Soleus m.

Tibial crest

Peroneal mm.

Soleus m.

Ext. dig. long. m.

Ext. hall. long. m.

Tibialis ant. m.

Lateral malleolus

Medial malleolus

ANTERIOR VIEW

Gluteus medius m.

Gluteus maximus m.

Iliotibial tract

Vastus lateralis m.

Gracilis m.

Flexor mm.

Semimembranosus and semitendinosus mm.

Semimembranosus, gracilis and sartorius mm.

Popliteal fossa

Biceps femoris m.

Gastrocnemius m.

Soleus m.

Medial malleolus

Tendo calcaneus (Achilles)

Calcaneus

Lateral malleolus

POSTERIOR VIEW

Gluteus medius m.

Ant. sup. iliac spine

Inguinal ligament

Tensor fasciae latae m.

Gluteus maximus m.

Sartorius m.

Greater trochanter

Rectus femoris m.

Flexor mm.

Vastus lateralis m.

Iliotibial tract

Biceps femoris m. (short head)

Tendon of quadriceps femoris m.

Tendon of biceps m.

Patella

Head of fibula

Tibial tuberosity

Gastrocnemius m.

Tibialis ant. m.

LATERAL VIEW

Soleus m.

Peroneal mm.

Tendo calcaneus (Achilles)

Ext. digitorum longus m.

Lateral malleolus

Ext. digitorum brevis m.

Tendons of peroneal mm.

Pansky

FIGURE 59. **Surface anatomy of lower extremity.**

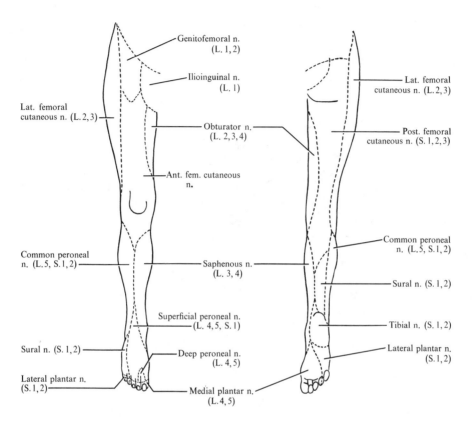

CUTANEOUS NERVES - SEGMENTAL DISTRIBUTION

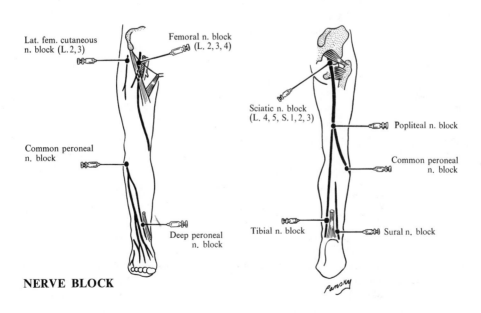

NERVE BLOCK

FIGURE 60. **Nerve blocks of lower extremity.**

I.

Name	Origin	Insertion	Action	Nerve
Flexor digitorum brevis	Medial process, tuber. calcaneus	Middle phalanx, lateral 4 toes	Flexes toes	Medial plantar
Quadratus plantae				
Lateral head	Lateral process, tuberosity calcaneus	Lateral border, long flexor tendons	Flexes toes	Lateral plantar
Medial head	Med. side calcan.			
Lumbricales				
1st	Medial side of 1st longus flexor tendon	Proximal phalanx and extensor tendons, toe 2	Flexes proximal, extends distal phalanges	Medial plantar
2nd, 3rd, and 4th	Long flexor tendons 2, 3, 4	As above for toes 3, 4, 5	See above	Lateral plantar
Abductor hallucis	Medial process, tuberosity calcaneus	Medial side, base, proximal phalanx of big toe	Flexes and abducts big toe	Medial plantar
Flexor hallucis brevis	Cuboid and 3rd cuneiform	Abductor tendon Base, proximal phalanx	Flexes proximal phalanx, big toe	Medial plantar
Adductor hallucis				
Oblique head	Peroneus long. sheath Base 2, 3, 4, metatarsals	With flexor hallucis brevis	Adducts and flexes proximal phalanx, big toe	Lateral plantar
Transverse head	Capsules 3, 4, 5, metatarso-phalang. joints	Lateral side of base of proximal phalanx, big toe	See above	See above
Abductor digiti minimi	Lat. and med. processes, tuberosity calcaneus	Lateral surface, proximal phalanx, small toe	Flexes and abducts proximal phalanx, small toe	Lateral plantar
Flexor digiti minimi brevis	Peron. long. tendon Base, metatarsal 5	Base of proximal phalanx, small toe	Flexes proximal phalanx, small toe	Lateral plantar
Plantar interossei (3)	Proximal shaft and base, metatarsals 3, 4, 5	Base of prox. phalanx of toes 3, 4, and 5 extensor tendons	Adducts toes Flexes proximal, extends distal part of toes 3–5	Lateral plantar

II. Special features

  A. THE MUSCLES AND TENDONS of the plantar surface of foot are arranged in 4 layers

     1. Layer 1: abductor hallucis, flexor digitorum brevis, and abductor digiti minimi

     2. Layer 2: quadratus plantae, lumbricales, tendons of flexor digitorum longus and flexor hallucis longus

     3. Layer 3: flexor hallucis brevis, adductor hallucis, and flexor digiti minimi brevis

     4. Layer 4: the interossei, tendons of tibialis posterior and peroneus longus

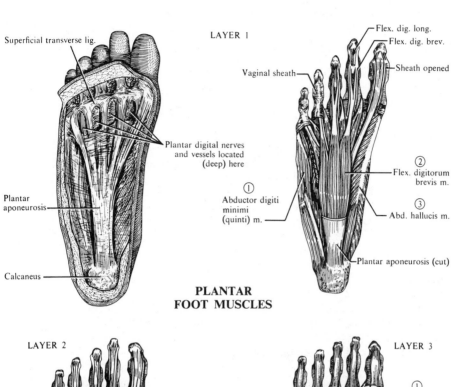

Superficial transverse lig.

LAYER 1

Flex. dig. long.
Flex. dig. brev.

Vaginal sheath

Sheath opened

Plantar digital nerves
and vessels located
(deep) here

② Flex. digitorum
brevis m.

① Abductor digiti
minimi
(quinti) m.

③ Abd. hallucis m.

Plantar
aponeurosis

Plantar aponeurosis (cut)

Calcaneus

**PLANTAR
FOOT MUSCLES**

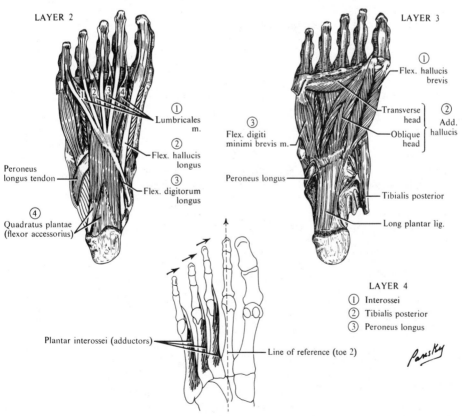

LAYER 2

LAYER 3

① Flex. hallucis
brevis

① Lumbricales
m.

② Flex. hallucis
longus

Transverse
head
② Add.
hallucis
Oblique
head

③ Flex. digiti
minimi brevis m.

Peroneus
longus tendon

③ Flex. digitorum
longus

Peroneus longus

Tibialis posterior

④ Quadratus plantae
(flexor accessorius)

Long plantar lig.

LAYER 4
① Interossei
② Tibialis posterior
③ Peroneus longus

Plantar interossei (adductors)

Line of reference (toe 2)

-547-

240. PLANTAR APONEUROSIS, THE VESSELS AND NERVES OF THE SOLE OF THE FOOT

I. Plantar aponeurosis: thickened deep fascia of the sole. Extends from the tuberosity of the calcaneus to end in 5 slips attached to the skin over the heads of the metatarsals and into the flexor tendon sheaths (bands 2, 3, and 4 are stronger than 1 and 5)

II. Arteries

A. POSTERIOR TIBIAL runs beneath the origin of the abductor hallucis muscle and divides into 2 branches
   1. Medial plantar along the medial side of foot and medial border of big toe
   2. Lateral plantar goes obliquely lateralward to reach the base of fifth metatarsal, then arches medially to a point between the bases of the first and second metatarsals. It joins the deep branch of the dorsalis pedis artery to form the *plantar arch*. Branches:
      a. Perforating to dorsal metatarsal arteries
      b. Plantar metatarsals (4) in the interosseous spaces where they divide into *proper plantar digital arteries* to the adjacent sides of toes. Send *anterior perforating branches* to the dorsal metatarsal arteries
      c. First plantar metatarsal supplies adjoining sides of big and second toes
      d. Proper plantar digital to the lateral side of the small toe

III. Nerves

A. CUTANEOUS
   1. The saphenous nerve (from the femoral) and the sural nerve (from both the tibial and peroneal nerves) supply small areas of skin on the medial and lateral parts, respectively, of the sole
   2. Plantar surface of the heel is supplied by the tibial nerve
   3. Most of the medial sole, including the toes to the middle of the fourth toe, are supplied by the medial plantar nerve. The medial plantar nerve accompanies the medial plantar artery and gives off branches
      a. Plantar cutaneous branches to medial side of sole
      b. Proper digital nerve to medial side of big toe
      c. Three common digital nerves, which split to give 2 proper digital nerves to adjacent sides of toes 1–4
   4. The lateral half of the fourth toe and lateral side of the foot are supplied by the lateral plantar nerve. The lateral plantar nerve is accompanied by the lateral plantar artery. The nerve gives off a superficial branch, which sends proper digital nerves to the lateral side of the little toe, and common digital nerves, which divide to supply adjacent sides of toes 4 and 5

B. MUSCULAR
   1. Medial plantar nerve to the abductor hallucis, flexor digitorum brevis, flexor hallucis brevis, and first lumbrical muscle
   2. Lateral plantar to the quadratus plantae and abductor digiti minimi muscles
      a. Superficial branch to flexor digiti minimi and the 2 interossei muscles of the fourth space
      b. Deep branch to adductor hallucis muscle, interosseous muscles of spaces 1, 2, and 3, and lumbrical muscles 2, 3, and 4

IV. Clinical: cuts of the feet may bleed profusely. Due to the numerous anastomoses between vessels, both ends of a severed artery must be ligated

A. PLANTAR SKIN REFLEX: when skin of sole is slowly stroked along outer border from heel forward, the toes flex. In patients with certain disorders of motor pathways, similar stimulation of sole results in a slow dorsiflexion of big toe and a spreading of other toes—known as the *Babinski reflex*

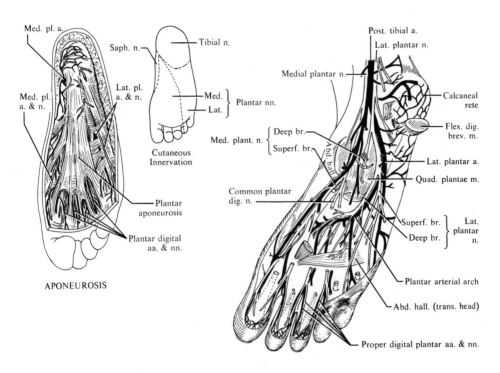

Med. pl. a.

Saph. n.

Tibial n.

Med. pl. a. & n.

Lat. pl. a. & n.

Med.
Lat.

Plantar nn.

Cutaneous
Innervation

Plantar
aponeurosis

Plantar digital
aa. & nn.

APONEUROSIS

Post. tibial a.

Lat. plantar n.

Medial plantar n.

Deep br.
Superf. br.

Med. plant. n.

Common plantar
dig. n.

Superf. br.
Deep br.

Lat.
plantar
n.

Calcaneal
rete

Flex. dig.
brev. m.

Lat. plantar a.

Quad. plantae m.

Plantar arterial arch

Abd. hall. (trans. head)

Proper digital plantar aa. & nn.

**VESSELS AND NERVES
OF PLANTAR FOOT**

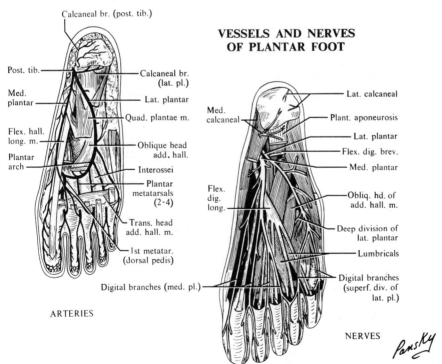

Calcaneal br. (post. tib.)

Post. tib.

Med.
plantar

Flex. hall.
long. m.

Plantar
arch

Calcaneal br.
(lat. pl.)

Lat. plantar

Quad. plantae m.

Oblique head
add. hall.

Interossei

Plantar
metatarsals
(2-4)

Trans. head
add. hall. m.

1st metatar.
(dorsal pedis)

Digital branches (med. pl.)

ARTERIES

Med.
calcaneal

Flex.
dig.
long.

Lat. calcaneal

Plant. aponeurosis

Lat. plantar

Flex. dig. brev.

Med. plantar

Obliq. hd. of
add. hall. m.

Deep division of
lat. plantar

Lumbricals

Digital branches
(superf. div. of
lat. pl.)

NERVES

Pansky

241. SURFACE ANATOMY OF LOWER EXTREMITY

I. Arteries

A. FEMORAL: with the thigh slightly flexed and laterally rotated, its course is indicated by a line beginning at a point halfway between the anterior superior iliac spine and pubic symphysis, extending downward to a point just above the adductor tubercle

B. POPLITEAL: beginning at a point on the back of the thigh just above the adductor tubercle, the line curves slightly laterally to the middle line of the thigh and then extends downward through the middle of the popliteal fossa to the apex of its inferior angle

C. POSTERIOR TIBIAL: its position can be indicated by a line drawn from the middle of the popliteal fossa to a point midway between the medial malleolus and the medial border of the tendo calcaneus

D. PERONEAL (FIBULAR): begins on posterior tibial line about 3 cm below the inferior angle of popliteal fossa, the line curving obliquely lateralward to reach the medial border of the fibula at the junction of its upper and middle thirds, thence continuing downward along the fibula, behind the lateral malleolus to end on the lateral side of the calcaneus

E. ANTERIOR TIBIAL: indicated by a line beginning at a point medial and just below the head of the fibula extending downward to a point midway between the malleoli at the front of the ankle

F. DORSALIS PEDIS: from a point midway between the malleoli at the front of the ankle to the base of the first intermetatarsal space on the dorsum of the foot

G. LATERAL PLANTAR AND PLANTAR ARCH: from a point midway between the medial malleolus and the medial border of the lower end of the tendo calcaneus, across the sole of the foot to the proximal end of the fourth intermetatarsal space; the arch continues medially across the sole of the foot to the proximal end of the first intermetatarsal space

II. Nerves

A. SCIATIC (IN THIGH): its course may be indicated by a line drawn from a point midway between the great trochanter of the femur and the ischial tuberosity to the superior angle of the popliteal fossa

B. TIBIAL: from the superior angle of the popliteal fossa down the midline of the fossa, thence following the line of the posterior tibial artery through the rest of its extent

C. COMMON PERONEAL (FIBULAR) diverges from the line of the sciatic at the superior angle of the popliteal fossa, continuing downward along the lateral boundary of the fossa (the biceps femoris) to the level of the head of the fibula where its terminal branches are given off. It is palpable against the biceps and the fibula

D. FEMORAL enters thigh about 1.25 cm lateral to the line of the femoral artery and extends only about 2.5 cm below the inguinal ligament where it breaks up into several branches

III. General relations

A. IN THE ERECT POSITION, the coccyx, pubic crest, middle of acetabulum, head of femur, and tip of greater trochanter are all in the same horizontal plane

B. THE FEMORAL ARTERY can be compressed by pressing directly backward at the midinguinal point

C. THE LOWEST LEVEL of the knee joint is at a level at the margins of the tibial condyles (about 1 cm below the apex of the patella)

D. THE HIGHEST LEVEL of the ankle joint is about 1 cm above tip of medial malleolus

E. MEDIAL EDGE of the sustentaculum tali is about 2–3 cm below tip of medial malleolus

F. THE TRANSVERSE TARSAL JOINT is indicated by a line from the back of the tuberosity of the navicular to a point halfway between lateral malleolus and tuberosity of the 5th metatarsal bone

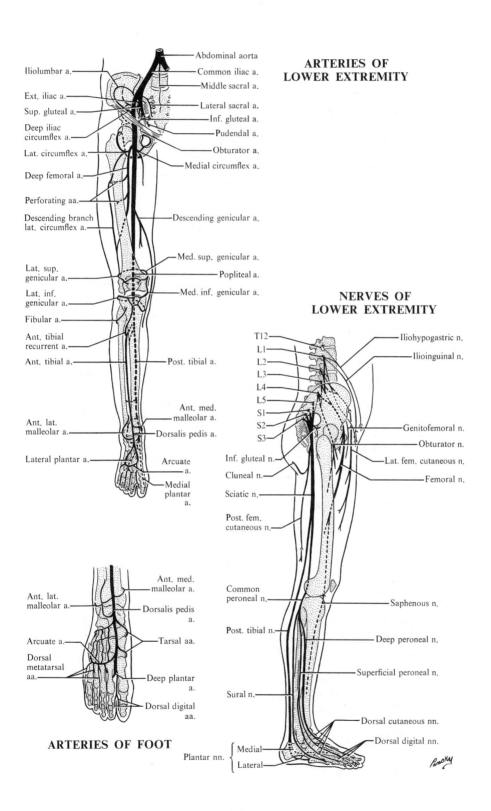

ARTERIES OF
LOWER EXTREMITY

Iliolumbar a.
Ext. iliac a.
Sup. gluteal a.
Deep iliac circumflex a.
Lat. circumflex a.
Deep femoral a.
Perforating aa.
Descending branch lat. circumflex a.
Lat. sup. genicular a.
Lat. inf. genicular a.
Fibular a.
Ant. tibial recurrent a.
Ant. tibial a.
Ant. lat. malleolar a.
Lateral plantar a.

Abdominal aorta
Common iliac a.
Middle sacral a.
Lateral sacral a.
Inf. gluteal a.
Pudendal a.
Obturator a.
Medial circumflex a.
Descending genicular a.
Med. sup. genicular a.
Popliteal a.
Med. inf. genicular a.
Post. tibial a.
Ant. med. malleolar a.
Dorsalis pedis a.
Arcuate a.
Medial plantar a.

NERVES OF
LOWER EXTREMITY

T12
L1
L2
L3
L4
L5
S1
S2
S3
Inf. gluteal n.
Cluneal n.
Sciatic n.
Post. fem. cutaneous n.
Common peroneal n.
Post. tibial n.
Sural n.
Plantar nn. { Medial / Lateral }

Iliohypogastric n.
Ilioinguinal n.
Genitofemoral n.
Obturator n.
Lat. fem. cutaneous n.
Femoral n.
Saphenous n.
Deep peroneal n.
Superficial peroneal n.
Dorsal cutaneous nn.
Dorsal digital nn.

Ant. lat. malleolar a.
Arcuate a.
Dorsal metatarsal aa.

Ant. med. malleolar a.
Dorsalis pedis a.
Tarsal aa.
Deep plantar a.
Dorsal digital aa.

ARTERIES OF FOOT

Panoky

242. JOINTS OF THE ANKLE AND FOOT

I. Ankle joint (talocrural)
A. TYPE: ginglymus (hinge)
B. BONES: medial malleolus of tibia, lateral malleolus of fibula, and talus
C. MOVEMENTS: dorsiflexion (extension) and plantar flexion (flexion)
D. LIGAMENTS
    1. Articular capsule
    2. Medial (deltoid) consists of 4 ligaments—the anterior and posterior tibiotalar, the tibiona-vicular, and the tibiocalcaneal. The anterior and posterior fibers check plantar and dorsi-flexion, respectively. The anterior fibers also limit abduction
    3. Lateral ligaments: the anterior and posterior talofibular and calcaneofibular

II. Subtalar (talocalcaneal) joint (only 3 more movable intertarsal joints mentioned)
A. TYPE: arthrodial
B. BONES: talus and calcaneus
C. MOVEMENTS: gliding
D. LIGAMENTS
    1. Anterior, posterior, medial, and lateral talocalcaneal
    2. Interosseous talocalcaneal

III. Talocalcaneonavicular joint
A. TYPE: arthrodial
B. BONES: talus, navicular, and calcaneus
C. MOVEMENTS: gliding and some rotation
D. LIGAMENTS
    1. Articular capsule
    2. Dorsal talonavicular
    3. Plantar calcaneonavicular supports the head of the talus (*spring ligament*)

IV. Calcaneocuboid joint
A. TYPE: arthrodial
B. MOVEMENTS: gliding and slight rotation
C. LIGAMENTS
    1. Dorsal and plantar calcaneocuboid
    2. Bifurcated: from the calcaneus to the cuboid and navicular
    3. Long plantar

V. Tarsometatarsal joints between the bases of the metatarsals, the 3 cuneiforms, and the cuboid. Are arthrodial joints with slight gliding movements. Each has dorsal, plantar, and interosseous ligaments

VI. Metatarsophalangeal joints between the heads of the metatarsals and the bases of the proxi-mal phalanges. Are condyloid joints, permitting flexion, extension, abduction, and adduc-tion. Each has plantar and 2 collateral ligaments

VII. Interphalangeal joints between bases and heads of adjoining phalanges. Are ginglymus joints permitting flexion and extension. Each has plantar and collateral ligaments. The exten-sor aponeurosis acts as dorsal ligaments

VIII. Clinical consideration
A. SPRAINS: a sprained ankle usually is an inversion injury in which, as a result of rocking the foot toward the midline, some part of a ligament is torn

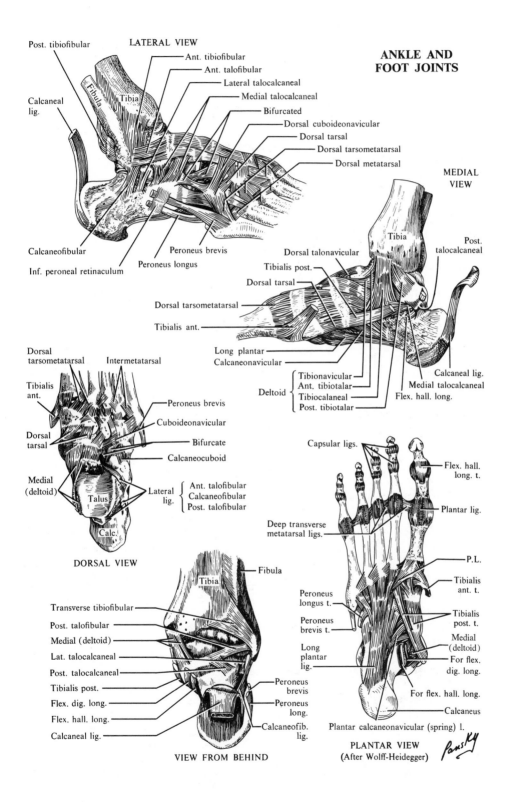

ANKLE AND
FOOT JOINTS

LATERAL VIEW

Post. tibiofibular
Ant. tibiofibular
Ant. talofibular
Lateral talocalcaneal
Medial talocalcaneal
Bifurcated
Dorsal cuboideonavicular
Dorsal tarsal
Dorsal tarsometatarsal
Dorsal metatarsal
Calcaneal lig.
Fibula
Tibia
Calcaneofibular
Inf. peroneal retinaculum
Peroneus brevis
Peroneus longus

MEDIAL VIEW

Tibia
Post. talocalcaneal
Dorsal talonavicular
Tibialis post.
Dorsal tarsal
Dorsal tarsometatarsal
Tibialis ant.
Long plantar
Calcaneonavicular
Tibionavicular
Ant. tibiotalar
Deltoid
Tibiocalcaneal
Post. tibiotalar
Calcaneal lig.
Medial talocalcaneal
Flex. hall. long.

DORSAL VIEW

Dorsal tarsometatarsal
Intermetatarsal
Tibialis ant.
Peroneus brevis
Cuboideonavicular
Dorsal tarsal
Bifurcate
Calcaneocuboid
Medial (deltoid)
Talus
Lateral lig.
Ant. talofibular
Calcaneofibular
Post. talofibular
Calc.

VIEW FROM BEHIND

Fibula
Tibia
Transverse tibiofibular
Post. talofibular
Medial (deltoid)
Lat. talocalcaneal
Post. talocalcaneal
Tibialis post.
Flex. dig. long.
Flex. hall. long.
Calcaneal lig.
Peroneus brevis
Peroneus long.
Calcaneofib. lig.

PLANTAR VIEW
(After Wolff-Heidegger)

Capsular ligs.
Flex. hall. long. t.
Plantar lig.
Deep transverse metatarsal ligs.
P.L.
Tibialis ant. t.
Peroneus longus t.
Peroneus brevis t.
Tibialis post. t.
Medial (deltoid)
For flex. dig. long.
Long plantar lig.
For flex. hall. long.
Calcaneus
Plantar calcaneonavicular (spring) l.

Pansky

– 553 –

243. ARCHES OF THE FOOT

I. Functional considerations
A. LARGE NUMBER OF BONES allow for both great mobility and absorption of shock
B. THE TOES are important in balancing, running, and climbing

II. Structure of the arches
A. LONGITUDINAL consists of 2 columns of bones, a medial and a lateral
  1. Medial: calcaneus, talus, navicular, 3 cuneiforms, and 3 medial metatarsals
    a. Most important: it bears most of the weight
    b. Talus and navicular are most subject to injury
  2. Lateral: calcaneus, cuboid, and 2 lateral metatarsals
    a. Balances weight. Calcaneus most subject to injury
B. TRANSVERSE: due to the fact that the second and third cuneiforms and the second through fourth metatarsals are wedge-shaped, with their base directed dorsally

III. Weight distribution and support
A. DISTRIBUTION (percentage of total body weight): 25% to calcaneus and 25% to heads of metatarsals (10% to head of first and 3.75% to the more lateral metatarsals)
B. SUPPORT: 80% stress on plantar ligaments; 20% on such muscles or tendons as tibialis posterior, peroneus (fibularis) longus, abductor and adductor hallucis for the medial longitudinal arch and adductor hallucis for the transverse arch

IV. Muscles acting on the ankle joint

Dorsiflexion	Plantar Flexion
Tibialis anterior	Gastrocnemius
Extensor digitorum longus	Soleus
Peroneus (fibularis) tertius	Plantaris
Extensor hallucis longus	Flexor hallucis longus
	Peroneus (fibularis) longus and brevis
	Tibialis posterior

V. Muscles acting on subtalar and transverse tarsal joints

Inversion and Adduction	Eversion and Abduction
Tibialis ant. and post.	Peroneus (fibularis) longus and brevis
Flexor hallucis longus	Peroneus (fibularis) tertius
Flexor digitorum	Extensor digitorum longus (lateral part)

VI. Clinical considerations
A. FLATFEET (pes planus): a condition in which the medial arch of the foot is nonexistent due to the misalignment of the bones and a stretching of the ligaments
  1. In infants, the flat appearance of the foot is normal and is due to the subcutaneous fat pads in the soles. This may persist for two years
  2. Flatfeet in adolescents and adults is caused by ''fallen arches,'' usually the medial longitudinal arches
B. FLATTENING OF THE TRANSVERSE ARCH of the foot may occur, and callus formation takes place under the heads of the lateral four metatarsal bones as a protective measure for the abnormal pressure that is exerted
C. THE SESAMOIDS of the big toe take the weight of the body, especially during the latter part of the walking phase. Their mechanism may be altered in bunions and in hallux valgus

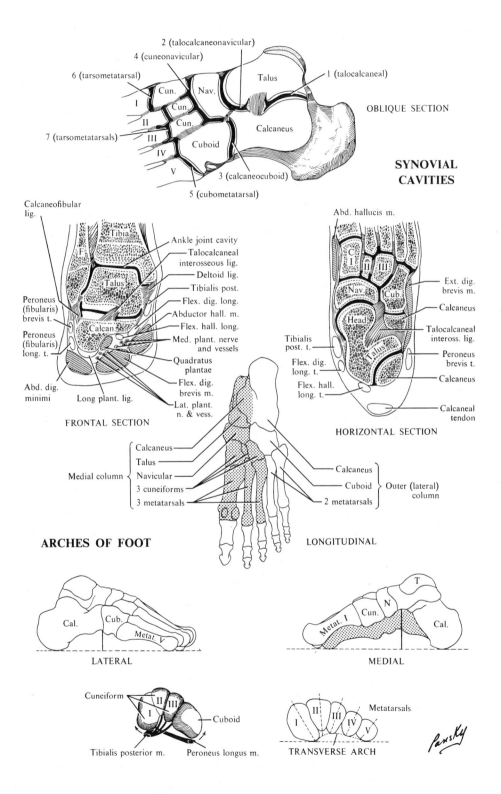

SYNOVIAL CAVITIES

OBLIQUE SECTION

2 (talocalcaneonavicular)
4 (cuneonavicular)
6 (tarsometatarsal)
1 (talocalcaneal)
7 (tarsometatarsals)
3 (calcaneocuboid)
5 (cubometatarsal)

Cun.
Nav.
Talus
Cun.
Cun.
Calcaneus
Cuboid

I
II
III
IV
V

FRONTAL SECTION

Calcaneofibular lig.
Tibia
Ankle joint cavity
Talocalcaneal interosseous lig.
Deltoid lig.
Talus
Tibialis post.
Flex. dig. long.
Peroneus (fibularis) brevis t.
Abductor hall. m.
Flex. hall. long.
Peroneus (fibularis) long. t.
Calcan.
Med. plant. nerve and vessels
Quadratus plantae
Abd. dig. minimi
Flex. dig. brevis m.
Long plant. lig.
Lat. plant. n. & vess.

HORIZONTAL SECTION

Abd. hallucis m.
C I
II
III
Nav.
Cub.
Head
Talus
Ext. dig. brevis m.
Calcaneus
Talocalcaneal inteross. lig.
Peroneus brevis t.
Calcaneus
Calcaneal tendon
Tibialis post. t.
Flex. dig. long. t.
Flex. hall. long. t.

ARCHES OF FOOT

LONGITUDINAL

Medial column {
Calcaneus
Talus
Navicular
3 cuneiforms
3 metatarsals
}

Outer (lateral) column {
Calcaneus
Cuboid
2 metatarsals
}

LATERAL

Cal.
Cub.
Metat. V

MEDIAL

T
N
Cun.
Metat. I
Cal.

Cuneiform
II III
I
Cuboid
Tibialis posterior m.
Peroneus longus m.

TRANSVERSE ARCH
I II III IV V
Metatarsals

244. POSTURE AND DISORDERS OF THE FEET

I. Posture of normal foot: the long-axis direction of the normal postured foot is determined by the rotation of the femur at the hip, since the talus is firmly held in the mortise of the ankle. The long axis may be directed inward (in-toeing) up to 15° or outward (out-toeing) up to 25°. The average posture when standing is a position of out-toeing of about 15°

II. Body weight distribution is transferred through the legs to both feet, each receiving half the load. Weight in the foot itself is supported like a tripod, whose legs are the calcaneus, the 2 sesamoids under the 1st metatarsal, and the heads of the 2 lateral metatarsals. Thus, the axis of balance, *when standing,* passes through the center of the heel and forward between the 2nd and 3rd metatarsals

A. THE BALL OF THE GREAT TOE sustains twice the pressure of that over each of the lateral metatarsal heads

III. Congenital anomalies

A. TALIPES OR CLUBFEET: used in describing the deformities of both ankle and foot (*pes* is term used when disorder is limited to foot alone)
  1. Simple types
    a. Talipes equinus: foot is in marked plantar flexion (like horse foot) with heel abnormally elevated and weight placed on balls of toes
    b. Talipes calcaneus: dorsiflexion of foot and prominence of heel with weight of the body resting on the heel alone
  2. Combined deformities
    a. Talipes equinovarus: most common deformity; usually seen at birth; the heel is in varus (bent inward), the ankle is in equinus (plantar flexed), and forefoot is in varus and supination. There is also an equinus of forefoot producing a cavus or increase of the longitudinal arch
    b. Talipes equinovalgus: same as 2a except heel is in valgus (bent outward) and longitudinal arch is flattened (pes valgus or planus, flatfoot)
    c. Talipes calcaneovarus: t. calcaneus and t. varus combined; the clubfoot is dorsiflexed, inverted, and adducted
    d. Talipes calcaneovalgus: t. calcaneus and t. valgus combined; the clubfoot is dorsiflexed, everted, and abducted
B. PES CAVUS: very disabling deformity of foot characterized by an exaggeration of the longitudinal arch (converse of flatfoot). One sees inversion at subtalar joint, possible calcaneal deformity, dropping of forefoot (equinus) at midtarsal joint, and clawing of all the toes
C. FLATFEET: weakening of the supporting muscles and plantar calcaneonavicular ligament removes support for head of talus which descends, stretching the ligament and flattening the longitudinal arch of foot

IV. Disorders of the toes

A. HALLUX VALGUS (bunion): long axis of big toe is deviated laterally and rotated internally in various degrees. Condition may be either congenital or acquired
B. CLAWING OF TOES: common; imbalance of short intrinsic mm.; consists of hyperextension or dorsiflexion of proximal phalanx with plantar flexion of distal phalanges
  1. Result of overactivity of long and short extensors with reciprocal inhibition of corresponding flexors and weakness of interossei and lumbricals. High heels may be a cause
C. HAMMER AND MALLET TOES: when clawing is confined to but one toe
  1. Hammer type: marked dorsiflexion of proximal phalanx with plantar flexion of 2nd phalanx. Distal phalanx may be neutral, hyperflexed, or hyperextended
  2. Mallet type: deformity consists only of a marked plantar flexion of distal phalanx

V. Nails

A. INGROWING TOENAIL: common, congenital or traumatic, usually involves great toe. Due to tight shoes, shrunken socks, or improper paring of nail corners
B. ONYCHOGRYPHOSIS: overgrowth of the nail (like a ram's horn)

DISORDERS OF FEET

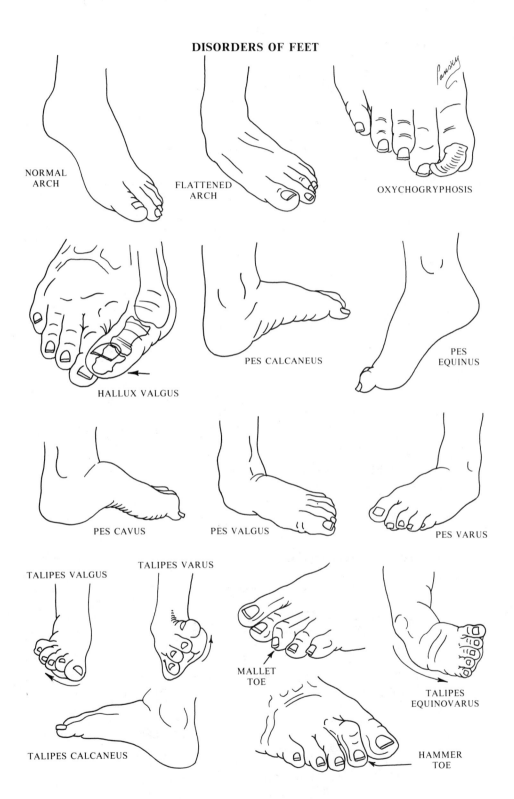

NORMAL
ARCH

FLATTENED
ARCH

OXYCHOGRYPHOSIS

HALLUX VALGUS

PES CALCANEUS

PES
EQUINUS

PES CAVUS

PES VALGUS

PES VARUS

TALIPES VALGUS

TALIPES VARUS

MALLET
TOE

TALIPES
EQUINOVARUS

TALIPES CALCANEUS

HAMMER
TOE

245. MUSCLE-NERVE RELATIONSHIP: FUNCTIONAL

I. Specific cases of nerve injury

A. NERVES to iliacus and psoas muscles: no superficial signs, but, in walking, pelvis does not swing when the extremity is fixed

B. FEMORAL: signs and symptoms are wasting of the anterior thigh; when the thigh is flexed, there is "leg drop"—the inability to extend the leg; loss of sensation on the anterior thigh and the medial leg; weakness in flexion of the thigh due to loss of action of rectus femoris, sartorius, and pectineus muscles

C. OBTURATOR: the thigh cannot be adducted, and there is some loss of sensation on its upper medial part

D. SUPERIOR GLUTEAL: there is weakness in abduction of thigh

E. INFERIOR GLUTEAL: there are wasting of the buttock and weakness in extension of thigh, as in climbing

F. SCIATIC: wasting of all hamstrings and muscles of leg and foot; except for flexion of thigh and adduction of thigh, no other activity in the lower extremity would be possible. Sensation loss on leg and foot

   1. Tibial division: if severed in the upper popliteal fossa, there would be loss of flexion of toes and foot, no inversion of foot, wasting of calf, loss of sensation on sole of foot and terminal parts of both surfaces of toes

   2. Common peroneal (fibular): if damaged near point of division from sciatic, will result in footdrop, loss of extension of all toes, and a loss of sensation on the side of the leg

II. Anatomy of the straight-leg-raising test

A. THE INVESTING SHEATHS OF THE SCIATIC NERVE (like all nerves) are continuous with the coverings of the CNS. Thus, the epineurium of the sciatic nerve is continuous with the dura mater. If the epineurium is pulled upon, some degree of tension will be transmitted to the dura mater and to the nerve roots. Clinically, one method of exerting tension on the nerve roots and the dura is to exert a pull on the sciatic nerve by means of the straight-leg-raising test. The nerve is a close posterior relation of the hip joint, and the 2 branches of the nerve are close posterior relations of the knee joint. Furthermore, the sciatic nerve is 2 cm broad and its supporting sheaths are strong. Thus, when the hip is flexed while the knee is extended, as in the straight-leg-raising test, tension is transmitted back to the L4, L5, S1, S2, and S3 nerve roots and to the dura mater. The clinician may elicit pain not only from the nerve roots and dura but also from other structures. Thus, clinical experience is necessary for the interpretation of the test results

III. Nerve-root alignment: arrangement of spinal nerve roots in relationship to vertebrae is of clinical significance. In the lumbar region, upward migration of the cord during development causes the lumbar nerve roots to slant downward at an acute angle. Thus, although the 4th lumbar nerve root exits between the 4th and 5th lumbar vertebrae, a disk herniation at this level will not affect the 4th lumbar root because it exits above the disk. Instead, it would cause pressure on the 5th nerve root. A similar situation exists with subsequent disks and nerves

IV. Segmental innervation of joint movements: in general, 2 adjoining spinal segments govern a specific joint movement. The 4 segments concerned in 2 opposing movements in a specific joint are in numerical sequence (e.g., hip flexion: L2, 3; hip extension: L4, 5)

A. IN THE LOWER LIMB, joint innervation is segmental, with each joint innervated by nerves arising one segment lower in the spinal cord than those that innervate the next proximal joint. Four spinal segments control hip, knee, and ankle

   1. Hip: L2, 3 govern hip flexion, internal rotation, and adduction

       L4, 5 control hip extension, lateral rotation, and abduction

   2. Knee: L3, 4 control extension; L5, S1 govern flexion

   3. Ankle: L4, 5 control dorsiflexion; S1, 2 mediate plantar flexion

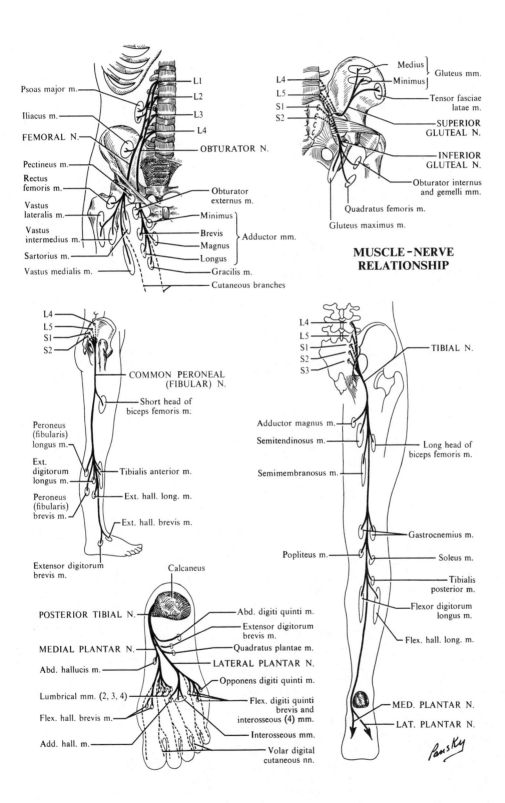

Psoas major m.
Iliacus m.
FEMORAL N.
Pectineus m.
Rectus femoris m.
Vastus lateralis m.
Vastus intermedius m.
Sartorius m.
Vastus medialis m.

L1
L2
L3
L4
OBTURATOR N.
Obturator externus m.
Minimus
Brevis
Magnus
Longus
Adductor mm.
Gracilis m.
Cutaneous branches

Medius
Minimus
Gluteus mm.
Tensor fasciae latae m.
SUPERIOR GLUTEAL N.
INFERIOR GLUTEAL N.
Obturator internus and gemelli mm.
Quadratus femoris m.
Gluteus maximus m.

L4
L5
S1
S2

MUSCLE-NERVE RELATIONSHIP

L4
L5
S1
S2
COMMON PERONEAL (FIBULAR) N.
Short head of biceps femoris m.
Peroneus (fibularis) longus m.
Ext. digitorum longus m.
Peroneus (fibularis) brevis m.
Tibialis anterior m.
Ext. hall. long. m.
Ext. hall. brevis m.
Extensor digitorum brevis m.

L4
L5
S1
S2
S3
TIBIAL N.
Adductor magnus m.
Semitendinosus m.
Long head of biceps femoris m.
Semimembranosus m.
Gastrocnemius m.
Popliteus m.
Soleus m.
Tibialis posterior m.
Flexor digitorum longus m.
Flex. hall. long. m.
MED. PLANTAR N.
LAT. PLANTAR N.

Calcaneus
POSTERIOR TIBIAL N.
Abd. digiti quinti m.
Extensor digitorum brevis m.
MEDIAL PLANTAR N.
Quadratus plantae m.
LATERAL PLANTAR N.
Abd. hallucis m.
Opponens digiti quinti m.
Lumbrical mm. (2, 3, 4)
Flex. digiti quinti brevis and interosseous (4) mm.
Flex. hall. brevis m.
Add. hall. m.
Interosseous mm.
Volar digital cutaneous nn.

Pansky

246. SUMMARY OF MUSCLE-NERVE RELATIONSHIPS IN LOWER EXTREMITY

Segment	Muscle	Named Nerve
L2–L3	Adductor brevis	Obturator
L2–L3	Adductor longus	Obturator
L2–L3	Gracilis	Obturator
L2–L3	Sartorius	Femoral
L2–L4	Iliacus	Femoral
L2–L4	Psoas	Nn. to psoas
L3–L4	Obturator externus	Obturator
L3–L4	Pectineus	Femoral
L3–L4	Quadriceps femoris	Femoral
L3–S1	Adductor magnus	Obturator and sciatic
L4–L5	Tibialis anterior	Deep peroneal
L4–S1	Gluteus medius	Superior gluteal
L4–S1	Gluteus minimus	Superior gluteal
L4–S1	Inferior gemellus	N. to quadratus femoris
L4–S1	Lumbricales	Medial and lateral plantar
L4–S1	Plantaris	Tibial
L4–S1	Popliteus	Tibial
L4–S1	Quadratus femoris	N. to quadratus femoris
L4–S1	Tensor fascia latae	Superior gluteal
L4–S2	Biceps femoris	Sciatic
L4–S2	Semimembranosus	Sciatic
L4–S2	Semitendinosus	Sciatic
L5–S1	Extensor digitorum longus	Deep peroneal
L5–S1	Extensor hallucis longus	Deep peroneal
L5–S1	Peroneus brevis	Superficial peroneal
L5–S1	Peroneus longus	Peroneal
L5–S1	Tibialis posterior	Tibial
L5–S2	Gluteus maximus	Inferior gluteal
L5–S2	Obturator internus	N. to obturator internus
L5–S2	Soleus	Tibial
L5–S2	Superior gemellus	N. to obturator internus
L5–S2	Biceps femoris	Sciatic
S1–S2	Abductor digiti minimi	Lateral plantar
S1–S2	Abductor hallucis	Medial plantar
S1–S2	Adductor hallucis	Lateral plantar
S1–S2	Extensor digitorum brevis	Deep peroneal
S1–S2	Extensor hallucis brevis	Deep peroneal
S1–S2	Flexor digiti minimi	Lateral plantar
S1–S2	Flexor digitorum brevis	Medial plantar
S1–S2	Flexor digitorum longus	Tibial
S1–S2	Flexor hallucis brevis	Medial plantar
S1–S2	Flexor hallucis longus	Tibial
S1–S2	Gastrocnemius	Tibial
S1–S2	Interossei	Lateral plantar
S1–S2	Piriformis	N. to piriformis
S1–S2	Quadratus plantae	Lateral plantar

APPENDIX I. DEFINITIONS OF ANATOMIC TERMS

Alveolus: bony socket of a tooth; pouch in lung air sac.

Ampulla: dilatation of a canal or duct.

Antebrachium: forearm.

Antrum: any nearly closed cavity, particularly one with bony walls.

Anulus: ring-shaped opening.

Aponeurosis: expanded tendon for the attachment of a flat muscle.

Artery (a.): vessel carrying blood from the heart through the body.

Articulation: connection between bones.

Autonomic nervous system: for the innervation of smooth muscle, heart muscle, and glands, consisting of a craniosacral (parasympathetic) and thoracolumbar (sympathetic) portion.

Axilla: arm pit.

Azygos: without a yoke; certain vessels or nerves not in pairs.

Belly: fleshy part of a muscle.

Body: broadest or longest mass of a bone.

Bone: inflexible structure composing skeleton.

Brachium: arm.

Bursa: literally, a purse; a small sac with fluid found in the fascia under skin, muscles, or tendons.

Capillary: anatomic unit connecting the arterial and venous systems; minute vessel, functional unit of the circulatory system.

Cartilage: substance from which some bone ossifies; gristle.

Cell: structural and functional body unit.

Central nervous system (CNS): brain and spinal cord.

Condyle: polished articular surface, usually rounded.

Cornua: plural of cornu, a horn.

Costal: relating to a rib or costa. Costal cartilage.

Crest: ridge or border.

Cribriform: resembling a sieve.

Deltoid: shaped like the Greek letter delta.

Diaphysis: shaft of a long cylindrical bone.

Diarthrosis: movable joint.

Diverticulum: outpocketing or sac; the cecum.

Eminence: low convexity just perceptible.

Endocrine: internal secretion without the use of glandular ducts.

Epicondyle: elevation near a condyle.

Epiphyseal plate (line): growth center for elongation of bone, found between shaft and extremities of the bone.

Epiphysis: extremity or head of a long bone.

Exocrine: secretion discharged by way of a duct system.

Facet: small articular area, often a pit.

Falciform: sickle-shaped (*falx:* a sickle).

Fascia: fibrous envelopment of muscle structures and other tissues.

Foramen: hole, perforation.

Fornix: vaultlike space.

Fossa: shallow depression.

Fundus: base.

Ganglion: group of nerve cell bodies outside the central nervous system.

Genu: knee.

Gingiva: gum.

Glenoid: having the form of a shallow cavity.

Head: enlarged round end of a long bone; knob.

Inguinal: belonging to or near the thigh, or inguen.

Insertion: relatively movable part of a muscle attachment.

Joint: connection between bones.

Jugular: belonging to the neck, or jugulum.

Lacrimal: to do with tears, or lacrimae.

Lamella: little plate or thin layer.

Lamina: plate or layer.

Ligament: fibrous tissue binding bones together or holding tendons and muscles in place.

Linea: line.

Lingual: belonging to the tongue.

Lymph vessels: like veins, but walls are thinner and valves more numerous; drain tissue spaces.

Macula: spot.

Malar: belonging to the cheek.

Mastoid: shaped like a breast.

Mesentery: double layer of peritoneum (mesothelium).

Muscle: contractile organ capable of producing movement.

Neck: constriction of a bone near head; connecting head and body.

Nerve (n.): group of fibers outside the central nervous system.

Neuron: nerve cell body plus its processes.

Nucleus: group of nerve cell bodies within the central nervous system.

Omentum: fold of peritoneum connecting abdominal viscera with the stomach.

Organ: 2 or more tissues grouped together to perform a highly specialized function.

Origin: relatively fixed part of a muscle attachment.

Peripheral nervous system (PNS): cerebrospinal nerves and the peripheral parts of the autonomic nervous system.

Process: projection (can be grasped with fingers).

Protuberance: swelling (can be felt under fingers).

Ramus: platelike branch of a bone; branch of a vessel or nerve.

Ramus communicans: nerve branch from the anterior root of a spinal nerve to the sympathetic chain of ganglia; white—nerve to chain; gray—chain back to spinal nerve.

Raphé: seam; the union of two parts in a line.

Shaft: body of a long bone.

Sheath: protective covering.

Spine: pointed projection of sharp ridge.

Suture: interlocking of teethlike edges.

Symphysis: union of right and left sides in the midline.

System: group of organs acting together to perform a highly complex but specialized function, e.g., nervous, skeletal, muscular, circulatory, respiratory, digestive, urinary, endocrine, and reproductive.

Tendon: fibrous tissue securing a muscle to its attachment.

Teres: round.

Thenar: relating to the palm or sole.

Tissue: differentiation and specialization of groups of cells bound together to perform a special function, e.g., epithelial, connective, muscular, and nervous.

Trochanters: 2 processes on the upper part of the femur below its neck.

Trochlea: spool-shaped articular surface.

Tubercle: small bump (can be felt under finger).

Tuberosity: large and conspicuous bump.

Uncinate: hooked; process shaped like a hook.

Vein (v.): vessel returning blood to the heart.

APPENDIX II.   ESSENTIAL TERMS OF DIRECTION AND MOVEMENT

Abduction (abd.): draws away from midline.

Adduction (add.): draws toward the midline.

Anatomic position: standing erect with arms at the sides and palms of the hands turned forward.

Anterior (ant.) or ventral (vent.): situated before or in front of.

Corrugator: that which wrinkles skin, draws skin in.

Deep: farther from the surface (in a solid form).

Depressor: that which lowers.

Distal (dist.): farther from the root.

Dorsal (dors.): toward the rear, back; also back of hand and top of foot.

Erector: that which draws upward.

Evert (ever.): turn outward (as foot at ankle joint).

Extension (ext.): straightening.

External (extern.): outside (refers to wall of cavity or hollow form).

Flexion (flex.): bending or angulation.

Frontal (front.) or coronal (coron.): vertical; at right angles to sagittal; divides body into anterior and posterior parts.

Horizontal (horiz.): at right angles to vertical.

Inferior (inf.): lower, farther from crown of head.

Internal (int.): inside (refers to wall of cavity or hollow form).

Inverted (invert.): turned inward (as foot at ankle joint).

Lateral (lat.): farther from midline (or center plane).

Levator (lev.): that which raises.

Longitudinal (longit.): refers to long axis.

Medial (med.): nearer to midline (or center plane).

Median: midway, being in the middle.

Midline: divides body into a right and left side.

Midsagittal: vertical plane at midline dividing body into right and left halves.

Oblique: slanting.

Palmar (palm.) or volar (vol.): palm side of hand.

Plantar (plant.): sole side of foot.

Posterior (post.) or dorsal (dors.): rear or back.

Pronator (pronat.): that which turns palm of hand downward.

Prone: forearm and hand turned palm side down; body lying face down.

Proximal (prox.): nearer to limb root.

Rotator (rotat.): that which causes to revolve.

Sagittal (sagit.): vertical plane or section dividing body into right and left portions.

Sphincter: that which regulates closing of aperture.

Superficial (superf.): nearer to surface (refers to solid form).

Superior (sup.): upper, nearer to crown of head.

Supinator (supinat.): that which turns palm of hand upward.

Supine: forearm and hand turned palm side up; body lying face up.

Tensor (tens.): that which draws tight.

Transverse (trans.): at right angles to long axis; body divided into upper and lower parts.

Ventral (vent.) or anterior (ant.): situated before or in front of.

Vertical (vert.): refers to long axis in erect position.

Volar (vol.) or palmar (palm.): palm side of hand.

APPENDIX III. CLASSIFICATION OF JOINTS

I. Immovable joints (synarthroses)
A. SUTURE: interlocking of bones along their saw-toothed edges, e.g., joints of cranium
B. SCHINDYLESIS: a thin plate of bone is received into a cleft or fissure formed by the separation of 2 laminae in another bone, e.g., vomer in fissure between maxillae
C. GOMPHOSIS: insertion of a conical process into a socket, e.g., roots of teeth with alveolae of jaws
D. SYNCHONDROSIS: where the connecting medium is cartilage, e.g., sternocostal joint

II. Slightly movable joints (amphiarthroses)
A. SURFACES CONNECTED BY FIBROCARTILAGE (symphysis), e.g., joints of spine (intervertebral disks)
B. SURFACES UNITED BY AN INTEROSSEOUS LIGAMENT (syndesmosis), e.g., inferior tibiofibular articulation

III. Freely movable (synovial or lubricated, diarthroses)
A. GLIDING JOINT (arthrodia): bones glide face to face limited by restraining ligaments, e.g., intercarpal, intertarsal joints
B. HINGE JOINT (ginglymus): movement about a transverse axis only, e.g., elbow (humeroulnar) joint, knee joint
C. SADDLE JOINT: increases the extent of hinge-joint movement by adding an axis of movement perpendicular to transverse axis, e.g., joint of thumb with wrist bone (carpometacarpal)
D. CONDYLOID (ellipsoidal): a modified ball-and-socket joint in which the opposed surfaces are ellipsoidal and not spherical. Increases extent of saddle joint by permitting circular movement that describes a cone, but vertical rotation is impossible, e.g., wrist joint
E. PIVOT JOINT (trochoid): a cylindrical form moving within a complete or partial ring (or a ring moving about the cylinder); only a vertical axis is present, e.g., joints of the forearm (radio-ulnar)
F. BALL-AND-SOCKET JOINT (enarthroses): provides freest movement by means of a spherical head in a cuplike cavity; adds to wide play of condyloid joint, the vertical axis of rotation, e.g., hip joint (at acetabulum), shoulder joint

APPENDIX IV. REFERENCES

Anderson, J. E.: *Grant's Atlas of Anatomy,* 8th ed. Williams & Wilkins Company, Baltimore, 1983.

Basmajian, J. V. (ed.): *Grant's Method of Anatomy,* 10th ed. Williams & Wilkins Company, Baltimore, 1980.

Basmajian, J. V.: *Primary Anatomy,* 8th ed. Williams & Wilkins Company, Baltimore, 1983.

Bates, B.: *A Guide to Physical Examination,* 3rd ed. J. B. Lippincott Company, Philadelphia, 1983.

Bergamin, C. P. L.: CT scans of head: Basic interpretation. *Hospital Medicine,* Jan., 1982, pp. 21–42.

Crafts, R. C.: *A Textbook of Human Anatomy,* 2nd ed. John Wiley & Sons, New York, 1979.

Felson, B. (ed.): Computerized cranial tomography I and II. *Semin. Roentgenol.,* **12**:3, Jan., 1977; **12**:95, April, 1977.

Gardner, E.; Gray, D. J.; and O'Rahilly, R.: *Anatomy: A Regional Study of Human Structure,* 4th ed. W. B. Saunders Company, Philadelphia, 1975.

Gordon, R.; Herman, G.; and Johnson, J.: Image reconstruction from projections. *Sci. Am.,* **233**:56, Oct. 1975.

Goss, C. M. (ed.): *Gray's Anatomy of the Human Body,* 29th Am. ed. Lea & Febiger, Philadelphia, 1973.

Grossman, C. G.; Gonzalez, C. F.; and Palacio, E.: *Computed Brain and Orbital Tomography. Technique and Interpretation.* John Wiley & Sons, New York, 1976.

Hamilton, W. J. (ed.): *Textbook of Human Anatomy,* 2nd ed. C. V. Mosby Company, St. Louis, 1976.

House, E. L.; Pansky, B.; and Siegel, A.: *A Systemic Approach to Neuroanatomy,* 3rd ed. McGraw-Hill Book Company,.New York, 1979.

International Anatomical Nomenclature Committee: *Nomina Anatomica,* 5th ed. Waverly Press, Baltimore, 1983.

Jaffe, Carl C.: Medical imaging. *Am. Sci.,* **70**:576–585, Nov.-Dec., 1982.

Langman, Jan, and Woerdeman, M. W.: *Atlas of Medical Anatomy.* W. B. Saunders Company, Philadelphia, 1978.

McMinn, R. M. H., and Hutchings, R. T.: *Color Atlas of Human Anatomy.* Year Book Medical Publishers, Inc., Chicago, 1977.

Moore, K. L.: *Clinically Oriented Anatomy,* 1st ed. Williams & Wilkins Company, Baltimore, 1980.

Nolte, J.: *The Human Brain.* C. V. Mosby Company, St. Louis, 1981.

O'Rahilly, R.: *Basic Human Anatomy,* 1st ed. W. B. Saunders Company, Philadelphia, 1983.

Pykett, Ian L.: NMR imaging in medicine. *Sci. Am.,* **246**:78–88, May, 1982.

Ramsay, R. G.: *Computer Tomography of the Brain: With Clinical Angiographic and Radionucleotide Correlation.* W. B. Saunders Company, Philadelphia, 1977.

Romanes, G. J. (ed.): *Cunningham's Textbook of Anatomy,* 12th ed. Oxford University Press, London, 1981.

Snell, R. S.: *Clinical Anatomy for Medical Students,* 2nd ed. Little Brown & Company, Boston, 1981.

Wagner, M., and Lawson, T. L.: *Segmental Anatomy.* Macmillan Publishing Company, New York, 1982.

Williams, P. L., and Warwick, R. (eds.): *Gray's Anatomy,* 36th Br. ed. W. B. Saunders Company, Philadelphia, 1980.

Wilson, D. B., and Wilson, W. J. W.: *Human Anatomy,* 2nd ed. Oxford University Press, New York, 1983.

Woodburne, R. T.: *Essentials of Human Anatomy,* 7th ed. Oxford University Press, New York, 1983.

INDEX

Figures in **boldface** type refer to pages on which illustrations appear.

tibial, 516, **517**, 530, **531**
ulnar, 262, **263**, 268, **269**
renal, **419**, 422, **423**, 424, 426, **427**
retroduodenal, **399**, 402
sacral, lateral, 462, **463**
middle, 462, **463**
scalp, 18, **19**
scapular, circumflex, 248, **249**
descending, 248, **249**
dorsal (descending), 58, **59**, **249**
scrotal, 466, **467**
sigmoid, 410, **411**
spermatic, internal (testicular), 388, **419**
sphenopalatine, 42, **43**
spinal, 200, **201**
splenic, 402, **403**, 420, **421**
sternocleidomastoid, 55
striate, 112, **113**
stylomastoid, 12
subclavian, **55**, 58, **59**, **69**
sublingual, 72, **85**
submental, 42, **43**
subscapular, **241**, 244, **245**, 248, **249**
supraduodenal, **399**
supraorbital, 18, **19**, 144, **145**
suprarenal, **419**, **423**, 430, **431**
suprascapular, 58, **59**, 248, **249**
supratrochlear, 18, **19**, 144, **145**
temporal, deep, **39**, 42, **43**
superficial, 18, **19**, 36, **37**, 50, **51**
testicular, 388
thoracic, internal, 58, **59**, 320, **321**
lateral, **241**, 244, **245**, 248
superior, **241**, 244
thoracoacromial, **241**, 244, **245**, 248, **249**
thoracodorsal, **241**, 244, 248, **249**
thyrocervical trunk, **55**, 58, **59**, 60
thyroid, ima, 60, **61**
inferior, 60, **61**, 64, **65**
superior, 50, **51**, 60, **61**, 64, **65**
tibial, anterior, 516, **517**, 530, **531**
posterior, 516, **517**, **525**, 530, **531**
recurrent, 516, **517**, 530, **531**
tonsils, 82, **83**
transverse, cervical, 58, **59**, **61**, **249**
facial, 36, 42, **43**
tubal, 454, **455**
tympanic, anterior, 42, **43**
ulnar, collateral, 254, **255**, **259**, **263**, 268, **269**
forearm, 258, **259**, 262, **263**, **265**
hand, **263**, 286, **287**, 290, **291**, **297**
recurrent, **259**, 262, **263**, 268, **269**
wrist, 282, **283**, 286, **287**, 290, **291**, **297**
umbilical, 392
ureter, 428, **429**
uterine, 454, **455**

vaginal, 454, **455**
vertebral, 50, **51**, **55**, 58, **59**, 114, **115**, 200, **201**
vesical, 438, 446, **463**
Articulations. *See* Joint(s)
Ascites, 390
Astigmatism, 142
Atrioventricular node, 358, **359**
valves, 345, 346, **347**
Atrium, heart, 344–346
nose, **159**
Auditory tube, 86, **87**, **151–153**
Auricle, ear, 150, **151**
heart, **341**, **343**, 344
Autonomic nervous system,
abdomen, 458–**461**
head and neck (sympathetic), 68, **69**
pelvis, 458–**461**
thorax (sympathetic chain), 366, **367**, **369**
Axilla, 242, **243**, 245
Axillary fascia, 242
Axis, 186, **187**
Azygos system, 366, **367**, **369**

Babinski test, **221**
Bachmann's bundle, of heart, 354, **355**
Back (dorsum), **179**
Bartholin's cyst, 464
Bell's palsy, 26
Berry aneurysm, 118, **119**
Biceps reflex, **221**
Bicipital aponeurosis, **253**, **259**, **261**
Bifurcation, carotid, 54, **55**, **61**
Bladder, gall, **411–414**, 416, **417**
urinary, **393**, **429**, 438–**441**, **443**, **453**, 457
Bleeding, intracranial, 100, **101**
Blepharitis, 132, **133**
Blind spot of retina, 144
Blood-brain barrier, 114
Body, carotid, 54
ciliary, 140, **141**, **145**, 148, **149**
mammillary, 102, **103**, **107**
perineal (central point), **441**, **445**, **465**, 466
vitreous, 140, **141**
Bone(s), calcaneus, **535**, **553**, **555**
capitate, 278, **279**, **303**
carpal, **225**, 278, **279**, 302, **303**
cervical vertebrae, **181**, 186, **187**, 204, **205**
clavicle, 232, **233**, **235**
coccyx, **185**, 190, **191**
concha, inferior nasal, **7**, 160, **161**
coxal (hip), 488, **489**
cuboid, 534, **535**, **553–555**
cuneiform, 534, **535**, 554, **555**
ethmoid, **15**, 136, **137**
femur, **481**, 494, **495**, **525**
fibula, **481**, 514, **515**, **525**
foot, **481**, 534, **535**, 554, **555**
frontal, **3–11**, 14
hamate, 278–280, **303**
hand, **225**, 278–**281**
hip, 377, 488, **489**, **524**

humerus, **225**, 236, **237**
hyoid, 6, **7**, **65**, **67**
ilium, 377, 488, **489**, **524**
incus, **151–153**
ischium, 377, 488, **489**, **524**
lacrimal, **7–9**, 136
lumbar vertebrae, **181**, 190, **191**
lunate, **225**, 278, **279**, **281**, **303**
malleus, **151–153**
mandible, 6, **7**
maxilla, **3–9**, **161**
metacarpal, **225**, 278, **279**, **281**, 302, **303**
metatarsal, **481**, 534, **535**, 554, **555**
nasal, **3**, 6, **7**, **9**, 158, 160, **161**
navicular, **225**, 534, **535**, 554, **555**
occipital, **8–15**
orbit, **15**, 136, **137**
palatine, 12, **13**, 160, **161**
parietal, **3**, 6, **7**, **9–11**, 14
patella, **481**, 514, **521**, **525**
pelvis, 377, 488, **489**, **524**
phalanges, fingers, **225**, 278, **279**, **281**, 302, **303**
toes, **481**, 534, **535**
pisiform, **225**, 278, **279**, **303**
pubis, 377, 488, **489**
radius, **225**, 256, **257**, 272, **273**, 302, **303**
ribs, 323, 325–**327**
sacrum, **185**, 190, **191**, 377, **489**
scaphoid (navicular), **225**, 278, **279**
scapula, 232, **233**, 234, **319**
sesamoid, **279**, **280**, **535**
skull, **3–15**
sphenoid, **7–9**, **12–15**, 126, **161**, 164
stapes, **151–153**
sternum, 322, **323**, **325**
talus, **481**, 534–**537**, **555**
tarsal, **481**, 534, **535**, **555**
temporal, **3**, **7–9**, 12, 14, **15**
thoracic vertebrae, **181**, 186, **187**
tibia, **481**, 514, **515**, **525**, 522, **523**, **525**
trapezium (greater multangular), **225**, 278–**280**
trapezoid, **225**, 278, **279**, **281**
triangular (triquetrum), **225**, 278–**280**
ulna, **225**, 256, **257**, 272, **273**, 302, **303**
vertebrae, **181**, **185–187**, **189–191**, 209
vomer, 12, 160, **161**, **163**
zygomatic, **3–9**, **13**
Brachial plexus, 240, **241**, 245, 306, **307**
injury of, 308, **309**
Brachioradialis reflex, **221**
Brachium pontis, 102, 108, **109**
Brain, 102–106
venous drainage of, 94–**97**
Breast. *See* Glands, mammary
Breathing, 370
Bregma, **9**, 10
Broad ligament, **451–453**
Bronchi, 332, 334, **335**, 362, **363**

Muscle(s) [*Cont.*]
  rectus abdominus [*Cont.*]
    capitis, anterior, 70, 71, 204, 218
      lateral, 70, **71**, 218
      posterior major, **69**, 198,
        199, 204
      posterior minor, **69**, 198,
        199, 204
    eyeball, **137–139, 141, 147**
    femoris, 496, **497**, 512, **559**
    rhomboideus, major, 194, **195**,
      232, 370
    minor, 194, **195**, 232, 234, 370
    risorius, 26, **27**
    rotatores, vertebral column, 196,
      218
    salpingopharyngeus, 62, **83, 85,**
      87
    sartorius, 496, **497, 499, 503**,
      512, 520
    scalene, anterior, **59**, 70, **71**, 218,
      370
      middle, **59**, 70, **71**, 218, 370
      posterior, 70, **71**, 218, 370
    semimembranosus, 508, **509**,
      512, **517**, 520
    semispinalis, **195–197, 199**, 204,
      218
    semitendinosus, 508, **509**, 512,
      517
    serratus, anterior, 234, 238, **239**,
      370
      posterior, inferior, **195**, 328,
        370
      superior, **195**, 328, 370
    soleus, 532, **533**, 554
    sphincter, ani, externus, **441,**
      445, 447, 449
      internus, **445, 447**
    pupillae, 140, **141, 149**
    urethrae, female, 464
      male, 466, **467**
    spinalis, 196, **197, 199**
    splenius, **195**, 196, **199**, 204, 218
    stapedius, 152, **153**
    sternocleidomastoid, **47**, 52, **53**,
      204, 218, 234, 370
    sternohyoid, 52, **53, 55, 71**
    sternothyroid, 52, **53, 71**
    styloglossus, 62, **63, 75, 77**
    stylohyoid, 52, **55, 73**
    stylopharyngeus, 62, **63, 77, 87**
    subclavius, 234, 238, **239**, 370
    subcostal, 328, **329**
    subscapularis, 238, **239**, 250
    supinator, **265, 267, 269**, 272
    supraspinatus, 246, **247**, 250
    temporalis, 30, **31**, 40, **43**
    tensor, fasciae latae, 504, **505**,
      512, 520
    tympani, 152, **153**
    veli palatini, 77, 84, **85, 87**
    teres, major, 238, **247**, 250, **255**
    minor, **243**, 246, **247**, 250
    thyroarytenoid, 64, **65, 67**
    thyroepiglottic, **65**
    thyrohyoid, **53**, 62, **63, 65**
    tibialis, anterior, 528, **529**, 540,
      541, 543, 554

  posterior, 532, **533**, 554
  tongue, intrinsic, 76, **77**
  transversus, abdominis, 382,
    383, 385, 389
    perinei, deep, 464–**467**
      superficial, 464–**467**
    thoracic, 328, **329**
  trapezius, 192, **193**, 204, 234, 370
  triceps brachii, 247, 250, 252,
    253, 255, 272
  uvular, **85, 87**
  vastus, intermedius, 496, **497**
    lateralis, 496, **497**
    medialis, 496, **497, 503**
  vocalis, 64
  zygomaticus, major, 26, **27**
    minor, 26, **27**
Musculi pectinati, 344, **345**
Myelencephalon, 102, 106, 110
Myocardium, **343, 345**
Myopia, 142

Nails, **301**
Nares, 158, **159**
Nasal, bones, 6, 7
  polyp, 160
Nasion, 6, **121**
Nasopharynx, 86, **87**
Nearsightedness (myopia), 140
Neck, humerus, 236, **237**
  radius, 256, **257**
Nephron, 422
Nerve(s), abdominal wall, 386, **387**
  abducens, 138, **139**, 146, **147**,
    170, 172
  accessory, **59, 61, 111, 171, 173**
  alveolar, inferior, 38, **39, 43**
    superior, 38, **39**
  ansa, hypoglossi, 52, 54, 68, **69**
    subclavia, 68, **69**
  antebrachial cutaneous, lateral,
    226, **227**, 254
    medial, 226, **227**, 240, **241**
    posterior, 226, **227**, 254
  anterior cutaneous of chest, 36,
    183
  auricular, great, 16, **17**, 58, **59**
    posterior, **19**, 28, **33**
    vagus, 150
  auriculotemporal, 16–**19**, 36–38
  axillary, 226, **227**, 240, **241**, 308,
    309
  to biceps femoris muscle, short
    head, 508, **559**
  brachial cutaneous, lateral, 226,
    227
    medial, 226, **227**, 240, **241**
    posterior, 226, **227**
  buccal, facial, 32, **33**
    trigeminal, 16, 38, **39, 43**
  to bulb of penis, 466
  calcaneal, 482, **483, 549**
  cardiac branches, sympathetic,
    358, **359**
    vagus, 358, **359**
  caroticotympanic, 36
  carotid, external, 36, **37**
    internal, 148, **149**, 168, **169**

  cauda equina, 214, **215**
  cervical, transverse, 16, **17**, 58, **59**
  chorda tympani, 32–**35**, 72–**75**,
    156, **157**
  ciliary, long, 146, **147**
    short, 148, **149**
  cluneal, 482, **483, 507**
  to coccygeus muscle, 448, **511**
  cochlear, 154, **155**, 170, **171**
  common palmar digital, 290, **291**
  cranial, summary, 170–**173**
    testing of, 172, **173**
  cutaneous, abdomen, 182, **183**
    back, 182, **183**
    buttock, **183**, 482, **483, 507**
    face, 16, **17**
    head and neck, 16, **17**
    lower extremity, 482, **483**
    thorax, 182, **183**
    upper extremity, 226, **227**
  diaphragm, 436
  digital, palmar, 290, **291, 300, 301**
    plantar, 548, **549, 559**
  dorsal, clitoris, 462, **463**
    cutaneous, foot, 482, **483**, 542
    penis, **467**, 470, **471**
    scapular, 194, 240, **241**, 248,
      308, **309**
  esophageal, from vagus, 364, **365**
  ethmoidal, 168, **169**
  facial, face, 26, 32–**35**, 156, **157**,
    170, **173**
    skull, 32, **33, 107, 151**, 156, **157**
  femoral, 498–500, 502, **503**
    anterior cutaneous, 482, **483**
    lateral cutaneous, 482, **483**
    posterior cutaneous, 482, **483**
  frontal, 146, **147**
  to gemellus inferior and
    quadratus femoris, 506,
    510, **511**
  to gemellus superior and
    obturator internus, 506,
    510, **511**
  genitofemoral, 388, 482, **483**
  glossopharyngeal, 36, **37**, 74, **75**,
    111, 171, 173
  gluteal, inferior, 506, **507**, 510,
    511
    superior, 506, **507**, 510, **511**
  great auricular, 16, **17, 37**, 58, **59**
  hypogastric, 458, **459**
  hypoglossal, **69**, 74, **75, 103**,
    169, 173
  iliohypogastric, 386, **387**, 482,
    483
  ilioinguinal, 386–388, 482, **483**
  incisive, 38, 168
  infraorbital, 16, **17**, 38, **39**
  infratrochlear, 16, **17**, 146, **147**
  intercostal, thoracic, 182, **183**
    thoracoabdominal, 182, **183**
  intercostobrachial, 182, 226, **245**
  intermedius, 72, 74, **103**
  interosseous, anterior, 262, **263**
    posterior, 266, **267**
  lacrimal, 16, **17**, 146, **147**
  laryngeal, external, **61**, 64, **65**
    inferior, 64, **65**